The International Atlas of Mars Exploration

The First Five Decades

Volume 1: 1953 to 2003

PHILIP J. STOOKE
University of Western Ontario

CAMBRIDGE UNIVERSITY PRESS
Cambridge, New York, Melbourne, Madrid, Cape Town,
Singapore, São Paulo, Delhi, Mexico City

Cambridge University Press
32 Avenue of the Americas, New York, NY 10013-2473, USA

www.cambridge.org
Information on this title: www.cambridge.org/9780521765534

First published 2012

Printed in the United States of America

A catalog record for this publication is available from the British Library.

Library of Congress Cataloging in Publication Data

Stooke, Philip.
The international atlas of mars exploration : the first five decades : volume 1: 1953 to 2003 / Philip Stooke.
p. cm.
Includes bibliographical references and index.
ISBN 978-0-521-76553-4 (hardback)
1. Mars (Planet) – Remote-sensing maps. 2. Mars (Planet) – Exploration. 3. Mars (Planet) – Maps.
4. Space flight to Mars – Maps. 5. Space flight to Mars – History. I. Title.
G1000.5.M3A4S8 2012
912.99′23–dc23 2012007339

ISBN 978-0-521-76553-4 Hardback

Time and time again I repeat, “It’s incredible.” And it truly is. Nothing before or after can compare. It is transparent, brilliant, boundless. An explorer would understand. We have stood on the surface of Mars.

Thomas A. Mutch

Contents

Missions and events – chronological list *page* viii
Foreword by Matt Golombek xi
Preface and acknowledgements xiii

1 Chronological sequence of missions and events 1
2 Phobos and Deimos 325
3 Mars mission data 344

Bibliography 347
Index 357

Missions and events – chronological list

Mars

Mars at the Dawn of the Space Age *page* 1
1950s and 1960s: Humans to Mars: Early Thoughts 4
1959: Robotic Mission Planning at JPL 8
10 October 1960: Mars 1960A 8
14 October 1960: Mars 1960B 8
24 October 1962: Sputnik 22 8
1 November 1962: Mars 1 9
4 November 1962: Sputnik 24 9
1963: Voyager 9
5 November 1964: Mariner 3 12
28 November 1964: Mariner 4 12
30 November 1964: Zond 2 18
18 July 1965: Zond 3 18
1968: Mars Hard Lander Capsule Study 18
25 February 1969: Mariner 6 20
27 March 1969: Mariner 7 21
27 March 1969: Mars 1969A 25
2 April 1969: Mars 1969B 25
1960s: Earth-Based Topographic Mapping 25
8 May 1971: Mariner 8 29
10 May 1971: Cosmos 419 29
19 May 1971: Mars 2 29
28 May 1971: Mars 3 31
30 May 1971: Mariner 9 33
21 July 1973: Mars 4 39
25 July 1973: Mars 5 45
5 August 1973: Mars 6 45
9 August 1973: Mars 7 52
Viking Landing Site Selection 52
Viking 1 Site Certification 66
20 August 1975: Viking 1 74
Viking 2 Site Certification 117
9 September 1975: Viking 2 124
1970s: Pioneer Mars 164
1979: Viking Rovers 169
1980s: Mars 1984 Rover/Penetrator Mission 172
1987: Sample Return Planning 178
1980s: Mars Rover Sample Return 181
7 July 1988: Phobos 1 192
12 July 1988: Phobos 2 192
1990s: Vesta/Mars-Aster 195
1990s: Mars Science Working Group 195
1990: Mars Landing Site Catalog 196
1990s: Hubble Space Telescope 196
1990s: Mars Network Mission Plans 204
Early ESA Network 204
Mars Global Network Mission 204
MESUR 222
Marsnet 225
Intermarsnet 228
Micro-Meteorological Network 230
1990s: Future Exploration Studies 233
1992: Mars Rover Reference Mission 237
25 September 1992: Mars Observer 238
1990s: Mars Discovery Missions 238
1994: Mars 94 241
1995: Exobiology Site Study 243
7 November 1996: Mars Global Surveyor 247
1996: Marsokhod International Mission 250
16 November 1996: Mars 96 250
Mars Pathfinder Landing Site Selection 253
4 December 1996: Mars Pathfinder 253
1997: Mars Reference Mission 275
1998: Mars 98 285
3 July 1998: Nozomi (Planet-B) 287
24 October 1998: Deep Space 1 288
11 December 1998: Mars Climate Orbiter 288
Mars Polar Lander Landing Site Selection 289
3 January 1999: Mars Polar Lander 289
3 January 1999: Deep Space 2 290

1999: Mars Stratigraphy Mission 294
1999: Mars Network Design 294
2000: Hydrothermal Sites 294
Mars Surveyor 2001 Landing Site Selection 298
2001: Mars Surveyor 2001 298
7 April 2001: 2001 Mars Odyssey 308
2000s: Seeking Water 315
2 June 2003: Mars Express 315
Beagle 2 Landing Site Selection 318
2 June 2003: Beagle 2 320
2003: Mars Bioinspired Aircraft 324

Phobos and Deimos

Introduction 325
1969: Mariners 6 and 7 325
1971–1972: Mariner 9 325
1976: Viking 330
1970s: Viking Phobos Mission Studies 331
1989: Phobos 2 331
1997: Mars Pathfinder 334
1990s: Aladdin 336
1997: Mars Global Surveyor 339
Russian Missions 339
2003: Mars Express 343
Human Exploration of Phobos and Deimos 343

Mars mission data

Mars Impact and Landing Events 344
Mars Flyby and Orbital Events 344
Mars Exploration Chronology by Mars Year 345

Foreword

Many geologists love maps. Maps show the spatial distribution of features, which is central to exploring, mapping surface materials and rocks and understanding how these rocks and materials are arranged at depth. Geologists map the distribution of rocks and other materials at the surface to understand what processes led to their formation.

Geologists who study the surfaces of other planets (planetary geologists) have an even more unique relationship with maps. Images acquired from Earth or from orbiting spacecraft of other planets and satellites are commonly made into maps and show most of what planetary geologists get to study. The outermost layer of the planet is what can be imaged, and placing these images into maps is essential for understanding where features and terrains (i.e., places) are on the planet and how the surface evolved.

For planetary geologists, like myself, who work on where surface landers and rovers will land, maps are the prism through which we view the planet. Where will the spacecraft land, what will it do when its on the ground, what will it look like when it gets there, is it safe to land on or rove over, what science will it do when it is there and how will the location of rocks and features seen at the surface be located on the planet? Considering and answering these questions require maps – lots of maps. Maps to show the landing ellipse on the planet; maps to show potential hazards such as craters, scarps and large rocks; maps to show the materials of interest to study; maps to plan the rover traverse.

The revelation in reading Philip J. Stooke's *The International Atlas of Mars Exploration*, which chronicles humankind's first five decades of Mars exploration, is that the maps of possible landing sites provide a fascinating look at how our knowledge of the surface has evolved and changed through time, with the earliest maps based on albedo features that could be seen with telescopes. These maps are overlain by early low-resolution flyby images of the Mariner 4, 6, and 7 spacecraft. Mariner 9 produced the first coarse global dataset of Mars surface features, and eventually the Viking orbiters created a Mars global image map at a couple of hundred meters per pixel that set the stage for our modern exploration of the Red Planet.

By turning the pages in this book, you will also get a view of how scientists thought about landing sites at different times, the scale of their ambitions and some of the gritty detail of actually selecting and certifying landing sites. The real treat is to see where scientists thought the landings sites were located on the planet versus where they actually are, as they have all been subsequently imaged directly by the sub-meter per pixel HiRISE camera. Finally, this book will show you exactly where all the features seen by the landers and the Pathfinder rover actually are on the planet. This book is a treat for anyone who loves maps.

Matt Golombek
La Cañada Flintridge, CA

Preface and acknowledgements

Preface

This is a book about the exploration of the planet Mars, presented in atlas format – in other words, primarily a book of maps, a book about places. It covers a period of five decades conveniently bracketed by Wernher von Braun's Mars Project of 1953 and the arrival at Mars of the European Space Agency's Mars Express in December 2003. NASA's Mars Exploration Rovers and later missions will be described separately. Like my earlier *International Atlas of Lunar Exploration*, this book is not about the technology of spacecraft or instruments, the geology or other scientific aspects of Mars, the people involved or the social and political background of space exploration. Its focus is squarely on those aspects of the history of Mars exploration which can be portrayed on maps and annotated images in this visual format.

Those other aspects of exploration are covered elsewhere, but details of the events and places involved are often lacking. Which areas were imaged by orbital or flyby missions? How were landing sites chosen? What activities took place at each site? What missions have been planned but not flown, what would they have done and where? For example, consider Viking landing site selection, a process described by Ezell and Ezell (1984). Their description omits many specific details such as those found here in Table 5, but the missing details turned up in an unpublished document by the same authors, as described later in the atlas. Or consider the Viking 2 sample area maps, which were drawn before the last sampler arm operations and never updated, or the Mars Pathfinder rover route map, which has only been published in a simplified form. Here, mission events are set out step by step in tables, text and illustrations, as completely as the space permits. The intention is to collect material from scattered and often obscure or unpublished sources into one convenient reference and to portray it visually where possible. For every mission actually launched to Mars there have been many proposals which were not fulfilled, and some are included here. It is impractical to try to cover them all, but a reasonable number are presented, with emphasis on those for which specific locations can be illustrated.

I have used several different global maps of Mars to illustrate the planet as our knowledge of it evolved over time. They indicate the level of knowledge and the kinds of maps available to planners and researchers at the time and help illustrate the rich history of planetary cartography. Apart from those global maps, contemporary maps of smaller regions are used when appropriate to locate and depict various sites, and the best modern images are used for the most detailed site characterization. As an example, Viking site selection is mostly mapped on Mariner 9 base maps, but the most detailed site maps make use of more recent data from Mars Odyssey and Mars Reconnaissance Orbiter. The maps follow the gradual exploration of the planet, revealing the growth of our knowledge as only time-series maps can.

The earliest stage of global mapping illustrated here is a set of older telescopic maps reprojected into the standard global mapping format adopted for this atlas. The global maps are presented in Azimuthal Equidistant projection, in hemispheres centred on the 0° and 180° longitudes. One of my goals here is to emphasize the role of William Herschel as the first cartographer of Mars. That title is often given to Wilhelm Beer and Johann Mädler, but their maps, drawn 50 years after Herschel's, are in some ways inferior. The unusual pole-centred presentation of Herschel's global drawing (Figure 5A) has caused it to be overlooked in some studies, but when reprojected it is seen to be a reasonable representation of the planet. Jürgen Blunck, the prominent German historian of Martian cartography, included Herschel's map at my suggestion in his definitive exhibit at the Staatsbibliothek in Berlin in 1993 and 1994 (Blunck *et al.*, 1993).

Apart from those historical maps, a variety of other representations of the planet are used throughout this book as backgrounds for various figures. The goal is to portray Mars at it was known at the time each event occurred. The first of these, a map used as a base for the early figures in this atlas, was derived from the Mars Experimental Chart 1 (MEC-1) drawn by the Aeronautical Chart and Information Center (ACIC) of the US Air Force for NASA in 1962 and used for Mariner 4 planning. When published this was already something of an anachronism, as it showed the network of linear 'canals' which astronomers had long since dismissed as optical illusions. Nevertheless, this map was used as a planning document for the first NASA missions to Mars, and it acts as the foundation on which this sequence of maps is built. The ACIC map was converted to the projection used throughout this atlas for global maps (Azimuthal Equidistant), a process which required some artistic license especially at the poles (Figures 1 and 2). It is used here to portray locations from the earliest Mars missions, including Mariner 4 and the first stages of Viking planning. Maps based on Mariner 6 and 7 images were combined to represent the next stages in knowledge (Figures 15, 16, 18 and 19). For each of those missions, far-encounter images were reprojected and combined to make a low-resolution global map, and high-resolution near-encounter images were superimposed. Maps from the year-long mission of Mariner 9 illustrate the next stage of knowledge, the first global view of the planet's exotic landforms. A US Geological Survey (USGS) map compiled from Mariner 9 images was stripped of its text and contour lines and reprojected to match the other views (Figures 28 and 29).

In much of the atlas, the base map is derived from Viking data. Global Viking image mosaics prepared by USGS were assembled and reprojected (Figures 49 and 50). The same mosaics, available through USGS's useful 'Map-a-Planet' website (www.mapaplanet.org), are frequently used for local and regional maps. Finally, later parts of the atlas make use of a new map derived entirely from Mars Global Surveyor (MGS) data. This began with a composite of two global images, one from the wide-angle camera on the spacecraft and another representing albedo with infrared data acquired by the Thermal Emission Spectrometer. Various defects were removed from both datasets before they were combined. Albedo markings changed significantly in the two decades between Viking and Mars Global Surveyor. This albedo composite was combined with MGS MOLA (laser altimeter) topography to show relief more clearly (Figures 3 and 4). The ACIC and MGS maps, bracketing the period of exploration covered in this atlas, are presented for reference at the beginning of the atlas. A similar set of maps of Phobos and Deimos, constructed specifically for this atlas, portray the exploration of the Martian satellites in the last section of the book.

The areocentric solar longitude (L_s, spoken as 'L-sub-S') identifies the season of the Martian year. Its value is 0° at the northern hemisphere's vernal (spring) equinox, 90° at the summer solstice, 180° at the autumnal equinox and 270° at the winter solstice. The perihelion of Mars's orbit (closest point to the Sun) occurs at L_s 250° and the aphelion (most distant point) at 70°. One degree of L_s varies from about 1.6 Earth days at perihelion to 2.3 at aphelion because the orbit of Mars is significantly elliptical. Seasons are identified in this way where appropriate in the atlas, especially in Tables 80 and 81.

Maps and images in this atlas are always shown with north at or very near the top, except where explicitly noted or indicated by a labelled grid, so north indicators are not added. Scales are indicated using scale bars. The system for measuring coordinates on Mars has changed, leading to potential confusion in Mars maps and literature. Planetocentric coordinates were in favour when this book was compiled in 2008–2011. In this system, longitude is measured from 0° to 360° East, and latitude is measured at the centre of the planet. The older planetographic coordinates were discontinued in the first few years of the twenty-first century. Planetographic longitude was measured from 0° to 360° West, so that it increased opposite to the direction of planetary rotation. This suited astronomers, as the longitude of the central meridian of the planet's disk would increase with time. Planetary scientists came to prefer planetocentric longitudes which increase in the same sense as x coordinates on conventional graphs. Planetographic latitude takes the planet's oblateness into account by measuring angles relative to ellipsoid surface normals (the local vertical). The change occurred after the time period covered by this atlas, and the older system is used here so that coordinates will correspond as much as possible to contemporary documents. It is always preferable to specify east or west in any statement of longitude, regardless of the prevailing convention, to avoid confusion.

Care must be taken with coordinates for another reason. On Earth we take it for granted that a point may be referred to in two distinct ways, by its geographical coordinates or by its position relative to surrounding features, such as a road intersection seen in a map or air photograph. If we know the coordinates, we can use a map to find that location in the landscape. If we know the location, we can use a map to find its coordinates. The two systems are interchangeable within the accuracy of our maps because two centuries of geodetic surveying have tied locations precisely to coordinates, and global positioning system (GPS) devices perform the same task today. For Mars and other worlds, we have no GPS or geodetic surveying, so there is always uncertainty about the locations of features with respect to the grid. Consider a typical Viking image of Mars and a cartographic grid. Where should the image be placed relative to the pure geometry of the grid? The problem is solved by laborious calculation based on large numbers of control points, small recognizable features in the images. The control network thus established ties features to the grid, but it is only as accurate as the calculations. Over the years our control networks naturally increase in accuracy, but this means that coordinates from older documents are difficult to compare with new data. Here, contemporary coordinates are used without any attempt to bring them into a unified system. This mismatch also explains why the Viking spacecraft landing coordinates were known with quite high precision from radio tracking (Mayo *et al.*, 1977), whereas their locations in spacecraft images were not certain until nearby landforms were identified (Parker *et al.*, 1999, for Viking 1) and the HiRISE camera on NASA's Mars Reconnaissance Orbiter obtained images of them (Figures 54, 87, 160; Parker *et al.*, 2007). Modern positional control is based on the MGS MOLA topographic map.

The exploration of Mars has been in progress for five decades, long enough that a consistent Martian dating system is becoming desirable. Consistent dates are useful for correlating events among simultaneously operating spacecraft, including Vikings 1 and 2 and later the Mars Exploration Rovers and Phoenix. They help to tie events to Martian seasons and to keep track of times between widely separated events. Many Martian calendars have been devised (Gangale and Dudley-Rowley, 2005), but in my view only one is sufficiently widely used that it can be considered established. This is the dating system of Clancy *et al.* (2000), initially devised for comparisons of meteorological data over several Mars years. In this system, each Mars year (MY) starts at the northern vernal equinox, or solar longitude (L_s) 0°, and

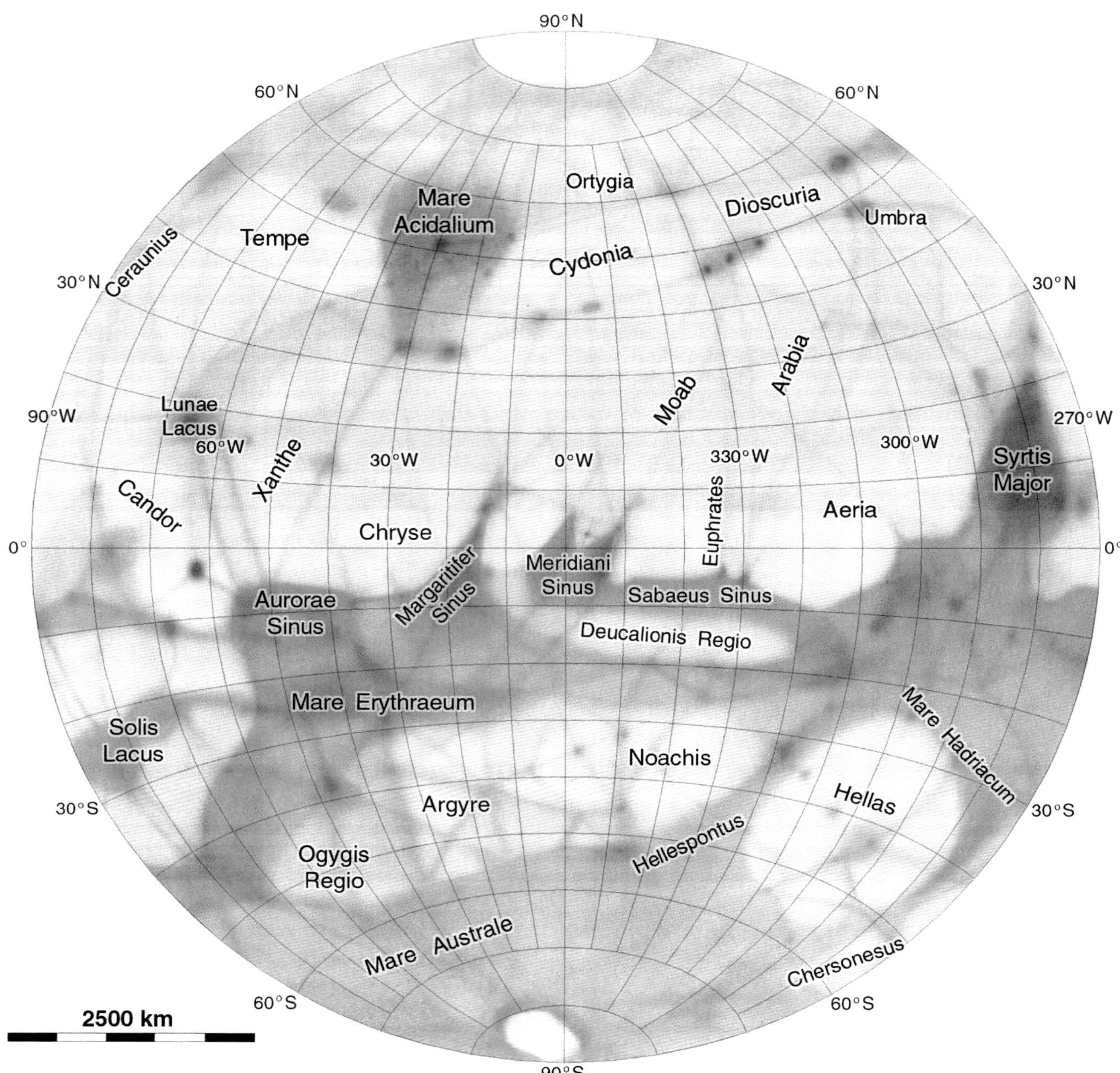

Figure 1 The Prime Meridian (0°) hemisphere of Mars with classical albedo feature names.
The map is a modified form of the Mars Experimental Chart MEC-1, published in August 1962 by the Aeronautical Chart and Information Center (ACIC) of the US Air Force.

the beginning of MY 1 corresponds to 11 April 1955, before the 'Space Age' began, and in particular before the global dust storm of 1956, the first to be well documented (McKim, 1996).

In this atlas Martian dates are given as a year and a 'sol', or Martian day. For instance, Viking 1 landed on MY 12, sol 209. When converting to Earth dates, no distinction is made here between the planet-wide sol, corresponding to local solar time at 0° longitude (*e.g.* the sol on which a spacecraft enters orbit), and the mission sol at a specific lander location, so discrepancies of up to a sol may occur here and there. Many other systems extend this concept to define

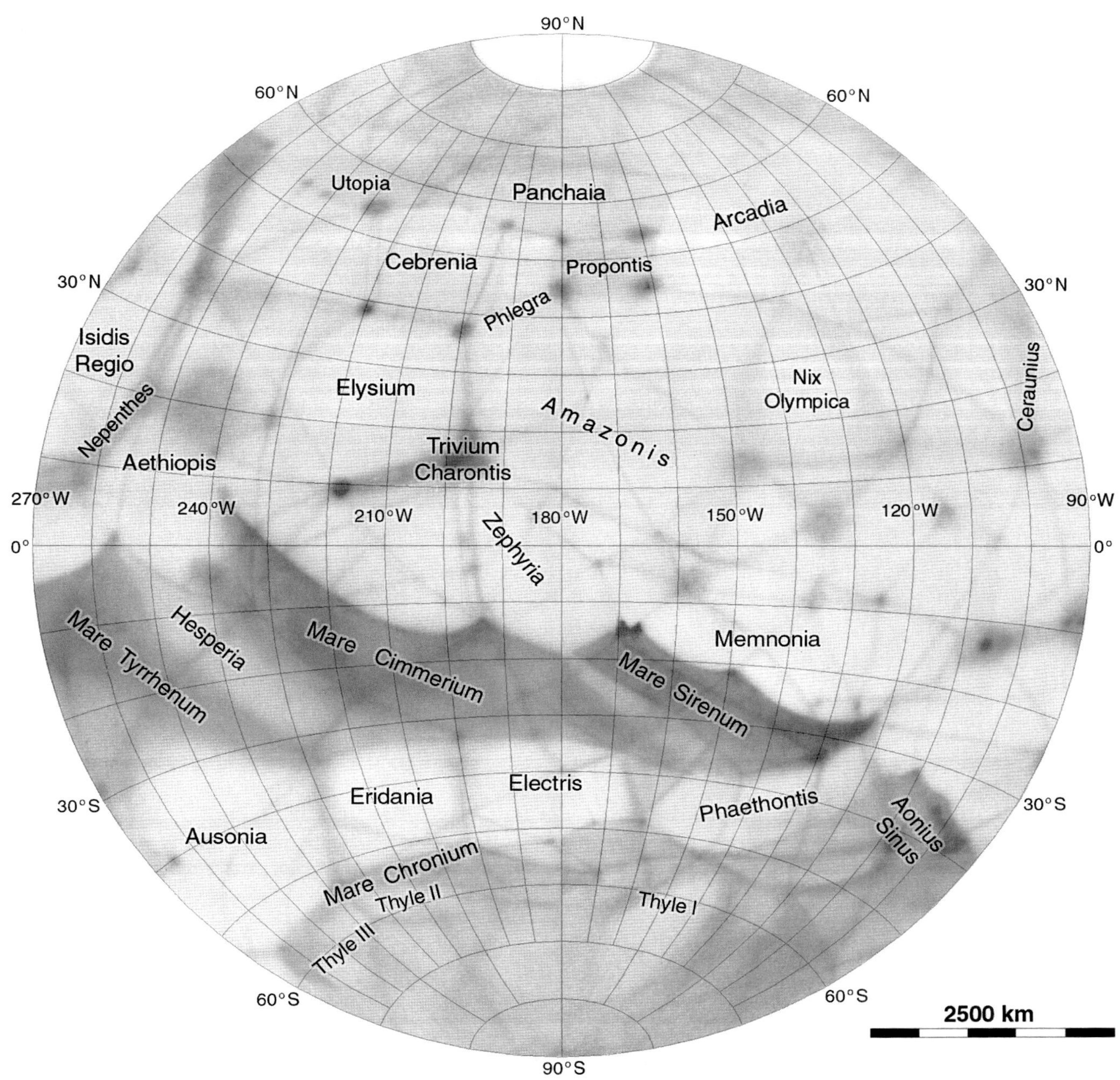

Figure 2 The 180° hemisphere of Mars, continued from Figure 1.

Martian months, but these often ingenious ideas have never been widely accepted. Tables of events in Mars exploration history are provided at the end of the atlas (Tables 80, 81 and 82). Calendar conversions can be made at a website provided by François Forget at the Laboratoire de Météorologie Dynamique, Université de Paris 6, which, at the time this was written, could be found at www-mars.lmd.jussieu.fr/mars/time/martian_time.html.

It is common to refer to events in space as 'making history', with good reason, but historians of the future cannot recount this history if its documents remain locked away. For NASA this problem is rare, but one concern arises with its competitive programs such as Discovery, for which mission proposals are considered proprietary. Some proposers release a public description, but many do not. A requirement for a public summary or a time limit on access restrictions would be helpful to

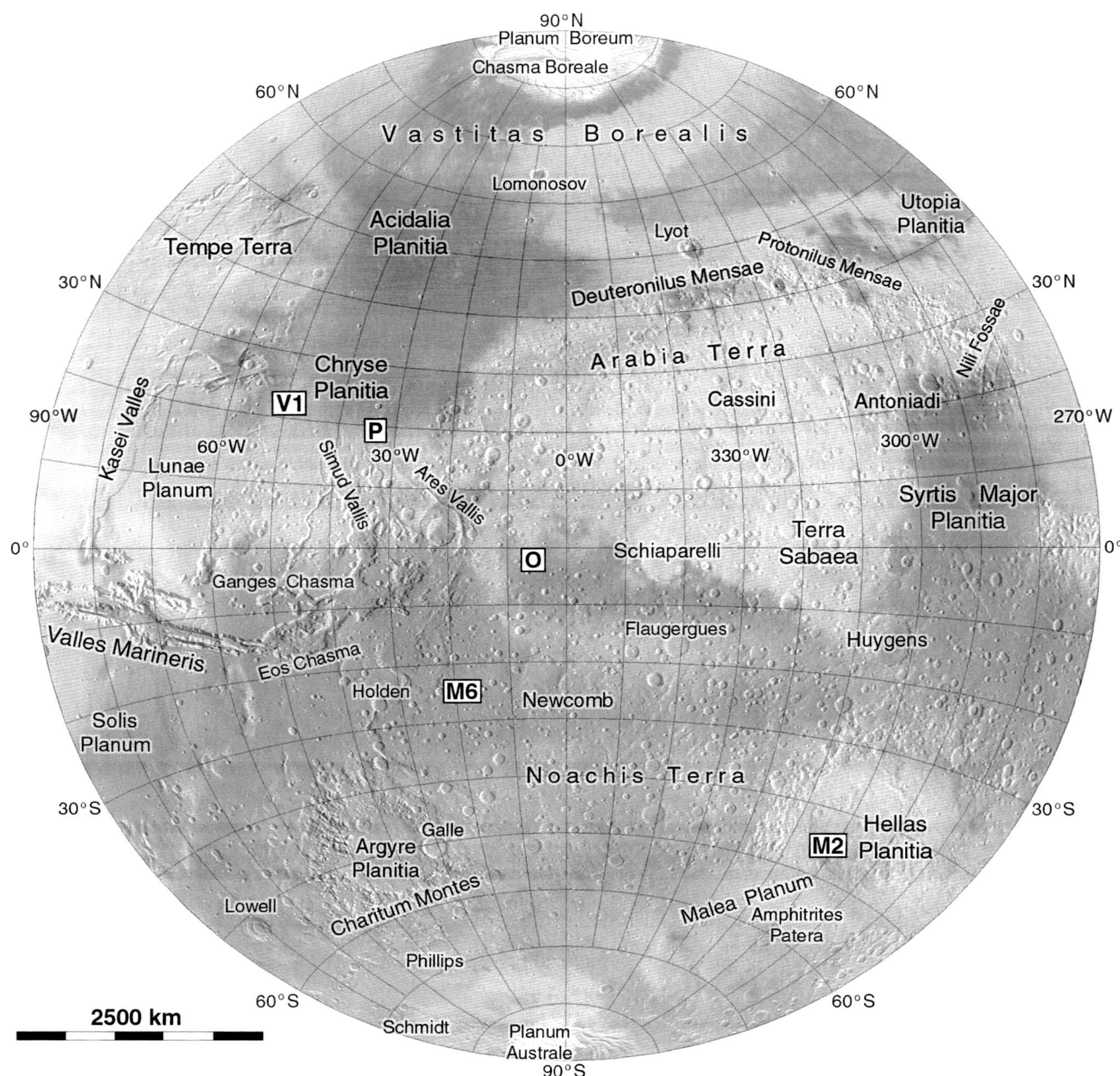

Figure 3 The Prime Meridian (0°) hemisphere of Mars with topographic feature names.
The map is a composite of Mars Global Surveyor albedo and topographic data. Artifact locations are indicated by initials: M2 – Mars 2; M6 – Mars 6; O – Opportunity; P: Pathfinder; V1 – Viking 1.

future historians. A more serious difficulty arises across the Atlantic, where the European Space Agency (ESA) has denied my requests for documents, notably for their Mars network mission proposals. Proprietary concerns aside, this shortsighted attitude serves only to diminish Europe's representation in any future histories, and as space exploration becomes more international, other nations with more restrictive cultures will present still greater difficulties. The best antidote to this may be the traditional openness of scientists, representatives of a truly international culture. Most of the people I acknowledge in this

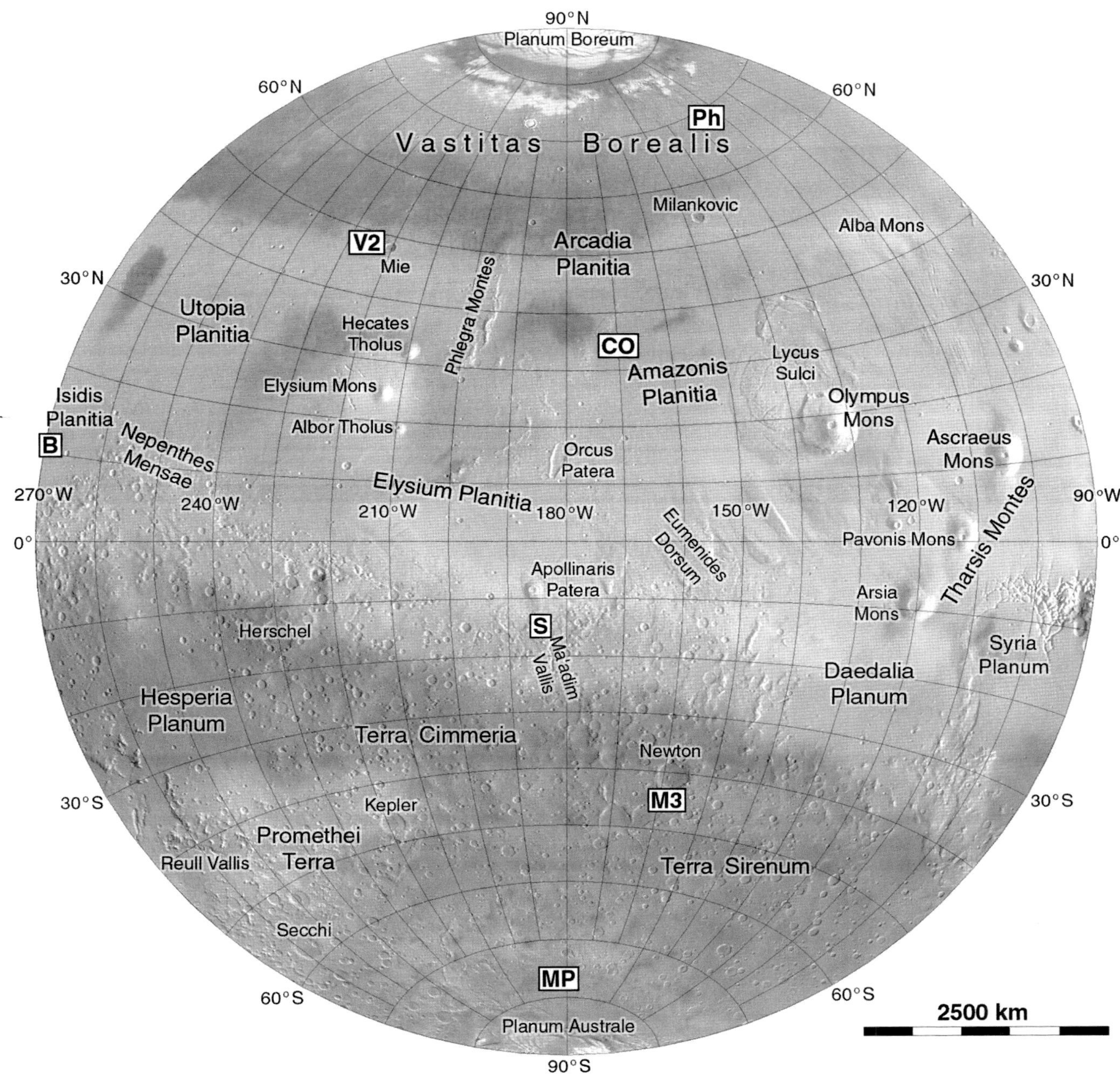

Figure 4 The 180° hemisphere of Mars, continued from Figure 3.
Artifact locations are indicated by initials: B – Beagle 2; CO – Mars Climate Orbiter (possible region of any fallen debris); M3 – Mars 3; MP – Mars Polar Lander (with Deep Space 2); Ph – Phoenix; S – Spirit; V2 – Viking 2.

book have provided information or access to documents in order that this history can be recorded, but sadly some people see little value in history, even when they are helping to make it. The penalty for secrecy is omission from history.

Above all I hope that this collection of maps and information will serve as a useful summary of the first half-century of Mars exploration. Despite all the details presented here, this is only a very condensed review of the subject, but the bibliography leads out into the broader literature. By summarizing widely scattered sources in one convenient volume and in a common format, I aim to create a useful reference and to help make some of the more obscure literature more accessible. In addition, substantial parts of the text go beyond what can be found elsewhere to fill gaps in that literature. Almost all of the illustrations were created specifically for this atlas. I hope others will find this volume as satisfying to read and use as I have found it satisfying to compile.

Acknowledgements

Many people and institutions have helped me as I compiled this atlas, and I gratefully acknowledge them here. The basic mission facts, such as launch or encounter dates and instrument lists, are largely based on the descriptions provided by the National Space Science Data Center, forming a framework on which I could construct the rest of the text.

Norman Crabill kindly provided the minutes of the Viking Landing Site staff meetings for the Viking landing site certification, a unique source not previously described in detail as far as I know. Mike Carr (US Geological Survey, retired), Matt Golombek (Jet Propulsion Laboratory, JPL) and Rick Kline (Cornell University) helped locate information. Matt Golombek also took time out from his busy schedule to write the foreword, for which I am extremely grateful. Andrew Ball (Open University) helped me track down European Space Agency (ESA) sources, and it is not his fault that they proved impossible to obtain in the end. David Portree (US Geological Survey Flagstaff) provided very useful documents, especially concerning the Mars 1984 mission plans. Ken Herkenhoff and Ron Greeley helped with my queries about mission plans. Gerald Gaidmore at the John Hay Library of Brown University provided access to the papers of Thomas Mutch, and Peter Schultz and Peter Nievert at the Northeast Planetary Data Center at Brown University gave me access to important items in their collection. Emily Lakdawalla (Planetary Society) very promptly found some information I needed. As always, the staff at the Lunar and Planetary Institute near Houston have been very helpful over the years, at first Stephen Tellier and more recently Mary Ann Hager, Linda Chappell and David Bigwood.

The University of Western Ontario (UWO) provided sabbatical time and research facilities, especially the Serge A. Sauer Map Library (formerly part of the Department of Geography). Cheryl Woods has been a constant friend and supporter of my work. I also acknowledge inspiring discussions with faculty and students at UWO's Centre for Planetary Science and Exploration. UWO students Jennifer Gill and Jennifer Stenson helped me gather material at an early stage in the preparation of this atlas. My Russian colleagues have also helped with the provision of data and facilities or other assistance while I was visiting Moscow. They include Kira B. Shingareva of the Moscow State University of Geodesy and Cartography (MIIGAiK) who first invited me to Moscow, her daughter Tanya and her colleague Bianna V. Krasnopevtseva of MIIGAiK, and the Nyrtsov family who made my visits so enjoyable and productive: Tamara P. Nyrtsova of MIIGAiK, Valery V. Nyrtsov and Maxim V. Nyrtsov. Alexander T. Basilevsky (Vernadsky Institute of Geochemistry) has helped me frequently over many years, and Vladislav V. Shevchenko and Jeanna F. Rodionova (Sternberg State Astronomical Institute) offered hospitality and provided data. I offer them my deepest thanks.

Other colleagues have helped me by providing images. Olivier de Goursac provided the cover image, for which I am extremely grateful, as well as information concerning informal place names at the Viking 1 landing site. William Farrell (NASA Goddard) kindly allowed me to use his MARSIS radargrams of the south polar region, though in the end I took others from ESA's Planetary Science Archive. Ted Stryk and Mike Malaska have provided me with their uniquely processed versions of images as noted next. Both are fellow contributors to www.unmannedspaceflight.com, a stimulating community of experts and enthusiasts devoted to the robotic exploration of the solar system which I am pleased to have been associated with since 2005. This forum was founded by Doug Ellison and is now run by the Planetary Society. The quote from Thomas Mutch on the frontispiece is from NASA (1978).

Last, I extend my warmest thanks to Vince Higgs and his colleagues at Cambridge University Press, including Claire

Poole and Louis Gulino, and to Brigitte Coulton at Aptara. Their confidence in me and their help at every stage of editing and production has made this atlas possible.

Data Sources

Most of the data used in creation of the illustrations in this atlas come from NASA planetary missions, and in most cases I have processed raw data from NASA's Planetary Data System rather than relying on press release images (*e.g.* Figure 109). Some figures also make use of maps prepared by the US Geological Survey (*e.g.* Figure 45), the US Air Force's Aeronautical Chart and Information Center (*e.g.* Figure 23B) and the Army Map Service (Figure 17C). I make frequent use of the web-accessible global photomosaics produced by USGS (*e.g.* Figure 125) and Arizona State University (*e.g.* Figure 187). All of these raw materials are in the public domain and are credited in captions.

A small number of images require specific credits. I especially thank Ted Stryk of Roane State Community College for allowing me to use his specially processed images from Mars 3, Mars 4 and Mars 5; Deep Space 1; and Phobos 2, as credited in the appropriate captions. I have compiled my own special mosaics for this atlas from other Soviet images and have made new multi-image composites from the Phobos 2 Termoscan images. Jeanna Rodionova (Sternberg State Astronomical Institute) kindly provided prints of those images which I scanned and then merged the visible and infrared channel data to prepare these illustrations (Figures 121D, 121E, 122B). Mike Malaska processed a mosaic of images from the Visual Monitoring Camera on Mars Express (Figure 190C).

All other credits below, listed by mission in roughly chronological order, are for specific spacecraft instruments and their Principal Investigators. Without their dedicated work, none of this exploration would be possible.

Mariners 4, 6 and 7: NASA/JPL and Robert Leighton.

Mariner 9: NASA/JPL and Harold Masursky.

Viking Orbiters: NASA/JPL and Michael Carr.

Viking Landers: NASA/JPL, Thomas Mutch, Raymond Arvidson.

Hubble Space Telescope: NASA, ESA, the Space Telescope Science Institute, the Hubble Heritage Team, and J. Bell, T. Clancy, D. Crisp, P. James, R. Kahn, S. Lee, A. Lubenow, L. Martin, J. Neubert, R. Singer, M. Wolff and R. Zurek.

Mars Observer: NASA/JPL/Malin Space Science Systems and Michael Malin.

Mars Global Surveyor: Mars Orbiter Camera: NASA/JPL/Malin Space Science Systems and Michael Malin. Mars Orbiter Laser Altimeter: NASA/Goddard Space Flight Center and David Smith.

Mars Pathfinder: Imager for Mars Pathfinder: NASA/JPL and Peter Smith. Rover Imaging Cameras: Henry Moore.

Mars Climate Orbiter: Mars Color Imager: Mars Observer: NASA/JPL/Malin Space Science Systems and Michael Malin.

2001 Mars Odyssey: Thermal Emission Imaging System: NASA/JPL/Arizona State University and Philip Christensen.

Mars Express: High Resolution Stereo Camera (HRSC): ESA and G. Neukum. Mars Advanced Radar for Subsurface and Ionospheric Sounding (MARSIS): ESA and G. Picardi, J. Plaut and R. Orosei. Visual

Monitoring Camera (VMC): ESA.

Mars Reconnaissance Orbiter: High Resolution Imaging Science Experiment: NASA/JPL/University of Arizona and Alfred McEwen.

1 Chronological sequence of missions and events

Mars at the Dawn of the Space Age

Mars entered the Space Age as a remote target of astronomy and became an explored world of geology and meteorology during the five decades covered by this atlas. The planet, long known for its faint reddish colour, is 6792 km in diameter or 53% of the diameter of Earth. It orbits the Sun at a mean distance of 228 million km with a period, a Martian year, of 687 days. Mars rotates about its axis with a mean solar period of 24 hours 39.6 minutes, a period now known as one sol, a Mars day. The sidereal rotation period, relative to the stars, is 24 hours 37.4 minutes. Its axis is tilted 25.2° to its orbit plane, causing seasons similar to those of Earth, though its eccentric orbit makes the seasons unequal in length. The northern hemisphere spring is the longest season (194 sols), and northern hemisphere autumn is the shortest (142 sols). One Mars year contains 668.6 sols. As Earth and Mars move in their respective orbits, opportunities to launch spacecraft to Mars recur at 26-month intervals. Because this period is not commensurate with the orbit of either planet, each mission arrives at a different Mars season and accessible latitudes vary accordingly, as can be seen by comparing Figures 156 and 180B, for example.

The history of Mars observation and mapping is described by Flammarion (1892, 1909), Blunck (1977, 1982) and Blunck *et al.* (1993). In 1659 Christiaan Huygens observed surface markings on Mars (Syrtis Major and the south polar cap) and deduced the rotation period. The earliest cartographic representation of the planet dates from 1783, when Sir William Herschel (1784) made a composite of seven drawings of different faces of Mars to create a map centered at the south pole (Figure 5A, 5C). Herschel observed only the southern hemisphere and low northern latitudes, but in the 1830s Wilhelm Beer and Johann Mädler (1831, 1841) combined observations from several oppositions to produce a map of the whole planet (Figure 5D). They also established an arbitrary prime meridian, which is the basis of that used today. Maps made over the next few decades showed broad regions of light and dark material, to which the names of lands and seas were applied by later astronomers, notably Richard Proctor (Figures 6A, 6B). When Mars made an unusually close approach in 1877, a new type of feature began to appear on maps, the famous canals of Mars. First drawn by Giovanni Schiaparelli (1877) and named 'canali', or channels, they were interpreted as rivers (Figure 6B) by Proctor (1888) but as artificial irrigation canals by Percival Lowell (1895, 1906, 1910). Many maps of the early twentieth century contained canals.

Most astronomers eventually concluded that canals in the form of thin straight lines were optical illusions, and they had largely disappeared from maps by mid century. Typical maps of the 1950s, including the International Astronomical Union (IAU) map drawn by Glauco de Mottoni y Palacios in 1957 (IAU, 1960) as a key to nomenclature (Figure 6C), showed broad light and dark areas similar to the cartography of the 1850s.

Despite this, canals were still being drawn by some cartographers in the middle of the twentieth century and are shown on the first maps made for the National Aeronautics and Space Administration (NASA) by the US Army and Air Force in 1962. The Air Force map (Figures 1, 2) made by the Aeronautical Chart and Information Center (ACIC) was drawn in a cylindrical projection, the Army map in several orthographic hemispheres. The ACIC map misplaced Syrtis Major, shown here at about 280° W. It should really be 10° farther west (Figure 14). The error is retained here for historical accuracy and to reflect the difficulty of comparing old and current maps, and it is repeated in all later uses of the ACIC map (Figures 5B, 7 and 8, for instance). The region is shown correctly on every other map used in this atlas. The canals were already anachronistic, but the general pattern of albedo markings reflects knowledge of Mars at the beginning of the Space Age.

No relief features were visible through telescopes. Claims to have seen craters were made in 1915 by John Mellish, and even as early as the 1890s by Edward Emerson Barnard (Sheehan and McKim, 1994), but it seems

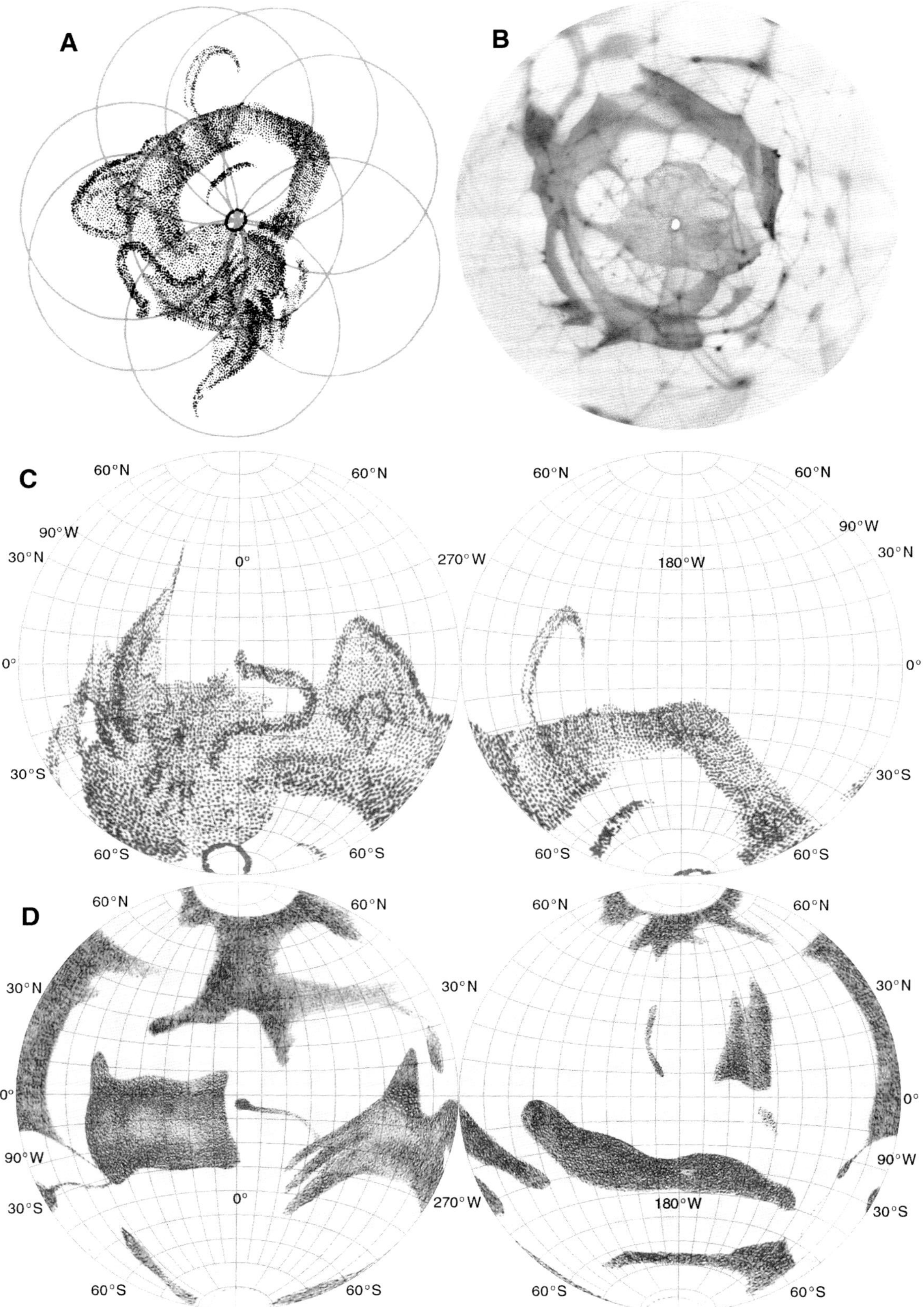

Figure 5 A: The first map of Mars, by William Herschel (1784). **B:** The ACIC map of Figure 1 reprojected to compare with Herschel's map. Syrtis Major is shown 10° too far east by ACIC and would match Herschel's map better if corrected. **C:** Herschel's map modified to match the geometry of other global maps in this atlas. **D:** A composite of two maps by Beer and Mädler (1831, 1841), also reprojected to the common atlas format and correcting an error in the original.

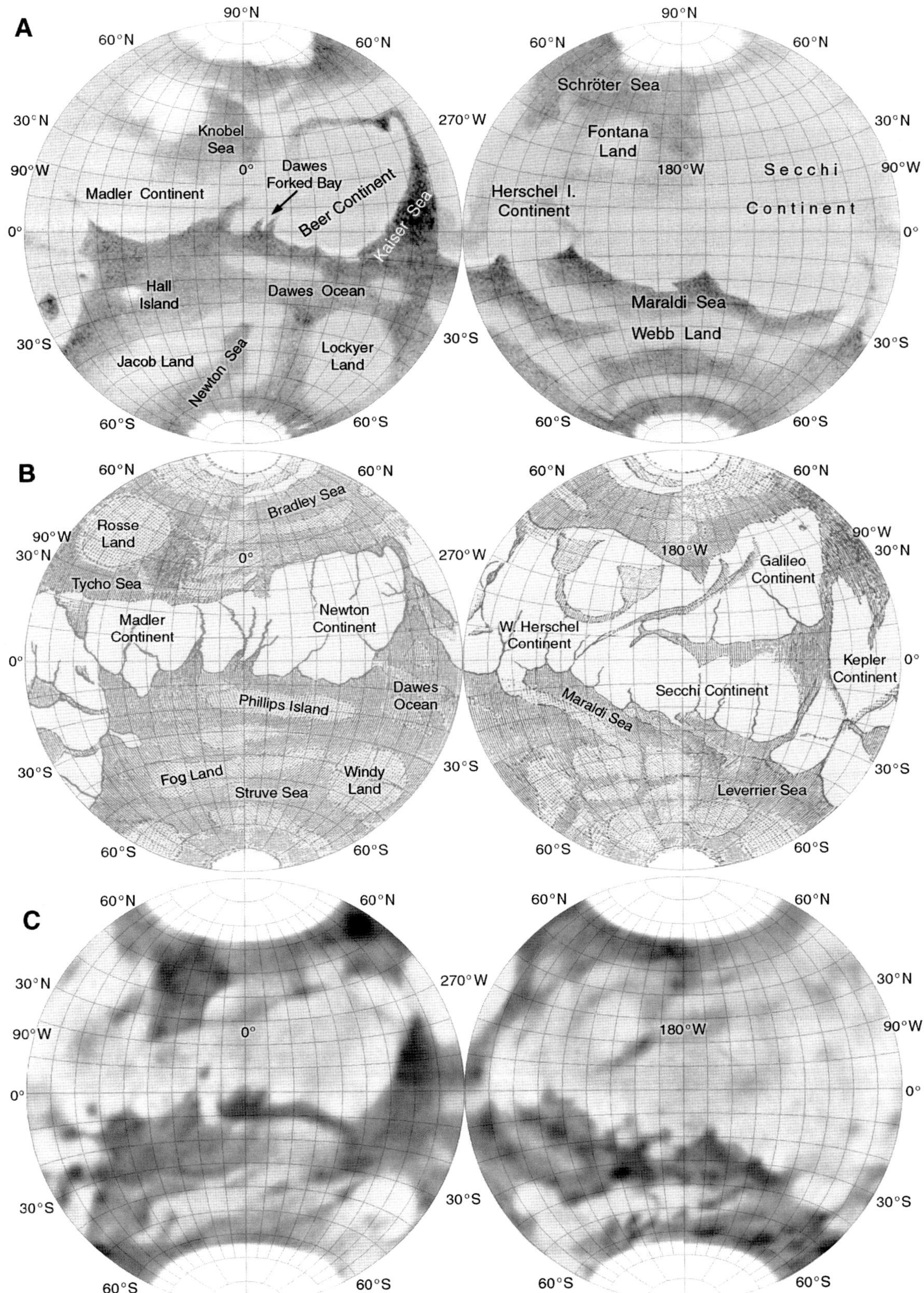

Figure 6 Early maps of Mars, modified to match the geometry of other global maps in this atlas.
A: Map of Mars by Nathaniel Green (1879). **B:** Map by Richard Proctor (1888). (A) and (B) include versions of Proctor's feature names. **C:** Albedo map of Mars based on a 1957 drawing by Glauco de Mottoni y Palacios (IAU, 1960), typical of most Mars maps from the half century preceding the Space Age.

clear that they were based only on glimpses of small albedo markings. A few of these may coincide with craters, but many more do not. A patch of dark sand dunes inside a crater is no more a crater than, on the Moon, Mare Crisium is the Crisium impact basin.

No unambiguous detection of a depressed circular area, a true crater, was ever made before the first spacecraft images were taken in 1965 by Mariner 4 (Figure 8). A related issue is the misrepresentation of Olympus Mons and Kasei Valles as craters in a 1970 map by the US Army Map Service (Figure 17C), based on roughly circular albedo markings seen by Mariners 6 and 7 (NASA, 1969). Another example is the overinterpretation of faint markings as craters in ACIC maps of some areas imaged by Mariner 4 (Figures 13C, 13D and 23B). Figure 13B shows another crater concept, that the small spots called 'oases' at canal intersections might be craters.

Astronomical observations had established the basic size, rotation state and surface features of the planet, but the nature of the surface remained unclear. The growth and decay of the white polar caps suggested Earth-like seasons. Interest in biology overshadowed geology at this time, and much of the focus was on the 'wave of darkening'. This phenomenon begins each Martian spring when a dark area forms around the shrinking polar cap and expands towards and across the equator during the summer season, alternating between the northern and southern hemispheres. It had been monitored by telescopic observers on Earth for decades and was widely interpreted as evidence of seasonal plant growth on Mars, fed by water released by the melting polar cap. Today it is attributed to atmospheric clearing as the winter carbon dioxide cap disappears. The smaller water ice cap does not melt.

Place names on Mars are based on the system of names from classical geography devised by Schiaparelli beginning in 1877, as described by Blunck (1977, 1982). These names referred to albedo markings only and replaced the names used by Proctor and others (Figure 6), which commemorated astronomers. Other astronomers added hundreds of additional names for small markings and canals. An entirely new set of names became necessary when spacecraft revealed the topographic features of Mars. Large craters were named for prominent scientists who had studied Mars or had worked in related fields. Smaller craters have since been named for small towns on Earth. Other features – the volcanoes, mountains, valleys and other landforms of this complex planet – took their names from nearby albedo markings, in combination with Latin descriptive terms to identify the type of landform (*mons* for mountain, *vallis* for valley and so on). Detailed information on place names and current maps of them are available at the US Geological Survey (USGS) Planetary Nomenclature website (planetarynames.wr.usgs.gov).

1950s and 1960s: **Humans to Mars: Early Thoughts**

As with the Moon, thoughts of human Mars exploration predate the dawn of the Space Age. Fictional accounts by Edgar Rice Burroughs, C. S. Lewis, Arthur C. Clarke and others helped to inspire space engineers and scientists in the mid twentieth century. Early Mars plans are summarized by Portree (2001) and Platoff (2001) and are only very briefly commented on here, with emphasis on suggested landing sites. Additional details are found in NASA (1964), including the statement (p. 193) that 'In general, man will play a much more dominant role in the exploration of Mars than in the exploration of the Moon because the planet is more complex and more remote, and is less amenable to comprehensive exploration by unmanned systems'. The last point turned out to be false, but early planners did not predict the dramatic developments in computing, automation and robotics which enabled later exploration by orbiters, landers and rovers. Mission studies dating from the 1950s and 1960s, as described by Portree (2001), are briefly summarized in Table 1.

The earliest detailed plan for a Mars expedition was described by Wernher von Braun (1953). His 'Mars Project' was a massive expedition involving ten spacecraft of 4000 tons each, a crew of 70 and no robotic precursor missions. These enormous spacecraft would be assembled in Earth orbit from parts launched by a fleet of fully reusable shuttles, requiring nearly 1000 launches. The journey to Mars would take eight months. Seven of the ten spacecraft would carry only fuel and crew quarters, but three would include winged gliders for access to the surface of Mars. Two landers would wait in Mars orbit while the crew used telescopes to seek a smooth landing site near the equator. The third crew would fly one of the large gliders to a landing on ski-like skids on one of the polar ice caps. The site would be chosen before launch from telescopic studies, based on the assumption of a very smooth snow-like surface. When a suitable location was found, the polar crew would drive 6000 km to the equator to build a smooth landing strip for the remaining landers. Operating from an inflatable habitat, the crew would spend 400 days on the surface, studying Martian life and canals, before leaving for the eight-month journey back to Earth. The surface activities are illustrated schematically in Figure 7.

Ernst Steinhoff, another of the original German rocket engineers, also suggested large winged shuttles for Mars exploration (Steinhoff, 1963). His mission would operate from a base on Phobos, using shuttles at first to drop supplies and crew by parachute before climbing back to orbit, and only later landing. The Mars base would be within 25° of the equator, and both Mars and Phobos bases would make use of local resources as much as possible. Another report from Space Technology Laboratories in NASA (1964) described possible Mars surface operations. A landing site would be studied from orbit. The crew would land and study their surroundings, deploying instrument stations nearby. Next they would fly an aircraft to a remote site, undertaking aerial reconnaissance on the way, and then study the remote site before returning to the landing area.

Avco Corporation conducted a study of human expeditions to Mars under contract to NASA Marshall Space Flight Center's Advanced Systems Office (Swan *et al.*, 1966). Avco also prepared studies of Venus and Mars robotic landers for the Voyager program (Tables 2 and 3, Figure 8). Avco described two types of human Mars missions. The first would be an extension of Apollo, carrying a crew of six. Four people would explore the surface of Mars for 21 days while the remaining two stayed in orbit in the command module. The expedition would deliver three modules to the surface: a shelter, a pressurized garage, and a two-person, 4000-kg mobile laboratory (MOLAB) able to travel up to 800 km. The astronauts would seek evidence of life (present or past), study Mars itself, and study natural resources to be used by the next mission. This first visit would land at a location where the so-called wave of darkening was in progress. Depending on the date of arrival, the target site would be adjusted to fall within the latitude affected by the wave at the time. A flight in 1978 would spend 21 days in Syrtis Major at the start of northern hemisphere summer.

The second class of mission, called a synodic (extended) expedition, would establish three Mars surface bases, each with a crew of 14, in the 1980s. They would stay on the surface for 300 days. Each of the three bases would include two seven-person shelters, two nuclear power plants, two garages (one of them pressurized) and a pair of two-person MOLABs able to make traverses up to 900 km long and lasting 30 days. A total of 8000 kg of scientific equipment would be carried. Four people would stay in Mars orbit, giving a full crew size of 46 people.

The three base sites would be chosen to provide access to all the major types of Martian feature, as understood at the time. A base in the northern dark area Syrtis Major would allow traverses to the bright 'desert' areas Libya and Aeria. A base anywhere in Hellas, a confined bright southern desert region, would be near Zea Lacus, where five canals intersect in the centre of Hellas. The third base would be located in the Mountains of Mitchel near the south pole. This area is seen as an isolated white spot as the cap recedes in the southern summer (2003 image in Figure 141) and was thought to be a frost-covered mountain range. It is now seen to be part of the rim of a large impact basin and an adjacent ridge (Figure 17). These Avco sites are shown in Figure 7.

An early Mars Surface Sample Return (MSSR) mission was described by Northrop Space Laboratories (NASA, 1967) as a way to increase the scientific value of a human Mars flyby. This study was done for the Advanced Studies Office at NASA's Marshall Space Flight Center after NASA's Planetary Joint Action Group (NASA, 1966) proposed MSSR as part of a piloted Mars flyby mission (Table 1). The MSSR probe would separate from the piloted vehicle; land; collect samples of dirt, dust, rock, air and (possibly) life; and launch them back to the passing flyby spacecraft for immediate analysis. Northrop's study assumed a launch in 1975, 1977, or 1979, with an approach over the planet's sunlit side, closest approach near the terminator and departure over the night side. The probe landing site would be in darkness and out of contact with the flyby spacecraft. Landing site latitude would be dictated mainly by communication requirements during the launch from Mars and the date and time of Mars

Table 1. *Early Plans for Human Mars Missions (Portree, 2001)*

Date	Description
1952	The Mars Project: ten spacecraft and 70 crew. The first crew land on one of the polar ice caps and undertake a traverse to the equator to build a landing strip for two more landers at a site chosen from orbit. Total of 400 days on the surface (von Braun, 1953).
1956	A considerably reduced version of von Braun's 1953 plan with two spacecraft and 12 crew. Landing sites, near the equator for maximum warmth, would be chosen using telescopes on an Earth-orbiting space station. Margaritifer Sinus was suggested, and observations from orbit would refine the site. Nine of the crew would spend a year on the surface (Ley and von Braun, 1956).
1960	NASA's first Mars study by the Lewis (now Glenn) Research Center. A crew of seven would travel to Mars in 1971 on a 420-day round trip using nuclear rockets. Two would land and spend 40 days on the surface. For the first time, radiation shielding received serious consideration.
1962	A 1980s Mars expedition described by Ernst Stuhlinger (ABMA), using electric propulsion with five spacecraft and 15 crew. A cargo lander would touch down first, and if successful a crew would follow for a 29-day stay. If the cargo lander failed, a backup could be used.
1962	EMPIRE (Early Manned Planetary-Interplanetary Roundtrip Expeditions) studies were made by Aeronutronic (Ford), Lockheed and General Dynamics for NASA's Marshall Space Flight Center. They involved a 450-day mission launching in 1975, with observations of Mars by the crew and by probes. One version (hot) included both Venus and Mars flybys, the other (cold) had aphelion in the asteroid belt after the Mars flyby, followed by the return to Earth. Another version was a Mars orbit mission with soft landers, balloon probes, Mars sample return probes, Phobos and Deimos hard landers similar to the early lunar Rangers, and a deployed remote-sensing orbiter. A small lander might put two people on the surface for seven days.
1963	Marshall Future Projects Laboratory studies (NASA, 1964). Multiple mission descriptions. In one, a crew flies to Mars in 120 days and spends 40 days on the surface as their spacecraft flies into solar orbit. When a second spacecraft, launched first on a 200-day trajectory, flies past Mars, the crew launch and meet it for the return trip. Another option included orbital missions with landers capable of 10- to 40-day surface excursions. Aerobraking might be used to enter orbit. A landing site suggested by Aeronutronic was a dark 'mare' region at Cecropia (65° N, 320° W), where organisms might make use of water from the retreating polar cap. One mission was dubbed UMPIRE (U = unfavourable, for the poor orbital geometry during the late 1970s and increased radiation from the more active Sun at that time). Its first landing should establish a long-term base.
1966	The Planetary Joint Action Group (Bellcomm and NASA centers) studied a variety of Mars and Venus flyby missions using upgraded Apollo hardware to follow the Mariner and Voyager robotic missions. A 1975 mission with a crew of four would deploy three penetrators, a polar orbiter, a small lander and a large sample return lander. The flyby included remote sensing and a 200-km close approach. Balloons might be deployed in some versions of missions like these. Planetary Joint Action Group suggested that crews might land on Mars and orbit Venus when nuclear rockets were available.
1968	A Boeing plan for a large nuclear rocket to send crews to Mars or Mars and Venus between 1978 and 1998, with various mission designs lasting between 460 and 900 days. North American Rockwell described a lander which could fly in 1982, a light version putting two people on Mars for four days, a heavier version placing four people on Mars for 30 days.
1969	NASA's Space Task Group described a mission involving nuclear rockets, with twin 1981 launches and 1982 Mars orbit operations. During two days in orbit, the crew would select landing sites for 12 robotic sample return probes. Samples would be brought back to the orbiters and studied for biological hazards. If they were safe, a crew of three would land, keeping a second lander on standby for a rescue. The crew would spend 30 to 60 days on the surface studying geology and biology and searching for water for future In Situ Resource Utilization planning. A Venus swingby on the return trip could include radar mapping and the release of four probes. The crew would be quarantined in Earth orbit and the spacecraft refurbished and reused. Repeat flights would occur every 2 years, leading to a 50-person surface base in 1989.

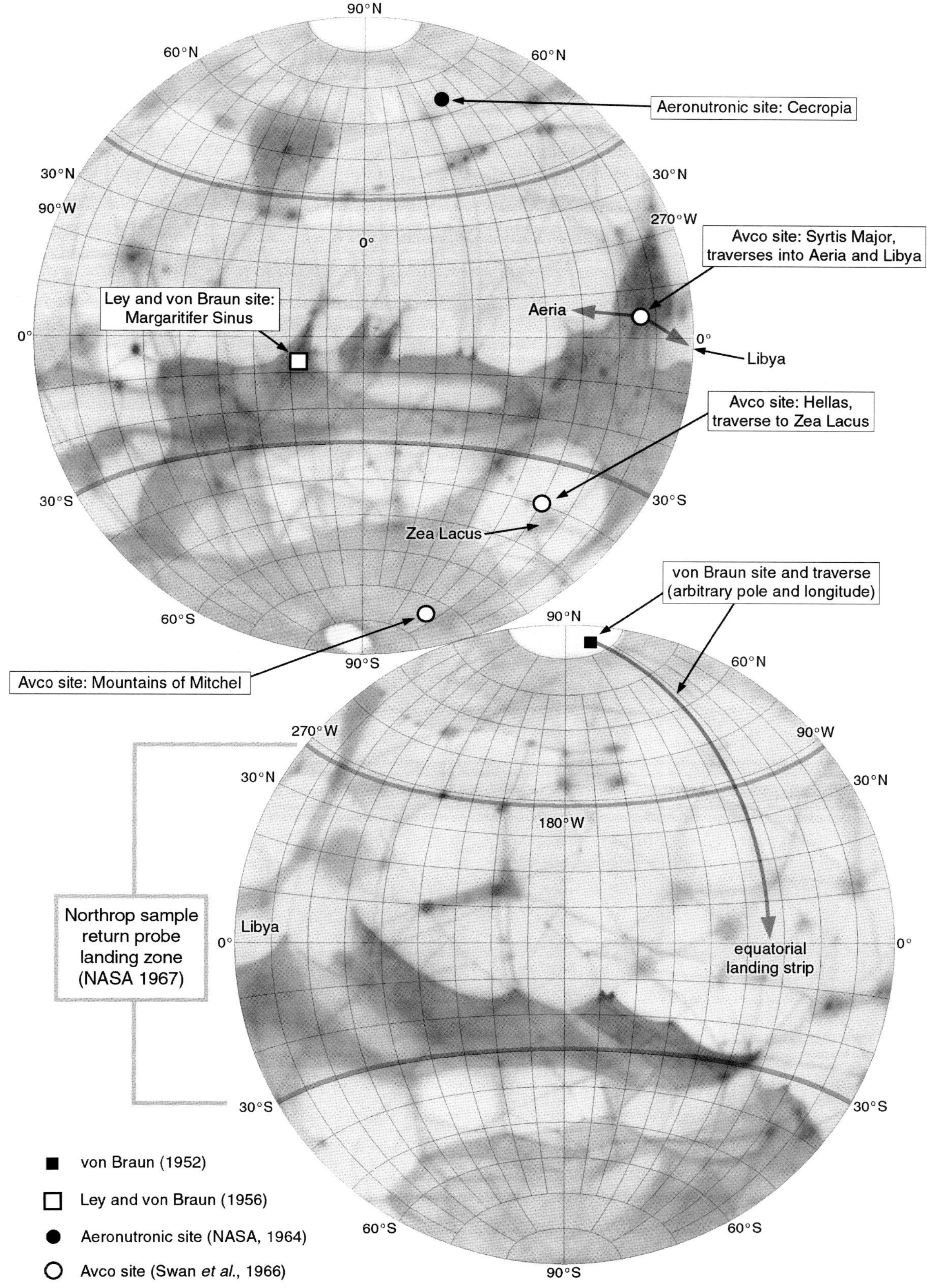

Figure 7 Landing sites suggested for early human missions to Mars, from Table 1.

arrival. In 1975, for example, the probe could land between 39° N and 30° S latitude (Figure 7). Liwshitz (1966) suggested trying to land within one of the many dark spots, roughly 100 km wide, glimpsed from Earth or seen by Mariner 4.

1959: **Robotic Mission Planning at JPL**

Very early in NASA's existence plans began to be prepared for spacecraft to explore the Moon and planets. An early example (Hibbs, 1959) from the Jet Propulsion Laboratory (JPL), prepared at NASA's request and drawing in part on earlier studies for the US Army, proposed a series of missions to the Moon, Mars and Venus between 1959 and 1964. The plan combined scientific goals with engineering feasibility to identify necessary technical advances and to plan a reasonable series of missions capable of satisfying both scientific goals and public interest. Instruments would include both film and television cameras, magnetometers, particle and micrometeorite detectors and spectrophotometers for compositional studies.

The report called for three or four launches per year from 1960 to 1964 beginning with a lunar flyby in August 1960, mainly to test equipment. The second mission, a Mars flyby in October 1960, would pass within 1.5 million kilometres of Mars to obtain infrared scans, images and particles and fields data. This would be followed by a similar Venus flyby, a lunar hard landing mission and lunar orbiter, and then in 1962 a Venus orbiter and entry probe. Mars would be the target again in November 1962 with an orbiter and entry probe. The orbiter would investigate the ionosphere, atmosphere and surface with radar, optical, ultraviolet and infrared sensors and particles and fields instruments. The surface would be mapped to look for future landing sites. The entry probe would measure temperature, surface hardness and illumination levels and atmospheric dust, and it would transmit surface images. The remaining missions in the JPL plan were landers for the Moon and Venus, but a Mars soft lander would follow in late 1964 as a backup to the Venus mission if it proved too demanding. This would include a rover and a communication relay satellite if needed, and biological studies would be undertaken on the surface. The JPL plan was eventually followed in its broadest outlines, but spread over a much longer period and with the more advanced Venus missions cancelled in favour of additional Mars exploration.

10 October 1960: **Mars 1960A (Soviet Union)**

This first attempt to send a spacecraft to Mars began with a launch at 14:28 UT from the Baikonur Cosmodrome. The Soviet designation was Korabl 4. The mission objective was to fly an 850-kg probe past Mars at close range, but the third stage of the launch vehicle failed, and the spacecraft was unable to reach orbit. A film camera similar to that on Luna 3 was to be carried, but weight problems eventually required that it be removed. Other instruments were to measure fields and particles, and a spectrometer would attempt to detect methane in the planet's atmosphere.

14 October 1960: **Mars 1960B (Soviet Union)**

This second mission of the first Mars opportunity, referred to as Korabl 5, was identical to Mars 1960A. It was launched at 13:51 UT from Baikonur, but a launch vehicle failure prevented it from reaching its parking orbit.

24 October 1962: **Sputnik 22 (Soviet Union)**

At the next Mars launch opportunity the Soviet Union attempted three Mars launches. The first spacecraft of the three, then known as Korabl 11, was an 894-kg probe designed to fly past Mars. It was launched at 17:55 UT and entered a 196 by 485 km parking orbit inclined 65° to the equator, with a period of 89 minutes. A stage of the launch vehicle failed as the spacecraft was entering or leaving the parking orbit, fragmenting the vehicle. The main fragments re-entered the atmosphere on 29 October and were destroyed. The Sputnik designation was intended to conceal the nature of the failed mission. The 1962 Mars missions were to carry the first cameras to Mars. These were improved versions of the Luna 3 cameras with colour filters and the capacity to record 112 frames.

The International Atlas of Mars Exploration

The First Five Decades
Volume 1: 1953 to 2003

Covering the first five decades of the exploration of Mars, this atlas is the most detailed visual reference available. A wealth of information from diverse sources is brought together for the first time, featuring annotated maps, photographs, tables and detailed descriptions of every Mars mission in chronological order, from the dawn of the Space Age to Mars Express. Special attention is given to landing site selection, including some missions that were planned but never flew. Phobos and Deimos, the tiny moons of Mars, are covered in a separate section. Contemporary maps reveal our improving knowledge of the planet's surface through the latter half of the twentieth century. Written in nontechnical language, this atlas is a unique resource for anyone interested in planetary sciences, the history of space exploration and cartography, whatever their background, whereas the detailed bibliography and chart data are especially useful for academic researchers and students.

PHILIP J. STOOKE is a cartographer and imaging expert at the University of Western Ontario, whose interest in mapping the Moon and planets began during the Apollo missions. He has developed novel methods for mapping asteroids, and many of his asteroid maps are now accessible from NASA's Planetary Data System. He has studied spacecraft locations on the Moon and Mars, especially attempting to locate Viking 2 on Mars. He is the author of many papers and articles on planetary mapping, planetary geology and the history of cartography and planetary science. His book *The International Atlas of Lunar Exploration* was published by Cambridge University Press in 2008.

1 November 1962: **Mars 1 (Soviet Union)**

Mars 1 was the first successful launch into a Mars-bound trajectory, so the 894-kg spacecraft received the first official 'Mars' designation. The mission was intended to fly past Mars at about 11 000 km range, photographing the planet on film which would be developed and scanned for transmission. It was also intended to measure the reflection spectrum of the surface, the magnetic and radiation environment of Mars, and cosmic rays and micrometeoroids. The radio signal would have probed the atmosphere to measure its density. The cylindrical spacecraft was 4 m long and up to 1.0 m wide, with a 1.7-m dish antenna. Power was provided by solar panels roughly 1 m square, one on each side of the spacecraft.

Mars 1 was launched at 16:14 UT from Baikonur and entered a 238 by 157 km parking orbit inclined 65° to the equator with a period of 88 minutes. An upper stage burn placed it in a heliocentric orbit with an aphelion at 1.604 Astronomical Units (AU, the mean distance of Earth from the Sun). This orbit had an inclination of 2.68° to the ecliptic and a period of 519 days and intersected the orbit of Mars, leading to a flyby of the planet at a range of 193 000 km on 19 June 1963 (MY 5, sol 237). Communications failed on 21 March 1963, possibly because of an attitude control problem. The last transmissions were from a distance of 106.8 million km, a record at the time. Particles, fields and micrometeorite data were transmitted successfully by Mars 1, but no data were obtained from Mars.

4 November 1962: **Sputnik 24 (Soviet Union)**

Sputnik 24 was another attempted Mars flyby mission, also known as Korabl 13 in the Soviet Union. It was launched at 15:35 UT and entered a 197 km by 590 km parking orbit inclined 65° to the equator, but its upper stage failed during the burn to send it on to Mars, fragmenting the vehicle. It re-entered the atmosphere on 19 January 1963.

1963: **Voyager (United States)**

The Avco Corporation of Wilmington, Massachusetts, under contract to NASA's Office of Space Sciences, prepared a study of a series of missions to Venus and Mars. This Voyager program involved multiple orbiters and landers for both planets, to be launched between 1969 and 1975 (Avco Corporation, 1963a, b).

The best opportunity for a study of the wave of darkening in the two decades covered by the study was in 1969. A 1969 lander could set down just before the maximum darkening and observe it through its peak and decline in less than two months of operation. When the 1971 landers arrived at Mars, the peak of the wave of darkening would be over, and in 1973 landing in a different season would require three or four months of observation. In 1971 the south polar cap would be accessible, but given the expected targeting accuracy, it was unlikely that the lander could observe the dark collar around the cap as it retreated past the landing site. Still, this was the best chance to study the south polar cap, and the region was given high priority. The bright 'desert' areas and supposed canals would be investigated in 1973, and in 1975 the north polar cap and its dark collar. When the 1975 landers arrived at the planet, the edge of the cap at 220° longitude would be near 78° N and receding northwards at about 1 km per day. Avco urged that efforts be made to achieve a more accurate landing, to increase the chance of observing the narrow collar. Releasing the lander from its carrier spacecraft a little later during the approach might accomplish this.

Table 2 lists the chosen landing sites and alternative selections for each flight year, as given in Table 2.2 of the Voyager Design Study summary volume (Avco Corporation, 1963a). Table 3 includes more information and some differences in coordinates from the full mission analysis (Avco Corporation, 1963b, section 7.3). These sites are also described by Swan and Sagan (1965), and they are mapped in Figure 8.

The General Electric Co. also designed a Mars mission for the Voyager program, involving two 660-kg landers carried on one 930-kg orbiter (General Electric, 1963, 1964). The fully fueled spacecraft mass would be 3190 kg. The landers would use rockets to brake during final descent, like the later Vikings, and would use protruding rods to orient themselves on the surface. Their landing targets for 1969 were Syrtis Major (10° N, 285° W) and Pandorae Fretum (24° S, 310° W) (Figure 7). There had been other studies of landers, including a 1961 study by the Aeronutronic Division of Ford Motor Co. (Kemmerly, 1961) of a camera-carrying probe which

Table 2. *Avco (1963a) Voyager Study: Mars Landing Sites*

Year	Lander	Landing site	Location	Notes
1969	1	1. Solis Lacus	28° S, 90° W	Dark spot
	2	2. Syrtis Major	15° N, 286° W	The 'wave of darkening' crosses this site 30 days after landing: very high-priority observations
1971	1	3. South polar cap	83° S, 30° W	
	2	4. Mare Cimmerium	18° S, 235° W	Dark regions where the 'wave of darkening' has recently passed
	3	5. Lunae Palus	15° N, 65° W	
	4	6. Aurorae Sinus	15° S, 50° W	
1973	1	7. Propontis	45° N, 185° W	'Canal' area
	2	8. Elysium	25° N, 210° W	'Pink desert' area
1975	1	9. North polar cap	78° N, 220° W	Goal is to intercept the 'dark collar' as the cap recedes
	2	10. Nepenthes-Thoth	25° N, 255° W	Canal area: nominal target – subject to retargeting based on earlier orbital mapping

Table 3. *Avco (1963b) Voyager Study: Mars Landing Sites and Alternatives*

1969. Primary Objective: Wave of Darkening. Two Landers.		
Solis Lacus	25° S, 90° W	History of surface changes at landing season, linear features.
Syrtis Major	15° N, 286° W	Very dark, frequent changes, circular boundary with Isidis Regio.
Mare Sirenum	30° S, 140° W	Alternative: reported changes, nearby Memnonia has unusual colour.
Lunae Palus	15° N, 65° W	Alternative: large recent changes, double cycle of seasonal change.
Trivium Charontis	20° N, 200° W	Alternative: unusual color changes, linear edge of Elysium desert.
1971. Primary Objectives: Polar Cap and Dark Areas. Four Landers.		
South Polar Cap	82° S, 40° W	Only chance for south pole, last of four landers directed here if first lander fails. Moisture measurements, permafrost features.
Mare Cimmerium	18° S, 235° W	Changes reported. Darkening from north can be seen in 1975.
Lunae Palus	15° N, 65° W	Wave of darkening present but waning; also see 1969 comments.
Aurorae Sinus	15° S, 50° W	Reported seasonal colour changes, crossing linear features.
Solis Lacus	20° S, 90° W	Alternative: see 1969, candidate for successive landings at one site.
Mare Serpentis	30° S, 315° W	Alternative: changes reported, darkening via Hellespontus channel?
1973. Primary Objectives: Deserts and Canals. Two Landers.		
Propontis	45° N, 185° W	Typical 'canal', striking changes, edge of desert hemisphere.
Elysium	25° N, 210° W	Circular pink desert, high priority if normal deserts seen already. Off-centre landing may permit a broad panoramic view.
Hellas	40° S, 290° W	Alternative: anomalous yellowish colour and high albedo.
Nix Olympica	20° N, 130° W	Alternative: bright area, center of desert hemisphere, many clouds.
Ismenius	40° N, 330° W	Alternative: striking color changes reported during landing season.
1975. Primary Objective: Polar Cap. Two Landers.		
North Polar Cap	78° N, 220° W	If possible, land so dark cap collar passes lander during its lifetime.
Nepenthes-Thoth	25° N, 255° W	Unusual changes in 1940s. Former canal features. This lander may be redirected to unexpected features discovered by earlier orbiters.
Mare Cimmerium	18° S, 235° W	Alternative: Best site to observe wave from north pole, see 1971.
Mare Acidalium	40° N, 40° W	Alternative: Biology suggested by infrared spectroscopy and polarimetry.

Note: Sites in each year are listed in priority order.

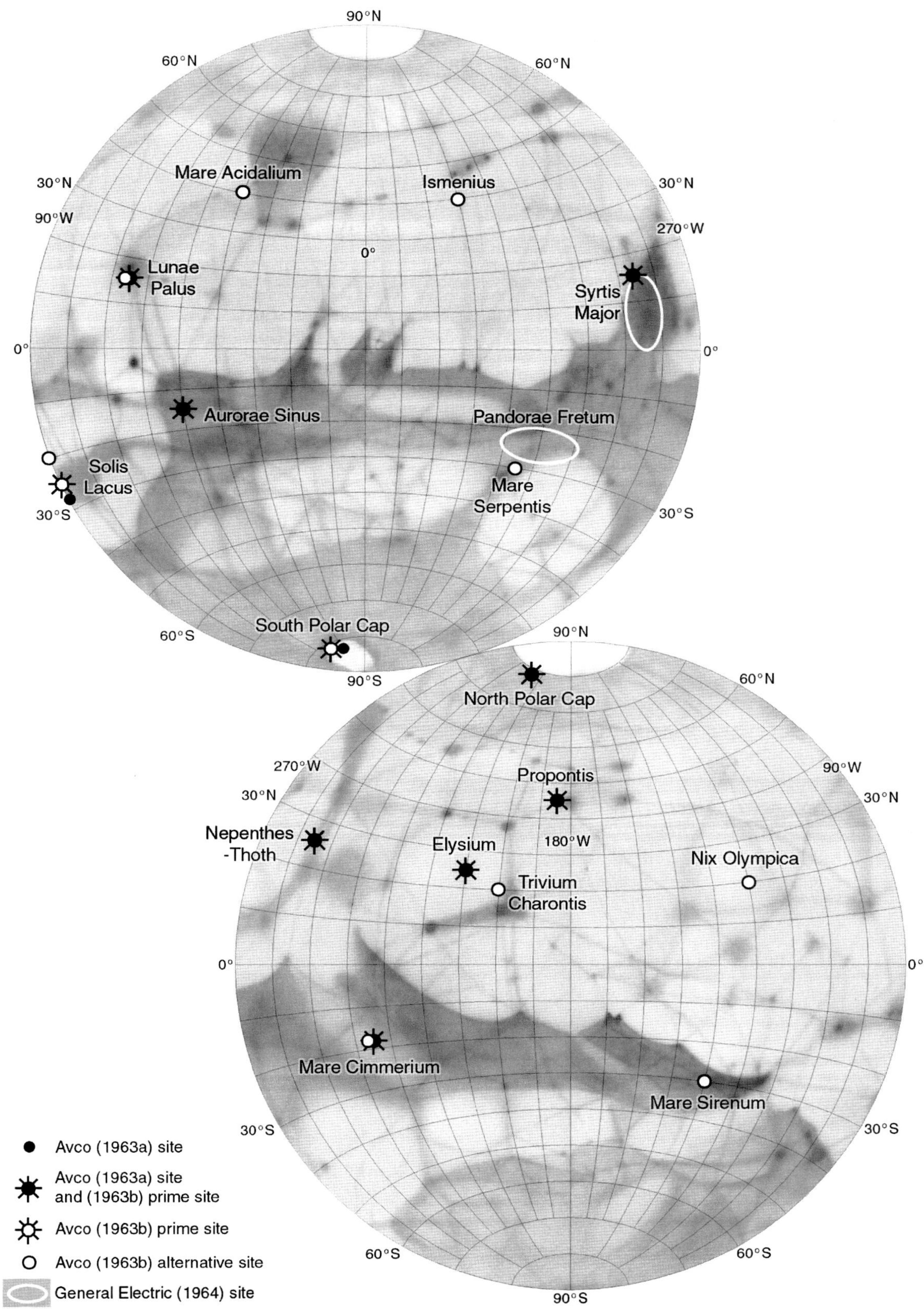

Figure 8 Voyager landing sites suggested by Avco (1963a, 1963b) and General Electric (1964).

might have been dropped into the atmosphere of Mars after a late 1964 launch. During its parachute descent it would obtain small colour images of the surface before impacting in a 'sub-solar equatorial region'. No specific landing site was identified, but if the probe could be designed to survive the impact, it might search for signs of life.

5 November 1964: **Mariner 3 (United States)**

The first NASA Mars probe was Mariner 3, identical to Mariner 4 in its design and mission plan. The goal was to fly past Mars at close range, taking images and studying the atmosphere. The 261-kg Mariner 3 was launched at 19:22 UT from Cape Canaveral and entered a parking orbit. The payload shroud did not jettison as intended. An upper stage burn placed the spacecraft in a solar orbit, but the extra mass of the shroud compromised the trajectory, so no Mars flyby was possible. The shrouded probe could not open its solar panels, so the battery power failed after 8.7 hours. Mariner 3 remained in an 0.82 by 1.615 AU orbit, inclined 0.52° to the ecliptic with a period of 449 days. At this early stage in planetary navigation, it was difficult to predict the exact time of flyby before launch, or the sub-spacecraft latitude and longitude, so the areas on Mars which would have been imaged by Mariner 3 were not identified. Images of Syrtis Major were considered important and might have been targeted, if possible.

28 November 1964: **Mariner 4 (United States)**

Mariner 4 was the first successful Mars flyby mission. The 261-kg spacecraft consisted of a 1.38 by 0.46 m octagonal frame with high and low gain antennae mounted on the top and four solar panels on the sides, spanning 6.9 m. Attitude control thrusters were mounted on the ends of the solar panels. The spacecraft was 2.9 m high. The scientific payload consisted of meteoroid, charged particle and cosmic ray detectors, a magnetometer, and a single TV camera on a pointable scan platform. It was launched at 14:22 UT from Cape Canaveral and entered a 184 by 172 km parking orbit inclined 28.3° to the equator and was then placed on a Mars-bound trajectory by its upper stage. After 228 days it flew past Mars on 15 July 1965 (MY 6, sol 305) at 01:00 UT, and at a distance of 9844 km.

Mariner 4 returned the first close-up photographs of the Martian surface (Figures 9 through 12). The 22 images revealed numerous craters (Leighton *et al.*, 1965) and a long valley now called Sirenum Fossae. Although there was no reason to suppose that craters would be absent, proof of their existence was regarded as a major discovery and was a surprise to some observers. One of the largest was named Mariner in its honour (Figure 11). Mariner 4 also studied the ionosphere and atmosphere, finding that carbon dioxide was the major constituent and that the surface pressure was only about 5 mb, less than a tenth of the value commonly expected. For the radio occultation experiment, radio signals from Mariner 4 passed through the atmosphere at the limb and were cut off by the body of the planet for 54 minutes. The surface atmospheric pressure and altitude could be estimated at the entry and exit points of the occultation as Mariner 4 passed behind the planet. Entry was between Electris and Mare Chronium near 55° S, 183° W, and exit in the Mare Acidalium region (Figure 13A) near 60° N at longitude 44° W (NASA, 1968b) or 34° W (Kliore *et al.*, 1965). Local topography makes the exact location uncertain. Pressures of between 4.1 and 7.0 mb and temperatures of about 173 K were estimated from the occultations, indicating that Mare Acidalium was as much as 5 km lower in elevation than Electris. Mariner 4 also found that any magnetic field was no more than about 0.3% as strong as Earth's (Nicks, 1967).

Figure 9 shows the Mariner 4 image area, chosen to include a variety of features, including, canals which might have been ridges or fractures. An early wish to include Syrtis Major in the imaged area could not be satisfied with the given trajectory and flyby time. Figures 10 through 12 show the Mariner 4 images and outline the areas covered by them on modern maps for comparison. The images were used by ACIC to create maps of the newly revealed features (Figures 13 and 22). Just before the flyby, Clyde Tombaugh, the discoverer of Pluto, produced a map (Figure 13B) of the Mariner 4 image area (Tombaugh, 1965). This included canals and also illustrated a hypothesis that oases (spots, mainly at canal intersections) might be impact craters. This pre-mission map and Figures 1and 2 contrast strongly with the post–

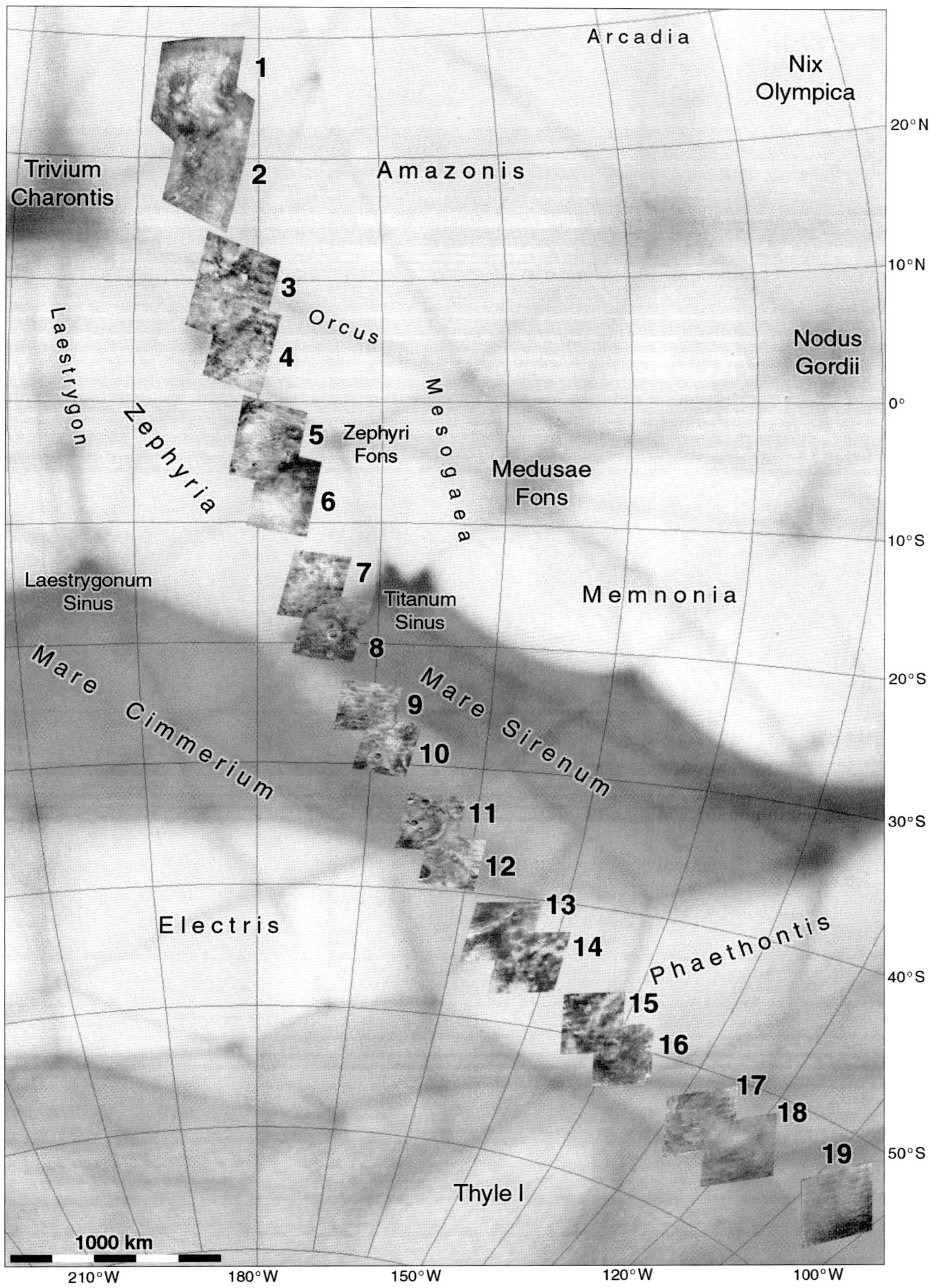

Figure 9 Mariner 4 images superimposed on the contemporary ACIC map, part of Figure 1.
Locations are matched to modern topography and are more accurate than contemporary maps.

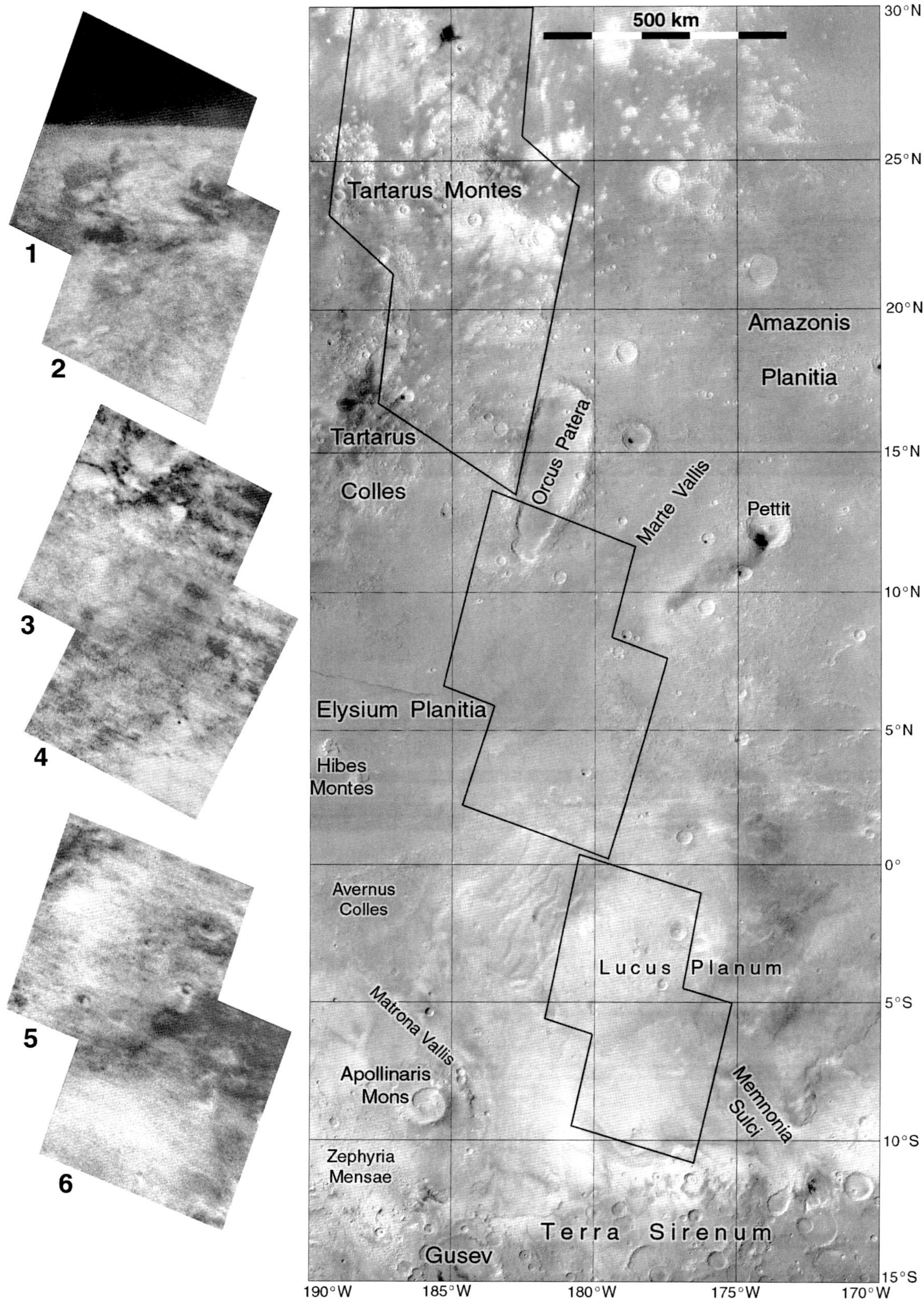

Figure 10 Mariner 4 images and locations.
The background is a mosaic of Mars Global Surveyor images.

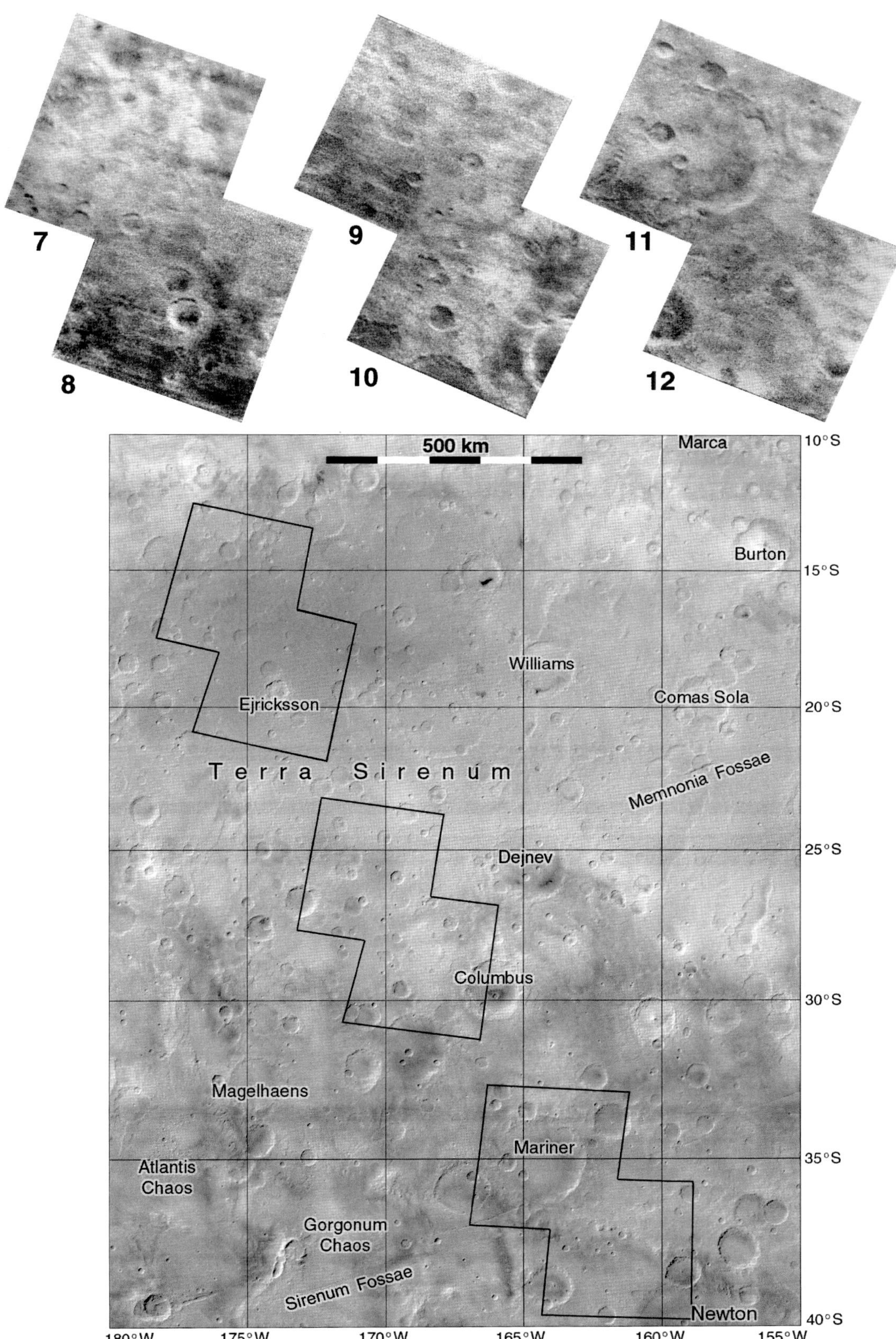

Figure 11 Mariner 4 images and locations, continued from Figure 10.

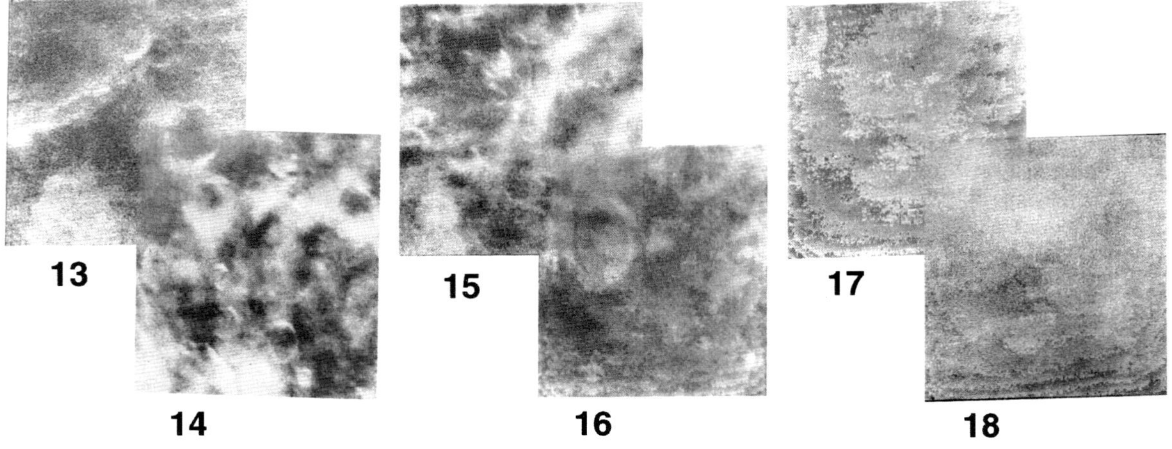

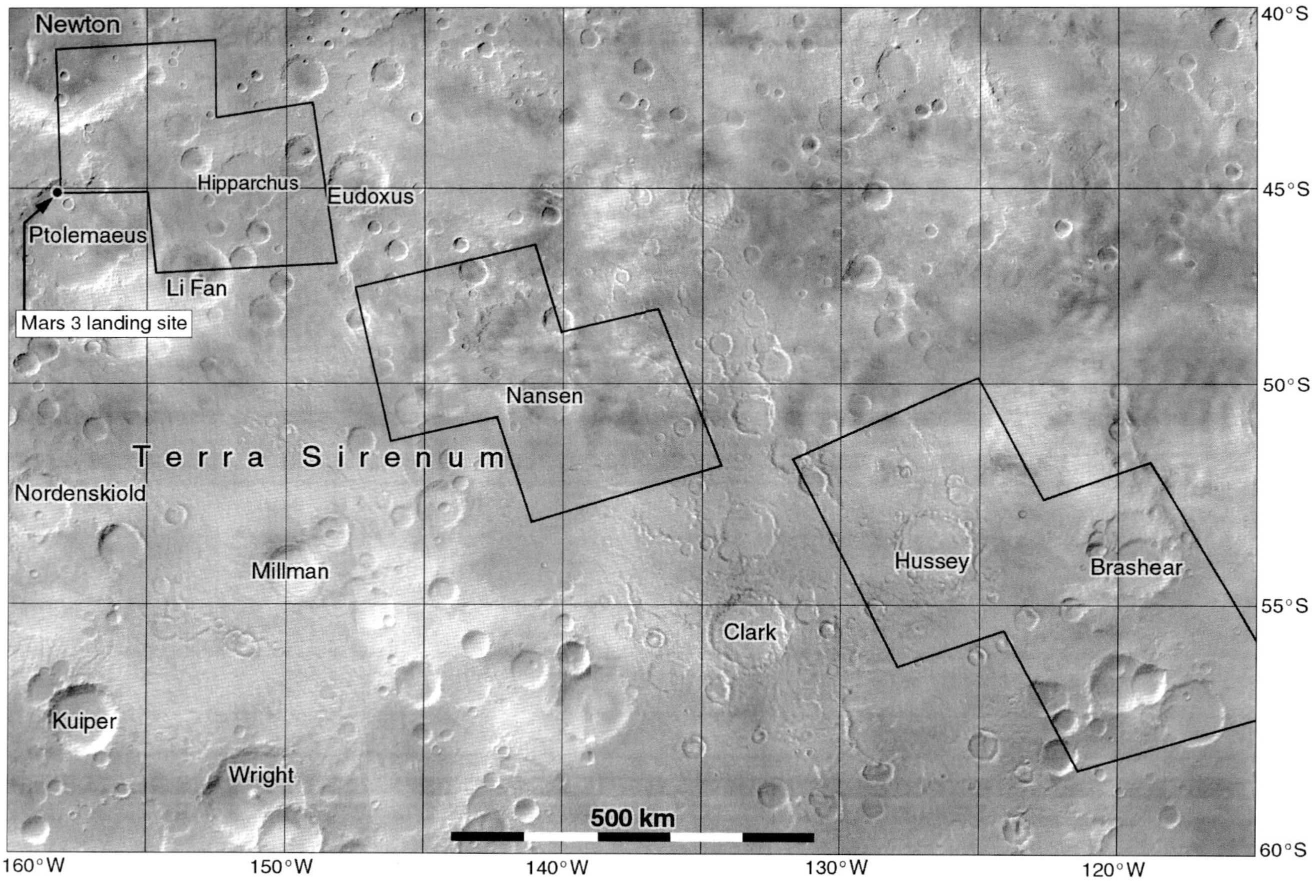

Figure 12 Mariner 4 images and locations, continued from Figure 11.

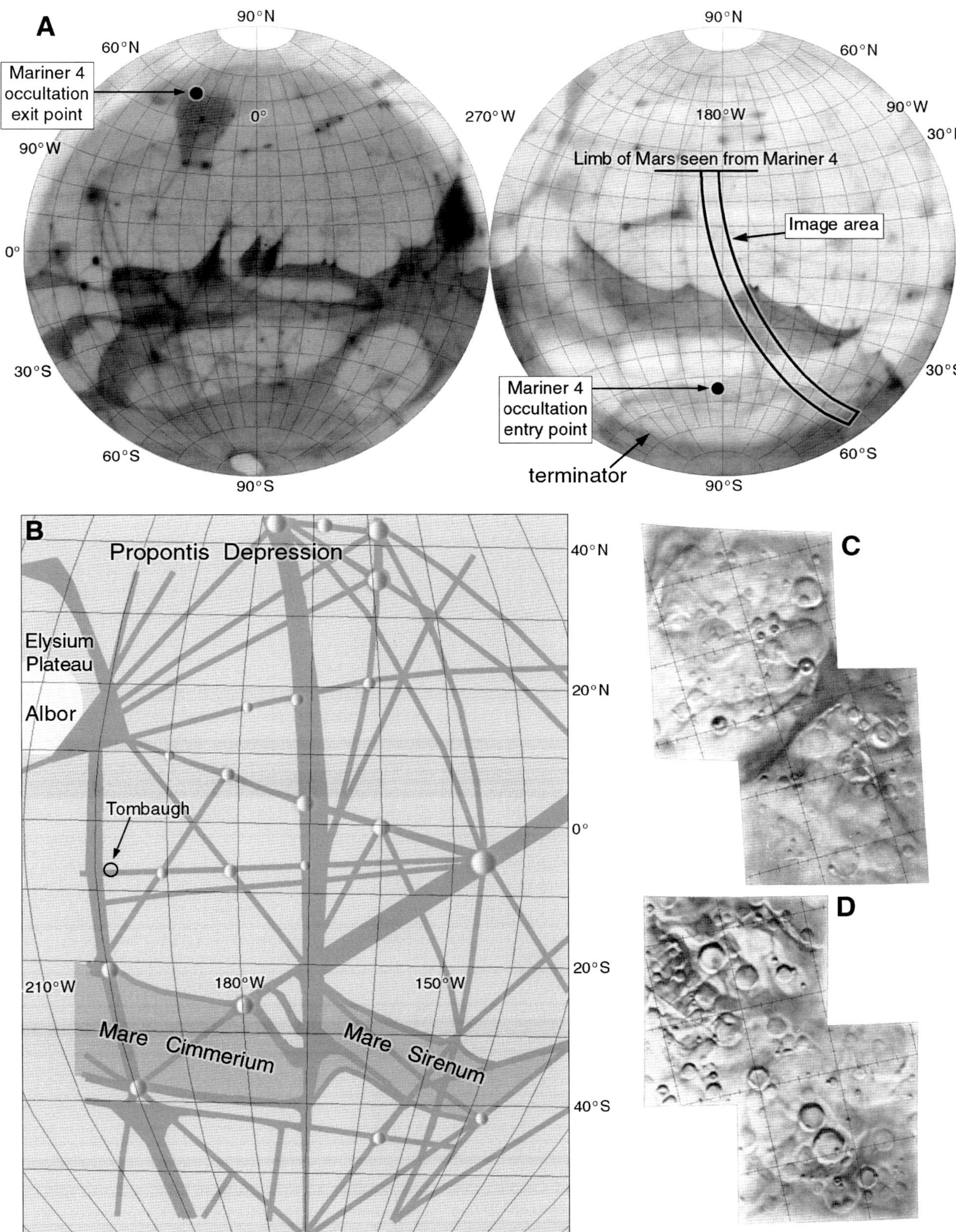

Figure 13 **A:** Mariner 4 image swath, terminator position and occultation points. **B:** Map redrawn from an original by Clyde Tombaugh, showing a pre-Mariner interpretation of the Mariner 4 imaging region (Tombaugh, 1965). It shows canals and hypothetical impact craters in the 'oases' at canal intersections. A crater at 3.5° N, 162° E in this region commemorates Tombaugh. **C and D:** ACIC maps made from Mariner 4 frames 5, 6, 7 and 8.

Mariner 4 maps, beginning a pattern of cartographic evolution which would pervade planetary exploration. Planning would be done on the best maps available before the mission, and subsequent maps would record the mission results. Often the final maps from one mission became the planning maps for the next.

Mariner 4 continued in its cruise until contact was lost on 1 October 1965, but its signal was regained late in 1967. Several micrometeoroid strikes were recorded on 15 September and 10 December of that year, energetic enough to change the spacecraft orientation and affect communications. Attitude control gas was exhausted on 6 December, and all communication ceased on 21 December 1967.

Other missions had been proposed for this period, notably a study by the TRW Inc. subsidiary Space Technology Laboratories (1962). This mission would have delivered an atmospheric entry probe to Mars in 1964 or 1967, with two launches in 1964 and three in 1967. The earlier launches would build confidence in the technology but might not be expected to succeed. This mission's science goals were to improve knowledge of the size of the Astronomical Unit (distance between Earth and the Sun) and the diameter of Mars.

30 November 1964: **Zond 2 (Soviet Union)**

The Soviet Union planned two Mars probes for the 1964 launch window with the goal of landing an instrumented probe on the planet. The 960-kg Zond 2 was launched at 13:12 UT from Baikonur and entered a 153 by 219 km parking orbit inclined 65° to the equator with a period of 88.2 minutes. It was injected into a heliocentric orbit which would have arrived at Mars on 6 August 1965 (MY 6, sol 326), 21 sols after Mariner 4, but communications were lost early in May 1965. Zond 2 used experimental ion engines for attitude control and carried a probe which would have been deployed on an atmospheric entry trajectory by the carrier spacecraft. After the probe was released, the carrier was intended to have been deflected so as to pass 1500 km from the planet's limb. Murray *et al.* (1967) argued that if the spacecraft had made a trajectory correction in the first 11 weeks of flight, as suspected but never confirmed, it would probably have been placed on an accurate planetary impact trajectory. Because communications were lost, the probe deployment and carrier deflection burn may not have taken place, and the vehicle possibly burned up in the atmosphere of Mars, becoming the first human artifact to reach that planet's surface. Murray *et al.* (1967) suggested that this would have delivered microbes to Mars from the unsterilized carrier. This is very uncertain, but if it happened, the location on Mars of any debris is unknown, as is the landing site targeting strategy for the mission.

18 July 1965: **Zond 3 (Soviet Union)**

Zond 3 was similar or identical to Zond 2, but a delay meant that it missed its 1964 launch window to Mars. Instead it was launched as a technology test vehicle and passed the Moon to image part of its far side, which had not been seen previously. The launch from Baikonur was at 14:38 UT, and Zond 3 entered a 209 by 163 km parking orbit inclined 64.78° to the equator with a period of 88.4 minutes. It was injected into a heliocentric orbit which intersected the orbit of Mars, but at a time when Mars was far from that point. It successfully transmitted high-quality images of the lunar surface and was still transmitting as it crossed the orbit of Mars as a test of navigation and communications at planetary distances. Zond 3 carried ultraviolet and infrared spectrometers, charged particle and meteoroid detectors, a magnetometer and a radiotelescope system. Zond 2 probably carried similar instruments.

1968: **Mars Hard Lander Capsule Study**

General Electric Re-entry Systems performed a study for NASA on a hard lander concept, designed to launch in the summer of 1973 and land in the spring of 1974 (NASA, 1968a). The mission would include an orbiter and might deploy its lander directly from the approach directory or later from orbit. The lander's goals would be to study the atmosphere during its decent, to take images on the surface, measure its composition and make meteorological observations. The orbiter would serve as a communication relay, image the surface for regional mapping, especially over the northern hemisphere where

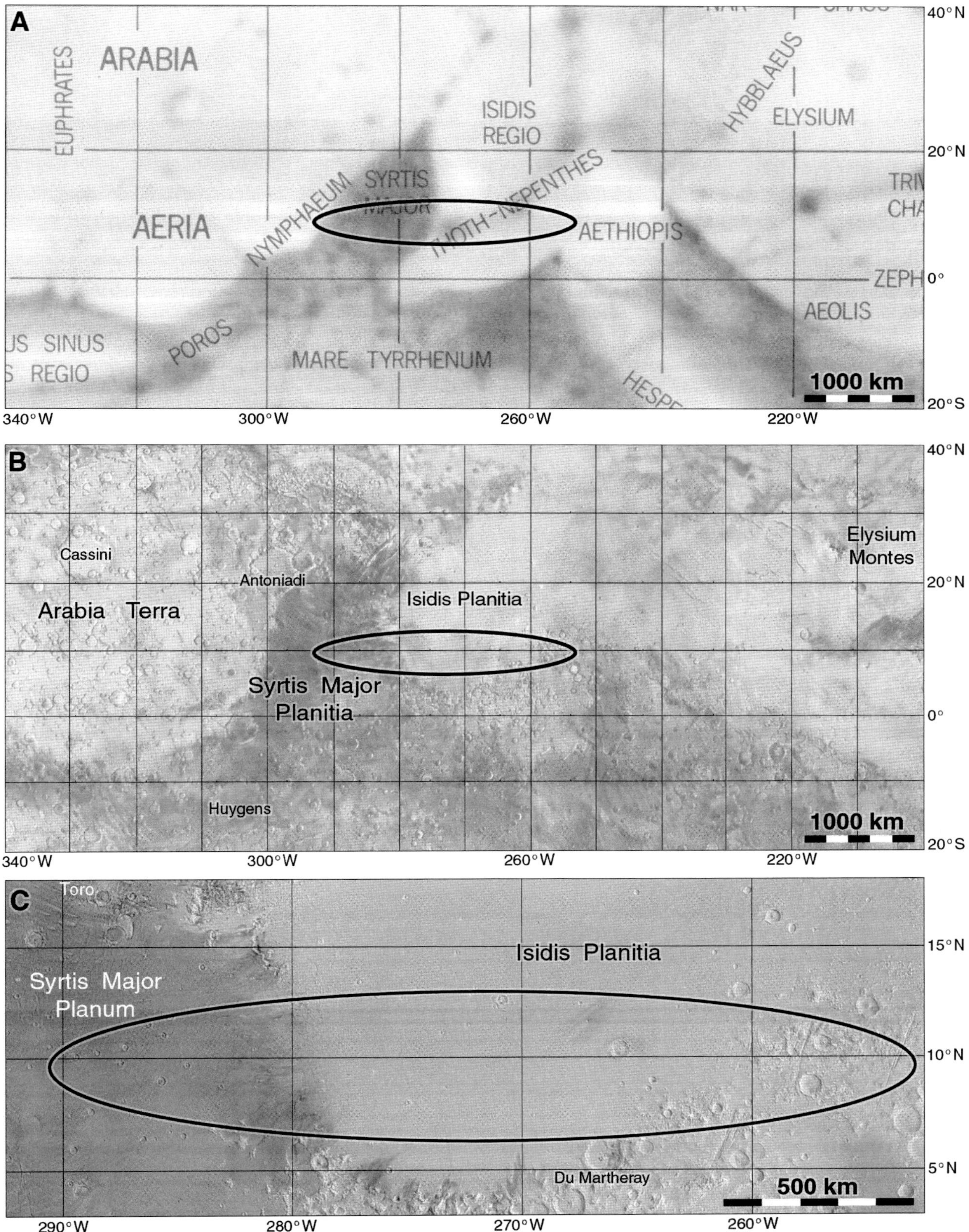

Figure 14 A: The General Electric Hardlander Study landing ellipse plotted on the General Electric Co. Mars map. **B:** The same area portrayed on a USGS map based on Mariner 9 images. **C:** Enlargement of the ellipse area from the USGS Viking global mosaic with relief enhanced using Mars Global Sorveyor MOLA data.

illumination would be most favourable, take high-resolution images of selected areas, including the landing site, and measure surface temperatures, meteorological characteristics, seasonal variations and gravity. It was assumed that this mission would follow a flyby mission in 1969 (Mariners 6 and 7) and an orbiter in 1971 (Mariners 8 and 9).

The orbiter would have a 60° inclination with 1000-km periapsis and 33 100-km apoapsis, giving it a period of one Mars day. It would be synchronous over the lander, but not stationary because of the elliptical and inclined orbit. If the lander was deployed from orbit, the orbiter would image the landing site before landing, then stay in its synchronous orbit for a few days to relay surface images from the lander. Later its periapsis longitude would be allowed to move around the planet to permit broad regional mapping while the lander transmitted smaller amounts of meteorological data directly to Earth. If the lander was deployed before orbit insertion, the landing site imaging would be taken later. The first images taken immediately upon landing would be transmitted to the orbiter for relay in case the lander did not survive the first cold night on Mars.

Landing sites would be in the northern hemisphere to suit the seasons in 1974. Latitudes of 10° N and 20° N were studied for the mission analysis, but latitudes farther north at 25° N and 50° N were also considered. A reference mission described in the report placed the landing near the prominent 'mare' (dark area) of Syrtis Major, where the supposed 'wave of darkening' could be studied. This was still considered a possibly biological phenomenon, and the surface imaging was designed to be able to identify lichen-like organisms if they existed. The reference mission's landing ellipse was centred at 10° N, 273° W, extending 20 east and west of that point so that it included Syrtis Major to the west and Isidis to the east. At that time nothing was known about these areas. Figure 13 shows the ellipse on the contemporary map prepared by General Electric, and a comparison map showing the surface with topographic details revealed in 1972 by Mariner 9, along with a high-resolution view of the landing ellipse. The General Electric map was based on the ACIC chart (Figure 1), with canals de-emphasized and a few spurious craters sketched in to reflect the recent Mariner 4 discovery of impact craters. In Figure 13A, Syrtis Major is displaced about 10° east of its correct position, a mistake copied from the ACIC map. The name of the topographic feature associated with Syrtis Major changed from Planitia (low-elevation plain) to Planum (high plain) when its elevation became clear in Mars Global Surveyor data in the 1990s.

25 February 1969: **Mariner 6 (United States)**

Mariner 6 was an upgraded 413-kg version of Mariner 4 intended to fly past Mars with greatly improved instruments. A version which could deploy a landing capsule was initially considered for a 1966 launch (Arnold *et al.*, 1964), but was not developed. Ten days before launch, while Mariner 6 was mounted on its Atlas/Centaur rocket on the launch pad, valves on the Atlas stage were accidentally opened, causing a loss of pressure which threatened to deflate the flimsy structure. Two members of the ground crew turned on pumps which saved the rocket from collapsing. Mariner 6 was moved to another rocket and launched on schedule, and the two men who saved the mission were awarded NASA Exceptional Bravery Medals. Mariner 6 was launched at 01:29 UT, and after 156 days it flew past Mars at a distance of 3431 km on 31 July 1969 (MY 8, sol 405), only a week after the Apollo 11 astronauts returned to Earth from the first Moon landing.

Mariner 6 had a 138-cm-wide octagonal body, 46 cm deep, with four solar panels spanning 5.8 m, a dish antenna on top of the frame and a low-gain antenna mounted on a 2.2-m tubular mast. Attitude control thrusters were located on the ends of the solar panels. Its 59 kg of instruments consisted of infrared and ultraviolet spectrometers, an infrared radiometer and wide- and narrow-angle cameras mounted on a scan platform underneath the body. The entire spacecraft was 3.3 m high. The spacecraft used an analog tape recorder system which compromised the image quality and was never used again.

The Mariner 1969 planning maps were versions of the ACIC map (Figures 1 and 2) augmented with features seen by Mariner 4. As the spacecraft approached Mars, a series of 49 images were taken over nearly two rotations, showing the whole planet up to 70° N more clearly than telescopic views of the time (Figures 15 and 16). The nearly full phase and low contrast concealed relief in these images but showed albedo variations well. Twenty-six more images taken during the close-encounter phase of the flyby showed topography clearly, with resolutions as good as 1 km/pixel (wide angle) and 100 m/pixel (narrow angle). Imaging commenced over Lunae Palus and ended in Sinus Sabaeus, a strip spanning 6000 km. A scan platform slew in the middle of the sequence allowed some additional coverage north of the main strip. The area covered by these images is mapped in Figure 17. Apart from the numerous craters, a few more enigmatic features were also revealed. Valleys were also seen, but not clearly enough to be unambiguously identified as fluvial channels rather than tectonic features. A large area of complex depressions filled with jumbled hills, referred to as chaotic terrain, was seen at high Sun angles, but its regional context between giant canyons and vast outflow channels was not made clear until Mariner 9 extended the imaged area.

Post-encounter cruise operations continued until 21 December 1970, involving tracking, the collection of 10.5 hours of ultraviolet spectra of the sky including the galactic plane made on 12–14 August 1969, and a test of communication ability when Mariner 6 was at superior conjunction (on the far side of the Sun as seen from Earth) between late April and mid May 1970. Published suggestions that Mariner 6 observed Comet 1969-B are incorrect. The post-encounter heliocentric orbit was 217 by 212 million km, inclined 1.8° to the ecliptic, with a period of 636 days. In a final engineering experiment on 18 December 1970, the course correction rocket was fired, and the ultraviolet spectrometer studied the exhaust plume.

Mariner 6 passed behind Mars after closest approach, and its radio signal was cut off by the planet (Kliore *et al.*, 1969). Just before the loss of signal, the radio transmission was affected by its passage through the atmosphere, providing information about atmospheric pressure and temperature and surface elevation at the occulting point. The same information was obtained 20 minutes later as the spacecraft emerged from behind the planet and its signal was reacquired. Night surface temperatures as low as –73° C at the equator were measured. The surface atmospheric pressure was 6 to 7 mb, and carbon dioxide was confirmed as the main constituent of the atmosphere. The inbound occultation point was in sunlight at 4° N, 5° W in Sinus Meridiani, and the outbound point at 79° N, 276° W (Figure 15). The latter point was in darkness where the long polar night was beginning, and measurements suggested that carbon dioxide would condense there at all altitudes.

Figure 17A shows the area covered by Mariner 6 near-encounter images. The topographic feature names were not assigned until 1973. Small white rectangles in that figure show the locations of narrow-angle frames. Mariner 6 and 7 also obtained topographic information, as the ultraviolet spectrometers on both spacecraft measured ultraviolet radiation scattered by the atmosphere (Barth and Hord, 1971). The amount of scattering varied with surface pressure and inversely with elevation. This information was collected along swaths coinciding with the imaged areas (Figure 17B). For each swath in that figure, a topographic profile is plotted against a vertical scale, with an assumed datum level indicated by a line. The low elevation of Hellas was revealed, as well as small low areas now called Juventae Chasma and Ganges Chasma. The data for Argyre are difficult to relate to topography, perhaps due to atmospheric effects and oblique viewing. Herr *et al.* (1970) obtained similar results from Mariner 6 and 7 infrared spectra, including data for Argyre with a better match to modern topography. They also described the complex topography now seen to be part of the great canyon system and associated chaotic terrain.

Figure 17C is a detail from a US Army Map Service map prepared from Mariner 4, 6 and 7 images. It shows Nix Olympica and Kasei Valles drawn as if they were large impact craters. This interpretation is made explicit by the camera team in NASA (1969).

27 March 1969: **Mariner 7 (United States)**

Mariner 7 was identical to Mariner 6. It was launched at 22:22 UT, and 133 days later it flew past Mars at a distance of 3430 km on 5 August 1969 (MY 8, sol 410), five sols after Mariner 6. As it approached the planet it suffered a loss of control, perhaps due to a venting battery, but recovered in time to operate successfully during the flyby.

Mariner 6 had flown over the equatorial region of Mars, but Mariner 7 was directed to a southern hemisphere flyby. Imaging commenced over Meridiani Sinus, giving an opportunity to search for changes in the week since Mariner 6 imaged the same area. The 33 images

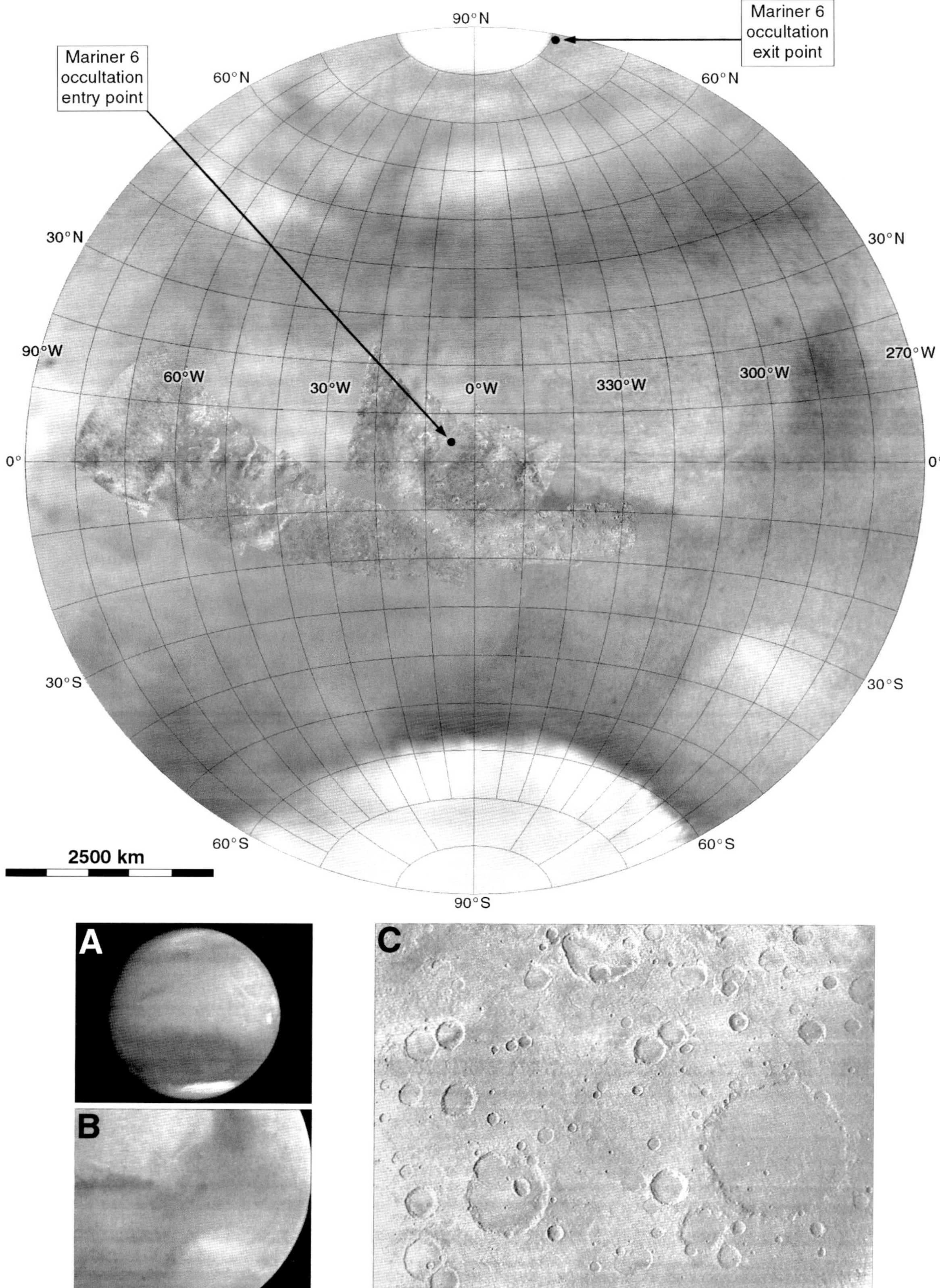

Figure 15 Top: The 0° hemisphere of Mars as it was known after the Mariner 6 flyby. The Mariner 6 near encounter images run east to west across the middle of the map. Mariner 6 did not observe the north polar region, which is shown here as in Figure 2. **Bottom, A:** Mariner 6 far-encounter frame 6f38 showing clouds over Tharsis. **B:** Far-encounter frame 6f49, showing Iapygia and the large crater Huygens. **C:** Near encounter frame 6n21 showing craters in Sabaeus Sinus.

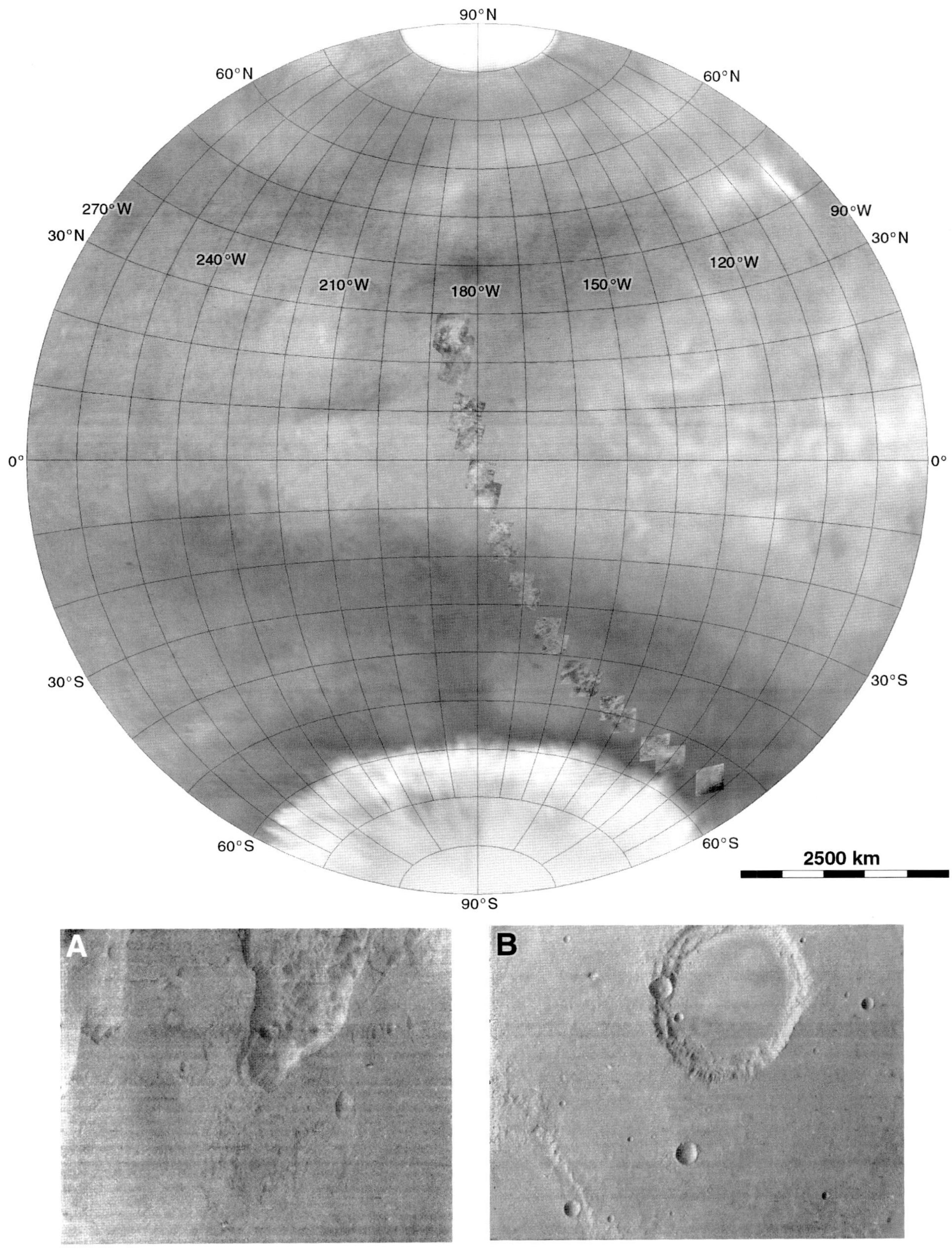

Figure 16 Top: The 180° hemisphere of Mars as it was known after the Mariner 6 flyby. The Mariner 4 images run north to south across the middle of the map. Many of the small bright spots are clouds. **Bottom:** Mariner 6 near-encounter image 6n14 **(A)** showing chaotic terrain at 12° S, 35° W, and image 6n18 **(B)** showing craters at 16° S, 4° W.

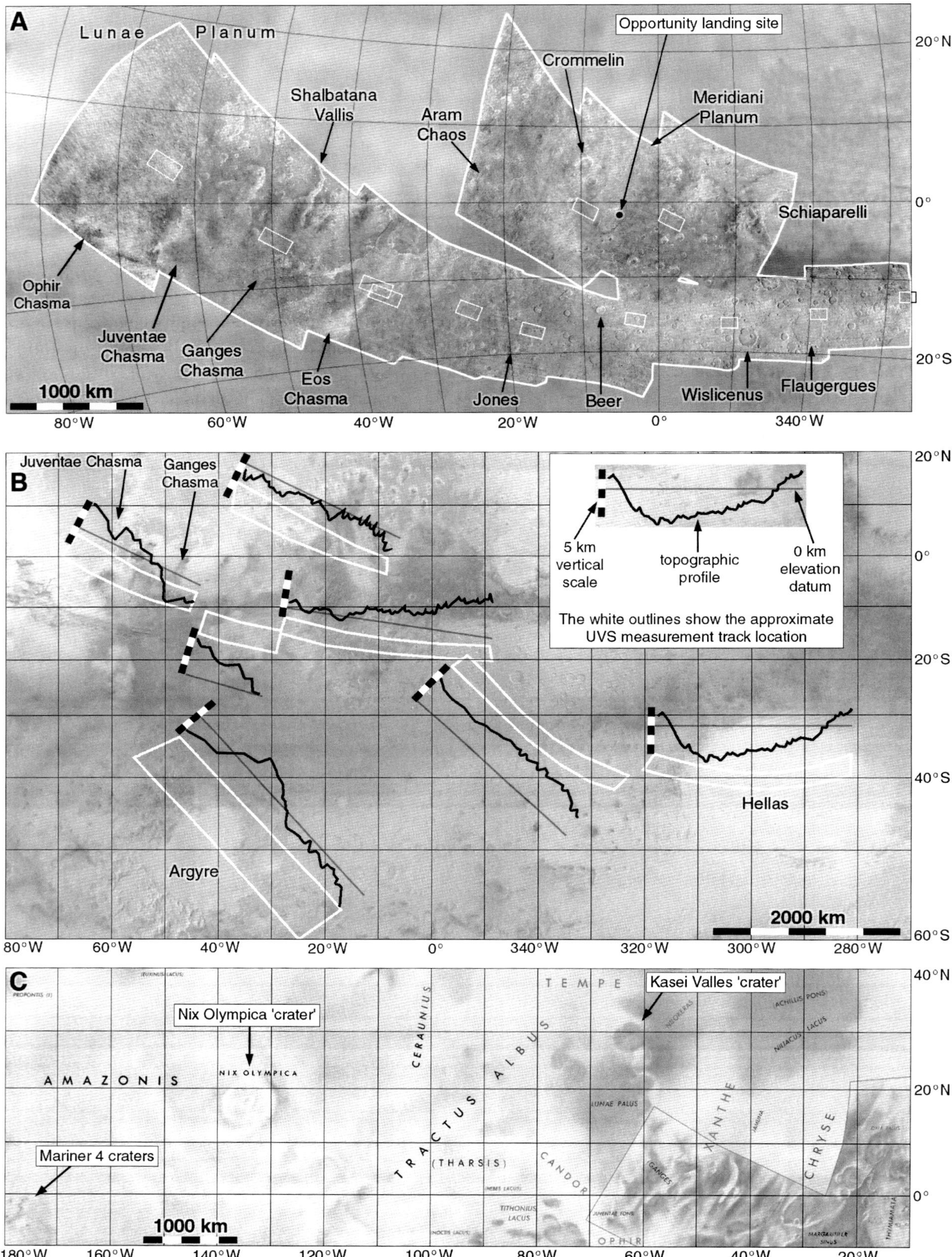

Figure 17 A: Mariner 6 near-encounter images. Small white rectangles show the locations of the narrow angle frames. **B:** Mariner 6 and 7 combined ultraviolet spectrometer topographic information (Barth and Hord, 1971). **C:** Part of an Army Map Service map drawn in 1970, showing two large features misinterpreted as craters.

covered a strip curving southeastwards into Hellas, with a separate swath enabled by scan platform slews which crossed the boundary of the south polar cap. The image coverage is illustrated in Figures 18 and 19, and images are shown in Figure 20. Apart from the expected craters, two anomalous areas were observed. Hellas was effectively featureless, though whether this was due to smooth topography or clouds was unclear. The south polar cap exhibited strange curvilinear ridges and irregular depressions, but resolution was insufficient to explain their origins. The cap boundary showed frost deposits on pole-facing slopes and dark frost-free areas where slopes faced the Sun. Ice on the bright ridge at the top of Figure 18C, Pityusa Rupes, is visible from Earth as a small detached bright spot as the shrinking spring ice cap recedes (Figure 141). This is the Mountains of Mitchel area targeted by Avco in 1966 (Figure 7). Mariner 7 also took 93 images over 2.4 rotations during the approach to the planet, which have been combined to form the background of Figures 18 and 19. Phobos was observed in some of these images, in enough detail to resolve its size and shape, but not to reveal surface features (Figure 196A).

The combined Mariner 6 and 7 images were used to create new maps of Mars. The US Army created a global map, part of which is shown in Figure 17C, and British planetary cartographer Charles Cross drew detailed maps of the areas imaged at high resolution for the RAND Corporation, where geodesist Merton Davies used the new images to improve cartographic control for the planet (Davies, 1972). The Mariner 7 occultation points were at 58° S, 330° W in Hellespontus (inbound, day side) and 38° N, 149° W in Amazonis (outbound, night side) (Kliore *et al.*, 1969). The occultation lasted for 30 minutes and the results suggested that the Hellespontus site was several kilometres higher than the other five Mariner occultation points. These points are illustrated in Figures 18 and 19.

The spacecraft returned to its interplanetary cruise after the encounter and data playback. The post-encounter heliocentric orbit was 206 by 210 million km, inclined 1.8° to the ecliptic, with a period of 609 days. Operations included engineering and communications tests, stellar imaging tests and ultraviolet scans of the sky, including the galactic plane. On 15 December 1970 the hydrazine-fueled course correction rocket was fired for about 80 seconds, and the ultraviolet spectrometer studied the exhaust plume. The gases released were ammonia, nitrogen and hydrogen, and the measurements were intended to simulate future studies of the atmospheres of the outer planets. Both Mariner 1969 spacecraft performed this experiment. The spacecraft failed due to the loss of attitude control gas just before a final tracking attempt on 28 December 1970.

Figure 19B shows a pair of craters informally referred to at the time as the Giant's Footprint. The larger of the two is now called Vishniac, after Wolf Vishniac, a biologist at the University of Rochester who worked on the Viking missions and died on 10 December 1973 during field work in Antarctica.

27 March 1969: **Mars 1969A (Soviet Union)**

The 1969 Mars missions were to be orbiters, intended to carry film cameras which could take 160 high-quality images. Each 4850-kg orbiter carried three cameras with colour filters, two for wide-angle imaging, taking strips of images covering 1500 km on a side at 1.5 km/pixel resolution, and one for high resolution, covering 100 km on a side at 100 m/pixel resolution. Launch from Baikonur was at 10:41 UT, but the mission apparently ended when a pump in its launch vehicle failed 7 minutes into the ascent.

2 April 1969: **Mars 1969B (Soviet Union)**

This mission was identical to Mars 1969A and launched at 10:33 UT from Baikonur. One of six engines failed immediately, and the ascending vehicle lost control about 25 seconds later and struck the ground several kilometres from the launch pad.

1960s: **Earth-Based Topographic Mapping**

During the 1960s, Earth-based radar studies of Mars began to provide topographic information about the planet, which complemented the higher resolution imaging by Mariner spacecraft. Some earlier authors had frequently assumed that dark areas were lower, as is generally the case for the Moon (Otterman, 1967), and

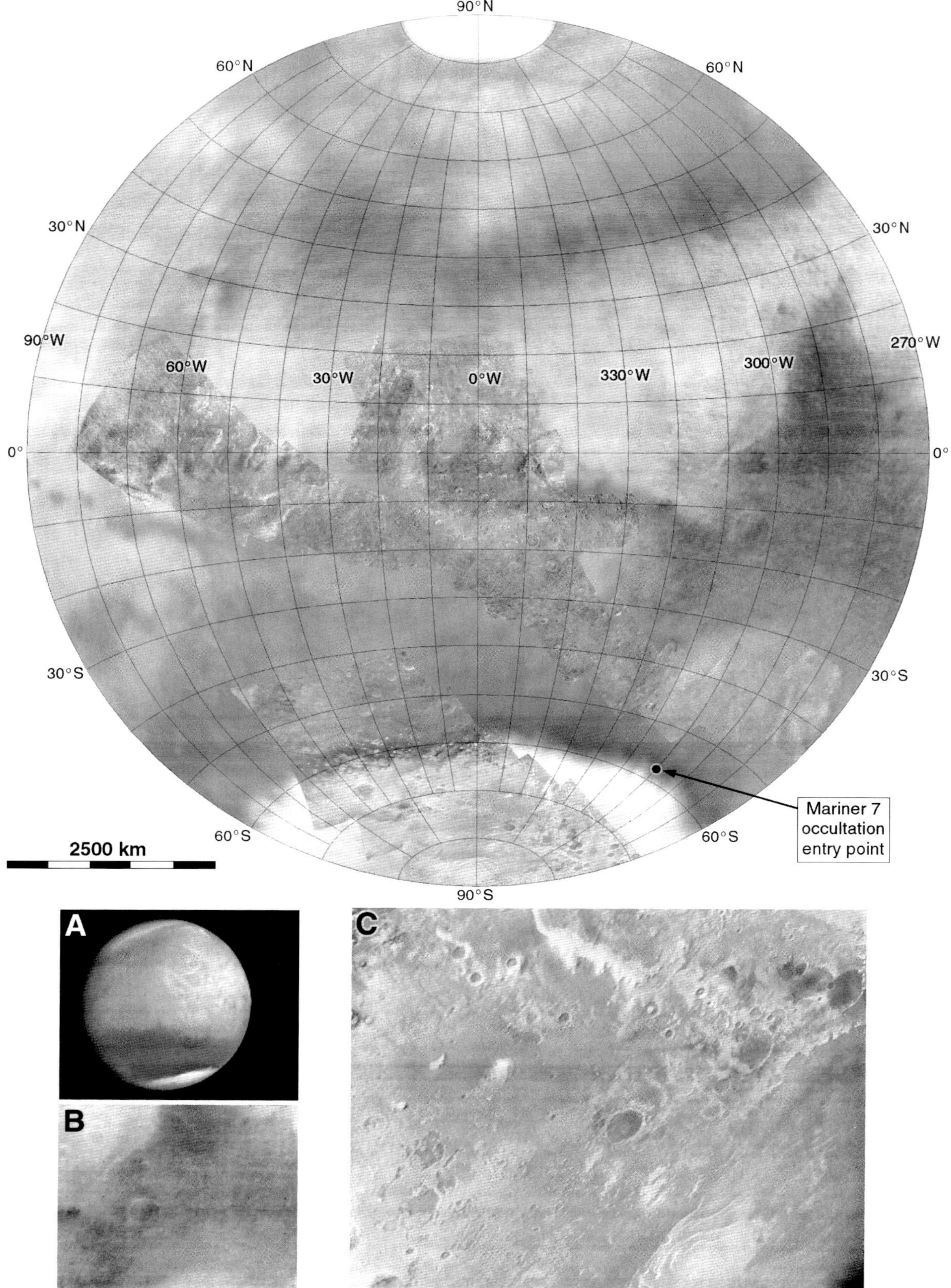

Figure 18 Top: Mars as it was known after Mariner 7. **A:** Mariner 7 far-encounter image 7f76 showing Tharsis, including the bright ring of Nix Olympica, interpreted at the time as a 500-km diameter crater. **B:** Image 7f90, showing Syrtis Major, the north edge of Hellas, and the large crater Huygens. **C:** Near-encounter image 7n17 covering the south polar carbon dioxide cap, with irregular erosion hollows (Sisyphi Cavi) at lower left and the ridges and troughs of the polar ice cap (Australe Mensa) at lower right.

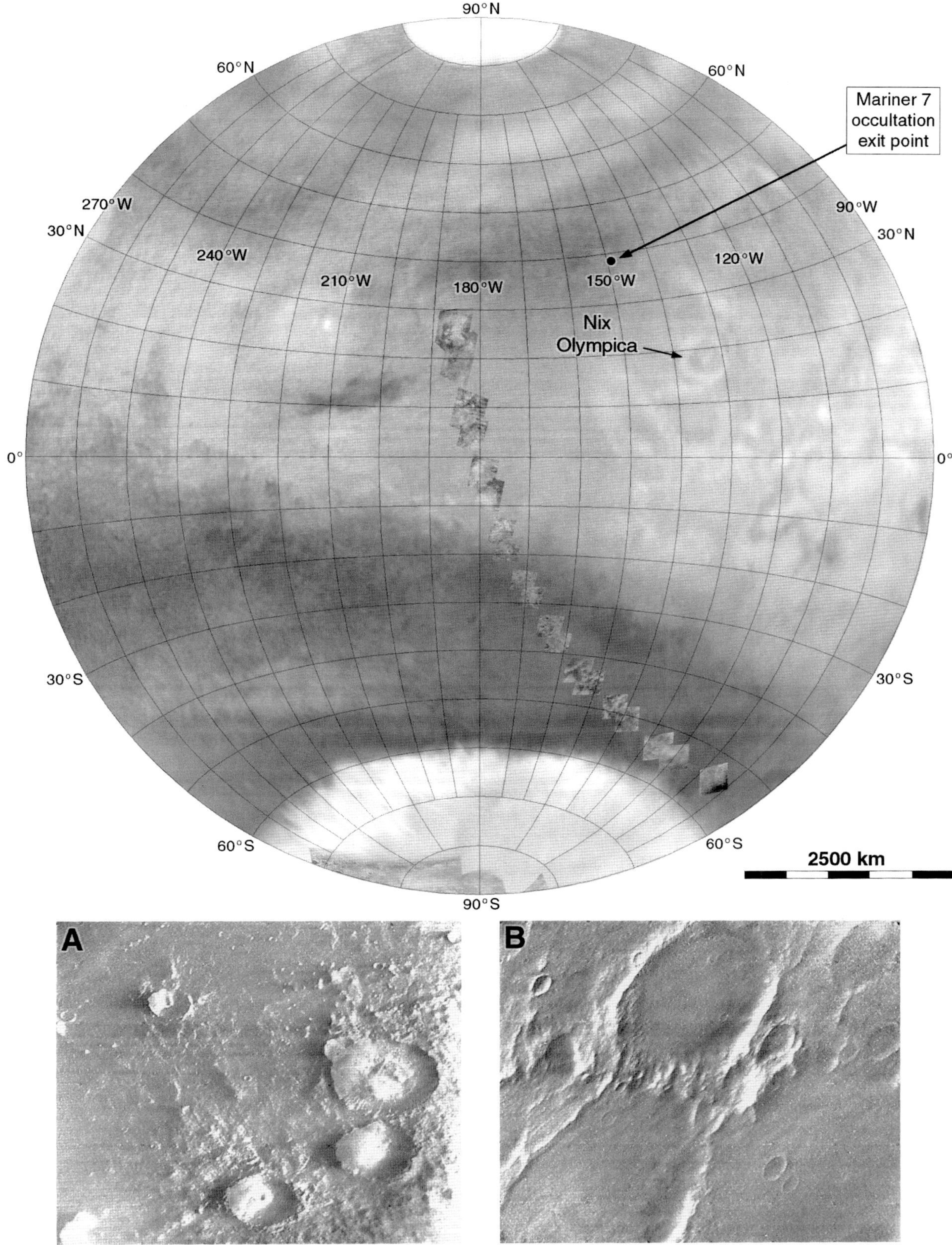

Figure 19 Mars as known after Mariner 7 (continued from Figure 17).
Top: The hemisphere centred on 180 longitude, showing Nix Olympica and other features more clearly than in Mariner 6 images. **A:** Mariner 7 near-encounter frame 7n12 showing craters on the edge of the carbon dioxide cap. Sunlight removes ice from north-facing slopes and shading preserves it on south-facing slopes, creating a reverse lighting effect. **B:** The 'giant's footprint' crater Vishniac (bottom left) near the south pole in image 7n20.

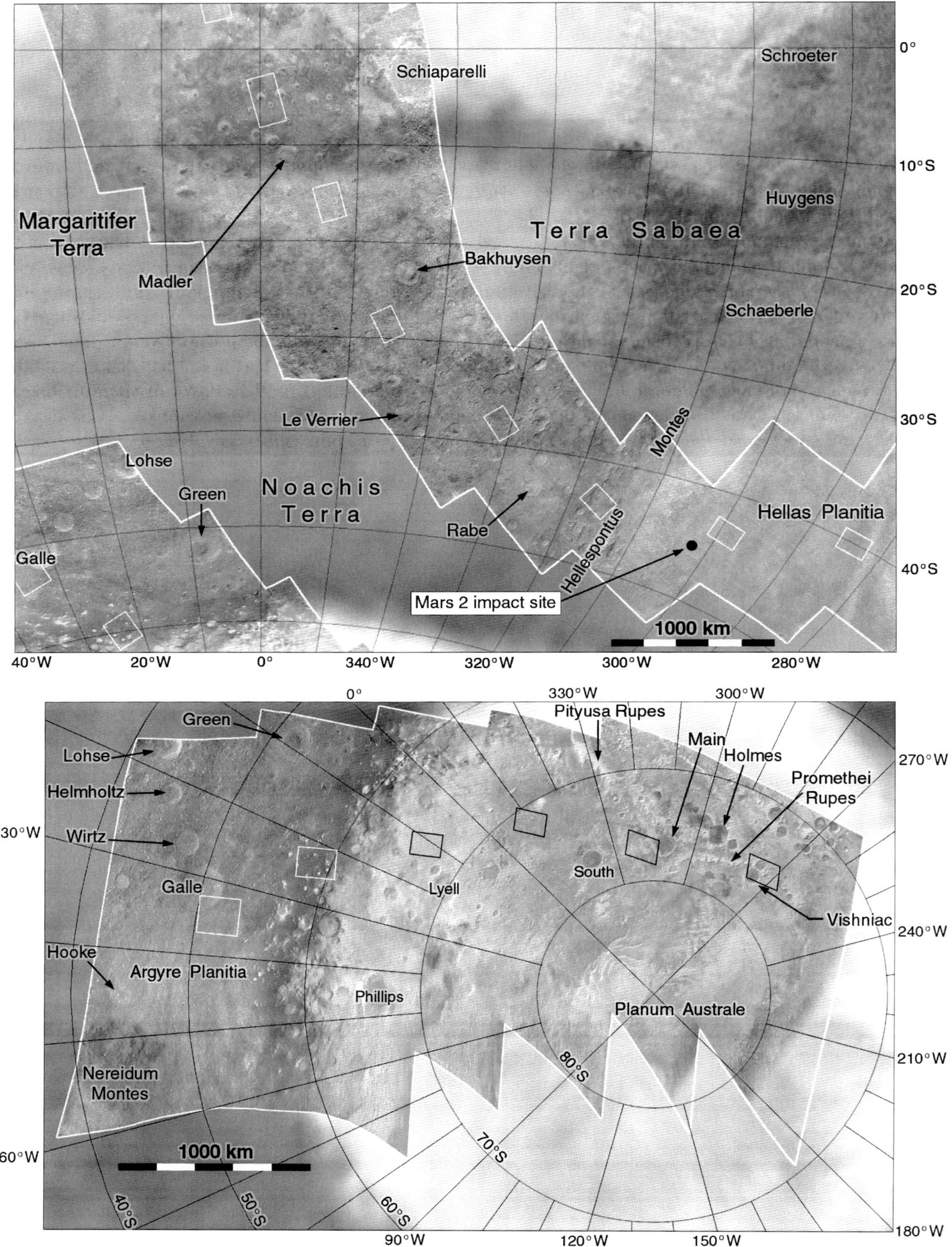

Figure 20 Mariner 7 near encounter images.
Small black or white rectangles show the locations of the narrow-angle frames. The Mars 2 impact site is shown in Hellas, on what appeared to be smooth plains. In fact the surface in that area was obscured by clouds. The actual target may have been the high-resolution frame.

areas affected by surface frosts were interpreted as higher despite some dissenting voices. The first radar reflections from Mars (Goldstein and Gilmore, 1963) suggested that dark areas such as Syrtis Major were more radar-reflective, and light areas less so, and Sagan *et al.* (1966) now argued that the dark areas were highlands. Further work showed exceptions but generally supported this observation. Topographic profiles were first obtained in 1967 (Pettengill *et al.*, 1969), and then at each subsequent opposition, gradually mapping topography in near-equatorial latitudes. Carbon dioxide spectra obtained from Earth (Wells, 1969; Belton and Hunten, 1969) indicated regions in which more or less of the gas was present, with lower values indicating higher terrain. Wells (1972) combined terrestrial radar and atmospheric spectra with Mariner occultation data and spectra to summarize topographic knowledge at that time. Figure 21B is a composite of several datasets showing the limited understanding of Martian topography prior to the Mariner 9 mission. Some features (Hellas, Chryse, Elysium, Isidis) are correctly represented, but others (Sabaeus Sinus, Syrtis Major) give poor matches to modern data. The dramatic relief of Tharsis was only hinted at by these early observations.

8 May 1971: **Mariner 8 (United States)**

Mariner 8 was part of a dual mission with Mariner 9 to orbit Mars and make the first global high-resolution studies of the surface. Mariner 8 was intended to provide near-global image coverage at about 1 km per pixel, with spot coverage at about 100 m/pixel, from a roughly 1800 by 17000 km orbit with an inclination of up to 80° and a period near 12 hours, imaging near the evening terminator (NASA, 1973). Mariner 9 would examine albedo and variable features from an orbit oriented to show the surface with higher Sun angles. After Mariner 8 was lost, Mariner 9's mission was adapted to obtain both types of data.

The 998-kg Mariner 8 spacecraft was launched from Cape Canaveral at 01:11 UT, but its upper stage failed and it fell into the Atlantic Ocean. The spacecraft design was similar to that of Mariners 6 and 7, but with its larger solar panels, each 215 by 90 cm, it spanned 6.9 m. Mariner 8 carried wide- and narrow-angle cameras, an infrared radiometer and ultraviolet and infrared spectrometers.

10 May 1971: **Cosmos 419 (Soviet Union)**

The Soviet Union launched three spacecraft towards Mars in 1971. This spacecraft, Cosmos 419, was intended to orbit Mars, but not to deploy a lander like the two spacecraft launched in the following days. It carried an enlarged fuel tank in place of a lander to allow it to brake into orbit from its faster approach to Mars. The 4650-kg spacecraft was launched at 16:59 UT into a 174 by 159 km parking orbit inclined 51.4° to the equator with a period of 88 minutes, but a command timing error caused its upper stage to fail, trapping it in the parking orbit. It reentered the atmosphere on 12 May. Because it did not leave Earth, its identity was concealed by giving it a Cosmos designation. This orbiter was intended to arrive at Mars before Mariner 8, to become the planet's first orbiter and to provide improved ephemeris data prior to the arrival of Mars 2 and Mars 3 (Perminov, 1999).

19 May 1971: **Mars 2 (Soviet Union)**

Mars 2 carried the first probe (after the uncertain fate of Zond 2) to enter the atmosphere of Mars and reach the surface, but the lander failed. Mars 2 weighed 4650 kg, of which 450 kg was the lander module and 385 kg the lander itself. It was launched from Baikonur at 16:23 UT into a 173 by 137 km parking orbit inclined 51.8° to the equator with a period of 88 minutes. It was successfully injected into a heliocentric orbit with an aphelion of 1.57 AU and a perihelion of 0.99 AU, inclined 2.2° to the ecliptic which would have had a period of 530 days, but after 192 days it arrived at Mars. The spacecraft deployed the lander about 40 hours before entering orbit and then made an orbit insertion burn on 27 November 1971 (MY 9, sol 563). The Mars orbit ranged from 24940 to 1380 km above the planet, inclined 48.9° to the equator, with a period of 17 hours, 58 minutes. Because Mariner 9 arrived at Mars 13 sols before Mars 2, the Soviet spacecraft became the planet's second orbiter.

The Mars 2 orbiter was intended to study Mars and serve as a relay for the lander's communications with Earth. It carried infrared and ultraviolet instruments to measure surface temperatures and atmospheric composition, a magnetometer and a film camera system to map the planet. Like Luna 3 and Zond 3, the spacecraft would

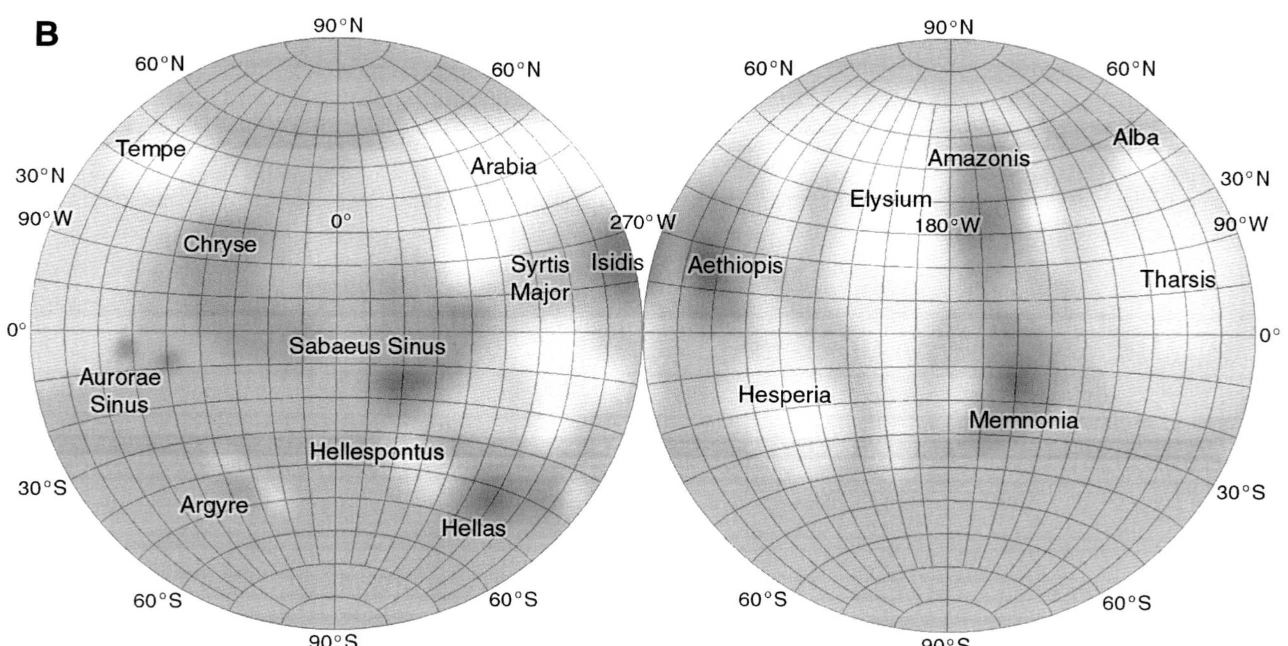

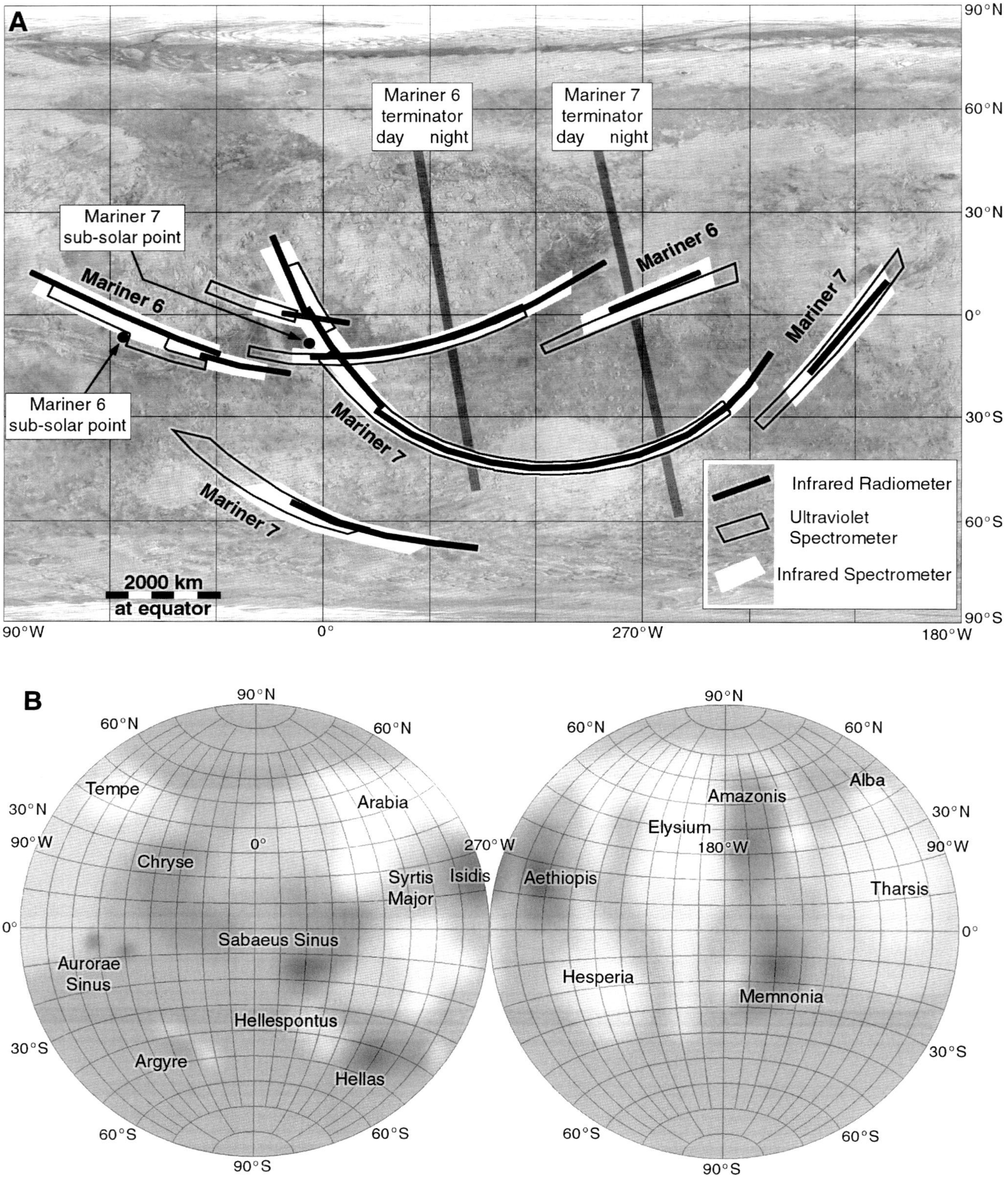

Figure 21 **A:** Mariner 6 and Mariner 7 instrument scans (Stallkamp *et al.*, 1971). The background map is from Mariner 9. **B:** Topography of Mars as it was understood before Mariner 9. Dark areas are lower than the average elevation and bright areas are higher. This map is compiled from data published by Wells (1969), Belton and Hunten (1969) and Wells (1972).

have recorded images on film, then processed and scanned the film and transmitted the image data to Earth. The camera system included both wide-angle and narrow-angle optics and apparently could resolve features between 10 and 100 m across. Images were to be taken in sets of 12, each 35 to 40 seconds apart, on each periapsis during the first 40 days in orbit, with enough film on board for 480 images. Imaging was delayed by a planet-wide dust storm and compromised by overexposure, and the spacecraft suffered telemetry problems which affected its images. It is not clear how many images were taken, but none have been released. Magnetic field data were obtained until as late as May 1972, and the mission ended on 22 August 1972 (MY 10, sol 156) after 362 orbits.

The Mars 2 lander with its sterilized scientific payload was designed to enter the atmosphere, then eject its heatshield and descend on a parachute. The spherical lander carried a dual-panoramic television system similar to that of Luna 13, the second Soviet lunar lander, as well as instruments to analyze the composition of the surface material and atmosphere and measure the temperature, atmospheric pressure and wind speed. It also carried a small rover, described with Mars 3. The probe carried the Soviet coat of arms to the surface, as was the custom with all Soviet lunar and planetary landers. About 30 meters above the surface, a braking rocket would reduce the velocity enough for a safe landing. The landing procedure failed, apparently because the vehicle entered the atmosphere too steeply. The spacecraft entered the atmosphere near 44° S, 313° W (Marov and Petrov, 1973), sometimes stated too precisely as 44.2° S, 313.2° W, but the parachute did not deploy and the vehicle crashed within about 150 km of 45° S, 302° W. The impact site falls within the area covered by Mariner 7 images, which gave the impression that the surface was extremely smooth. According to Alexander T. Basilevsky (T. Stryk, personal communication, 4 February 2009), this influenced the choice of target area. The latitude was determined by the spacecraft trajectory, but the longitude could be chosen to fall within the limited imaging coverage.

Figure 22 shows the impact area in Hellas Planitia. The circles do not reflect the true uncertainty in position, which is about 2.5° or 150 km radially from the given coordinates. The site at 302° W lies well within the smooth area seen by Mariner 7. The background map in Figure 22A is a composite of Mars Global Surveyor topography and wide-angle imaging. Figure 22B is a Mars Odyssey thermal emission imaging system (THEMIS) infrared mosaic of the landing region. The surface was much rougher than the Mariner 7 images suggested. If the atmospheric entry angle was too steep, the true landing target would have been farther east, closer to the centre of the Hellas basin.

An alternative location for Mars 2, at 4° N, 47° W, near Nanedi Valles, is mentioned in several places, including the website of NPO Lavochkin, the spacecraft builder, in 2009. This does not correspond with contemporary accounts or orbiter trajectories and is obviously incorrect. Because 47° W is 313° E the longitude at least appears to be a simple mistake.

28 May 1971: **Mars 3 (Soviet Union)**

Mars 3 was identical to Mars 2 in design and purpose. It was launched at 15:26 UT from Baikonur into a parking orbit of 234 by 140 km inclined 51.6° to the equator with a period of 88.2 minutes and then placed on a heliocentric Mars transfer trajectory with an aphelion of 1.57 AU and a perihelion of 0.99 AU, inclined 2.2° to the ecliptic with a solar orbit period of 530 days. After 188 days it reached Mars on 2 December 1971 (MY 9, sol 568), five sols after Mars 2, and was placed in a 211 400 by 1100 km orbit inclined 60° to the equator with a period of 12.7 days. The intended orbit should have reached no higher than about 25 000 km, but excessive fuel use during cruise prevented a full-orbit insertion burn, leaving the spacecraft trapped in a high orbit. Periapsis rose during the mission to more than 2250 km by mid April 1972.

The orbiter carried cameras like those of Mars 2, a French-built solar radio receiver (1 m wavelength) as part of a joint program called Spectrum 1, several photometers and radiometers to study atmospheric and surface composition and temperature, an infrared spectrometer, a magnetometer and particles and fields instruments. Mars 2 and Mars 3 provided data until May 1972 on atmospheric composition and pressure, surface temperatures varying from 286 K at noon to 163 K at night, and the weak magnetic field. The high-resolution imaging system on Mars 3 failed, so only low-resolution images with 250 scan lines could be transmitted. Sixty images

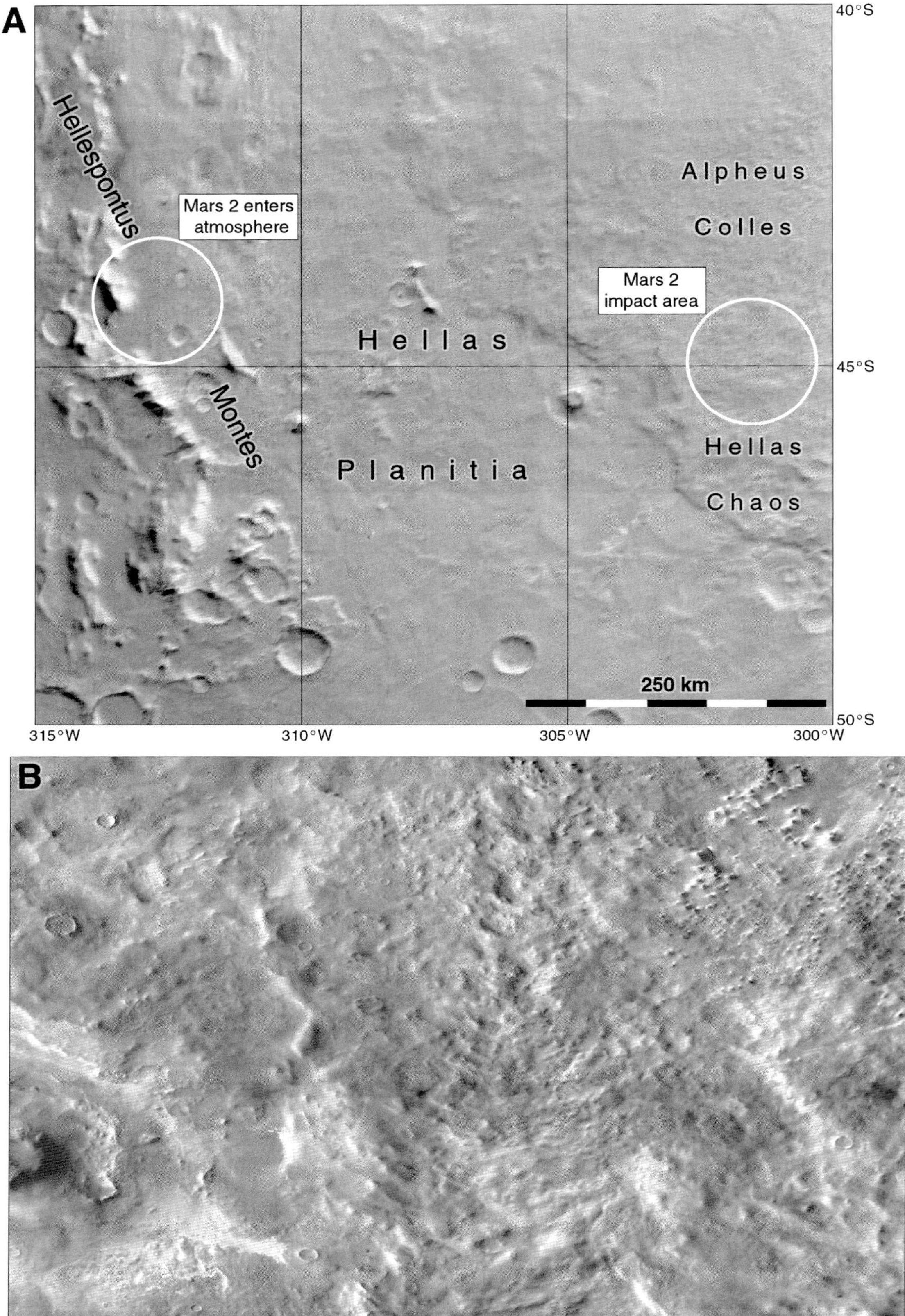

Figure 22 **A:** Mars 2 impact region in Hellas Planitia, with the atmospheric entry position in Hellespontus Montes at left. **B:** Mars Odyssey infrared mosaic of the Mars 2 impact area. The image is 250 km wide. The locations are centred on these circles but are uncertain by 150 km.

were said to have been returned from Mars 2 and Mars 3 together, most of them from Mars 3. Some were taken from near apoapsis on 10 and 12 December, 28 February and 12 March, and other data were collected at lower altitudes on 15 and 27 December 1971 and 16 and 22 February 1972 (Marov and Petrov, 1973). One of the early lower altitude images showed haze layers on the limb (Figure 25D). The two orbiter missions were declared over on 22 August 1972 (MY 10, sol 156).

The lander was released from the orbiter at 09:14 UT on 2 December 1971, shortly before orbit braking. It entered the atmosphere at about 13:49 UT and descended to the surface, making the first landing on the planet at 13:52 UT. A brief period of data transmission followed the landing, lasting only 20 seconds. A global dust storm in progress at the time of landing may have contributed to the loss of the lander. The signal included data from one of the dual cameras, but it could not be converted into an intelligible picture. It may have resembled the first lines of Luna 13 panorama 2 (Academy of Sciences of the USSR, 1969; Stooke, 2007), which showed the second camera housing rather than the lunar surface. It has been represented as a view of the horizon of Mars, but that cannot be correct. The lander carried a miniature walking rover on a 15-m tether. Its 'feet' were two ski-like horizontal rods, raised and lowered to step the rover forwards. Similar rovers were carried on Mars 2 and Mars 6.

The Mars 3 landing site was within 2.5° (150 km) of 45° S, 158° W. Ezell and Ezell (1984) say 45° S, 168° W, but the Mars 2 and Mars 6 trajectory data suggest this is the atmospheric entry location. The landing area is shown in Figure 23. The site is at the corner of Mariner 4 image 13 (Figure 11). This suggests that Mars 3 was targeted to fall in the imaged area, because absolutely nothing was known about other parts of the surface in this hemisphere at the time. A likely target point might have been the smooth level area shown at about 41° S, 155° W in an ACIC Mariner 4 map (Figure 23B). That location, just south of Newton crater, would be at 44° S, 156° W in current (2010) coordinates (Figure 22A). Similarly, Mars 2 was targeted for the area in Hellas imaged by Mariner 7 at this latitude (Figure 20). The only other region previously imaged at this latitude was Argyre (Figure 20), including a site sometimes mistakenly associated with Mars 7 (Figure 38). This may have been an alternative target for Mars 2 or Mars 3.

The Mars 3 orbiter obtained several crescent images of the planet, an image showing an elevated haze layer above the horizon, and at least two low-altitude surface images (Figure 25). Very few images have been released, and many were probably rendered nearly featureless by the dust storm, though Argyre was faintly visible in Mariner 9 approach images (Figure 27B). The two released low-altitude images were presumably taken as the dust storm subsided. Their locations are tentatively identified in Figure 24, which also shows the orbital ground tracks on three periapsis passes when passive X band microwave radiometry was used to estimate surface material density and electrical properties. These are taken from a map shown by Carl Sagan at a Viking Landing Site Staff meeting (LSS 45) on 20 August 1976. Marov and Petrov (1973) plot temperature, pressure and water data along these tracks, but show their locations slightly differently. Figures 25G and 25H show these tentative image locations in more detail.

30 May 1971: **Mariner 9 (United States)**

Mariner 9, the first spacecraft ever to be placed in orbit around another planet, was identical to Mariner 8, but originally had a complementary mission goal. As Mariner 8 mapped the planet, Mariner 9 was to observe areas at regular intervals to search for seasonal changes, from an orbit of roughly 1800 by 41 500 km inclined about 60° to the equator and a period of about 33 hours (NASA, 1973). After Mariner 8 was lost on 8 May 1971, Mariner 9 was assigned the global mapping goal, which was essential for Viking landing site selection. It was designed to operate for three months, but in fact lasted for 11 months and fulfilled many of the goals of both missions. Mission details and image coverage maps (Figure 26) are taken largely from Jet Propulsion Laboratory (1974) and Steinbacher and Haynes (1973).

The 998-kg spacecraft was launched at 22:23 UT from Cape Canaveral, cruised for 167 days, arrived at Mars on 14 November 1971 (MY 9, sol 550) and was placed in orbit with a periapsis of 1385 km and an inclination of 65° to the equator. Although it was launched later than Mars 2 and Mars 3, it arrived before them, becoming the first spacecraft to orbit Mars. Its instruments were infrared and ultraviolet spectrometers, an

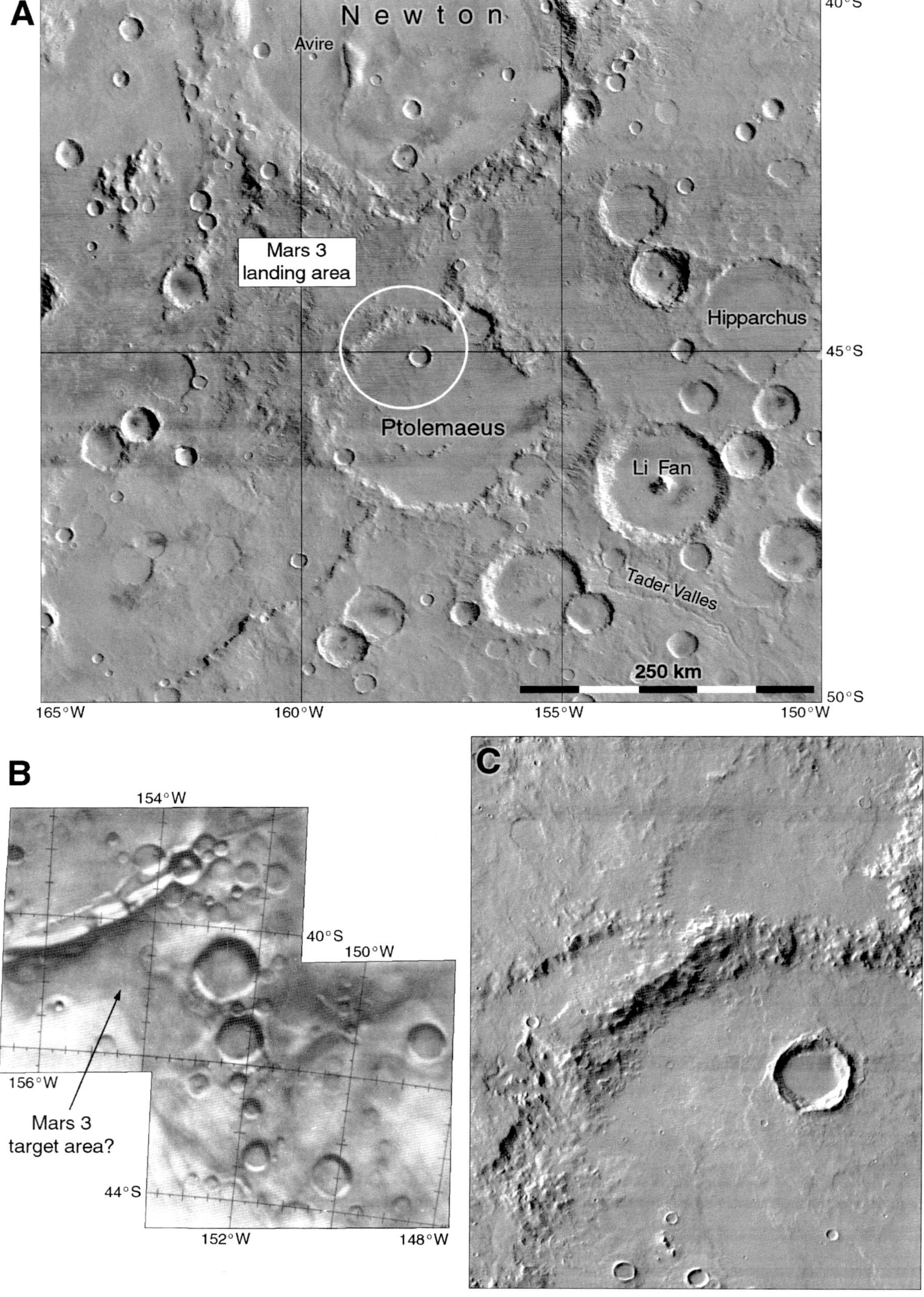

Figure 23 **A:** Mars 3 landing area, uncertain by 150 km. **B:** Probable Mars 3 target area on a US Air Force (ACIC) map based on Mariner 4 images. **C:** Mars Odyssey THEMIS infrared mosaic of the landing area.

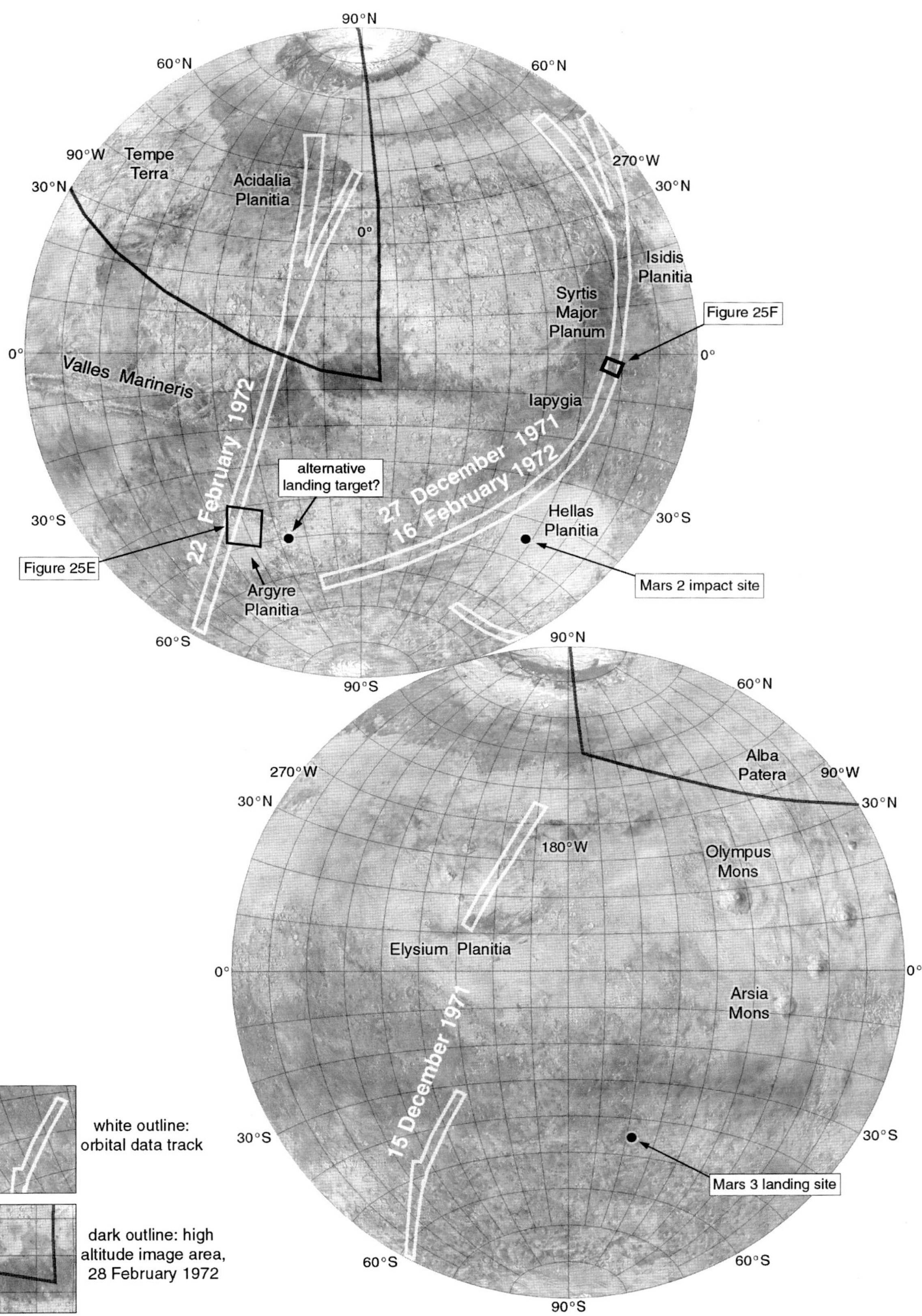

Figure 24 Mars 3 orbital remote sensing groundtracks (white strips) and identified image locations.

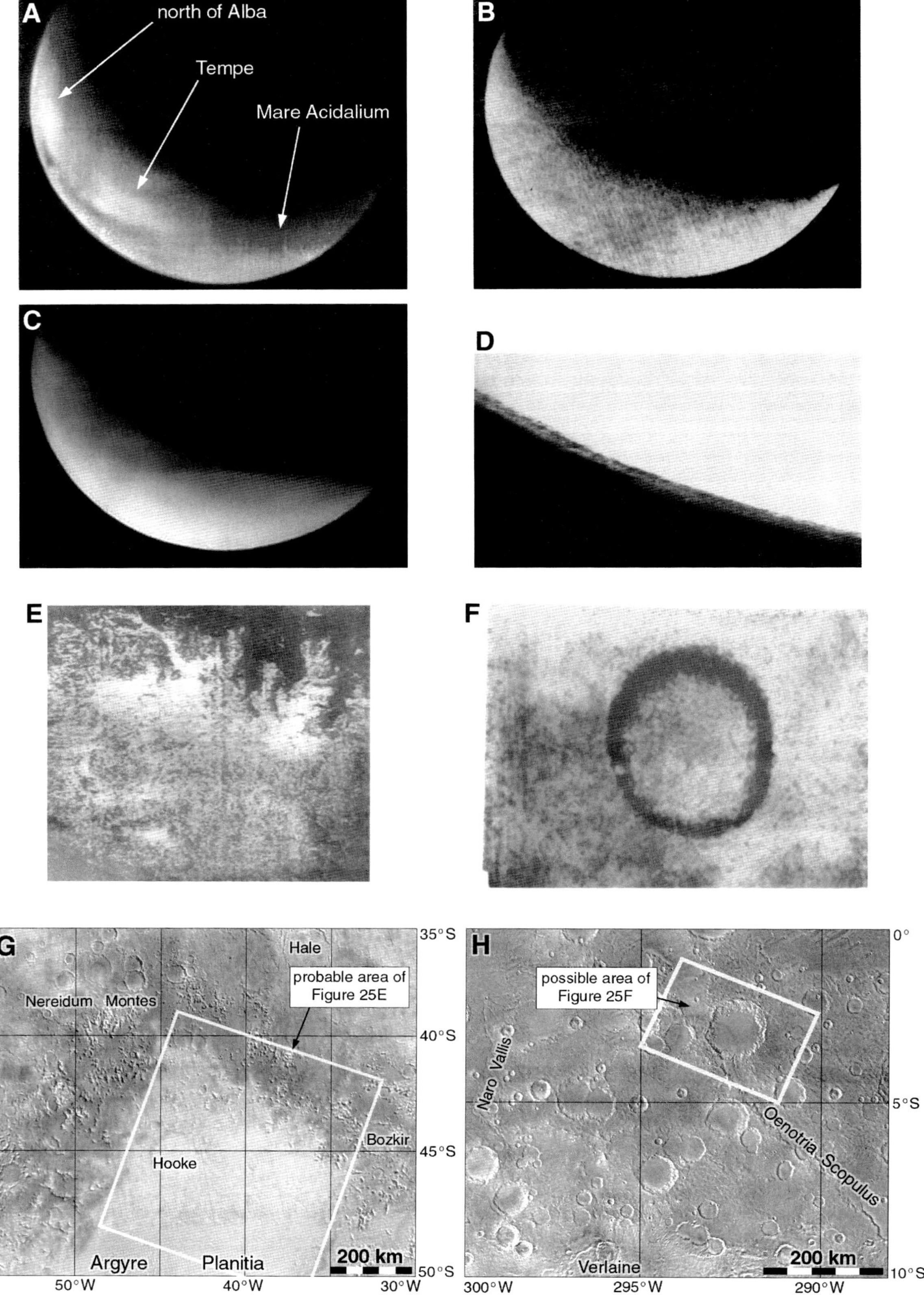

Figure 25 A, B, C, D, E and F: Mars 3 images provided by T. Stryk, modified by P. Stooke. **G and H:** locations of images (E) and (F), respectively. (A) is the image of 28 February 1972 in Figure 24.

infrared radiometer and two cameras for wide- and narrow-angle imaging, mounted on a scan platform. Mariner 9 performed well in orbit until contact was lost on 27 October 1972 (MY 10, sol 220). It is still in orbit, with an estimated de-orbit date of 2022. It obtained global image coverage at 1 to 2 km/pixel and spot coverage at 100 m/pixel. Only very small gaps in coverage remained after this mission, as indicated in Figures 42 and 46. The images provided data for selecting the Viking landing sites and established the general understanding of surface processes and geologic history for Mars. Seasonal changes were monitored for half a Mars year. In addition, the satellites of Mars, Phobos and Deimos were observed for the first time, with sufficient resolution to reveal surface features (Figures 196 and 197). Mariner 9 did not carry an altimeter, but its radio occultations on nearly every orbit and estimates of atmospheric pressure from infrared spectra provided rough elevation data which were used to create the first useful topographic maps of the planet.

After the loss of Mariner 8, the compromise orbit planned for Mariner 9 was referred to as '17/35', meaning that after 17 Mars days and 35 orbits, the groundtracks would repeat. Periapsis would be at 1250 km, to allow A-frame (wide-angle) images to overlap slightly. Over the course of one 17-day orbit cycle, a full latitude band would be imaged, and the next cycle would cover a new range of latitudes (Figure 26). However, as Mariner 9 approached Mars, a planet-wide dust storm arose, hiding the surface from view. According to Ezell and Ezell (1977), 'one scientist, in a bit of gallows humor, suggested they must have visited Venus by mistake. This humor was not well received.' The small south polar cap and a few dark spots were all that could be seen (Figure 27). The spacecraft entered orbit unable to begin its mapping sequences, so other observations were quickly devised. An orbit adjustment was made on the fourth orbit, and instruments were calibrated on the seventh orbit. The comparatively clear south pole was imaged, as well as dark spots in Tharsis, which turned out to be volcanic calderas elevated above the dust clouds on their gigantic mountains. Meanwhile, new imaging sequences were planned to obtain science data during the dust storm, including some limb images which revealed haze layers. These images were taken during the period which came to be called the reconnaissance phase.

By 17 November 1971 some craters began to appear in images as bright circles. The atmosphere cleared gradually during late November and early December, but clearing ceased later in December before resuming at the end of the year. Harold Masursky (USGS) was afraid that some regions might never be seen properly. Figure 26A shows the areas imaged during this phase of the mission.

In January 1972 the air was clear enough for mapping to begin. On orbit 96 the periapsis was raised to 1650 km, increasing the area covered by each wide-angle image to allow global mapping in the reduced time now available to the mission after the dust storm. The orbit period was 24 hours, so data transmission times were the same every Earth day, and the same ground station, Goldstone, could be dedicated to the mission. The periapsis longitude moved eastwards on each orbit, completing a full cycle each 38 orbits, a little different from the original plan for a 35-orbit repeat time. Table 4 lists the various mission phases, and imaging coverage is shown in Figure 26. Systematic imaging began on orbit 100. Three cycles of imaging during the primary mission covered most of the planet's surface up to 45° N, with some additional imaging farther north which was compromised by the hazy atmosphere (Figure 26A).

By 2 April 1972, Mariner 9 had taken 6876 images covering about 85% of the planet. The north pole was emerging from winter but was still largely shrouded by clouds, a phenomenon called the polar hood. The first extended mission phase included re-mapping the south pole before it disappeared into southern hemisphere winter night. A few images were taken in the following weeks to monitor clearing, but the north pole could only be properly mapped late in the mission. Several problems limited data return later in the mission, including communication conflicts at Goldstone due to the Apollo 16 lunar mission in the second half of April 1972. Operations were also interrupted when the orbit and attitude geometry turned the solar panels until they generated too little power. The bright star Canopus had been used for attitude control, but the orientation problem was overcome by switching to Arcturus and Vega. Between orbits 244 and 258, no data were returned due to computer problems, and after orbit 262, solar occultations began to limit power. At the same time, the distance to Mars increased as conjunction approached, reducing the data transmission rate.

Table 4. *Mariner 9 Mission Phases*

Phase	Orbits	Notes
Pre-orbit		30 images per day for 3 days
Calibration	1–15	Instrument calibration and initial imaging while planning imaging strategy during the dust storm
Interim cycle	16–23	Further imaging as planning continues
Reconnaissance 1	24–63	South polar mapping, imaging of dust-free areas
Reconnaissance 2	64–99	Further imaging and orbit adjustment on orbit 94
Mapping Cycle 1	100–138	Mapping, 65° S–25° S
Mapping Cycle 2	139–178	Mapping, 25° S–25° N
Mapping Cycle 3	179–217	Gap fill, 30° S–equator, and mapping, 25° N–45° N
Extended Mission Phase 1	218–262	Gap fill south of 45° N, repeat south polar imaging, some images of haze-covered north polar region
Conjunction	263–415	Relativity experiment, imaging hiatus
Extended Mission Phase 2	416–459	Mapping, 45° N–65° N (north polar collar)
Extended Mission Phase 3	613–673	No systematic imaging, targets of opportunity, Viking sites
Extended Mission Phase 4	674–676	Final imaging targets

On 7 September 1972 Mars moved through superior conjunction, and communication was again interrupted. A test of general relativity was undertaken before and after the time of conjunction, to examine the predicted bending of Mariner 9's radio transmissions by the Sun's gravity. This used a significant amount of attitude control fuel, and early maneuvers during the dust storm had also used more fuel than expected. The limited fuel budget available late in the mission reduced the planned imaging of Viking landing site candidates from 32 to 24, of which only 19 were imaged before the end of the mission (Table 6, Figure 40).

The second extended mission phase began after conjunction. The objectives of this mission phase were to monitor variable features, to complete the north polar mapping, and to image sites determined to be of interest to Viking, based on analysis of the primary mission imaging (Table 6). On orbit 458 a special task was performed, the imaging of a Soviet landing site at 44.9° S, 160.8° W (Jet Propulsion Laboratory, 1974). This was the contemporary estimate of the Mars 3 landing location (Marov and Petrov, 1973). The last images were obtained on orbit 676.

This pioneering spacecraft revealed Mars to be a world intermediate between Earth and the Moon in its geological complexity (Masursky, 1973), with a cratered southern hemisphere and generally smoother low-elevation plains in the north (Figures 28 and 29). Its surface was studded with volcanic shields, including some of the largest volcanoes in the solar system. It was cut by large tectonic structures, including a vast canyon or rift valley system (Valles Marineris, named for the Mariner series of spacecraft), and also by enormous valleys resembling dry river channels. The polar caps were revealed as water ice deposits several hundred kilometres across, covered by carbon dioxide frost caps several thousand kilometres across in winter, underlain in both hemispheres by extensive eroded layered deposits which appeared to preserve a long history of climate change.

Figure 27 illustrates some representative examples of Mariner 9 images. Figures 27A and 27B are approach images showing the dust-shrouded surface, with dark spots hinting at the presence of the Tharsis volcanoes, and bright streaks revealing the immense canyons of Valles Marineris. A bright spot at the bottom of each image is the south polar ice cap, much smaller than it was during the Mariner 6 and 7 missions. Figures 27C, D, E and F illustrate some of the dramatic discoveries of this first systematic exploration of a new planet. They show, respectively, the vast volcanic shield Arsia Mons, the meandering channel Shalbatana Vallis, a small section of Valles Marineris including Ius Chasma and Louros

Valles, and eroded polar deposits at Sisyphi Cavi. Figure 27F is a high-resolution frame.

Mars was found to be a world of superlatives, of geomorphological extremes, a 'wonder world' in Carl Sagan's words (Sagan, 1994). Though it lacked the canals of Percival Lowell, it was not the cratered clone of the Moon which Mariner 4 images had suggested. The question of life arose again as scientists contemplated the widespread evidence for water on the surface, so Mars became the target of a series of sophisticated robotic missions over the next four decades. USGS undertook a mapping program, producing the first global map of Martian landforms at 1:25 000 000 scale (Figures 28 and 29) and a series of maps at 1:5 000 000 scale covering Mars in 30 sheets (*e.g.* Figures 30, 38A, 42). Special maps at 1:1 000 000 scale were prepared for potential landing sites (*e.g.* Figures 37A, 38B, 45, 46A, 77). Despite the spectacular advances in knowledge of Mars made possible by Mariner 9, a comparison of Figures 30B and 51A shows how much detail north of 45° N was concealed by the poor atmospheric conditions during this mission.

Geologic mapping was also performed to establish the context of potential Viking landing sites, using methods developed for the Moon (Figure 30D). These maps were based on interpretations of surface morphology and made use of a historical framework similar to that developed for the Moon in preparation for Apollo missions. The old cratered terrain typified by the Noachis region formed in the Noachian era. More lightly cratered and ridged plains such as Hesperia Planum formed in an intermediate era called Hesperian, and the least cratered plains such as Amazonis Planitia formed in the Amazonian era. There was much controversy regarding the absolute timing of the transitions between eras, but the basic subdivision of time became entrenched in the literature of Martian studies.

The Infrared Spectrometer (Conrath *et al.*, 1973) measured thermal emissions from the surface and atmosphere, from which estimates of surface pressure and water vapor abundance were derived. These resulted in low-resolution topographic maps of much of the planet and suggestions that water was most abundant over the north polar cap during the mission. Dust particle sizes and atmospheric pressure variations during the dust storm were also measured.

Topographic mapping was a very important goal of the Mariner 9 mission because of its importance to Viking landing site planning. Low areas were desirable for landing because the denser atmosphere would make the parachute more effective and increase the likelihood of finding liquid water in the surface materials. Christensen (1975) combined results from Mariner 9 radio occultations and spectral data from the ultraviolet and infrared instruments with Earth-based radar to produce the best pre-Viking topographic map. Now it was clear that the cratered regions, mostly in the southern hemisphere, were higher than the northern plains, and that the volcanic provinces of Tharsis and Elysium were also elevated. There was no consistent relationship between albedo and topography. A preliminary version of this map (Viking Data Analysis Team, c. 1972) is included in Figure 29.

21 July 1973: **Mars 4 (Soviet Union)**

The next Mars launch window was in 1973, the original target date for Viking. When Viking was delayed until 1975, the Soviet Union had another chance to land the first operating science package on Mars. A four-spacecraft fleet was sent to Mars in 1973, two landers and two orbiters intended to image the surface and relay data from the landers. The landers had to be flown separately from the orbiters in this relatively unfavourable launch window. The first was the 3440-kg Mars 4 orbiter, launched at 19:31 UT from Baikonur into a 179 by 147 km parking orbit inclined 51.5° to the equator with a period of 87.5 minutes. An upper stage burn placed Mars 4 on a heliocentric orbit with an aphelion of 1.63 AU and a perihelion of 1.02 AU, inclined 2.2° to the ecliptic with a period of 556 days.

Mars 4 arrived at the planet after a 204-day cruise and should have braked and entered orbit on 10 February 1974 (MY 11, sol 10). The orbit injection rocket failed, and Mars 4 flew past the planet at a distance of 2200 km, taking images of an area north of Argyre including the probable Mars 7 landing target in northern Argyre Planitia. A scanning camera similar to those tested at the Moon by Luna 19 and Luna 22 transmitted two low-resolution image strips during the flyby (Figure 31), and another camera took 12 images on film (Sidorenko, 1980), which

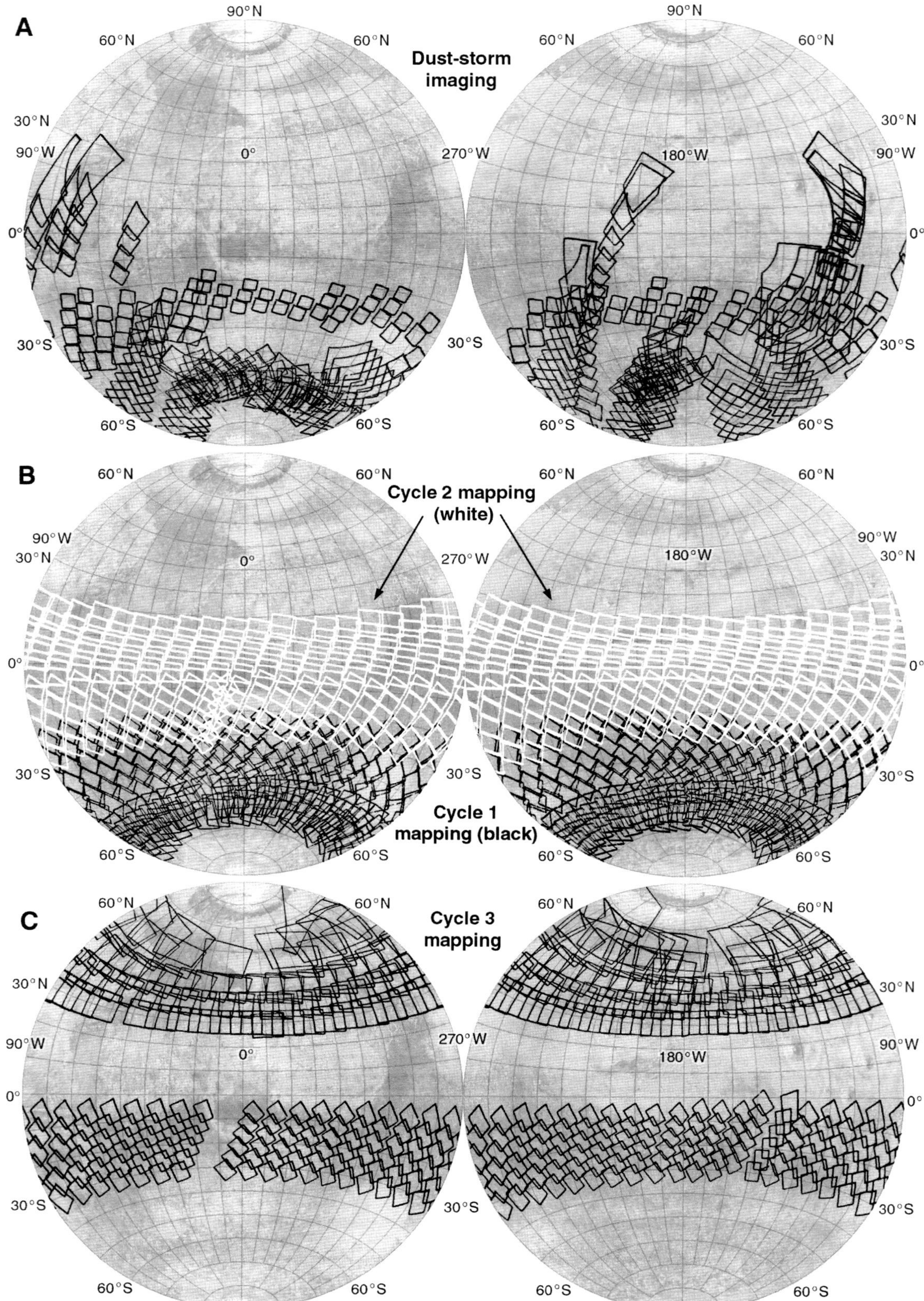

Figure 26 Mariner 9 imaging coverage during the primary mission.

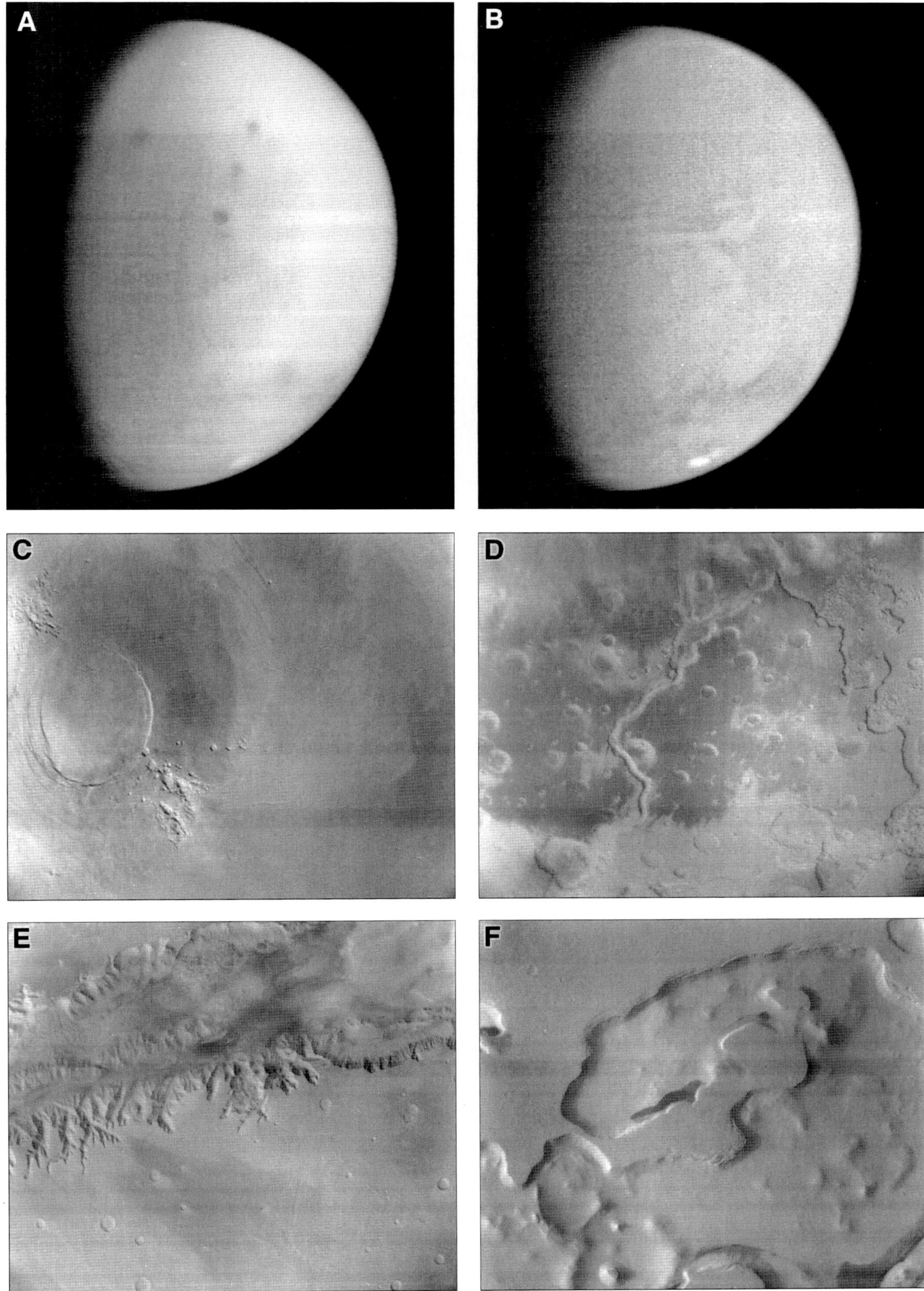

Figure 27 Mariner 9 images.
A: Image 01509677. **B:** Image 01497357. **C:** Image 08513514. **D:** Image 07615338. **E:** Image 07398688. **F:** Image 08763869.

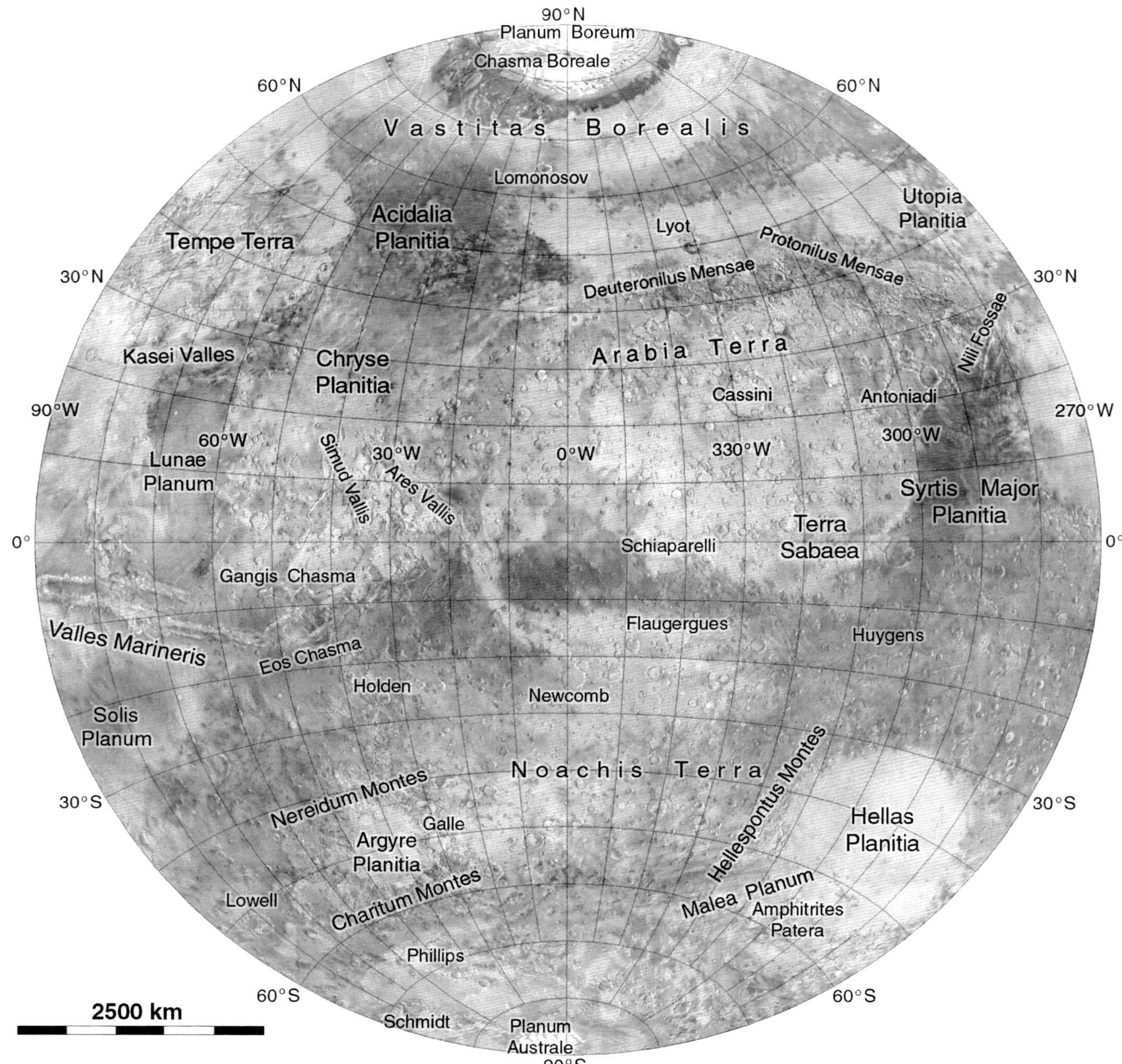

Figure 28 Mars as revealed by Mariner 9.
This hemisphere is centred on the 0° meridian and extends from Valles Marineris on the west to Syrtis Major Planitia (later referred to as Planum) on the east. The giant canyon system was hinted at in Mariner 6 images but first seen clearly by Mariner 9 and is named after the Mariner series of spacecraft. The map has been adapted from US Geological Survey map I-961 by removing some details and converting the map projection to Azimuthal Equidistant. The albedo markings are a composite of Mariner 9 observations made during 1972.

Figure 29 (opposite page): **Top:** Mars as revealed by Mariner 9, continued from Figure 28. This hemisphere, centred on the 180° meridian, includes the volcanic provinces of Tharsis and Elysium discovered in Mariner 9 images. **Bottom:** Topographic contours plotted on a cylindrical projection Mariner 9 base map. The topographic data are a combination of Earth-based radar and Mariner 9 occultation data (Christensen, 1975; Viking Data Analysis Team, c. 1972). Contours are 1 km apart with the zero contour shown as a heavier line.

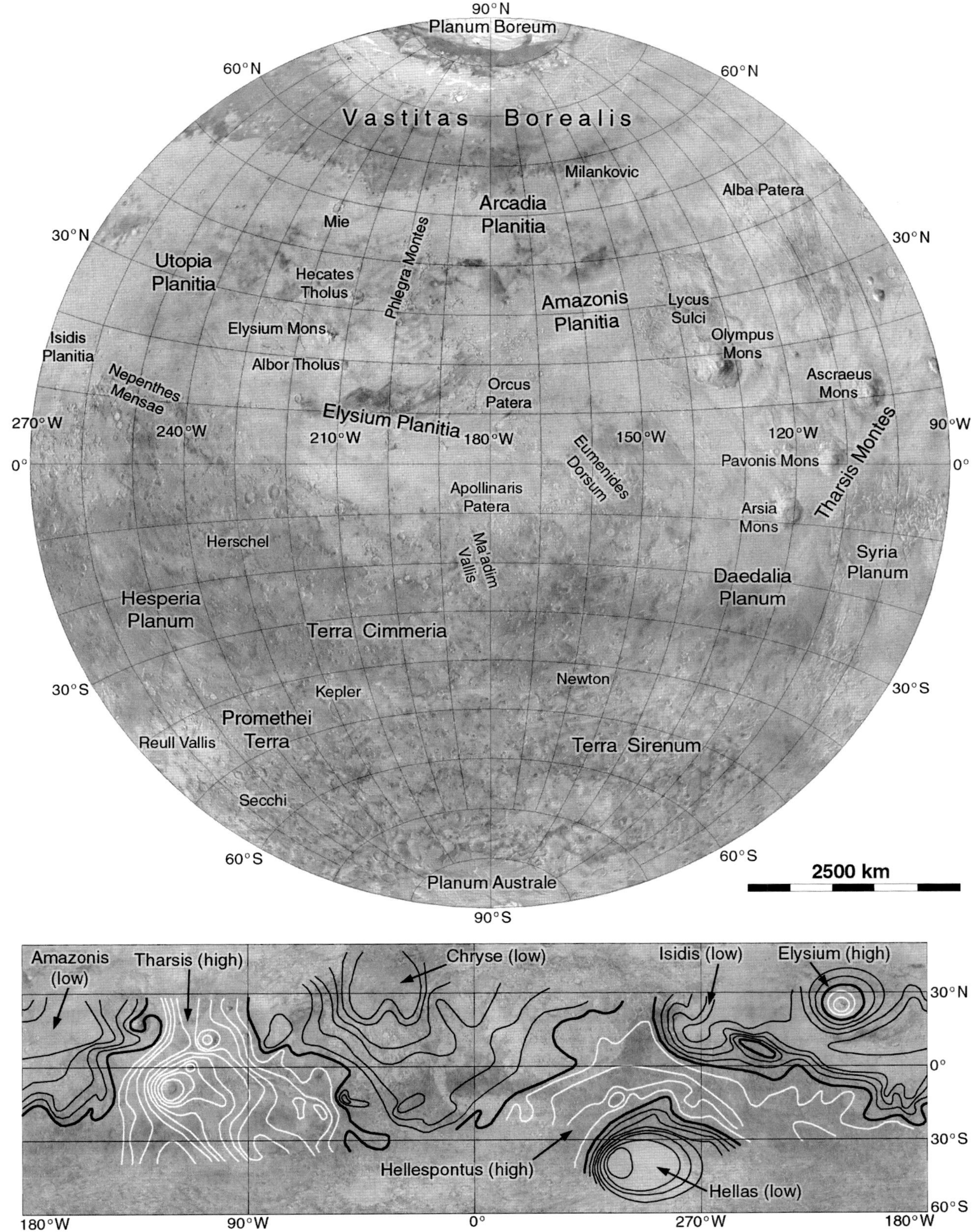
90°N
Planum Boreum
60°N
60°N
Vastitas Borealis
Milankovic
Alba Patera
Arcadia
Planitia
Mie
Phlegra Montes
30°N
30°N
Utopia
Planitia
Hecates
Tholus
Amazonis
Planitia
Lycus
Sulci
Olympus
Mons
Isidis
Planitia
Elysium Mons
Albor Tholus
Nepenthes
Mensae
Orcus
Patera
Ascraeus
Mons
270°W
240°W
Elysium Planitia
210°W
180°W
Eumenides
Dorsum
150°W
120°W
90°W
0°
0°
Pavonis Mons
Tharsis Montes
Apollinaris
Patera
Arsia
Mons
Herschel
Ma'adim
Vallis
Syria
Planum
Daedalia
Planum
Hesperia
Planum
Terra Cimmeria
Newton
30°S
30°S
Kepler
Promethei
Terra
Reull Vallis
Terra Sirenum
Secchi
60°S
60°S
2500 km
Planum Australe
90°S
Amazonis
(low)
Tharsis (high)
Chryse (low)
Isidis (low)
Elysium (high)
30°N
0°
30°S
Hellespontus (high)
Hellas (low)
60°S
180°W
90°W
0°
270°W
180°W

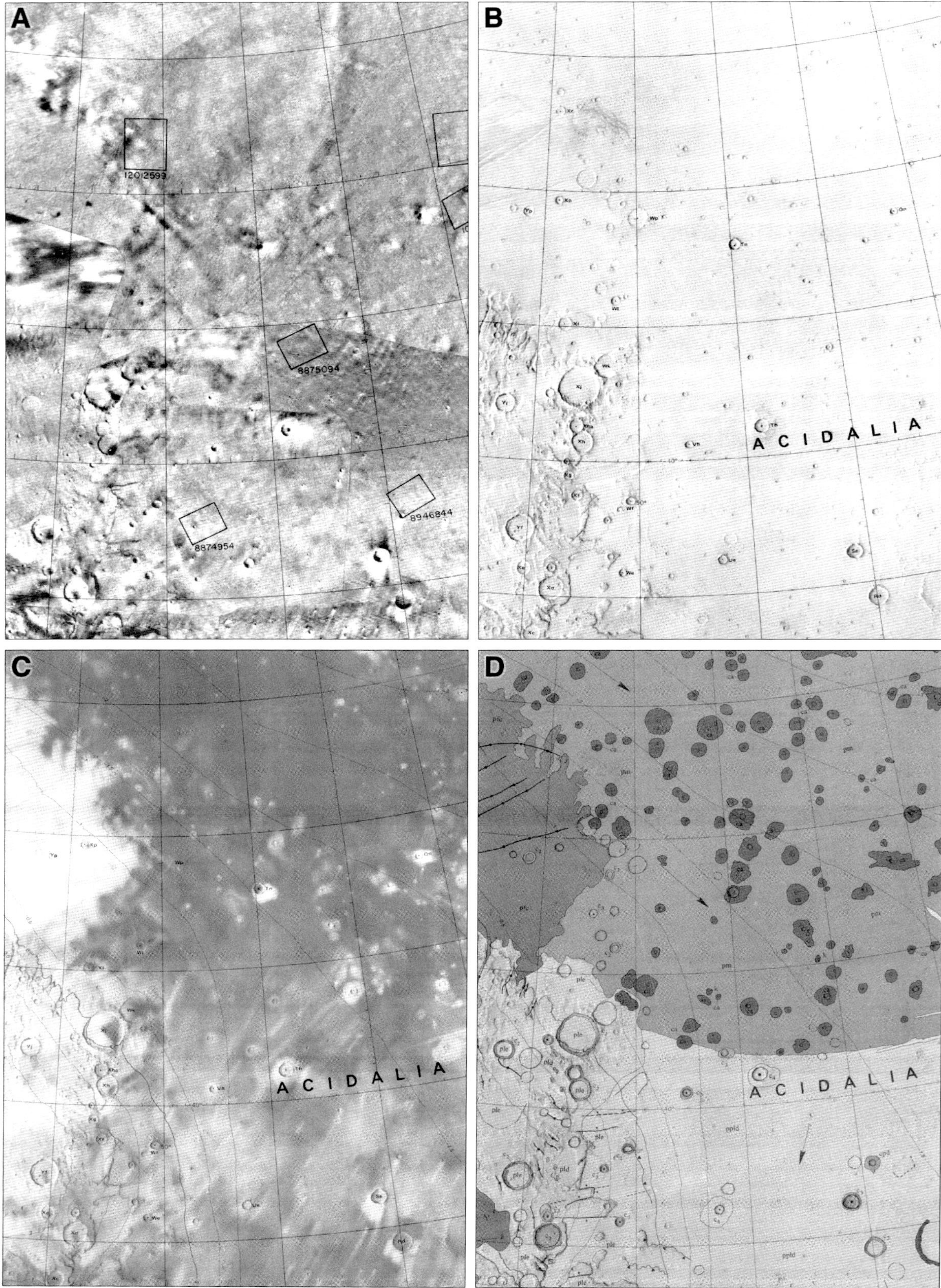

Figure 30 Maps made from Mariner 9 images.
The surface of the planet was covered by 30 sheets of USGS maps at 1:5 000 000 scale. Here, part of sheet 4 (Mare Acidalium) centred at 45° N, 50° W is shown as a photomosaic (A), a shaded relief map (B), a map showing topography and albedo markings (C) and a geological interpretation (D). 30A is an unpublished image index map, 30B is USGS map I-958, 30C is USGS map I-979 and 30D is USGS map I-1048.

were developed, scanned and transmitted after the flyby (Figure 32). The maps in Figures 31 and 32 show the area on Mars covered by these images, between Argyre and Valles Marineris. Mars 4 was similar in design to the Mars 2 and Mars 3 orbiters and carried the same instruments as Mars 5. During the cruise it operated a French-built instrument for solar radio astronomy and charged particle studies.

25 July 1973: **Mars 5 (Soviet Union)**

The second probe of the 1973 fleet, Mars 5, was launched at 18:56 UT from Baikonur and placed in a 174 by 159 km parking orbit inclined 51.6° to the equator with a period of 87.8 minutes. It then entered a Mars transfer orbit with an aphelion of 1.65 AU and a perihelion of 1.01 AU, inclined 2.2° to the ecliptic with a period of 567 days. After a 202-day cruise it arrived at Mars on 12 February 1974 (MY 11, sol 12), and this time the braking rocket worked successfully. Mars 5 entered an orbit ranging from 32550 to 1760 km above the planet, inclined 35° to the equator of Mars with a period of 24 hours, 53 minutes, and operated for only 22 orbits before failing due to a loss of pressure in the transmitter compartment.

The instruments on Mars 5 were a series of photometers and polarimeters for atmospheric studies, an infrared radiometer, a passive microwave system and a gamma ray spectrometer for surface studies, a magnetometer and the French radio astronomy and charged particle package also operated by Mars 4 during cruise. The atmospheric instruments detected ozone in the atmosphere and atomic hydrogen at very high elevations. Mars 5 also carried two cameras having resolutions from 50 m/pixel to about 1 km and a scanning camera with a resolution of 3 km/pixel. The cameras took images on film and scanned the resulting negatives for transmission as earlier Soviet probes had done. Figures 33 to 36 show the areas covered by all the imaging sequences, and the area for which surface data were obtained by the passive microwave system and the gamma ray spectrometer, including the Mars 6 landing site (Figure 37). The compositions inferred from the gamma ray experiment resembled basaltic rocks, with more radioactive elements detected over southern Tharsis than over the cratered southern highlands (Surkov, 1990). The two framing cameras each took 12 images at periapsis on five orbits, on 17, 21, 23, 25 and 26 February, covering the area from 0° S to 45° S and 0° W to 95° W. Some were scanned and transmitted at full resolution, others only at low resolution. Five panoramic swaths extended from 5° N to 45° S and from 10° W to 170° W. In all 108 images were received, 43 having reasonable quality (Sidorenko, 1980).

Scientists including Alexandr P. Vinogradov, Vasilii I. Moroz and Nadir B. O. Ibragimov, all involved in astronomy or planetary exploration in the Soviet Union, are commemorated by craters in the area imaged by Mars 4 and Mars 5 (Figures 32, 35, 37C). A crater named after the engineer Georgy N. Babakin is shown in Figure 35. The Soviet Union published maps of this area derived from their images (Blunck *et al.*, 1993).

5 August 1973: **Mars 6 (Soviet Union)**

The third and fourth spacecraft of the 1973 Soviet Mars fleet were the two landers, Mars 6 and Mars 7. The 3260-kg Mars 6 was launched from Baikonur at 17:46 UT into a 193 by 154 km parking orbit inclined 51.5° to the equator with a period of 87.9 minutes. It was injected into a 567-day, 1.67 by 1.01 AU heliocentric transfer orbit inclined 2.2° to the ecliptic. After two months a failure prevented further communication to or from Earth, and Mars 6 executed the rest of the mission in a preprogrammed mode. It arrived at Mars after 219 days, on 12 March 1974 (MY 11, sol 40), where the carrier spacecraft deployed its lander at a distance of 48000 km. The lander, now able to transmit its own data, entered the atmosphere at 09:06 UT, and the carrier vehicle flew on past the planet at a range of 1600 km. The lander carried an imaging system, atmospheric temperature, pressure and wind sensors, an altimeter and accelerometer and instruments to measure atmospheric and surface composition. The carrier spacecraft instruments were an imaging system, a magnetometer, solar wind, cosmic ray and micrometeorite detectors, and a French instrument for studying solar radio emissions and particles.

Zezin *et al.* (1975) illustrate the target, atmospheric entry and impact points for Mars 6. NASA was asked to provide Mariner 9 images of the Mars 6 landing target, and USGS produced the Mare Erythraeum map (Figure 37A) in support of this mission. The spacecraft

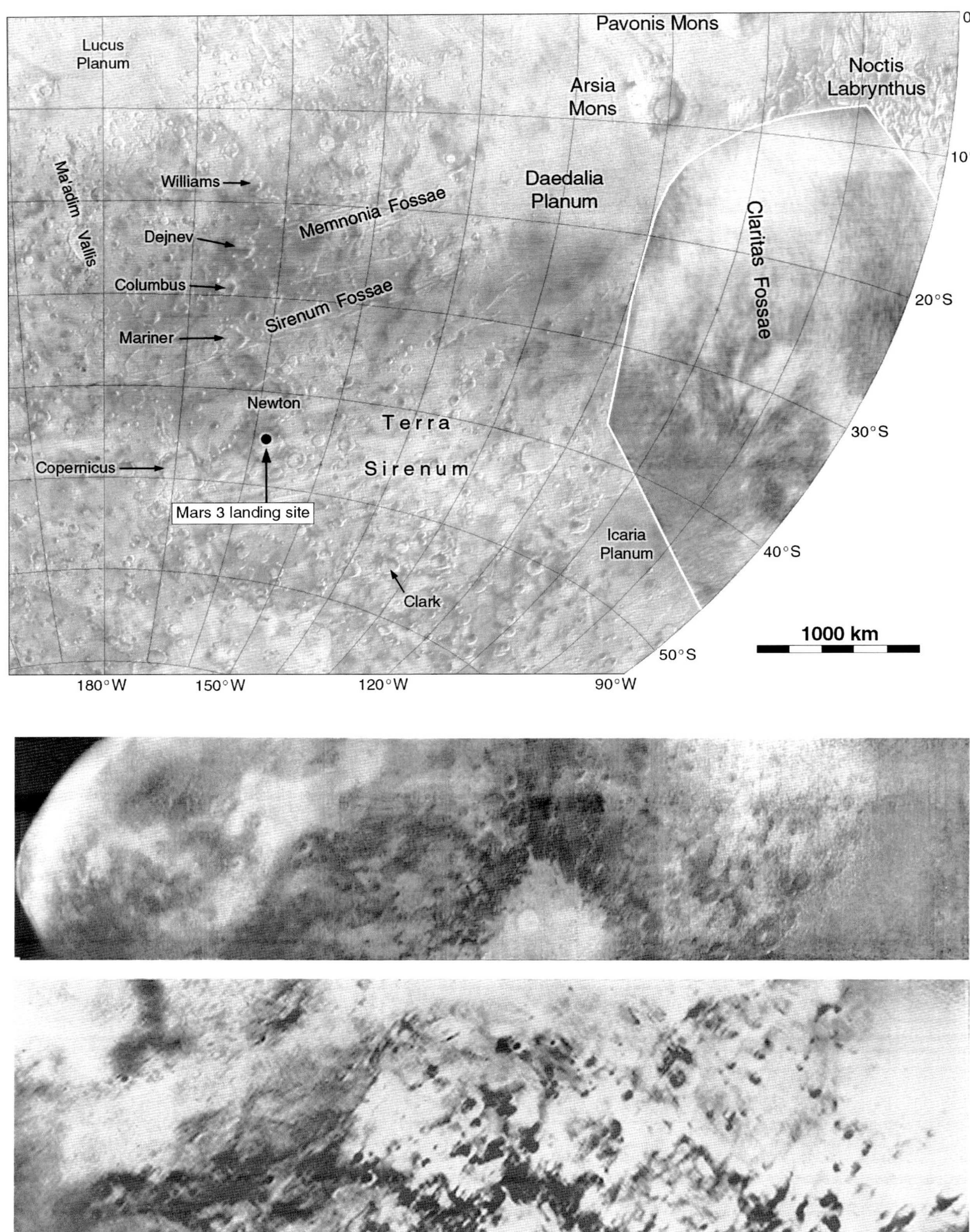

Figure 31 Top: Mars 4 image coverage in the 180° hemisphere. **Bottom:** Panoramic scanning camera images from Mars 4, provided by T. Stryk. The bright patch with a dark collar at bottom centre in each image is Argyre.

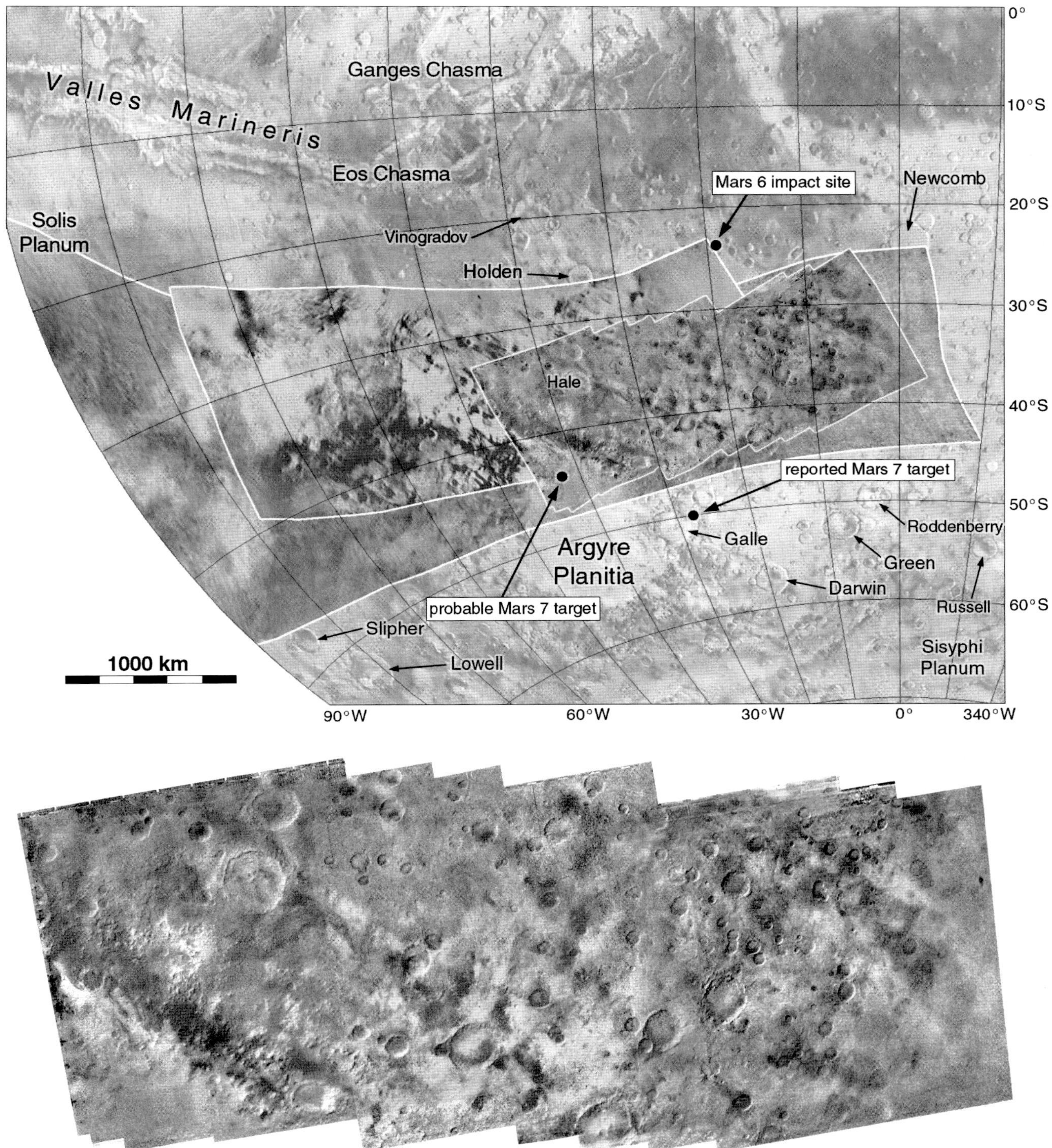

Figure 32 Top: Mars 4 image coverage, 0° hemisphere. **Bottom:** Mosaic of Mars 4 images (P. Stooke). Argyre is in the lower left corner. The image coverage maps are plotted on the USGS base maps made from Mariner 9 data, enlarged from Figures 28 and 29.

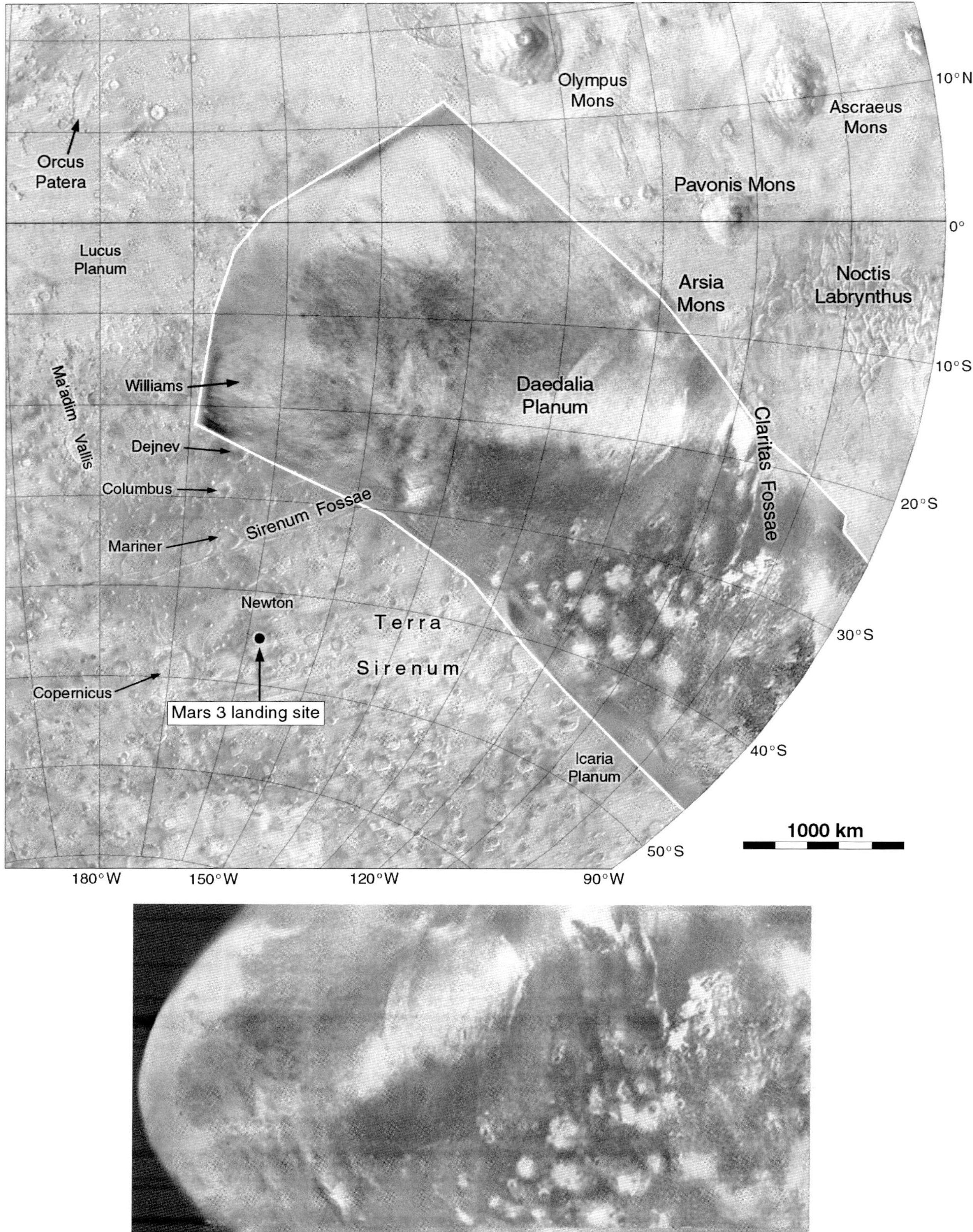

Figure 33 Mars 5 image coverage in the 180° hemisphere (top) and one of five panoramic scans (bottom), provided by T. Stryk.

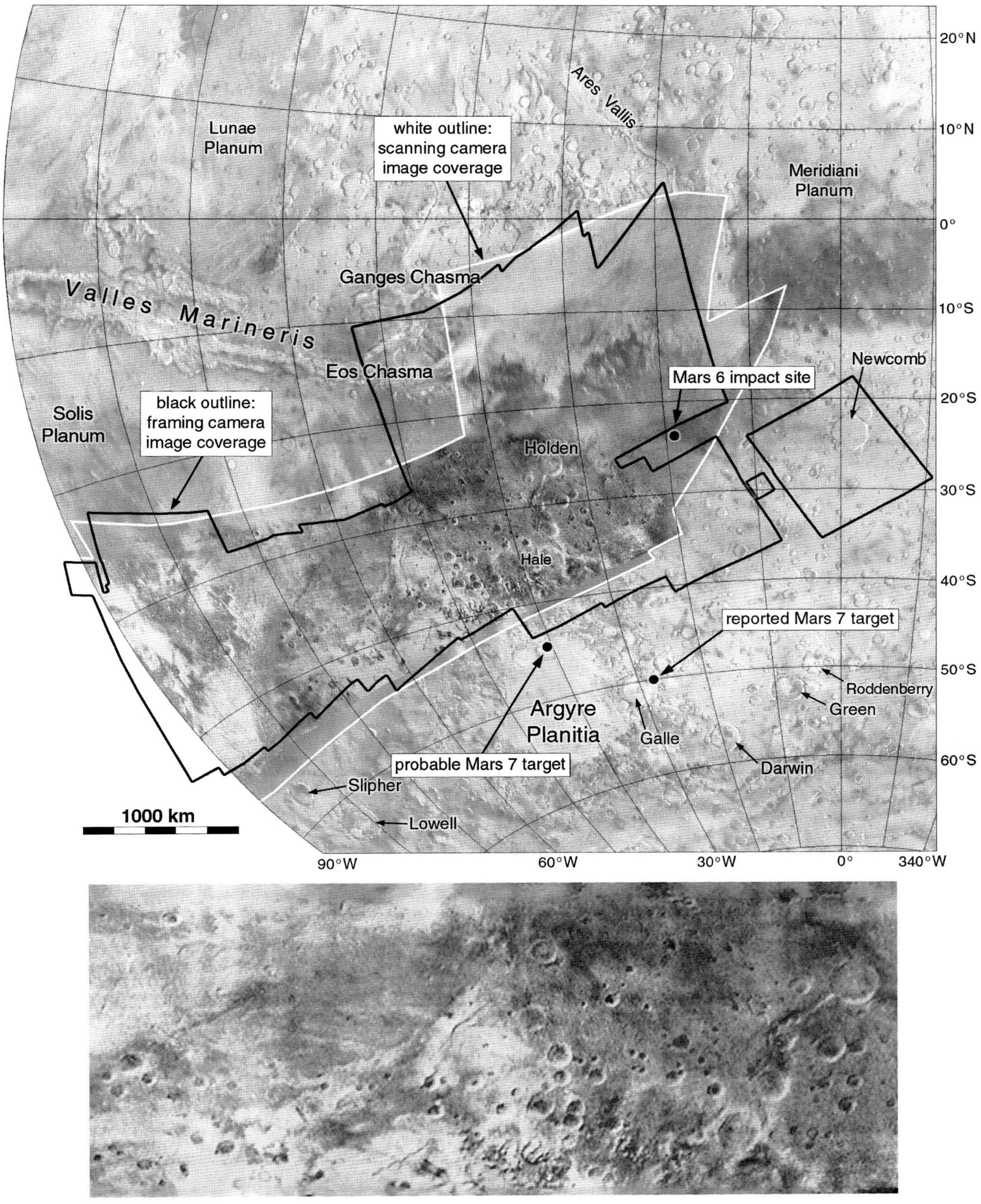

Figure 34 Mars 5 image coverage in the 0° hemisphere (top) and scanning camera panorama (bottom), provided by T. Stryk.

Figure 35 Mars 5 framing camera image coverage (top) and mosaic (bottom).
White outlines: Images transmitted only in low-resolution mode. Black outlines: Images transmitted in high-resolution mode. Only one image is not shown, an almost featureless view of the limb in the vicinity of 20° S, 110° W. The background map is USGS map I-810, based on Mariner 9 images. Mosaics by P. Stooke.

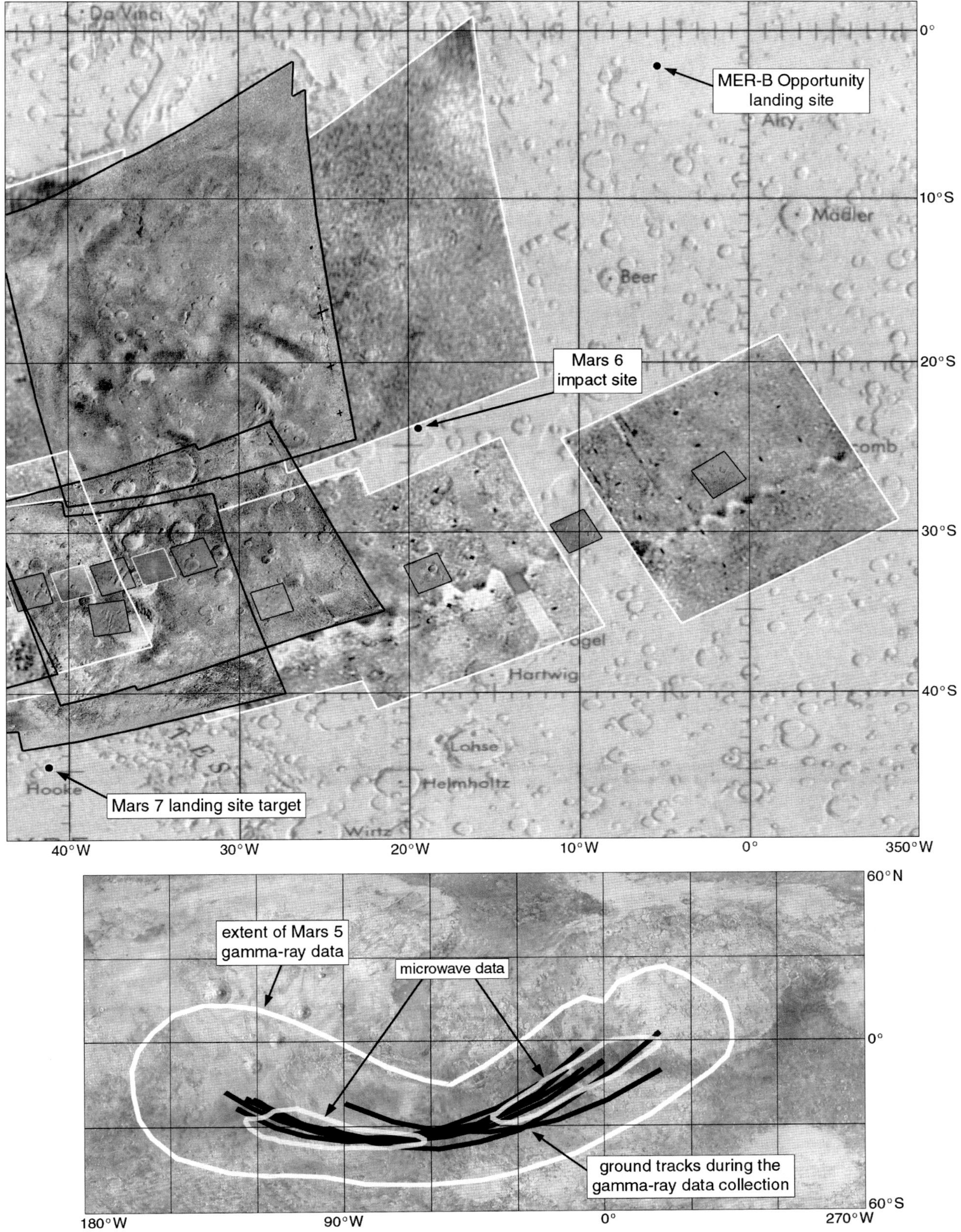

Figure 36 Top: Mars 5 image coverage, continued from Figure 35. **Bottom:** Areas of gamma-ray surface composition data and microwave surface density data.

entered the atmosphere near 23° S, 29.5° W, travelling eastwards. Its target was at 24° S, 25° W, but it overflew that point. The 1200-kg lander reached the surface at 09:11 UT at 24° S, 19.5° W (or 23.90° S, 19.42° W, but the location is uncertain by 150 km or 2.5° radially from that point). The site could lie anywhere within Figure 37C. The 10° difference between the longitudes of atmospheric entry and landing recalls confusion in the literature relating to the Mars 2 and Mars 3 locations, where the entry point is sometimes mistaken for the landing point.

The landing probe, which was similar to the Mars 3 lander, operated successfully during descent and provided the first direct measurements from the atmosphere of Mars, including pressure, temperature, deceleration and winds. About 2.5 minutes after the parachute opened, the spacecraft stopped communicating just before the expected landing time. The final measurements transmitted showed a temperature of 230 K (–43° C) and an atmospheric pressure of 6 mb. Mars 6 is assumed to have crashed. It carried temperature and pressure sensors and equipment to analyze the surface and atmosphere, as well as a dual-panoramic scanning camera system like those on Luna 13 and Mars 3 and a small tethered rover like that on Mars 3. Similar rovers were also considered later for the joint European/Russian Marsnet landers.

9 August 1973: **Mars 7 (Soviet Union)**

This, the last of the 1973 probes, was another lander. It was launched at 17:00 UT into a 193 by 154 km parking orbit inclined 51.5° to the equator with a period of 87.9 minutes. Its 1.69 by 1.01 AU transfer orbit was inclined 2.2° to the ecliptic with a period of 574 days. After a 212-day cruise, it reached Mars on 9 March 1974 (MY 11, sol 36), four sols before Mars 6. The carrier spacecraft flew past Mars but released its lander about 4 hours too early, and the entry capsule missed Mars by 1280 km.

The landing target has been identified as 50° S, 28° W, for example, on the National Space Science Data Center website in 2010. This site on the rim of the 200-km diameter crater Galle (Figure 38) seems an unlikely choice as its rugged topography was clear from Mariner 9 images, and other information suggests a more likely target. When USGS made the map for Mars 6 (Figure 37), they produced another sheet for an area called Nereidum Montes, 1200 km farther south. Figure 38A shows the area covered by this map, and Figure 38B is a detail of its central region, the same area imaged by Mars 3 (Figure 25E). The surface appears smooth in Mariner 9 images, making it a reasonable target (Figure 38D). This area at 45° S, 41° W was probably the Mars 7 target. The site near Galle may have been misidentified due to confusion between that crater and the 120-km Hooke, a simple map-reading mistake. Alternatively, it may have been the Mars 7 sub-spacecraft point at closest approach, either actual or targeted, or possibly a backup site for Mars 2 or Mars 3, as discussed in connection with Mars 3.

Viking Landing Site Selection

The Viking site selection process is described by Masursky and Strobell (1976a, b), Masursky and Crabill (1981), Ezell and Ezell (1984) and Moore (1994). This section also includes material from the papers of Thomas (Tim) Mutch in the Brown University archives and a draft report on Viking site selection (Ezell and Ezell, 1977) which contained details missing from other sources. The Viking mission was to search for evidence of present or past life, and water was considered the key. Attention was directed towards areas of water production (volcanic areas where venting might occur), areas of collection (basins receiving flow as suggested by dry channels on the surface) and areas of storage (polar caps or the subsurface).

Selecting landing sites was a difficult and time-consuming task, not unlike the selection of Apollo landing sites in the previous decade. Although the Apollo case involved risks to human life which were not present for Viking, Apollo had the advantage of very high-resolution images for site certification. Viking would land almost blindly in comparison. The fundamental problem, as for Apollo, was to find a compromise between safety and scientific interest. When the process began, only Mariner 4 images (Figures 8 through 11) were available, and they contributed little except to establish the general nature of the planet's surface.

The Viking Project Science Steering Group (VPSSG) first met in February 1969. The Viking Lander imaging team leader, geologist Thomas Mutch (Brown University), did not expect orbital images to reveal enough

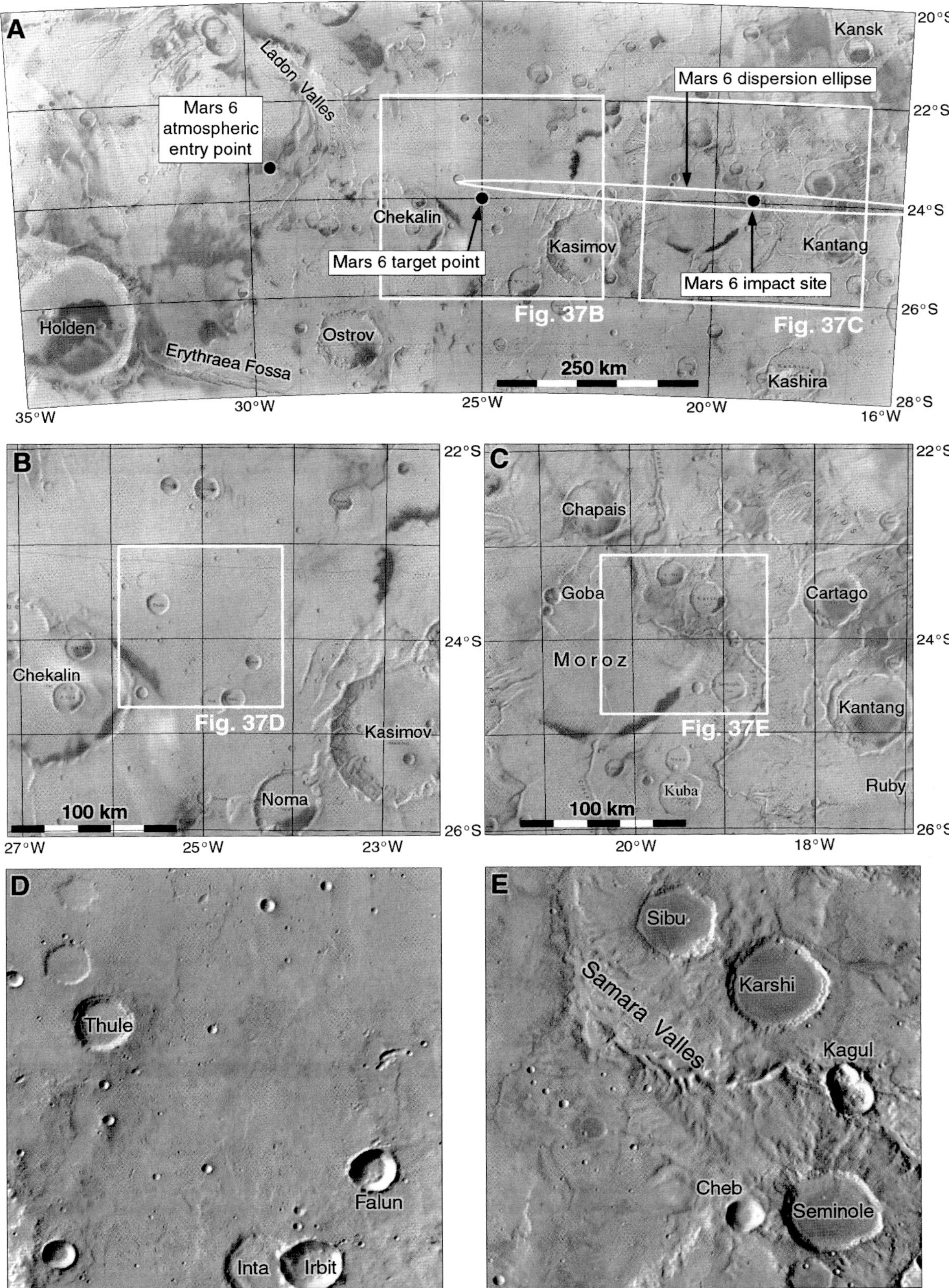

Figure 37 Mars 6 landing site.
A: USGS map I-986 showing the landing ellipse. **B:** Mars 6 target area. **C:** Mars 6 impact area.
D: Infrared mosaic of the target area, 50 km across. **E:** Infrared mosaic of the centre of the impact area, 50 km across.

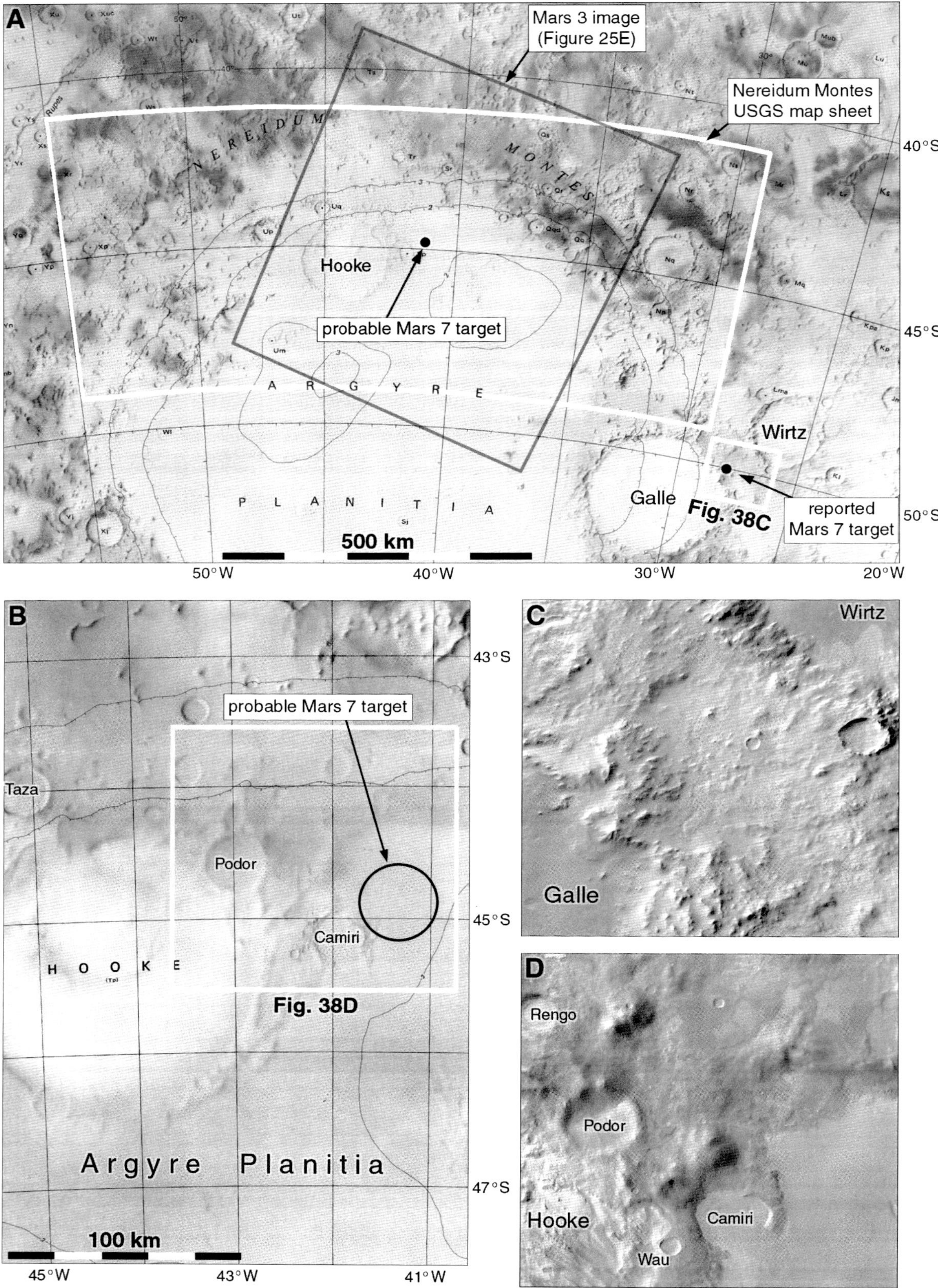

Figure 38 Mars 7 target point.
A: Part of USGS map I-985 (Argyre), 1976. **B:** Part of USGS map I-1002 (Nereidum Montes), 1977. **C:** Mars Odyssey daytime infrared mosaic of reported target point. **D:** Mars Odyssey daytime infrared mosaic of the more probable target in Argyre.

detail to certify sites as safe, as Lunar Orbiter had done for Apollo, but biologist Wolf Vishniac (University of Rochester) argued that lower resolution images could still establish the geological context of a site and suggest its biological potential. He also emphasized the value of landing in both a dark area and a light area as revealed by telescopic images (Figures 1 and 2), assuming these to be different geologically, as on the Moon, or perhaps biologically.

The VPSSG met again in March 1969 and heard new suggestions for landing sites. For meteorology a broad smooth plain would be desirable, whereas biology and chemistry considerations favoured landing in a warm, low spot. At that time neither type of site could be identified, though earth-based radar and spectroscopy were beginning to provide some data on topography (Figure 21). A 'preliminary reference mission' study used for planning purposes had described hypothetical sites in the 30° south latitude zone, but without any scientific justification. At this very early stage, the landing was scheduled for 1973, but in December 1969 NASA announced a delay until 1975.

In 1970 a Viking Landing Site Working Group (VLSWG) was established under Thomas Young to select the landing sites. It first met on 2 September 1970 to consider the needs of biologists, geologists and meteorologists. Although Mariners 6 and 7 had imaged some areas at reasonable resolution (Figures 15 through 20), the serious planning could not really begin until Mariners 8 and 9 provided global coverage and more detail. Then the superior Viking Orbiter cameras would inspect interesting sites at still higher resolution just before the landings. At the next meeting on 26 and 27 October 1970, the Working Group had to defend these cameras against severe budgetary pressure. Cameras might have been dropped altogether. Gerald Soffen said 'We have a 17–22 million dollar problem and the Orbiter Imaging System costs 25 million. Any suggestions?' An option to re-use the Mariner 9 camera design was rejected, as its results would be no better than telescopic images of the Moon. Eventually the better cameras were retained.

At a meeting on 2 and 3 December 1970, the VLSWG made a very preliminary recommendation for landing sites to give mission planners something to work with, accepting that these would probably change as new information appeared. An important engineering constraint was the landing site latitude range, between 30° N and 30° S. This was referred to as the 'Viking Zone' in some documents (Figure 39), by analogy with the lunar 'Apollo Zone'. Another requirement was to target lower elevations with higher atmospheric pressure, both to enhance chances for life and to allow maximum parachute braking, though regions of low elevation could not yet be reliably identified (Figure 21). An afternoon landing was preferred to allow some data to be transmitted via the orbiter before it dropped below the horizon. The two sites should be far enough apart to prevent radio interference during communications and to reduce the likelihood of simultaneous dust storms over both sites.

The size of landing ellipse used for initial planning would be 400 by 840 kilometres, reduced to 300 km in length in 1972 and to 100 by 240 km in 1976 (Masursky and Crabill, 1981). Ellipses of many different sizes were illustrated in landing site documents, generally getting smaller over time. Several lists of possible sites of interest were presented by Working Group members at this meeting (Table 5a and Figure 39). Sagan ranked sites by safety, Binder by priority, but their reasoning is not explained. Kieffer was honest about his choice – very little was known about the Hellas region. Baum's argument for geological activity assumed that the smooth surfaces seen in Mariner 4 and Mariner 7 images were equivalent to the lunar maria. Uncratered terrain observed in the first few Mariner 4 images (Figure 9) might be smooth enough to land on, and though Mariners 6 and 7 had shown mainly rougher terrain, Hellas appeared smooth (Figure 20). Masursky and Smith mentioned the likelihood of finding water at their sites, probably based on small channels seen in Mariner 6 images of Pandorae Fretum and fogs or clouds frequently seen at Nix Olympica and Phoenicis Lacus.

The Working Group recommended three sites from these lists for each lander (Table 5b), requiring that one landing site be north of the equator and one south. Mission analysts quickly assessed them and found no problems. On 7 December, Jim Martin chose two of these sites for further analysis and planning. They were Thoth-Nepenthes at 15° N, 275° W for the first landing and Hellas at 30° S, 300° W for the second (Figure 39). The VLSWG met on 21 April 1971 to discuss Viking planning and to prepare for the acquisition and use of Mariner 8 and 9 data. Masursky was concerned that Mariner data

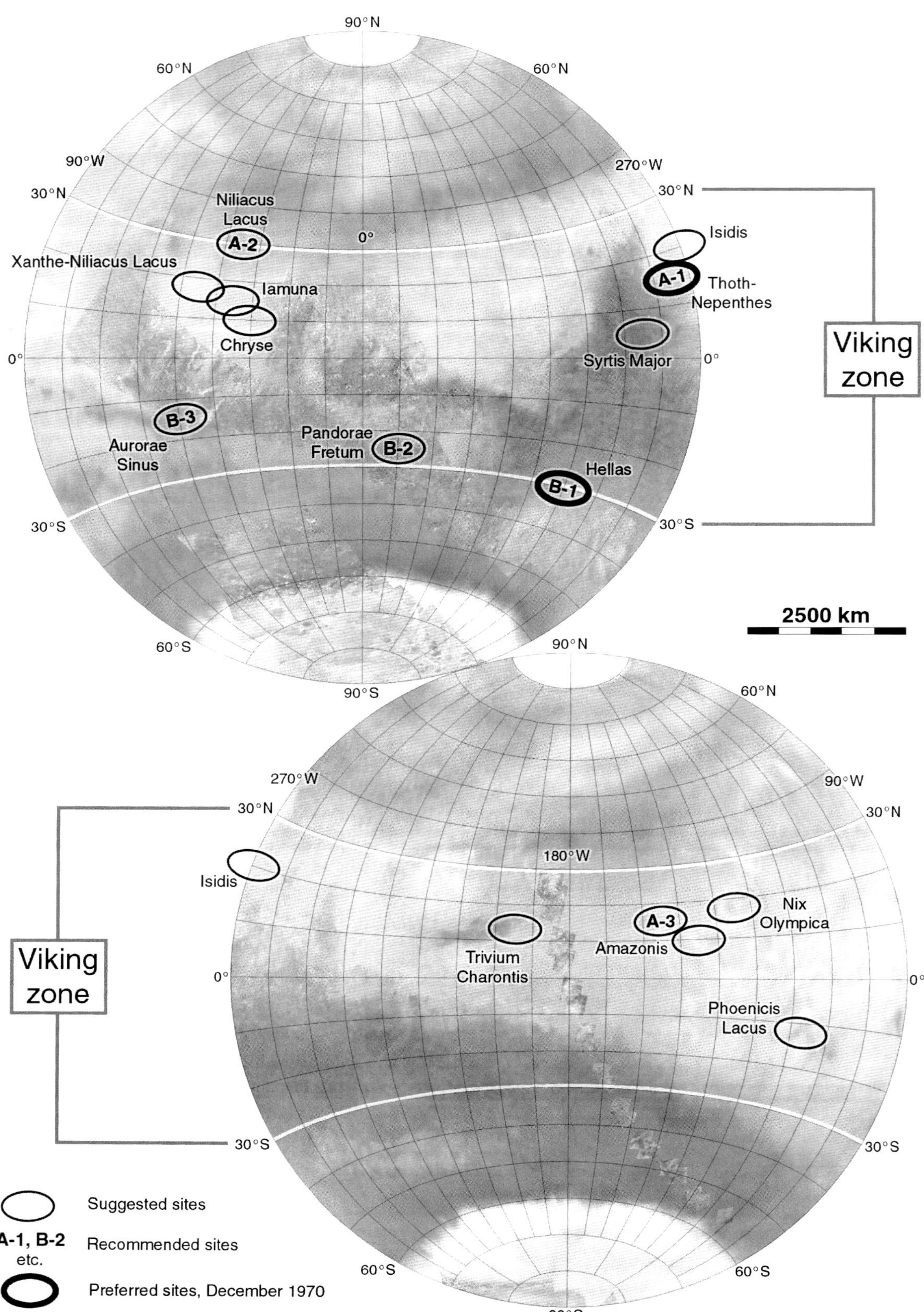

Figure 39 Early proposals for Viking landing sites, December 1970.
The base map includes Mariner 4, 6 and 7 images.

Table 5. *Candidate Sites Proposed by VLSWG Members, 2–3 December 1970*

5a. Individual member proposals

Proposer (comments)	Site name	Location
Carl Sagan (in order of decreasing safety)	Chryse	10° N, 30° W
	Xanthe-Niliacus Lacus	18° N, 45° W
	Thoth-Nepenthes	13° N, 275° W
	Trivium Charontis region	13° N, 195° W
	Amazonis	10° N, 145° W
	Aurorae Sinus	15° S, 50° W
Alan Binder (in order of decreasing priority)	Isidis region	20° N, 270° W
	Syrtis Major	5° N, 285° W
	Niliacus Lacus	30° N, 35° W
	Iamuna	15° N, 35° W
Hugh Kieffer (little known about this region)	Hellas	30° S, 295° W
William Baum (areas of known geological activity)	Amazonis	15° N, 155° W
	Hellas	30° S, 300° W
Hal Masursky, Brad Smith (maximum probability of water)	Nix Olympica	18° N, 140° W
	Phoenicis Lacus	12° S, 118° W
	Pandorae Fretum	25° S, 350° W

5b. Working Group recommendations

Mission	Site name	Location
A-1	Thoth-Nepenthes	15° N, 275° W
B-1	Hellas	30° S, 300° W
A-2	Niliacus Lacus	30° N, 35° W
B-2	Pandorae Fretum	25° S, 350° W
A-3	Amazonis	15° N, 155° W
B-3	Aurorae Sinus	15° S, 50° W

Notes: A, B: the two spacecraft. 1, 2, 3: three mission pairs in decreasing priority.

processing could take a year or more to complete, which would be too slow for effective Viking planning. Could Viking funds be spent to speed it up? The answer was yes. Tobias Owen mentioned that Soviet scientists were open to collaborating on site selection, a fact which saw some data sharing in time for Mars 6 and Mars 7 in 1973. Three weeks later, on 8 May, Mariner 8 was lost in a launch accident, but Mariner 9 was launched successfully on 30 May, arrived at Mars late in the year and performed well in Mars orbit, gathering images and atmospheric data to enable Viking planning to proceed. Another important source of new data was earth-based radar ranging, which allowed topography to be mapped at a regional level (Figure 21). Radar also provided estimates of surface roughness and texture and the degree of consolidation from the dielectric constant. Viking needed loose surface material, not solid rock, for its arm to operate successfully, and a relatively smooth but not dusty surface.

Mariner 9's mapping cycles are illustrated in Figure 26. The first cycle covered an area too far south for Viking, but the second was of interest to the mission. This took place between 22 January and 10 February 1972 (orbits 139–178). The third cycle, 11 February to 1 March, extended this, and the next cycle (2 to 22 March) completed the data collection needed for initial planning. At the same time Alan Binder combined the Earth-based radar data from oppositions in 1967, 1969 and 1971 with

Mariner 9 occultation measurements of atmospheric density to produce a preliminary topographic map of the Viking zone (Figure 29). Now surface features and elevations could be used to find safe landing sites for the first time. At a VPSSG meeting held on 16 and 17 February 1972, the available information was considered. Mars had been revealed as a dynamic world with volcanic, tectonic and possibly fluvial processes all contributing to a very varied landscape. Many areas looked too steep or rough to land in.

The VLSWG met again on 25 April 1972, when a revised latitude limit (25° N to 25° S) was presented. According to Ezell and Ezell (1977), 'The beginning of a rift between believers in radar and believers in photography began to show' at this meeting and would continue up to the time of landing. Thirty-five sites of interest to Viking planners were recommended for imaging by Mariner 9 during its extended mission by the Viking Data Analysis Team (VDAT) (1972), based on images taken during the main mapping phase of the mission (Table 6a). The list was presented to VLSWG by Howard Robbins, and the Working Group recommended that all 35 sites be imaged using three high-resolution B-frames and one A-frame at each site. Two of the sites had already been dropped by the time the VDAT report was completed, and their locations are not known. The constraints on landing site location at this time were the previously stated latitude limits, elevations between 0 and –5 km (using the very limited radar and occultation data then in hand) and surface roughness based on preliminary geological evaluations and radar data.

The candidate sites were evaluated after the meeting by George Recant, Tim Mutch, Bob Schmitz and Travis Slocumb, considering safety and science but with safety ranked far higher. Three of the sites were dropped as a result, though it is not clear whether these three included the two identified in Table 6. This revised list was passed on to the Mariner 9 project. A new version of the topographic map by Alan Binder suggested new areas of accessibility, and four new sites were submitted for addition to the list on 9 June by Masursky, Binder and James Gliozzi of VDAT (Table 6b). These were said to be 'of paramount geologic/biologic interest'. Mariner 9 mission constraints, especially the limited attitude control fuel, reduced the number of sites actually targeted to 24, of which 19 were eventually imaged. These are indicated with an asterisk in the first column of Table 6. All these sites are illustrated in Figure 40.

The next task was to identify 30° by 45° regions for topographic mapping by USGS Flagstaff. It was still not clear that either Mariner 9 data or USGS funding would allow full global mapping, so a more limited effort was anticipated. However, global cartography, both topographic and geological, was eventually completed by USGS along with detailed mapping of the most likely Viking landing sites. Eight regions were identified as being of most interest (Table 7a), and they became the USGS map quadrangles which were screened for landing sites (Figure 40). The quadrangle list given here was presented at a Mars Geologic Mapping Conference held at USGS Menlo Park on 20 September 1972.

VLSWG met again on 5 August 1972, when Masursky reported that topographic features were just starting to become visible on the floor of Hellas, the last place to be shrouded in dust following the great storm. Interest groups representing different disciplines indicated their primary choices for landing regions. In the words of Jim Porter, 'with maps, overlays, theories and opinions abounding, an hour was devoted to allow each group to indicate their primary regions' (Ezell and Ezell, 1977). The two most favoured latitude bands were the 15° N zone and 0° to 10° S. The first lander would be targeted for the northern zone, and the second for the southern zone. Eight smaller regions in the map quadrangles were identified as the candidate landing areas (Table 7b) and were examined in more detail in the coming month (Table 8). These map regions and landing areas are illustrated in Figure 41.

Influential biologist Joshua Lederberg now argued that higher latitudes should also be targeted, in an area exposed by the retreating polar cap, because water was expected to be released as the cap melted. Lederberg had been consulted by telephone on 5 August and had replied 'I am about to leave the U.S. for about a week, but on my return will prepare a statement of dismay (for the record). Biology is assertedly a prime goal of Viking. The M'71 [Mariner 9] data surprised us by indication [*sic*] that Mars's H_2O is principally poleward. Yet here is the box we are in for Viking '75: to be choosing "optimal" landing sites within the least promising zone'. Lederberg's expectation that the retreating cap would release water depended on the assumption that the seasonally varying part of the cap

Table 6. *Mariner 9 Imaging Targets for Viking Site Planning*

6a. Targets suggested by Viking Data Analysis Team (1972)

Target	Location	Comments
T-1*	15.0° S, 5.9° W	Optimum dielectric constant
T-2	22.0° N, 27.0° W	Smooth area adjacent to prominent scarp and cratered plateau
T-3	21.6° N, 40.0° W	High water vapor according to ground-based spectra
T-4*	18.6° N, 34.0° W	Plains region near plateau
T-5*	21.4° S, 44.5° W	Plateau near 'Red River' (Nirgal Vallis)
T-6	2.0° S, 63.7° W	Frequent white cloud occurrence according to ground-based data
T-7*	18.2° S, 76.3° W	Optimum dielectric constant
T-8	21.5° N, 75.5° W	Plains near mouth of large channel
T-9*	20.5° S, 142.5° W	Plains near normal faults
T-10	2.5° S, 147.0° W	Plains at mouth of braided channel
T-11*	4.0° N, 152.0° W	Frequent white cloud occurrence according to ground-based data
T-12*	23.0° N, 153.0° W	High water vapor according to ground-based spectra
T-13*	5.5° N, 164.3° W	Plains near mesas and channels
T-14	11.3° S, 169.0° W	Optimum dielectric constant
T-15	(deleted)	
T-16	7.0° S, 187.0° W	Caldera flank
T-17*	3.0° S, 191.0° W	Plains
T-18	2.0° S, 203.0° W	Plains
T-19	8.0° N, 207.0° W	Optimum dielectric constant, variable feature
T-20*	8.0° N, 215.0° W	High water vapor according to ground-based spectra, variable feature
T-21	1.3° N, 221.4° W	Smooth plains adjoining hummocky terrain
T-22*	14.0° S, 230.0° W	Optimum dielectric constant
T-23*	21.5° N, 232.5° W	Channel with 'islands' in plains
T-24	21.0° N, 245.0° W	High water vapor according to ground-based spectra
T-25	16.0° N, 250.0° W	High water vapor according to ground-based spectra
T-26	22.0° S, 251.0° W	Dandelion crater (Tyrrhena Patera) – volcanic caldera
T-27*	21.0° N, 251.0° W	High water vapor according to ground-based spectra
T-28*	1.9° N, 253.4° W	Frequent white cloud occurrence according to ground-based data
T-29*	22.0° N, 264.0 W	High water vapor according to ground-based spectra
T-30*	6.2° N, 268.8° W	Frequent white cloud occurrence according to ground-based data
T-31	10.0° N, 275.0° W	Optimum dielectric constant, variable feature
T-32*	17.5° S, 288.3° W	Upland, smooth intercrater area
T-33	(deleted)	
T-34*	16.0° S, 331.8° W	Smooth area with high dielectric constant
T-35*	17.4° S, 350.2° W	Smooth area with high dielectric constant, near variable feature
6b. Targets proposed for addition to VDAT list, 9 June 1972		
1	24.4° N, 59.1° W	Channel outlet
2	20.5° N, 210.0° W	Fractures and volcanic cones
3	9.8° S, 73.2° W	Channel (canyon) floor
4	11.2° S, 32.0° W	Upland cratered terrain adjacent to chaotic terrain (two options)
	8.5° S, 59.5° W	

Note: The 19 sites identified with an asterisk above and with white symbols in Figure 40 were eventually imaged by Mariner 9.

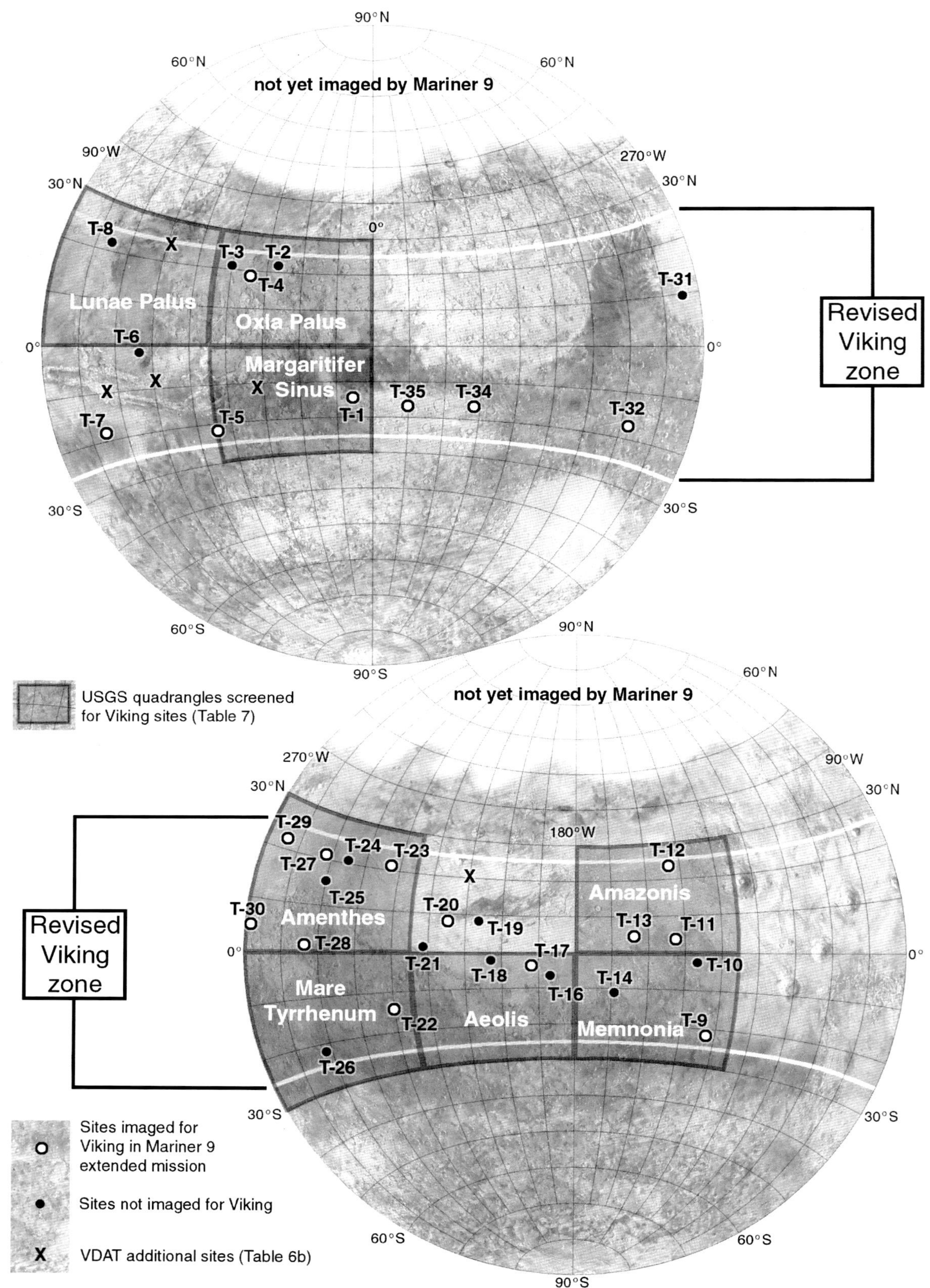

Figure 40 Mariner 9 extended mission imaging targets required for Viking planning (Table 6), and USGS quadrangles screened for landing sites (Table 7a).
The base map shows the extent of Mariner 9 imaging during the primary mission.

Table 7. *USGS Landing Region Quadrangles and Landing Site Areas, 1972*

7a. USGS quadrangles screened for Viking landing sites, 1972 (Figure 40)

Quadrangle number	Coordinate range	Name
8	0°–30° N, 135°–180° W	Amazonis
10	0°–30° N, 45°–90° W	Lunae Palus
11	0°–30° N, 0°–45° W	Oxia Palus
14	0°–30° N, 225°–270° W	Amenthes
16	0°–30° S, 135°–180° W	Memnonia
19	0°–30° S, 0°–45° W	Margaritifer Sinus
22	0°–30° S, 225°–270° W	Mare Tyrrhenum
23	0°–30° S, 180°–225° W	Aeolis

7b. Candidate landing regions selected by VLSWG on 4–5 August 1972 (Figure 41)

Region number	Coordinate range	Name
1	13°–24° N, 150°–175° W	Amazonis
2	13°–24° N, 33°–83° W	Chryse – Lunae Palus
3	13°–24° N, 237°–276° W	Amenthes
4	0°–10° S, 150°–165° W	Memnonia
5	0°–10° S, 180°–200° W	Aeolis
6	15°–20° S, 33°–38° W	Aureum
7	20°–25° S, 248°–253° W	Hesperia
8	17°–25° S, 185°–190° W	Ma'adim Vallis

consisted of frozen water, though it is now thought to be a layer of more volatile carbon dioxide ice. Nevertheless, a study was undertaken to explore the feasibility of landing between 65° N and 80° N. The low-latitude sites imaged by Mariner 9 were still being studied in detail as the high-latitude site study was undertaken. On 27 September 1972, the day before the next VLSWG meeting, Masursky faxed a list of these sites to Tom Young (Table 8). Geological maps of these areas were being compiled at USGS, regionally in the quadrangles listed in Table 7a, and at larger scale using the best available images. Masursky's sites included a prime (highest ranked) site in each of the two latitude bands and several alternatives. Those which differ from Table 9 sites are shown in Figure 41.

At the 28 September 1972 VLSWG meeting, the far northern site idea was rejected on technical and economic grounds. Sites in the 30° N to 30° S zone were discussed and ten were recommended: six for the first lander, four for the second (Table 9). The selection criteria mentioned at this time were evidence of water, low altitude, 'river delta' (depositional) sites, different geology at the two sites, unobstructed areas for meteorology, and relative proximity of the two sites to facilitate seismology (events detectable at both locations). Viking operational limitations now restricted the Viking zone to between 24° N and 25° S for the first mission, but the second mission was not restricted to this range. Then on 19 October Masursky and William Baum (Lowell Observatory) recommended minor changes in six of the ten sites (Table 9) to optimize Mariner 9 high-resolution (B camera) imaging coverage. The lists provided by Ezell and Ezell (1977) differ from these only in that Nepenthes was also designated 'lower priority'.

More details concerning these sites were provided by Masursky and Strobell (1976a). Site 4 (Uraniae) was to be moved to 9° N, 164° W to avoid a crater. Site 5 (Candor) had an alternative location at 14.7° N, 79.3° W, and both it and the site at 80° W might be moved 1° to the southwest to avoid obstacles. The Candor site was not in Candor Chasma, part of the giant canyon system, but north of it in the upper part of Kasei Valles. Site 10 was

Table 8. *Masursky's Landing Site List of 27 September 1972*

8a. Northern tier candidate sites (A mission)

Number	Site designation	Description	Location	Name
1	Prime	Alluvial	19.5° N, 34.0° W	Chryse
2	First backup	Volcanic plain	18.0° N, 65.0° W	Lunae Palus
3	Second backup	Alluvial	19.0° N, 233.0° W	Aethiopis
4	Third backup	Aeolian plain	20.0° N, 158.0° W	Eumenides

8b. Central/southern tier candidate sites (B mission)

Number	Site designation	Description	Location	Name
1	Prime if A lands	Volcanic plain	17.0° S, 136.0° W	Memnonia
2	Prime if A fails	Aeolian plain	3.7° S, 185.5° W	Zephyria
3	First backup	Alluvial – aeolian	2.0° N, 148.0° W	Amazonis
4	Second backup	Upland volcanics	16.0° S, 252.0° W	Hesperia (Dandelion)
5	Third backup	Terra plain	13.0° S, 33.0° W	Aureum

Table 9: *Candidate Viking Sites, September and October 1972*

			28 September		19 October	
Spacecraft	Site		Latitude	Longitude	Latitude	Longitude
Mission A	Eumenides	1	20° N	158° W	21° N	157° W
*Site 2: lower priority	Lunae Palus	2*	20° N	77° W	19° N	65° W
	Chryse	3	19.5° N	34.0° W	no change	
	Uraniae	4	12° N	158° W	8° N	163° W
	Candor	5	12° N	77° W	10° N	80° W
	Nepenthes	6	12° N	267° W	10° N	269° W
Mission B	Amazonis	7	2° S	148° W	no change	
	Zephyria	8	2° S	186° W	no change	
	Memnonia	9	9° S	144° W	9° S	141° W
	Aquae Apollinares	10	9° S	181° W	no change	

listed as 7.6° S, 177.6° W, and two ellipses were mentioned, the second presumably listed in Table 9. Sites 2 and 6 from Table 9 were dropped, and two more were added in November. These were Site 11, Amenthes at 20° N, 253° W, and Site 12, Hesperia (informally called the Dandelion, now Tyrrhena Patera) at 16° S, 251° W. Hesperia was soon dropped from consideration. Geologic maps were prepared of the remaining sites, and two landing site scenarios were described (Table 10). All these sites are shown in Figures 41 and 42.

The VLSWG met on 4 and 5 December 1972 to pick the primary and secondary landing sites for each lander. The prime and backup sites in the lower latitudes were chosen after much debate (Table 11). The site numbers refer to Table 9. The north polar region was again under consideration after forceful lobbying by Lederberg, so prime and backup sites would now be needed at high latitudes. The strategy would be to send the first lander to a low-latitude site and the second lander to a high-latitude site only if the first succeeded. NASA had wanted to

Table 10. *Viking Site Scenarios from Masursky and Strobell (1976a)*

Scenario	Mission	Site
I	Mission A prime site	3 – Chryse
	Mission B prime site	9 – Memnonia
	Mission A backup site	1 – Eumenides
	Mission B backup site	10 – Aquae Apollinares
II	Mission A prime site	7 – Amazonis
	Mission B prime site	(not chosen)
	Mission A backup site	5 – Candor (or 4 – Uraniae)
	Mission B backup site	7 – Amazonis (or 8 – Zephria)

announce the selected sites in December 1972, but the ongoing analysis delayed the choice until the following spring.

A backup target for Viking 1 and the question of going farther north were considered well into the early months of 1973. An ad hoc group met on 14 December 1972 to select high-latitude sites and chose five (Table 12, Figure 43) for further consideration by the VLSWG.

Lederberg's wish for a high-latitude landing site did not meet with much approval among members of the VLSWG, both because it affected orbital operations and because so little was known about the surface at those latitudes. Low-latitude sites had the advantage of Earth-based measures of radar reflectivity, an indicator of surface roughness. The strongest support came from biologists who expected to find water in the surface materials at the far northern latitudes. In the end a polar option was maintained in addition to the low-latitude target. After further consideration, six sites were approved by the Working Group at a meeting in February 1973 (Table 13, Figure 43). Site numbers above 10 in these lists are not always consistent. Ezell and Ezell (1977) reported that 'some unknown person suggested that when ultimately chosen the mission B site should be called "Crisis Continuum", but at higher levels that sense of levity was not shared'.

Table 12. *Viking High Latitude Sites, December 1972*

Site	Name	Location
12	Ortygia	73° N, 350° W
13	Uchronia	74° N, 225° W
14	Cedron	63° N, 0° W
15	Scandia	63° N, 85° W
16	(not named)	63° N, 160° W

The northern locations seemed favourable to water, but not water in liquid form because of the very low temperatures. That consideration eventually caused even Lederberg to abandon the high-latitude site he had promoted. A compromise was to move into northern midlatitudes, about 40° N to 50° N, where there was a chance that temperatures could rise high enough to melt water in the surface materials. The Working Group met again on 2 April and considered a list of midlatitude sites presented by Tom Young (Table 14).

The Working Group members eventually selected two sites in the 40° to 45° N latitude zone, Site 16 (Cydonia) and Site 17 (Alba) (Masursky and Strobell, 1976b). VPSSG were responsible for designating 16 as prime and 17 as backup. Eventually a region near site 20, first identified at this time, became the Viking 2 landing site. For now, though, the four Viking sites, two primary and

Table 11. *Viking Prime and Backup Landing Sites, December 1972*

Mission	Site	Comments	Location
Mission A prime	3	Chryse	19.5° N, 34° W
Mission A backup (no consensus)	12	Eumenides (weak surface?)	21° N, 157° W
		Lunae Palus (too high?)	19° N, 65° W
Mission B prime	10	Apollinaris	9° S, 181° W
Mission B backup	9	Memnonia	9° S, 144° W

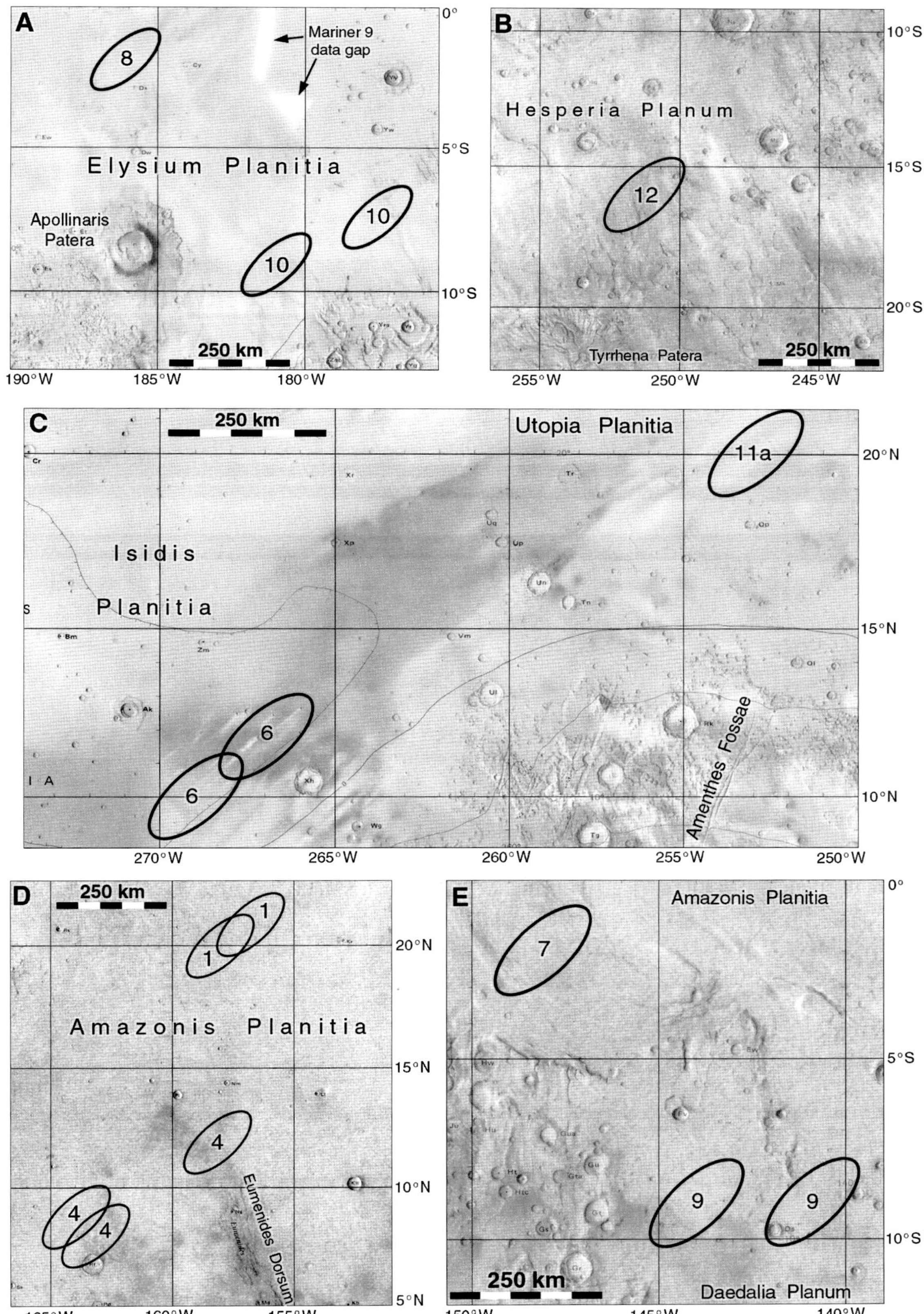

Figure 42 Viking sites (continued from Figure 41).
The base maps are USGS maps I-1000 (42A), I-1123 (42B), I-967 and I-1024 (42C), I-956 (42D), and I-1075 (42E). Maps 42B, D and E include a low-resolution albedo overlay from USGS map I-961.

Table 13. *Six Viking Landing Sites, February 1973*

Mission	Site	Location	
Mission A primary	3	19.5° N, 34.0° W	Chryse
Mission A backup	11	20.0° N, 252.0° W	Tritonis Lacus
Mission B, polar option, primary	12	73.0° N, 350.0° W	Ortygia
Mission B, polar option, backup	13	73.5° N, 221.5° W	Uchronia
Mission B, equatorial option, primary	10	9.0° S, 181.0° W	Apollinaris
Mission B, equatorial option, backup	9	9.0° S, 144.0° W	Memnonia

Table 14. *Midlatitude Sites Presented by Tom Young, 2 April 1973*

Latitude zone	Site	Location
40°–45° N	16	44.3° N, 10.0° W
	17	44.2° N, 110.0° W
	20	44.0° N, 256.0° W
50°–55° N	18	51.0° N, 7.8° W
	19	52.0° N, 135.0° W
	21	55.0° N, 73.0° W
	22	55.0° N, 5.0° W

two backup, had been decided. The selection allowed USGS to plan its mapping program and mission planners to define the trajectories, orbits and procedures necessary for each site. The sites were announced publicly on 7 May 1973 and are listed in Table 15 and portrayed in Figure 45. The site numbers used previously were now replaced with alphanumeric designations: A-1, B-2 and so on. Tom Young called the site selection process traumatic: 'We really thought we were embarking on a reasonably simple task ... and the one thing that we very rapidly found out was that there was ... divergence of knowledge and opinion on Mars' (Ezell and Ezell 1977).

After that selection was made, continuing analysis of the available data provoked some new thoughts about landing sites. Confidence in the Mission B sites at 44° N waned, and 'C sites' near and just south of the equator were studied again. The desire was to find extremely safe sites for use if radar data or Viking cameras revealed the A and B sites to be unusable. The C site subcommittee met in December 1974 and again on 6 February 1975 and recommended radar studies of three latitude zones (8.5° S, 4° S, and 4° N to 6° N) that could be observed from August to November 1975. The B sites were too far north for radar studies from Earth, so these C sites (Table 16, Figure 46) were now favoured for the second lander. Moore (1994) said that four equatorial sites were considered, perhaps including a second ellipse in C-1.

Viking 1 Site Certification

The Viking landing sites were chosen using Mariner 9 data which were not adequate to ensure a site's safety or to understand its geology. Viking images with greatly improved resolution, and other data such as Earth-based radar estimates of surface roughness, would be used to certify the sites as safe or to search for alternatives. This section is based on the minutes of the Landing Site Staff (LSS, the Viking team members responsible for site certification), kindly provided by Norman Crabill.

The process began as the Viking 1 spacecraft was approaching Mars. On 6 June 1976 the radar data for site A-1 were assessed at a Preliminary Radar Assessment meeting. Carl Sagan summed up the situation, saying 'we are fantastically ignorant of factors that could be fatal to a successful landing'. Mariner 9 pre-dawn temperatures were used to calculate thermal inertia and infer 'blockiness', the number of rocks in a given area on the surface. These new data suggested that C-1 was free of rocks. Mariner 9 images suggested that A-1 was covered with sand sheets and channel deposits, C-1 by small craters, and A-2 by knobby hills and craters. Radar data were considered for A-1, A-2, C-1 and a comparison site at 1.3° N, 90° W in Syria Planum. The southern half of the A-1 ellipse and the middle part of A-2 were crossed by radar ground tracks, and the A-1 radar showed a

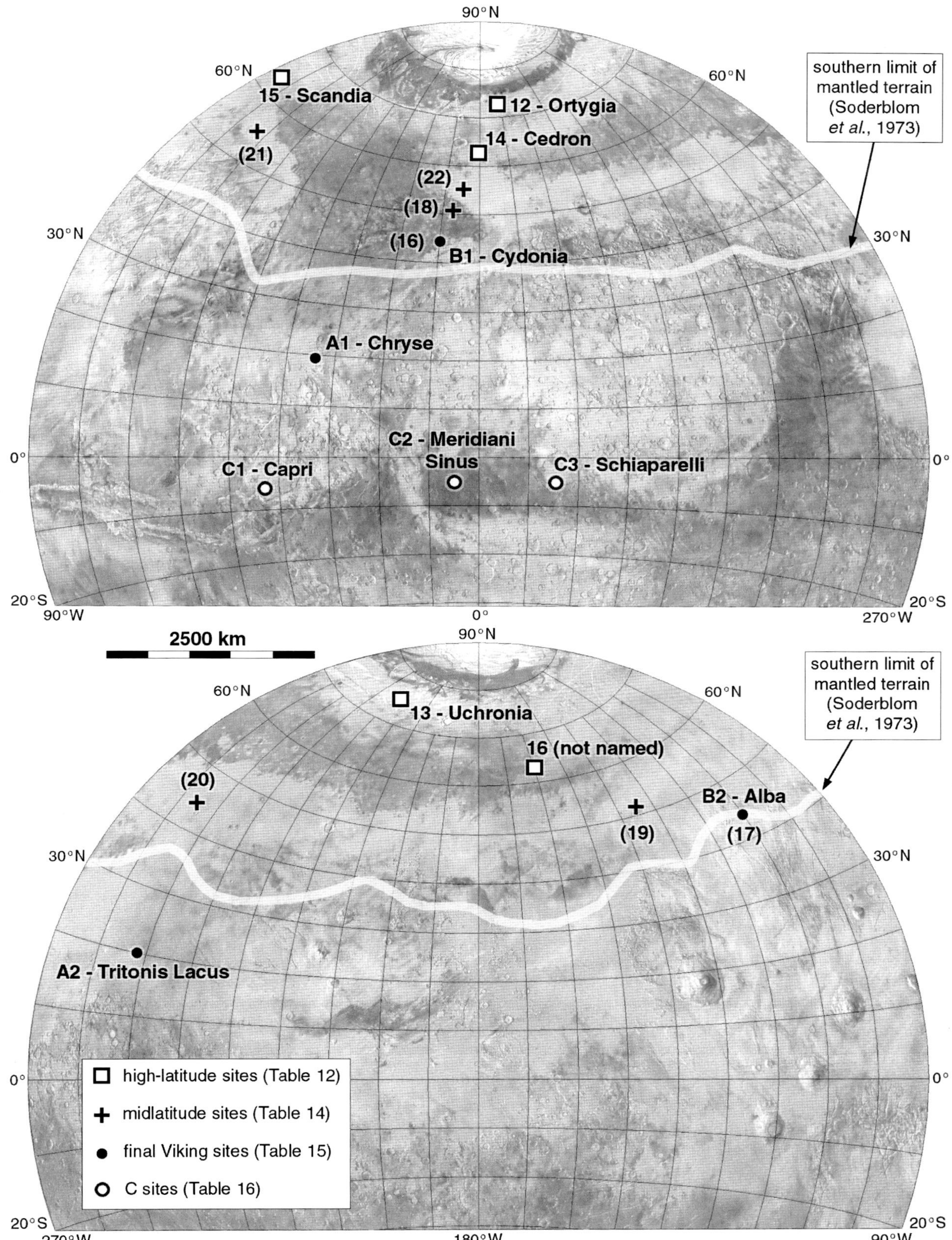

Figure 43 Potential Viking landing sites.
The high-latitude sites are numbered, and the midlatitude sites are numbered in parentheses. A and B sites are the prime and backup sites for the first and second missions, respectively. C sites are the radar-safe equatorial sites. The white line delineates the southern limit of the mantled region described by Soderblom *et al.* (1973).

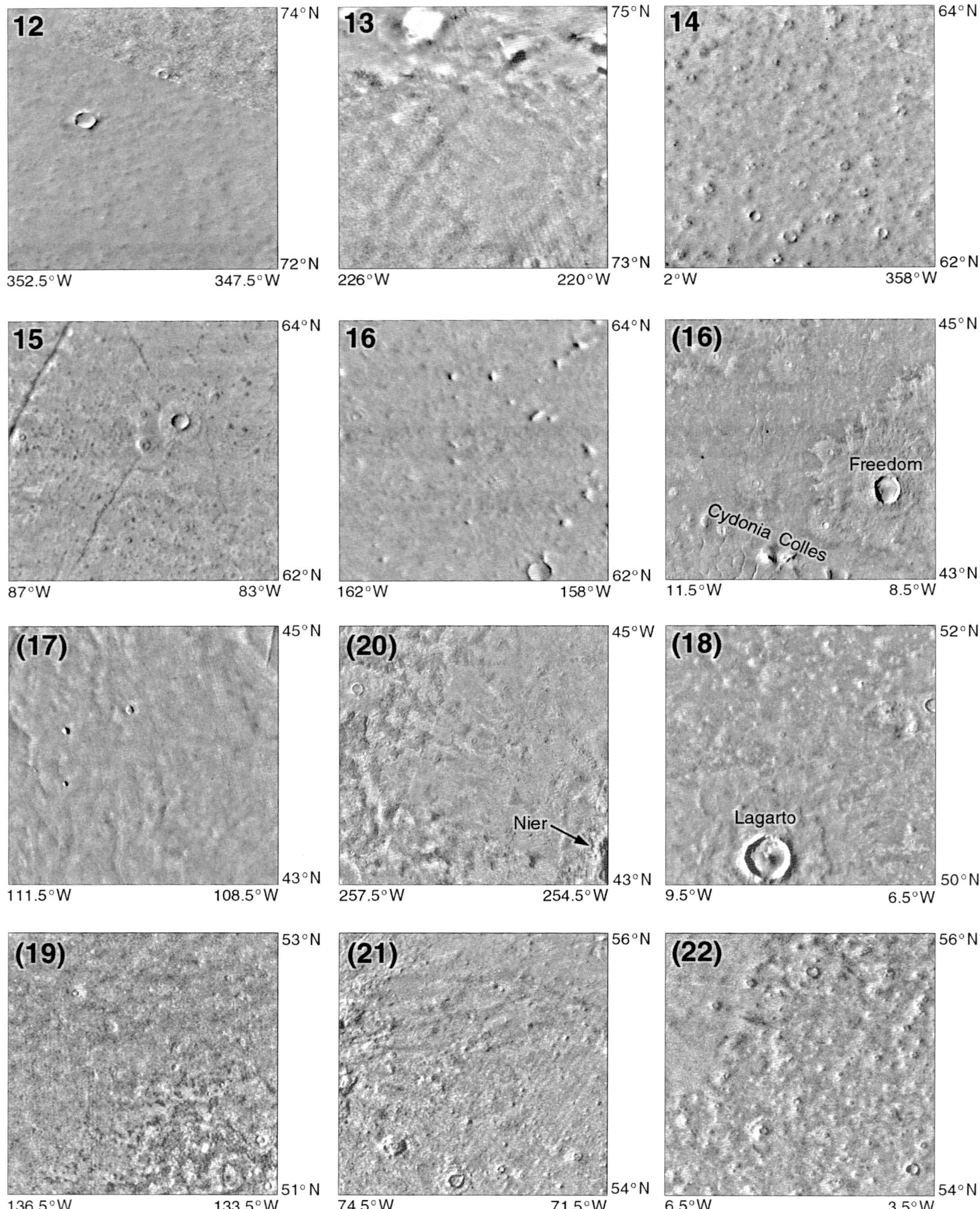

Figure 44 Details of high-latitude and midlatitude landing sites from Tables 12 and 14 (numbers in parentheses correspond to Figure 43 and are ordered as in Table 14).

Each view is a mosaic of Viking images showing more detail than was available to mission planners at the time. The mosaic squares are 120 km across, spanning 2° in latitude.

Table 15. *Viking Landing Sites, April 1973*

Mission	Site	Name	Location	
Mission A, primary	A1	3	Chryse	19.5° N, 34.0° W
Mission A, backup	A2	11	Tritonis Lacus	20.0° N, 252.0° W
Mission B, primary	B1	16*	Cydonia	44.3° N, 10° W
Mission B, backup	B2	17	Alba	44.2° N, 110° W

*Site 16 refers to Table 14, not Table 12.

Table 16. *Viking C Sites, 1975*

Site	Name	Location
C-1	Capri	6° S, 43° W
C-2	Meridiani Sinus	5° S, 5° W
C-3	Schiaparelli	5° S, 345° W

smooth patch on the floor of Ares Vallis south of A-1 (A1-SW in Figure 47). One point of concern was whether the C sites might be too high for effective use of the parachute, which needed an adequate depth of atmosphere to provide sufficient braking. New analyses of radar altimetry and Mariner 9 occultation data suggested that C-1 at least was adequate. The list of potential C ellipses is given in Table 17 and illustrated in Figure 46. A poll of the meeting showed that the A latitude (20° N) was preferred over the C latitude (5° S).

The first LSS meeting was held on 16 June as the first Viking Orbiter 1 approach images became available and atmospheric conditions could be monitored. Morning clouds were reported around Nix Olympica (Olympus Mons) and Chryse, but were expected to clear by the afternoon. By the second meeting on 17 June, these were shown to be artifacts of interpretation caused by incorrect brightness estimates, not real clouds, but changes probably due to real clouds were observed over South Spot (Arsia Mons) and fog or frost at Memnonia. Meetings were held nearly every day from now on, putting great pressure on the people and computer systems at JPL and USGS Flagstaff.

Another radar assessment meeting was held on 18 June to consider the A-1 and A-2 sites. The radar now suggested that the region just south of A-1 might be rougher than desired, but that a region northeast of there (A-1 R, Figure 47) was smoother. The sites discussed at the meeting, with their root mean square slopes, are listed in Table 18. Henry (Hank) Moore (USGS) argued that A-2 was preferred over A-1 from the radar results.

Viking 1 was now nearing Mars, and the third LSS meeting (LSS 3) on June 20 reviewed approach images and image-processing capabilities, which would be stretched to the limit by the coming weeks of activity. At LSS 4 the next day the final approach images, including colour composites, were discussed. Bradford Smith urged the group to use the official IAU names for features (Arsia Mons, Olympus Mons) rather than the informal or traditional names (South Spot, Nix Olympica) that most people were still more familiar with. The new IAU nomenclature was introduced in 1973 but was not yet entrenched in literature or memory. Viking 1 entered orbit safely on 19 June (MY 12, sol 178), and the first orbital images from the Viking cameras, a vast improvement over Mariner 9 data, were received on 22 June. At LSS 5 on that date the meeting began while the team were still awaiting those images and planning their processing and analysis. According to the minutes, 'the meeting was adjourned on a euphoric note due to the startling clarity of the P3 pictures coming in'. P3 (periapsis 3) was the imaging sequence which produced the first data from orbit, targeted for the A-1 ellipse (Figure 47).

The new images were clear, but they showed a complex and rough surface, etched by erosion in places and cut by fractures in others (Figure 48F). Was there a safe landing ellipse in the A-1 region? Some irregular pits suggested thermokarst, hollows produced by melting ground ice. At LSS 6 on 23 June, Carl Sagan asked whether Viking would melt its way down through an icy surface. Hugh Kieffer showed it would not, as the lander decreased solar heating by shading more than it added heat by thermal radiation.

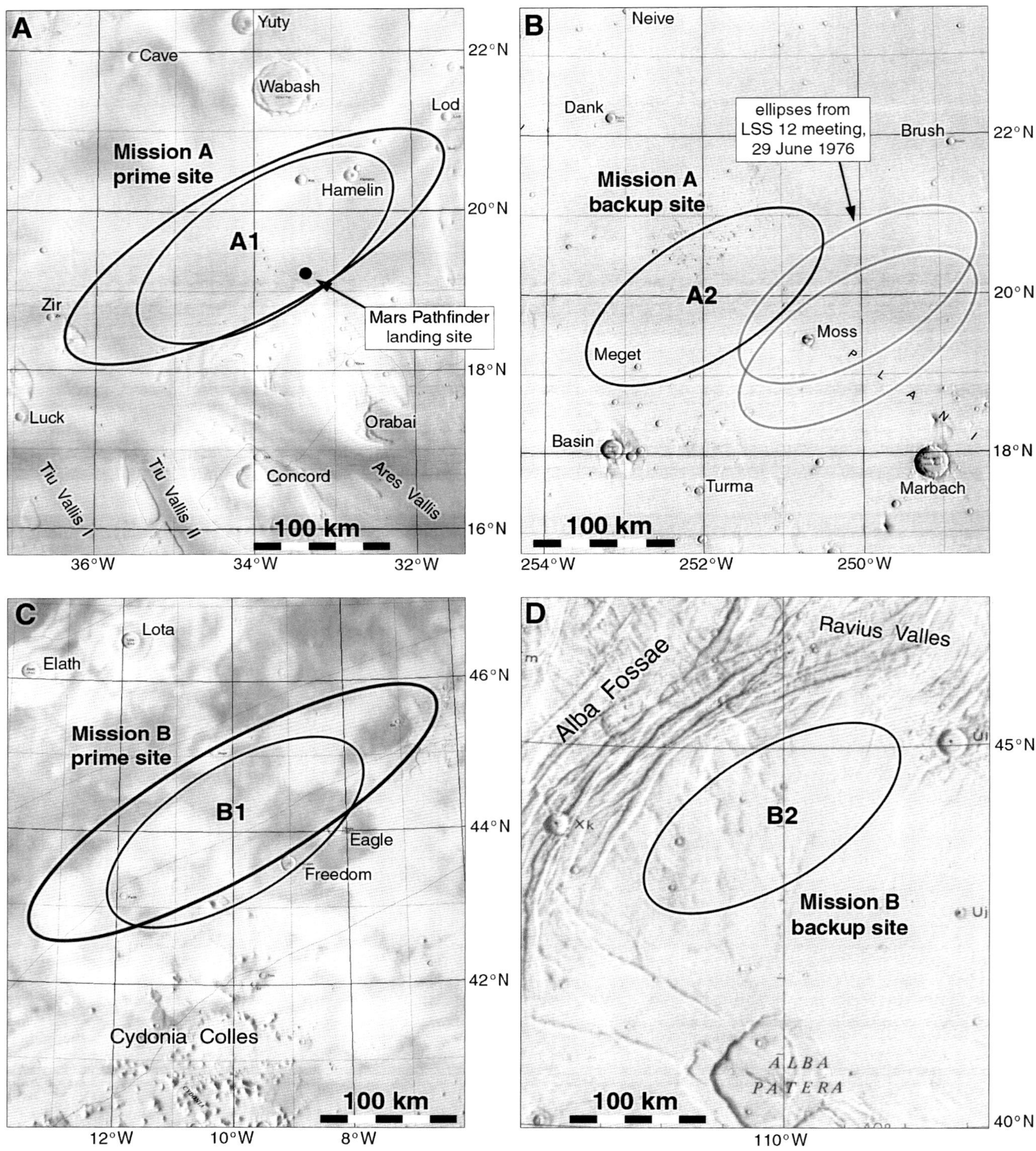

Figure 45 Viking primary (1) and backup (2) site ellipses for the A and B missions.
The smaller ellipses are located according to Table 15 and sized to match Masursky and Crabill (1981). The two larger ellipses are from Klein *et al.* (1976). The size differences reflect changing estimates of landing accuracy. Portrayals of Viking ellipses vary considerably from source to source. Two additional ellipses from the LSS 12 meeting are shown at A2. The base maps are USGS maps I-983 (45A), I-1055 (45B), I-988 (45C) and I-963 (45D). Masursky and Crabill (1976) show the B2 ellipse at Alba moved about 20 km to the northwest.

Table 17. *Viking C Site Altitude Update, June 1976*

C Site	Location	Best estimate of radius (km)	elevation (+/– 1.5 km)
C-1	6.25° S, 44° W	3396.45 (3395.25)	3.31 (2.11)
	4.10° S, 44° W	3396.3 (3394.6)	3.05 (1.35)
C-2	6.25° S, 4° W	3394.7	1.35
	4.10° S, 4° W	3393.9	0.43
	2.9° S, 4° W	3394.1	0.59
C-3	6.25° S, 346° W	3395.4	1.81
	2.9° S, 340° W	3395.9	2.06

Notes: Elevations are measured relative to a 6.1-mb areoid (planetary datum) calculated from Mariner 9 data. Any elevation below 4.25 km would be acceptable, but the lower the better. Most data are from Myles Standish except values in parenthesis at C-1, from Irwin Shapiro.

Table 18. *Average Root Mean Square (RMS) Slopes from Radar Data, 18 July 1976*

Site	Location	RMS slope
Syria Planum	1.3° N, 90° W	< 0.5°
Capri	4° S, 44° W	c. 2°
Tithonium Lacus	20° N, 251° W	c. 3°
Isidis	20° N, 272° W	< 2°
N. E. Chryse	22.5° N, 31° W	c. 3°
N. W. Chryse	22.5° N, 36° W	c. 8°
S. Chryse	17.5° N, 34° W	c. 6°

On 24 June at LSS 7, difficulties in receiving processed images quickly enough to work with were already being felt, and the team was feeling short-staffed. Hazard mapping in A-1 suggested it was too rough to land in, but better areas were available to the northeast and northwest.

At LSS 8 on 25 June, the team considered whether to move the orbiter to look at A-2 or to extend imaging around A-1 looking for a better site. The orbiter had a 24.6-hour period corresponding to one day on Mars, so it would always pass over the landing site at the same time each day to relay data from the surface to Earth after the landing. If it needed to image another area, its orbital period was adjusted slightly, and it would 'walk' through different periapsis longitudes until it arrived at its destination. That would take several days for a distant site. For A-1, the Viking images were far superior to Mariner 9 data. Harold Masursky described the process of trying to fit a landing ellipse among the various obstacles on photomosaics and hazard maps: 'We are moving the ellipse around to avoid islands, craters, ejecta blankets and etched terrain. The current location is about 19.35° N, 32.5° W, centered in a cratered lunar mare type terrain unit. Many successful landings have been made on the Moon in that kind of unit'. The new ellipse location shown in Figure 47 was called A-1 adjusted, A-1S or A-1R. The original site was now called A-1 nominal. Masursky called his process of sliding ellipses around the maps 'cosmic ice hockey'.

The next day, 26 June, at LSS 9, the group considered landing options. They could collect more data to the northwest of A-1, or move to A-2. A landing on the bicentennial holiday, 4 July, was rejected because the sites were rough and not understood, so lander-scale obstacles could not be extrapolated from orbital images with about 40 m/pixel resolution. The possibility of a 4 July landing had been suggested in 1970 when trajectory analysis suggested that period as a likely landing date. Despite an obvious desire to land on such an iconic date, Martin was afraid that the landing 'would have been lost among the tall ships', upstaged by all the other festivities. Also, images of B-1 and C-1 were needed in the near future, mitigating against A-2. This was needed because the Viking 2 periapsis latitude had to be decided soon. There was a risk of having two landers in orbit simultaneously if Viking 1 did not land before July 25, which made it likely one would have to land after solar conjunction on 25 November 1976. C sites could not be used after conjunction because surface temperatures would be too warm, but Viking 1 could go to C-1 first if necessary.

Meeting LSS 10 was held on 27 June, mainly to discuss images and mapping. A new dataset was presented, maps of clouds observed at Lowell Observatory from 1907 to 1958 and also seen in red light in 1969. They suggested both A-1 and A-2 would be mainly clear. At LSS 11 on 28 June, another source of data was mentioned, a stellar occultation by Mars observed the month before which provided an atmospheric density profile superior to the previous best by the Soviet Union's Mars 6. The options for landing near A-1 now included

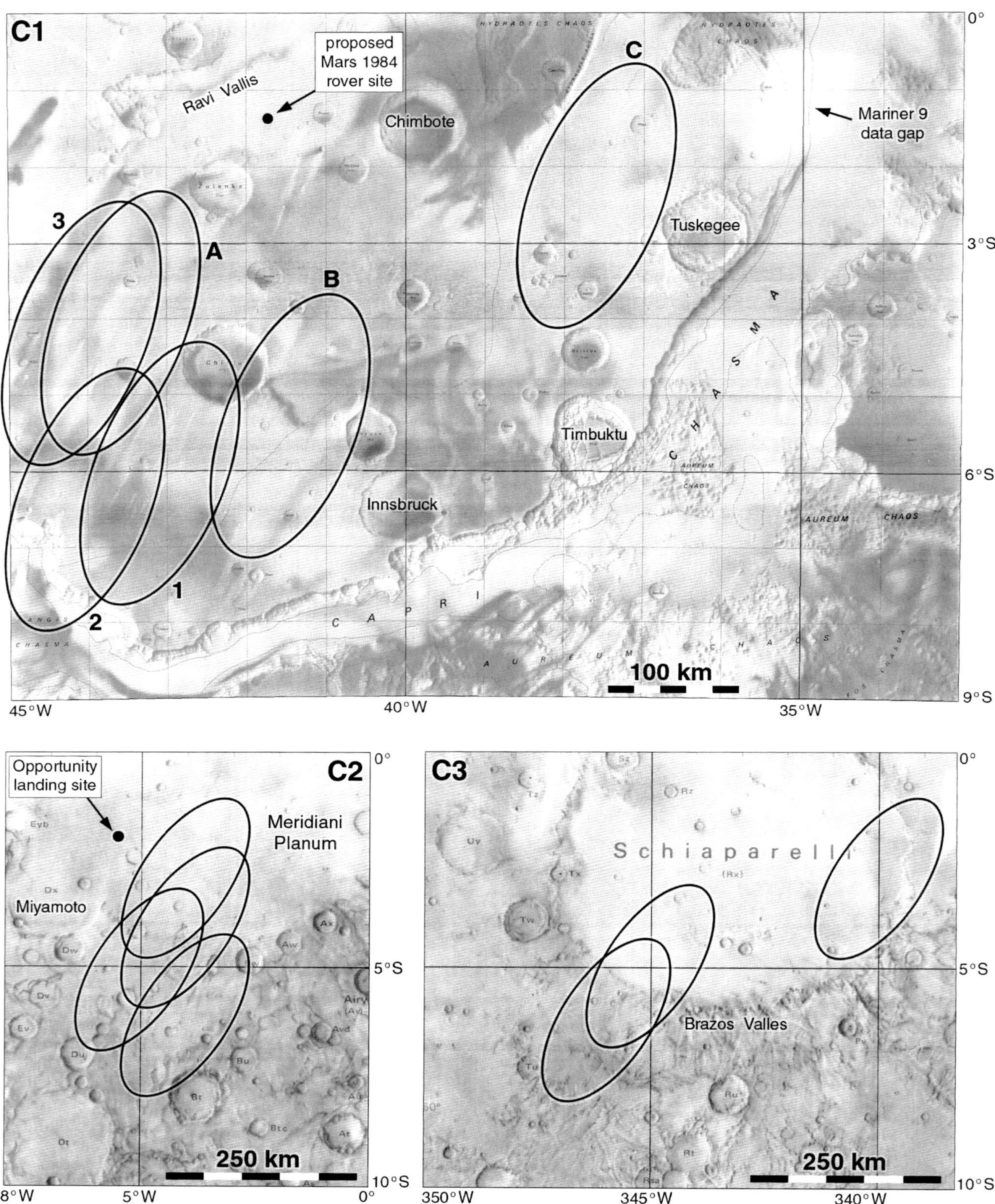

Figure 46 Viking C sites.

C1 sites A, B and C are from Masursky and Crabill (1981). All other sites are from Tables 16 and 17, though Table 16 sites may refer to regions rather than specific ellipses. A potential Mars 1984 rover site (Figure 112) is indicated at C1. The base maps are USGS maps I-1046 (site C1), I-927 (site C2) and I-1050 (site C3). The C2 and C3 maps include a low-resolution albedo overlay from USGS map I-961.

the area to the northwest (A-1 NW), the target for future imaging, and A-1 NE, an area northeast of A-1 also referred to as 'Plateau' (not specified but probably near 22° N, 29°). A-1 NW was called 'Chryse Phoenicia' in a Viking Mission Status Report on 27 June (Ezell and Ezell, 1984). Imaging of A-1 NE was rejected by the time of LSS 12, the next day, because it prevented imaging at A-1 NW. At LSS 12 the sites still considered for Viking 1 were A1-R at 19.5° N, 32.5° W, A-1 NW at 23.5° N, 43° W, or A-2 at 19.5° or 20° N, 250° W. A landing at A-1R on 9 July was considered possible. Images of B-1 were now available and were interpreted to show ejecta, lava flows and polygonal fractures similar to periglacial features seen on Earth. A region northeast of B-1, called B-1 NE (Figure 78), appeared mantled (buried) by sand or dust in Mariner 9 B-frames and might be smoother, according to Larry Soderblom. A map by Soderblom *et al.* (1973) showed the global extent of this mantling layer, inferred from scattered Mariner 9 high-resolution images (Figure 43).

LSS 13 was held on June 30 to consider the site options. Fatigue was affecting work, and the computers used for image processing could barely keep up with demand. Images were being processed in Flagstaff and flown to JPL. Multiple versions of each image were being made, and that requirement was dropped to reduce workload. The group was polled regarding sites: should they land at A-1R on 9 July, and if not, should they go to A-1 NW or A-2? A-1 was pitted with deflation hollows, A-1R was less well understood, A-1 NW looked more depositional than erosional, and A-2 looked more like a lunar surface. Henry Moore now waxed poetic, quoting Robert Service's 'The Ballad of Blasphemous Bill' (Service, 1909): 'North by the compass, North I pressed; river and peak and plain Passed like a dream I slept to lose and I waked to dream again'.

He said the A-1 NW area (24° N, 44° W) looked best, and that A-1 was a geologist's dream – too interesting to be safe. The assumption based on Mariner 9 images was that A-1 was a site of deposition from Ares Vallis, but Viking pictures suggested it was largely shaped by erosion. A-1 NW may be where the Ares sediments had been deposited, and so may be smoother. This suggestion was made by Bob Hargraves of Princeton University. Crofton (Barney) Farmer reported on water vapour mapping from the orbiter. It was at a maximum in Hellas due to high pressure and low temperatures, but low in the C-1 (Capri) area. In Martian terms Capri was 'hotter than Hellas' and too dry to be desirable. Masursky summarized their work: two years earlier they had been looking at the channel front area in the hope of landing on once-wet sediments. Now they had one ellipse to the northwest to avoid boulders, and one to the northeast to match the best radar data. In the end, A-2 was ranked the least desirable site and A-1 NW the best.

A Project Manager's meeting was held on 1 July, at which landing was approved at A-1 NW, though at a site still not determined, rather than moving to A-2. An argument raised in favour of A-2 was that if it proved unacceptable, there was plenty of room for alternative sites nearby, such as Isidis Planitia. In the end the delay caused by walking the orbiter to A-2 overruled that argument. A-1 NW coordinates at this time were given as 23.4° N, 43.4° W, with cartographic uncertainties of about 0.5°.

LSS meetings resumed on 2 July with the 14th meeting of the series, to consider what data Viking Orbiter 1 (VO1) could collect to help select the periapsis latitude for VO2. Further imaging northeast and west of B-1 were the highest priority, with temperature data at the B and C latitudes also desirable. The mantled area described by Soderblom at LSS 12 was too far north, but the smooth mantling material might be found closer to B-1 in new images. Mantling was also discussed at LSS 15, the next day. Henry Moore said the deposit at B-1 appeared to be eroded. The C-1 site was smooth, and its craters appeared partly filled with dust. C site images were also discussed at the LSS 16 meeting on 5 July. The day after that, at LSS 17, improved coordinates for site A-1 NW were presented, 23.36° N, 43.34° W based on the USGS maps and 23.42° N, 43.33° W from Viking pointing data and the site location within the images. Locations on Mars were still subject to significant uncertainties. Radar data were being obtained for this site, and Moore was becoming concerned that any fine debris might have been stripped away by the wind, leaving a rocky surface which the Viking arm could not collect samples from. A site nearby called A-1 SW (17.5° N, 33° to 37° W) was mentioned, but was not seriously considered. It included a very smooth area mentioned at the Preliminary Radar Assessment Meeting on 6 June 1976. Masursky was looking for ellipses in the C-1 area and asked to be allowed to move the ellipse closer to the big canyon rim than the previously established limit of 50 km to avoid large craters.

At LSS 18 on 7 July, a choice had to be made between going to A-1 NW or looking farther west again. Radar data improved towards the west, but were difficult to interpret. A landscape of mesas and buttes might look smooth to radar but be dangerous, whereas sand dunes might look bad to radar but be safe to land on. Von Eshleman said of the potential dangers 'We are like the Flying Dutchman getting close to the Bermuda Triangle'. Moore reiterated his concern about a lack of fine material for sampling and described one feature as looking like a volcanic neck with radiating dykes, like Ship Rock in New Mexico. The vote was for A-1 NW, which Moore called 'Northwest Territories'. Despite this, at the next meeting, LSS 19 on 8 July, a new site northwest of A-1 NW and referred to as A-1 WNW was considered. The WNW area spanned longitudes 46° to 56° W, with a specific site of interest at 22.7° N, 47.4° W. Because of the delays in landing, the minutes recorded that 'The SAMP Director noted that in keeping with the landing delay from the 17th to the 20th, the obstetrician now advises his wife's delay from July 14 to the 21st'. The SAMP (Science Analysis and Mission Planning) director was Gentry Lee, and Cooper Gentry Lee obliged by arriving on 31 July. Norman Crabill showed a cartoon of a giant eagle emblazoned with all the hazards Mars had to offer, being confronted with a rude gesture by a small rodent representing Viking, in a 'last great act of defiance'. Humour helped deal with the exhaustion setting in on the team.

Data needed for B site mapping were discussed at LSS 20 on 10 July. Imaging of areas west and northeast of the nominal B-1 site would be used to look for a safe B-1 location. New images just south of B-1 would be added as well. The landing strategy now was that if Viking 1 landed safely at A-1 WNW, the B options would still be the B or C latitude bands, but that might be reconsidered if Viking 1 failed. At LSS 21 on 11 July, attention returned to A-1 WNW, sometimes called W A-1 NW, whose moon-like appearance (Figures 47 and 51) was promising. Two tentative ellipses were identified, having azimuths of 58.5° clockwise from north and centres at 22.7° N, 47.5° W and 21.9° N, 47.3° W (Figure 47). The latter is very close to the Alpha ellipse noted below. At the next meeting the issue was effectively resolved. This was LSS 22, held on 12 July. The radar data were bad at 44° W, so that area was rejected. At 47.5° W, according to Tom Young, the images looked good, radar data were uncertain with 5° slopes, but the geology was thought to be understood, so probable rock size distributions could be estimated. Masursky presented three ellipses, Alpha at 22.4° N, 47.5° W, Beta at 22.5° N, 49.0° W and Gamma at 22.2° N, 51.0° W. Radar data suggested the Gamma site was best, but it had other hazards from nearby craters. Alpha was on balance better than Gamma and so became the preferred site. Jim Martin said it was 'the safest site we can get in a reasonable time'. Mike Carr said 'don't prolong the debate – the choice is clear'. This site was recommended to the Viking Project, and LSS turned its attention to Viking 2 until LSS 26.

On 18 July at LSS 26, the coordinates of the Viking Lander 1 target point were given as 22.4° N, 47.5° W. These were refined by new analyses reported at LSS 27 on 19 July. Two separate estimates gave values of 22.29° N, 47.71° W and 22.31° N, 47.68° W. These are not two different locations, but different estimates of the surface coordinates of the same target location seen in images. Also at this meeting William Baum reported on a cloud seen in recent images at the landing site. He concluded that similar clouds would have no effect on the landing.

20 August 1975: **Viking 1 (United States)**

Viking 1 placed the first fully successful lander on Mars, after Mars 3 which failed very soon after landing. The mission consisted of two spacecraft launched and flown together, an orbiter and a lander. The 2328-kg Viking Orbiter 1 was designed to obtain high-resolution images of the surface of Mars to help find a safe landing site for its lander and to function as a communication relay for the lander. It greatly exceeded these goals, operating for 4.1 years and returning 34 918 images as well as temperature data and atmospheric water vapour measurements. The 663-kg Viking Lander 1 was intended to image its landing site, collect seismic and meteorological data, study the composition of the surface material and atmosphere, and search for evidence of living organisms (Klein *et al.*, 1976). It operated for 6.3 years, fulfilling all its objectives except the seismic experiment, whose instrument failed to deploy.

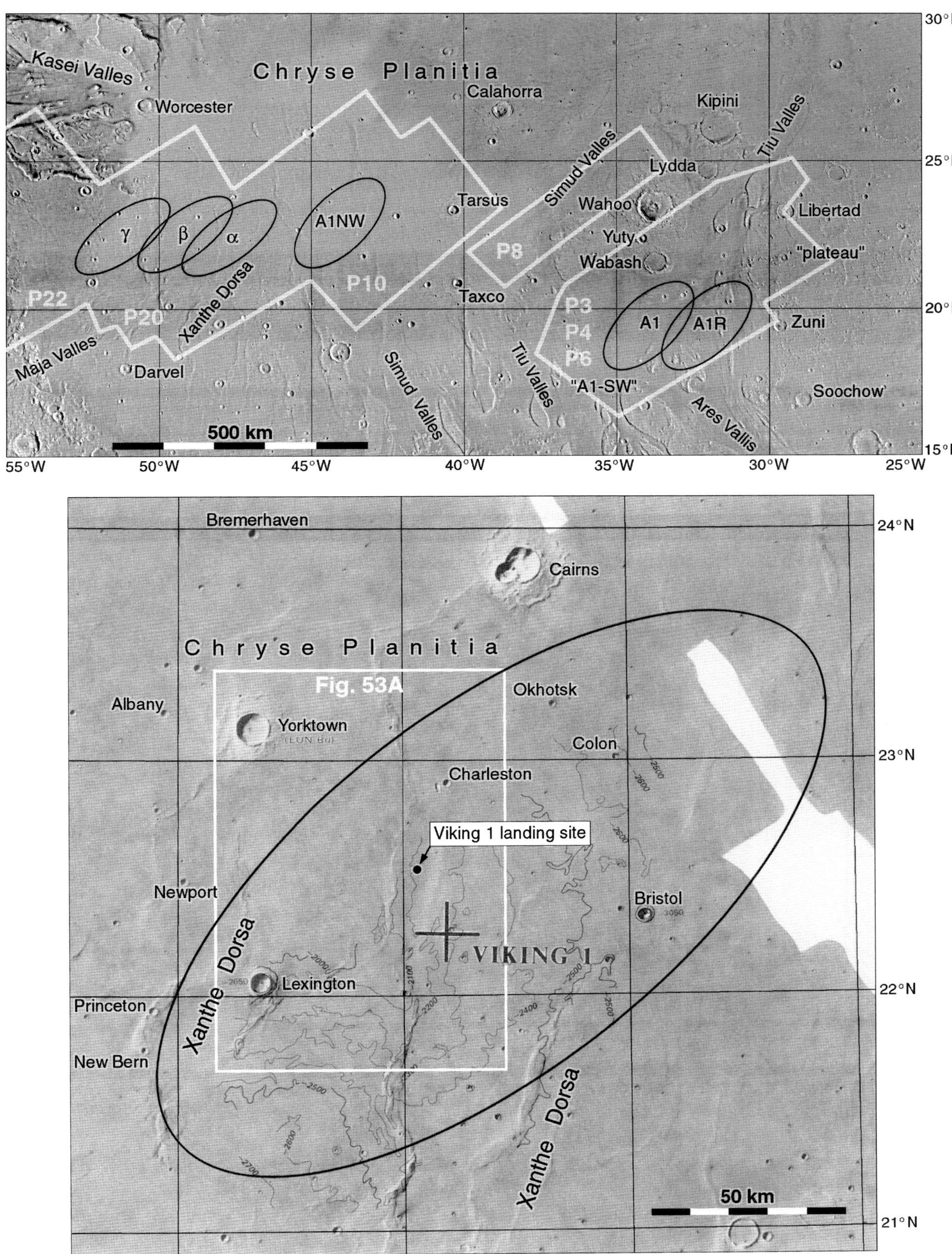

Figure 47 Top: The search for a new A site. The background is a mosaic of Viking images. White outlines show the areas imaged for site certification at periapsis on orbits between 3 and 22 (labelled P3, P22 etc.). **Bottom:** A1α, the final landing ellipse for Viking 1. The base map is USGS map I-1068 and the cross marks an early estimate of the landing site location, 20 km south of the correct position.

Viking Orbiter 1 was similar in design to the earlier Mariners, but its orbit insertion engine was larger to cope with the extra mass of the lander. The orbiter's octagonal frame, 46 cm high and 250 cm wide, supported four solar panels, each 320 cm long and 123 cm wide, with a total span of 9.75 m. The whole structure of the orbiter was 3.29 m high. It carried an atmospheric water detector and an infrared thermal mapper as well as two cameras. The lander had a triangular frame 102 cm high and 284 cm wide with three footpads, surmounted by antennae, two scanning cameras, a meteorology mast and other equipment including a robotic arm. The arm carried a sample scoop, a temperature sensor, a magnet and mirrors to permit small views of otherwise inaccessible areas. The meteorology mast held sensors for measuring air temperature and wind speed and direction. The lander instruments were a gas chromatograph mass spectrometer (GCMS) for organic and atmospheric chemistry analysis; an x-ray fluorescence spectrometer (XRFS) for inorganic chemistry analysis; the biology lab consisting of a pyrolitic release experiment, a labelled release experiment and a gas-exchange instrument; a seismometer; and an air pressure sensor. The whole lander fitted inside an aeroshell (heatshield and backshell) 3.6 m across. The aeroshell also carried instruments to provide data on the upper atmosphere.

Viking 1 was launched at 21:22 UT on 20 August 1975 from the Kennedy Space Center. A helium leak in the fuel tanks required unplanned rocket firings to reduce pressure, slowing Viking 1 so that it arrived at Mars seven hours later than planned. On 19 June 1976 (MY 12, sol 178) after a 304-day cruise, it entered a preliminary two-day orbit, which was soon adjusted to the planned 32 600 by 1500 km Mars orbit, inclined 37.9° to the equator with a period of 24.6 hours. This sequence put the mission back on its proper timeline after two days. The first task was to certify the landing sites chosen earlier (Figure 45) by taking high-resolution images.

The lander was released from the orbiter at 08:51 UT on 20 July 1976 (MY 12, sol 209), the seventh anniversary of the Apollo 11 landing on the Moon, and entered the atmosphere several hours later. Both Vikings approached their landing sites from the southwest. After the initial frictional braking, it discarded the heatshield at an elevation of 6400 m and descended on a parachute to an elevation of 1200 m, then discarded that and dropped to the surface braked by thrusters. Viking Lander 1 touched down at 11:53 UT (all times at Mars are Earth-received times) on Earth, and at 4:13 pm local time on Mars, and began transmitting its first image only 25 seconds later. Shadows cast by a cloud of dust raised by the landing appeared as streaks on the left side of the first image from the scanning camera (frontispiece, and also shown for Viking 2 in Figure 95). Craters near the Viking 1 landing site (Figures 52, 53) were named after places associated with the American Revolutionary War, as the landing took place in the bicentennial year of the Declaration of Independence.

The operations of the two Viking missions were conducted in several mission phases (Snyder, 1977; Snyder, 1979; Snyder and Evans, 1981, with contributions from a mission summary at NASA's Planetary Data System). The Orbiter and Lander primary missions commenced with orbit entry and landing, respectively, for the two spacecraft pairs and ended just before solar conjunction on 15 November 1976. An extended mission began after conjunction and continued until 31 May 1978, considerably extending the results of the relatively short primary mission. The extended mission was followed by a continuation mission which allowed the orbiters to continue mapping while the landers transmitted data in an automated mode. Viking Orbiter 2 failed during this phase, which ended on 26 February 1979 as Voyager 1 was approaching Jupiter.

The demands on communication and staffing at JPL during the Voyager encounters with Jupiter forced Viking to limit operations from 26 February to 19 July 1979, a period called the Viking interim period. Viking operations could have ended at any time with sudden equipment failures, so an effort was now made to obtain as much data from them as possible, beginning with the period from 19 July to 6 November 1979, which was called the survey mission. Viking Orbiter 1 obtained high-resolution images for future landing site selection while lander operations continued in automated mode. From 6 November 1979 to 23 April 1980, corresponding to the completion mission phase, gaps in medium-resolution mapping of the planet were filled in by Orbiter 1. The orbiter also undertook a search for unknown satellites in low orbits, without finding any (Duxbury and Ocampo, 1988), and monitored the north polar region. Viking Lander 2 failed during this phase of operations.

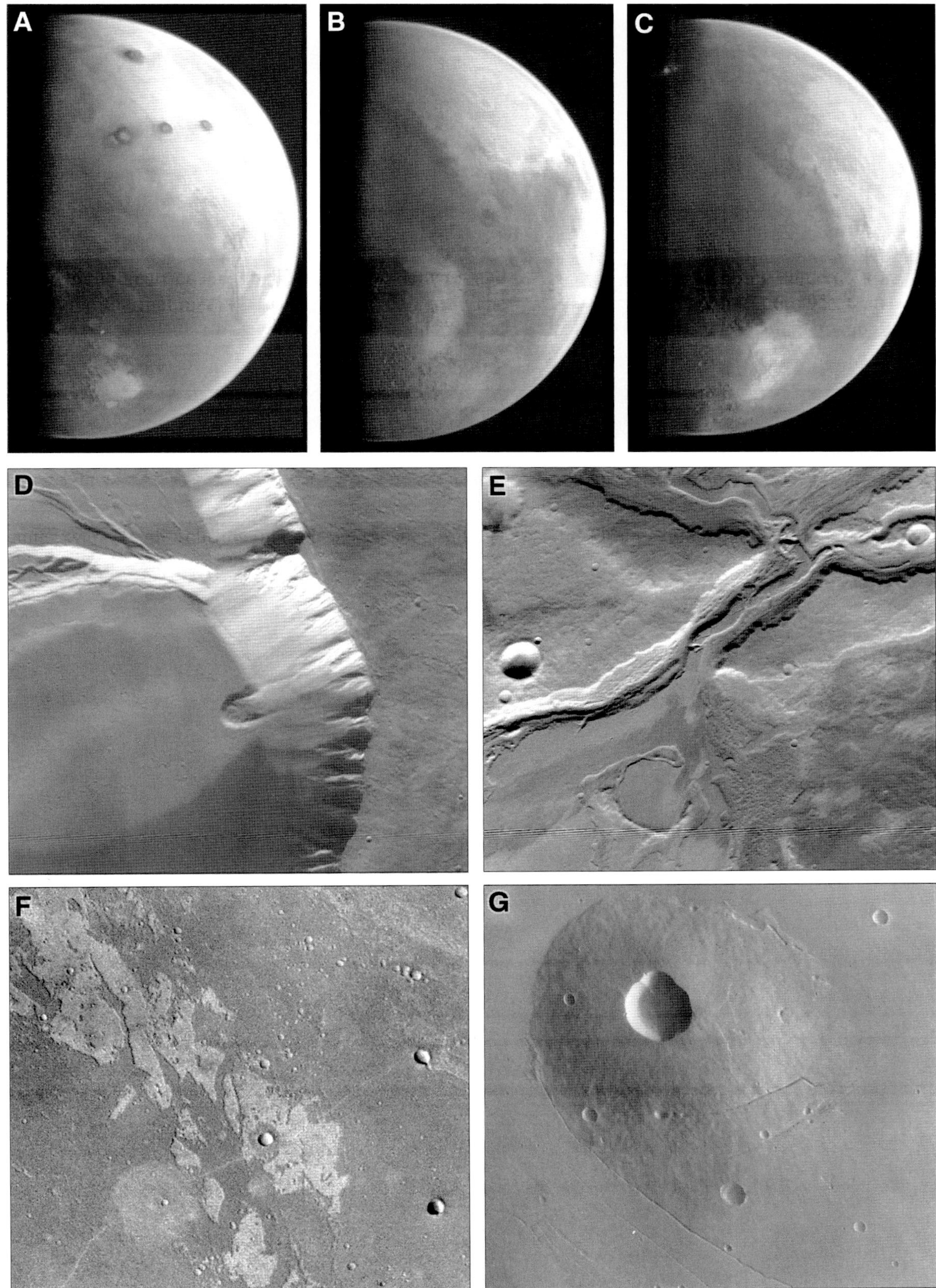

Figure 48 Viking Orbiter 1 images.
A: f169c23; **B:** f170c07; **C:** f170c12; **D:** 474S24, **E:** 457S02, **F:** 004A19, **G:** 846A21.

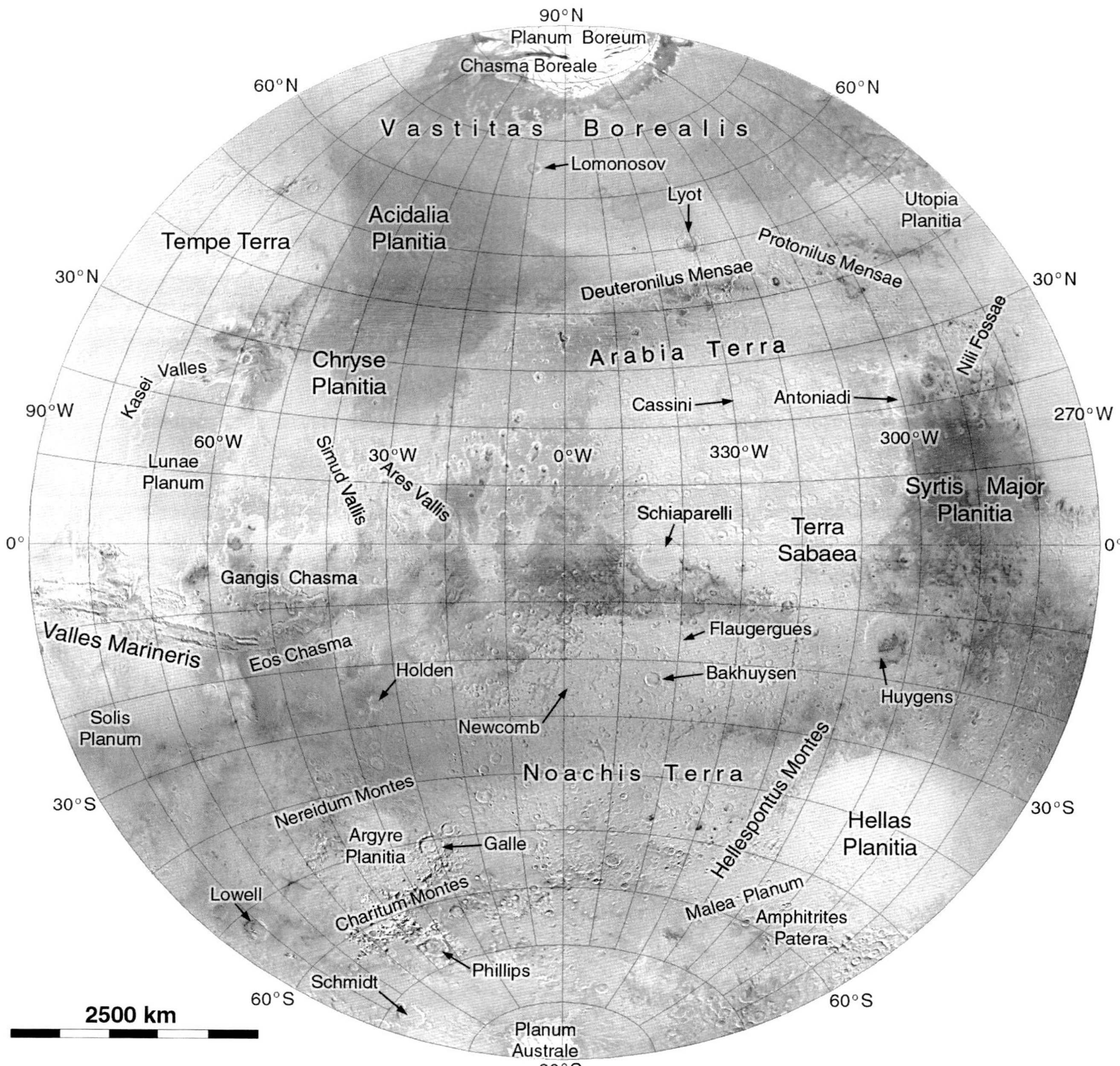

Figure 49 The 0° meridian hemisphere of Mars as seen by the Viking Orbiters.
Albedo changes had occurred in many areas since the Mariner 9 mission, especially near Kasei Valles and Utopia Planitia.

Figure 50 (opposite page): **Top:** The 180° hemisphere of Mars as seen by the Viking orbiters. Figures 49 and 50 are made from a global colour image mosaic assembled by USGS, with some modifications to remove clouds and artifacts. It was compiled from images taken over several years, so the albedo markings include inconsistencies caused by seasonal and longer term variations. The low volcanic shield at Alba was called Alba Patera at this time, but is now named Alba Mons (Figure 4). **Bottom:** Several people associated with Viking are commemorated by named craters in the area south of Chryse. The crater Barsukov commemorates Valeri Leonidovich Barsukov, a prominent Soviet planetary scientist and former director of the V. I. Vernadsky Institute of Geochemistry.

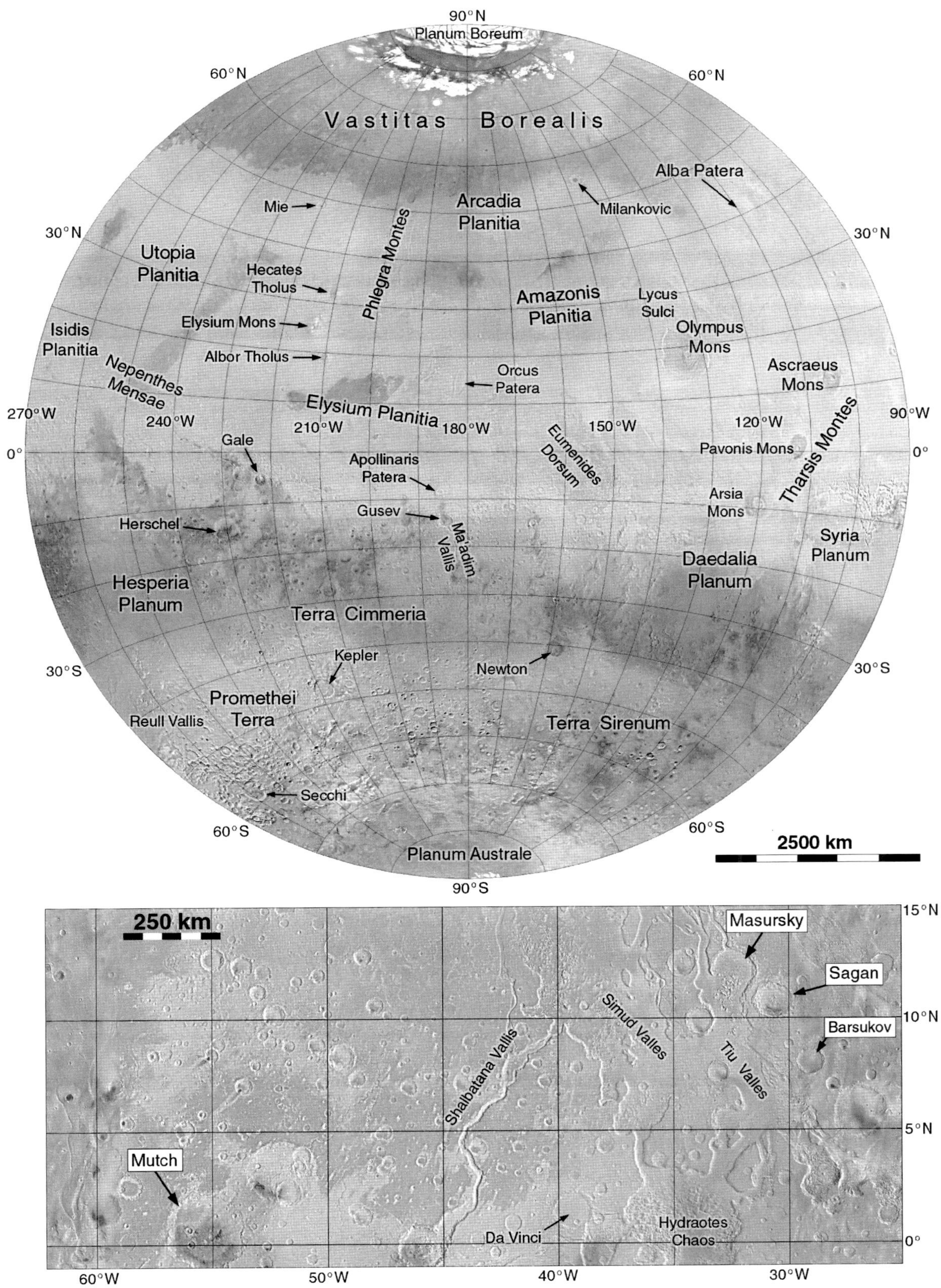
90°N
Planum Boreum
60°N
60°N
Vastitas Borealis
Alba Patera
Mie
Arcadia Planitia
Milankovic
30°N
30°N
Utopia Planitia
Hecates Tholus
Phlegra Montes
Amazonis Planitia
Lycus Sulci
Elysium Mons
Olympus Mons
Isidis Planitia
Albor Tholus
Orcus Patera
Ascraeus Mons
Nepenthes Mensae
Elysium Planitia
Tharsis Montes
270°W
240°W
210°W
180°W
150°W
120°W
90°W
0°
0°
Gale
Eumenides Dorsum
Pavonis Mons
Apollinaris Patera
Arsia Mons
Gusev
Herschel
Ma'adim Vallis
Syria Planum
Daedalia Planum
Hesperia Planum
Terra Cimmeria
Kepler
30°S
30°S
Newton
Promethei Terra
Reull Vallis
Terra Sirenum
Secchi
60°S
60°S
Planum Australe
2500 km
90°S
250 km
Masursky
15°N
Sagan
10°N
Barsukov
Simud Valles
Tiu Valles
Shalbatana Vallis
5°N
Mutch
Da Vinci
Hydraotes Chaos
0°
60°W
50°W
40°W
30°W

Finally the mission entered its final phases, with a survey mission II period lasting from 23 April to 14 July 1980 and a brief termination phase ending on 7 August 1980 (MY 14, sol 311) when Viking Orbiter 1 was shut down. In this period high-resolution mapping with some stereoscopic coverage was obtained in the Aeolis and Mangala areas, and Phobos and Deimos were imaged from a distance to refine knowledge of their orbits. The termination phase began on 15 July with the first of four orbit changes designed to exhaust the remaining fuel and delay impact on the planet. Estimates of remaining fuel were very imprecise, and any one of the burns might have been the last. The last image was taken on 30 July and transmitted on 5 August, and the last orbit change was on 31 July. The orbiter was now in a 47.6-hour orbit extending from 411 km to over 55 000 km altitude, an orbit which should delay atmospheric entry until about 2025. After August 1980 only Viking Lander 1 survived, continuing to operate in automated mode until November 1982.

Viking Orbiters 1 and 2 together imaged the whole surface of Mars in much greater detail than Mariner 9, returning 51 539 images between them with resolutions over most of the planet of about 250 m/pixel or better. Resolutions as high as 8 m/pixel were obtained in some areas. A total of 50 500 images were taken of the planet's surface, and the rest included some taken for calibration and others for a satellite and ring search, as well as Earth departure views. Samples of Viking 1 images are shown in Figure 48. Approach images in Figure 48 show the Tharsis volcanoes and Argyre (A), Syrtis Major (B) and Hellas (B and C). Orbital images in Figure 48 show a landslide in the Olympus Mons caldera (D), channels in Mangala Valles (E), eroded terrain in the A-1 ellipse in Chryse (F) and the volcano Albor Tholus (G). Preliminary image targets for Vikings 1 and 2 together are shown in Figure 82, and image coverage at different resolutions is shown in Figures 83 and 84.

The new images permitted major advances in mapping. Stereoscopic imaging and spacecraft occultations greatly improved knowledge of topography. The whole surface of Mars was remapped in a series of photomosaics at 1:2 000 000 scale and as greatly improved shaded relief maps (Figure 51). Mosaics of areas of special interest were made at 1:500 000, geologic mapping was greatly enhanced, and the new results formed the basis for future landing site studies for the next three decades. In the 1990s the production of paper maps slowed considerably with the introduction of new digital and online products. Only geological maps of Mars were routinely published on paper by USGS after 2000. These Viking-derived digital maps are used throughout this atlas, supplemented later by Mars Odyssey products with higher resolution and more uniform illumination. Both orbiters also made bistatic radar observations, reflecting their radio signals off Mars for reception on Earth to provide estimates of surface roughness (Simpson and Tyler, 1981). Earth-based radar had done this at low latitudes, but the orbiters could extend the data to higher latitudes. Phobos and Deimos were also imaged extensively during frequent distant flybys and several targeted close passes (Figures 198, 199). Their entire surfaces were mapped with much more detail than Mariner 9 had obtained. Viking Orbiter 1 activities are summarized in Table 19, based on JPL (1978), Snyder (1979) and Snyder and Evans (1981).

During the primary mission, Viking Lander 1's surface sampler collected 12 samples, three for the biology experiment, three for organic chemistry (GCMS) and six for inorganic chemistry (XRFS). The biology instrument performed four analyses, GCMS made two analyses of the first sample, three of the second and did not analyze the third, and XRFS made 73 analyses. The lander cameras and the meteorology instrument collected data daily. Twenty-four atmospheric analyses were performed, and the surface sampler conducted one soil test for physical and magnetic properties.

The backhoe was a protruding plate under the sample arm's scoop. It could be used to dig a trench by being pulled backwards through the regolith, but it was also used to estimate the cohesion of surface materials by being pressed down on them. These tests were called backhoe touchdowns (Moore *et al.* 1987), and all touchdowns for Viking 1 were made in the extended mission. All of these arm activities are listed in Tables 20 and 21 and illustrated along with other arm activities in Figures 60 through 73.

The samples were analyzed for their composition and introduced to components of the biology instruments. Results are generally interpreted as highly unfavourable for life, with a strongly oxidizing component in the samples which would quickly remove any carbon compounds.

Table 19. *Viking Orbiter 1 Orbital Events*

Date	Orbit	Events
19 June 1976	0	Mars orbit insertion (MY 12, sol 178)
21 June 1976	2	Adjust orbit for site certification
9 July 1976	19	Adjust orbit to move periapsis westward
14 July 1976	24	Adjust orbit to be synchronous over the landing site
20 July 1976	30	Viking Lander 1 separation and landing (MY 12, sol 209)
3 August 1976	43	Adjust orbit to remain synchronous over lander
11 September 1976	82	Decrease orbital period to move periapsis eastward
20 September 1976	92	Adjust orbit to approach synchronization over Viking Lander 2
24 September 1976	96	Synchronous orbit over Viking Lander 2
6 October 1976	107	First Earth occultation (last on orbit 133, 1 November 1976)
15 November 1976	117	End of primary mission (MY 12, sol 323)
25 November 1976	156	Conjunction (MY 12, sol 333)
14 December 1976	175	First commands of extended mission received
22 January 1977	213	Adjust orbital period to approach Phobos
4 February 1977	227	Approximately synchronize orbital period with Phobos
12 February 1977	235	Precise synchronization to Phobos orbit for close approaches
20 February 1977	243	Phobos closest encounter, altitude 89 km (MY 12, sol 418)
11 March 1977	263	Reduce periapsis to 300 km for higher resolution imaging
22 March 1977	275	Daily Earth occultations begin, used for topographic mapping
24 March 1977	278	Adjust orbital period to 23.5 hours
2 April 1977	289	Periapsis reaches most northerly latitude, 39.2° N (MY 12, sol 458)
20 April 1977	305	First of 554 daily Sun occultations
13 May 1977	329	Periapsis crosses terminator into dark, high-altitude mapping starts
15 May 1977	331	Maneuver to avoid Phobos impact, several flybys before and after
28 June 1977	376	Partial attitude control gas leak, one system turned off on 1 July
1 July 1977	379	Increase orbital period to 24.0 hours
24 September 1977	463	Orbiter images Phobos shadow over Lander 1 (MY 12, sol 628)
28 September 1977	467	Orbiter images Phobos shadow and dust storm over Lander 1
6 November 1977	507	Flyby of Deimos gives unique view of trailing side (MY 13, sol 1)
30 January 1978	593	Periapsis crosses equator from north to south
7 May 1978	689	One Mars year in orbit (MY 13, sol 178)
31 May 1978	713	End extended mission, begin continuation mission (MY 13, sol 202)
4 August 1978	778	Periapsis crosses evening terminator into the light
19 October 1978	854	Phobos flyby at a distance of 600 km (MY 13, sol 339)
23 October 1978	858	Last of 554 Sun occultations
10 November 1978	876	Last of 602 Earth occultations
2 December 1978	898	Adjust orbital period to 24.85 hours, start slow longitude walk
28 December 1978	923	Periapsis at maximum southern latitude (39.2° S)
20 January 1979	946	Solar conjunction. No communications, orbits 934–966
26 February 1979	984	Interim period begins – limited activity until 19 July 1979
19 May 1979	1061	Adjust orbital period to 25.0 hours, continue longitude walk
8 July 1979	1109	Second period of Earth occultations begins
19 July 1979	1119	Periapsis raised to 357 km, period 24.8 hours, start Survey Mission
31 July 1979	1131	Periapsis crosses morning terminator into dark
23 August 1979	1153	Second period of Sun occultations begins
6 November 1979	1226	Start of Completion Mission (MY 14, sol 43)
14 February 1980	1326	Satellite and ring search
23 April 1980	1396	Final phase of high-resolution mapping, Survey Mission II, begins
31 July 1980	1485	Orbit adjusted to 320 by 56000 km to delay impact until 2019
7 August 1980	1489	Orbiter 1 powered down, end of mission (MY 14, sol 311)

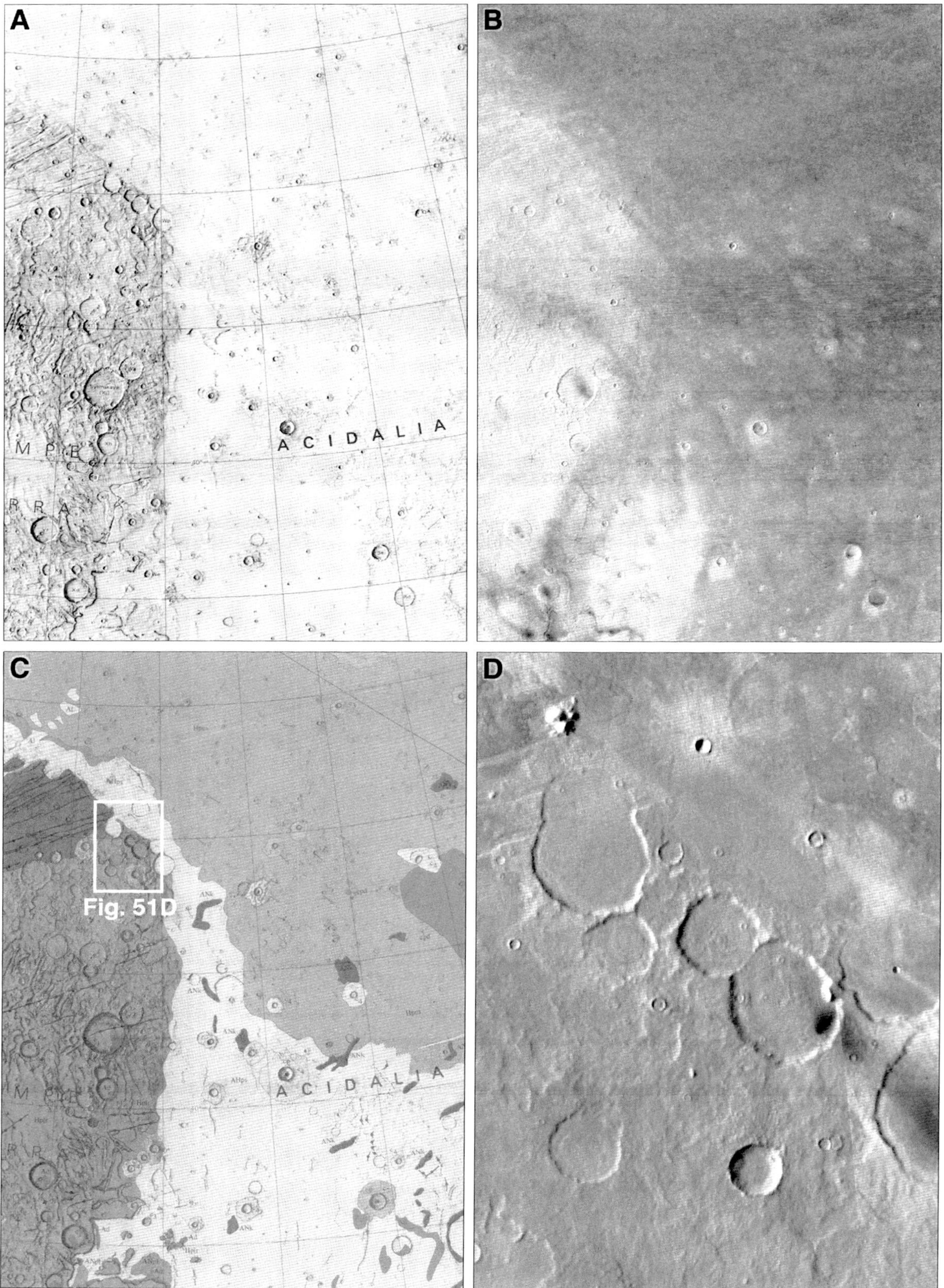

Figure 51. Viking maps.
A: Shaded relief (USGS map I-1476). **B:** Albedo (USGS map I-2574, Sheet 2). **C:** Geology (USGS map I-1614). **D:** Detail of the Mars Digital Image Model v. 2.1, a USGS global photomosaic. Figure 30 shows Mariner 9 maps of the same location at 45° N, 50° W, and Figure 188 shows Mars Odyssey images of the Figure 51D area.

No unambiguous evidence for growth of or metabolism by Martian organisms was found, though this interpretation is sometimes questioned. Viking Lander 1 completed its primary mission on 15 November 1976 but continued to operate until mid November 1982. It collected meteorological data and imaged the surface repeatedly to search for changes, which are described in association with Figures 74 and 75. The two Viking landers returned more than 4500 pictures between them.

Viking Lander 1 touched down within the A1-Alpha landing ellipse shown in Figure 47. Images of the surrounding landscape (Figures 55 and 56) revealed a very rocky landing site with patches of fine-grained material. Efforts began immediately to locate the lander relative to the craters and ridges seen in orbital images. The horizon showed many relief features (Figure 55), but most seemed to be too small and nearby to be visible in early Viking Orbiter images. The first attempts to locate the lander are illustrated in Figures 52A and 52B, which are details of two USGS maps showing different locations for Viking Lander 1.

At LSS 32 on 24 July, the post-landing position of Lander 1 from radio tracking during sols 1 to 3 was estimated as 22.27° N (areocentric) or 22.47° N (areographic, as shown on maps), with an uncertainty of 0.15°, and 48.00° W with an uncertainty of 0.5°. Bradford Smith was trying to locate the site by comparing horizon features with orbital images, without success. The pre-landing and post-landing ellipses are shown in Figures 47 and 52, respectively. This analysis was refined by the time of LSS 38 on 11 August. The new estimated areocentric site coordinates for the centre of the landing ellipse were 22.27° ± 0.02° N, 48.00° ± 0.07° W, with an ellipse size of 11 km north-south and 4 km east-west. The combined uncertainties were about 13 km north-south and 5 km east-west. The best effort to locate that position on images suggested that the site lay between two mare ridges (Figure 52), and one ridge might be visible as a light-toned hill east of the lander, a feature referred to as White Mesa. Further work on crater counts reported at LSS 40 on 13 August included another feature west of the landing site, Tara Vallis, an early reference to the feature now known as Maja Vallis, whose outflows eroded ridges near the landing site.

Morris *et al.* (1978) identified several features around the lander in Viking Orbiter 1 image 27A63, with a resolution of 40 m/pixel. The most important feature in this identification was Crater A (Figures 55, 56F), which lies southwest of the lander. Better images were obtained after the periapsis of Viking Orbiter 2 was lowered to 300 km from 1500 km, leading Morris and Jones (1980) to refine that location and identify several smaller ridges and crater rims nearby. A nearby dark spot in the highest resolution Viking image of this location was interpreted by Olivier de Goursac and James Garvin, very plausibly at the time, as the heatshield impact site (Anonymous, 1987). This location (Figure 52C) was accepted for many years and was the basis for a later rover mission design (Figure 140). Analysis of radio transmissions after landing had allowed Mayo *et al.* (1977) to locate the lander at 22.272° N, 47.94° W, accurate to within about 120 m (0.002°) in latitude and 12 km (0.2°) in longitude. Knowing the coordinates of the lander and the location in images gave an important tie between images and coordinates for cartographic control.

After Mars Pathfinder landed in 1997 at a site unambiguously determined from surrounding features (Figure 161), it was found that the image locations of Pathfinder and Viking 1 could not be reconciled simultaneously with their tracking coordinates. The Viking 1 location was recalculated and found to be farther east than Morris and Jones had suggested. Merton Davies of the RAND Corporation, a veteran planetary cartographic control analyst, suggested locations in Viking images (M. E. Davies, personal communications, April 1998) which were about 10 km east of the Morris and Jones (1980) site. Stooke (1999) proposed a topographic fit near Davies' suggested site in which boulder-covered ridges east of the lander were interpreted as parts of a wrinkle ridge. Slightly later analyses brought the calculated site to an intermediate location (Zeitler and Oberst, 1999), and here Parker *et al.* (1999) finally found a good match. These sites are shown in Figure 52C. Parker's match to topography was made using noisy low-contrast Viking images, but his result was confirmed when better images became available, first MOC images from Mars Global Surveyor (Malin Space Science Systems, 2004, 2006) and most spectacularly HiRISE images from Mars Reconnaissance Orbiter (Parker *et al.*, 2007). The latter images are used in Figures 53 and 54. Attempts to associate small-horizon features in panoramas with features seen from orbit are often ambiguous, and features

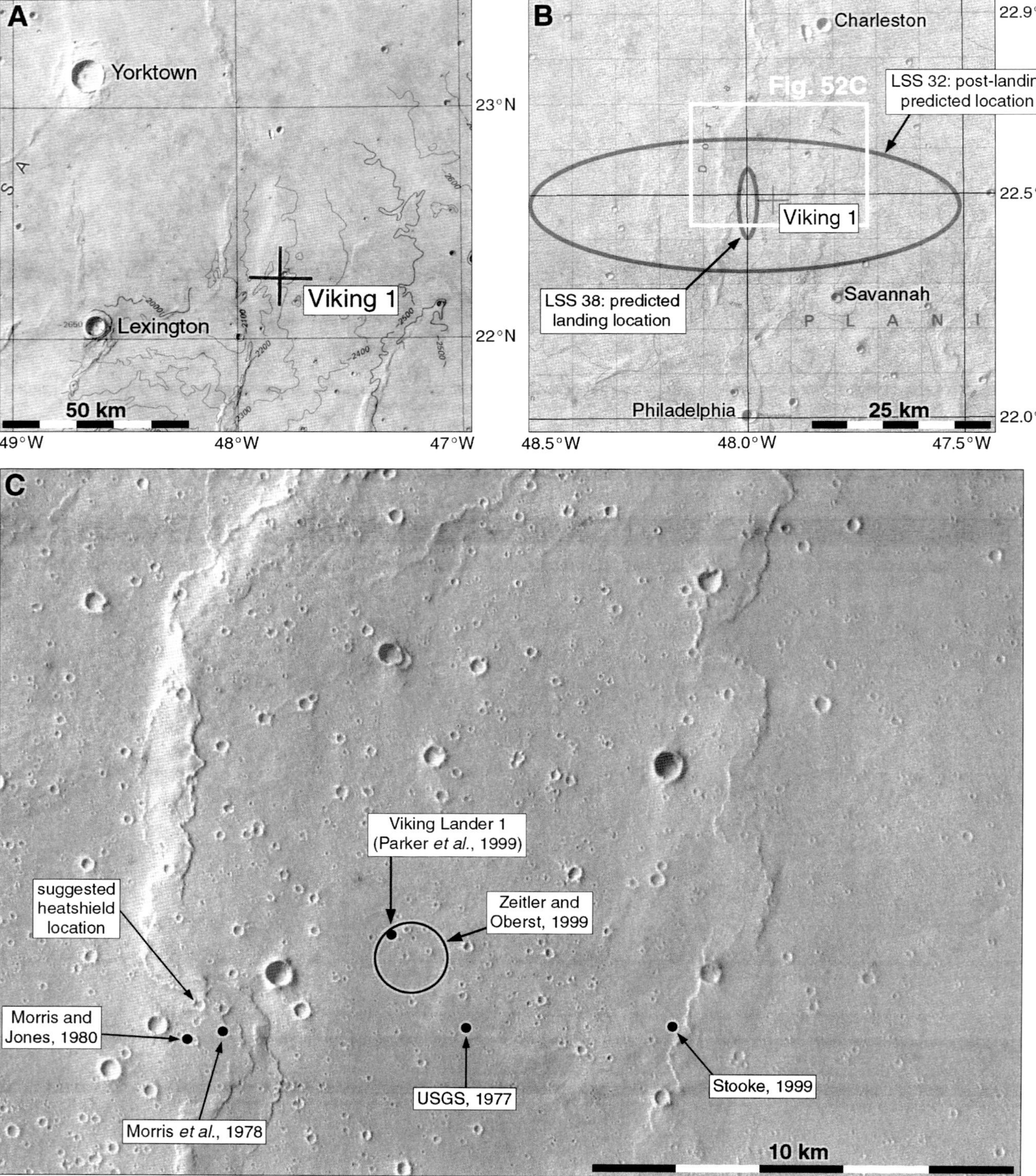

Figure 52 Attempts to locate Viking Lander 1.
A: USGS map I-1068. **B:** USGS map I-1059, with superimposed landing ellipse predictions. **C:** Mosaic of THEMIS images V06368024 and V27534024 showing various estimates of the lander location.

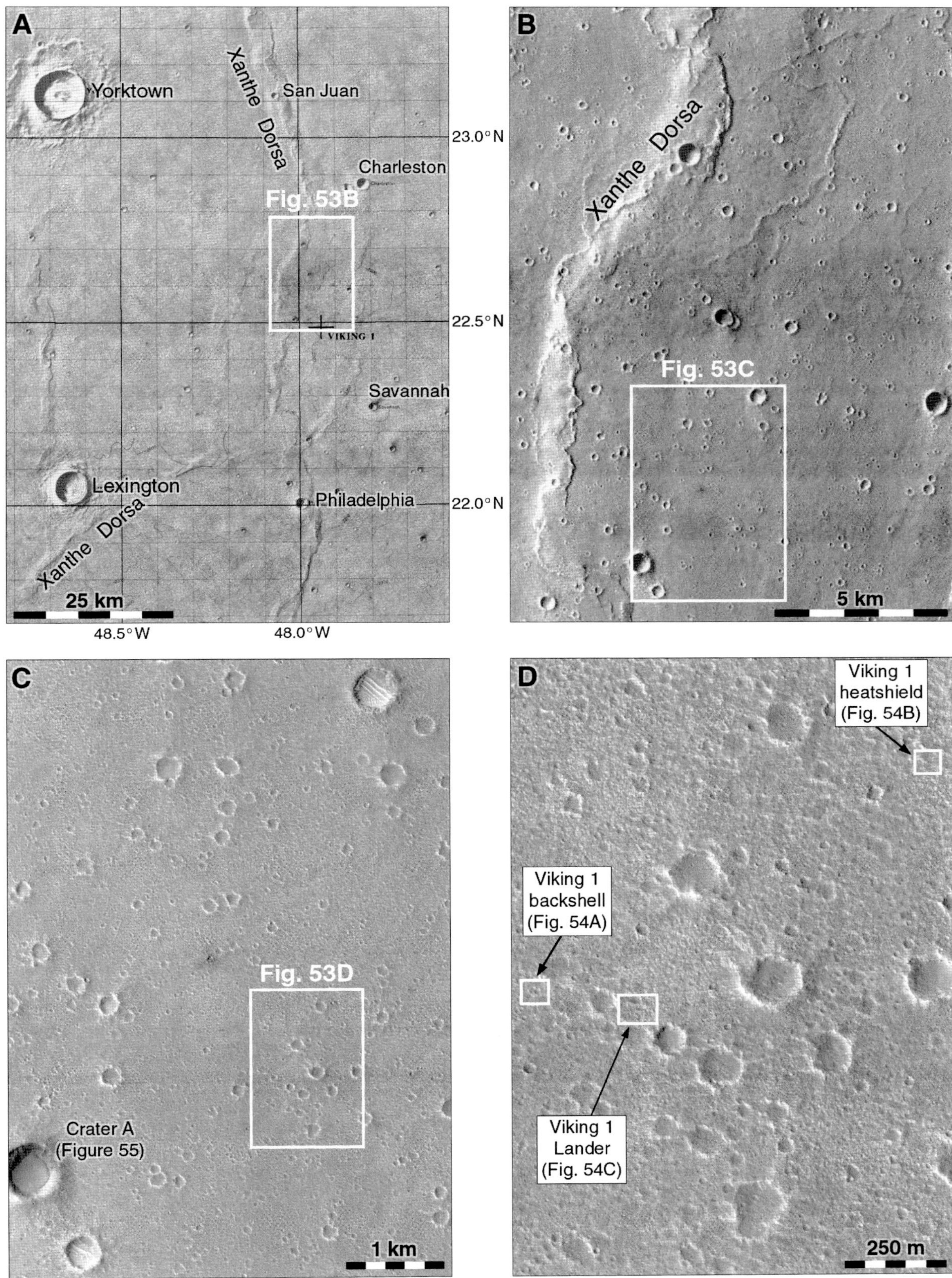

Figure 53 Viking 1 landing site.
A: Detail of USGS map I-1059, showing the estimated location of the landing site. **B:** Mars Odyssey THEMIS image V27534024.
C and D: Mars Reconnaissance Orbiter HiRISE image PSP_001521_2025.

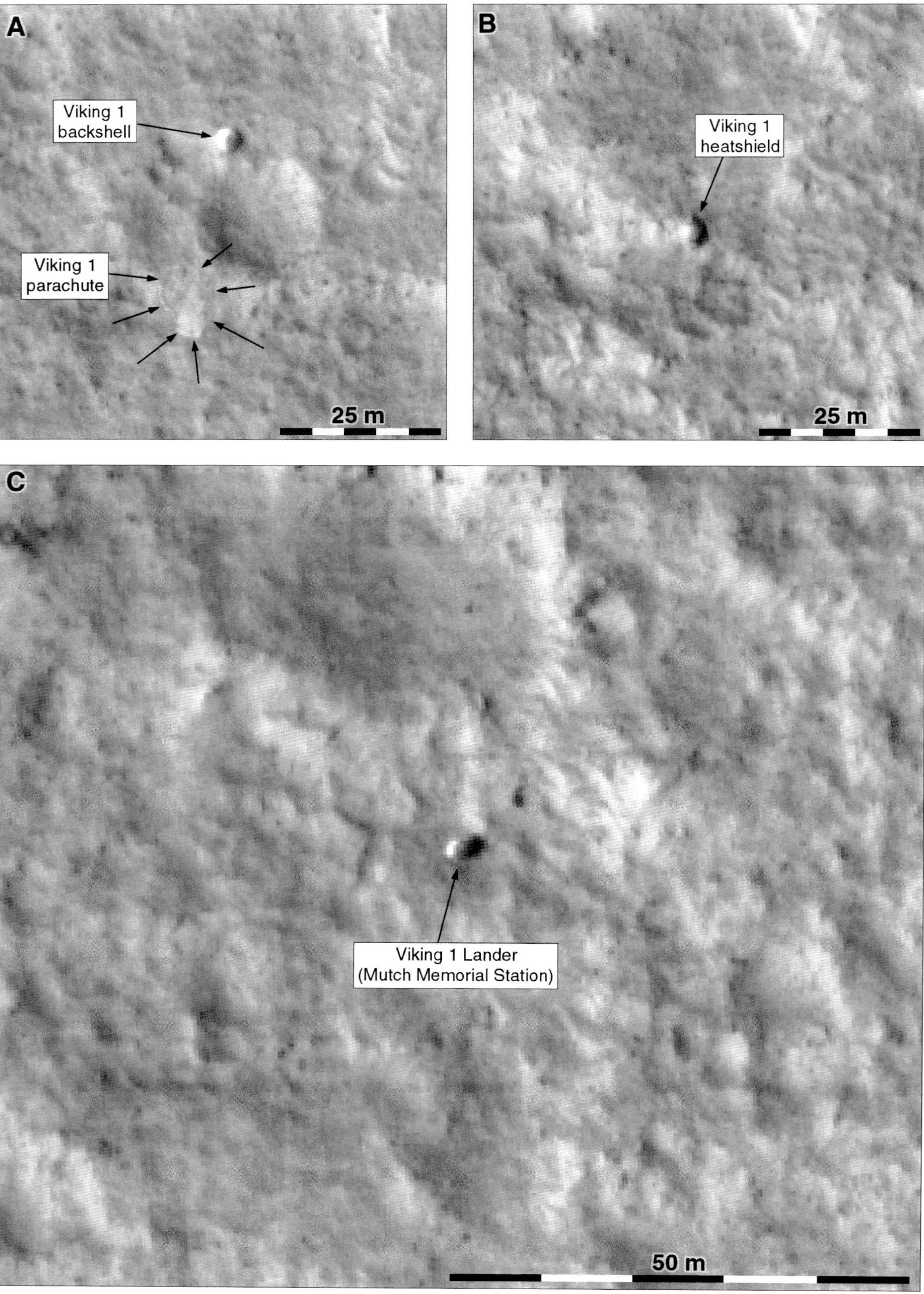

Figure 54 Viking hardware in HiRISE image PSP_001521_2025.

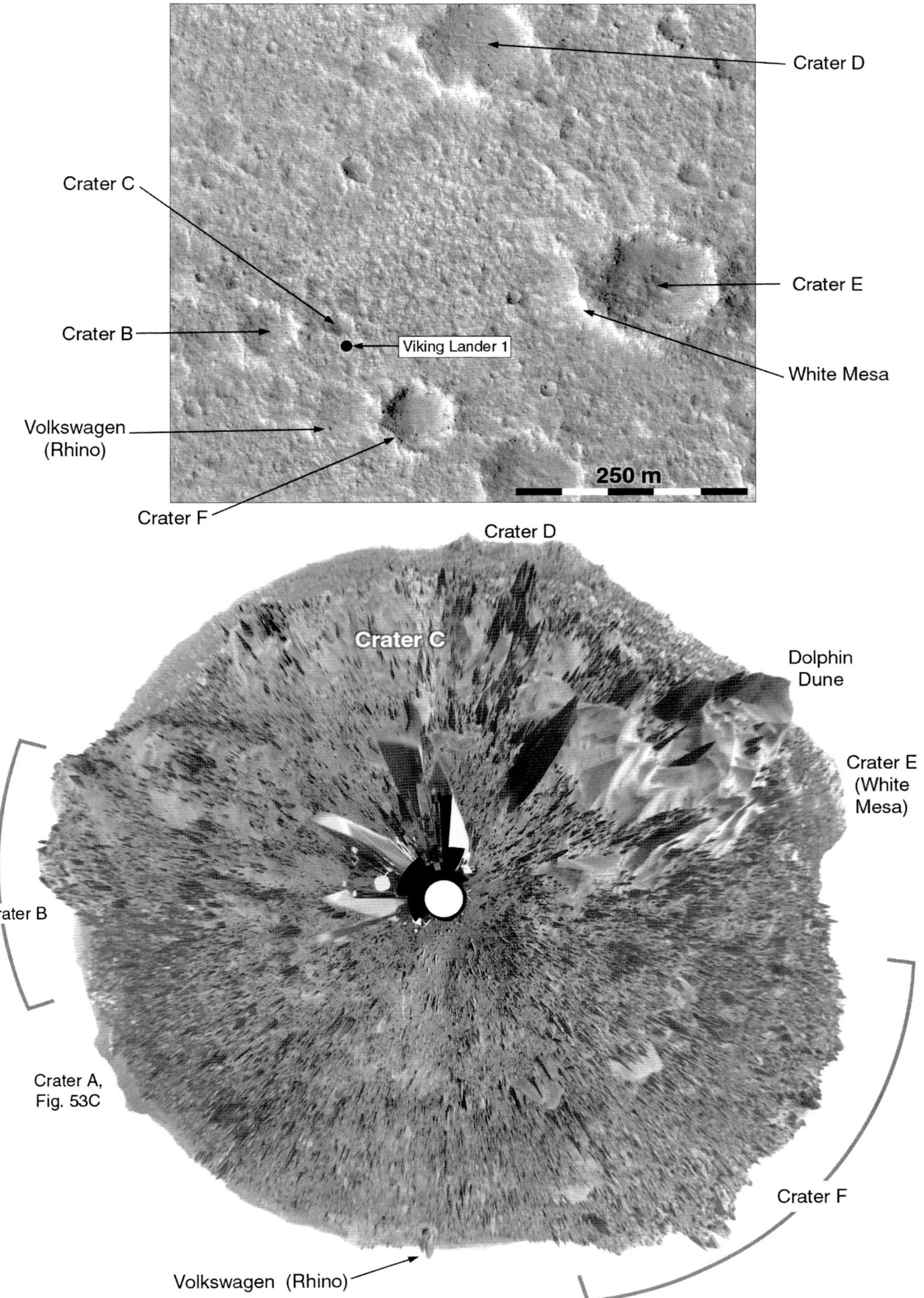

Figure 55 Comparison of HiRISE image PSP_001521_2025 and Viking 1 panoramas projected into a circular view with exaggerated horizon relief.
Volkswagen, also called Rhino, is a large rock.

expected but not seen can often be explained away, so it is perhaps not surprising that several locations were described by different analysts.

The Viking Lander 1 spacecraft was designated the Mutch Memorial Station soon after Thomas Mutch, the imaging team leader, died while climbing in the Himalayas on 6 October 1980. A plaque commemorating this designation is displayed at the National Air and Space Museum in Washington, DC, and is intended to be placed on Viking Lander 1 at some future date. Craters near the Viking 1 landing site were named after Mutch, Carl Sagan and Hal Masursky (Figure 50).

Immediately after landing, the high-gain antenna and the meteorology boom were deployed, and the analytical instrument cover was ejected. The seismometer failed to uncage and was never used. Then an image of a footpad on the ground and part of a panorama were transmitted via the orbiter, the first images successfully obtained from the surface of Mars. For 43 sols the lander followed its primary mission program of imaging the landing site, collecting and analyzing samples and obtaining weather data. A period of limited activity (the 'reduced mission') followed while Viking 2 landed and conducted its primary mission. The primary and reduced missions together were called the nominal mission. Mars passed through superior conjunction, behind the Sun as seen from Earth, on 25 November 1976, causing a short interruption to Viking operations, but both landers had long extended missions after that. Nominal mission activities are summarized in Table 20 and extended mission activities in Table 21. The sample area, the region within reach of the surface sampler, is shown in Figure 58B as it was before sampling commenced, with informal names for several areas and rocks. Figure 62A shows the area at the end of the nominal mission, and Figures 65A, 69A and 73A show it at intervals during the extended mission. The summaries of activities presented here are based on Soffen (1976), Shorthill *et al.* (1976), Snyder (1979) and especially Moore *et al.* (1987). The Viking Mission Status Bulletins released by JPL during the mission were also consulted, the last of the set (JPL, 1978) being particularly useful.

On sol 2 the shroud which protected the sampler arm during landing was ejected. It fell in front of the lander, bounced twice and landed about 1 m away, leaving visible pits where it bounced (Figures 58 and 59). A latch pin should have been ejected after that, but it stuck, temporarily raising fears of an immobile arm which would have been a major failure. Also on sol 2 a picture was taken of the area underneath engine 2 by using a small mirror on the sampler arm. Without the mirror that region would have been hidden by the lander body. The area was of interest because it would have been eroded by the engine exhaust, and the degree of erosion would indicate its hardness. As the lander was descending in the last few seconds before landing, it drifted about 1.5 m towards the northwest. Part of the area eroded by exhaust was visible as a shallow pit at the bottom of the panoramas, but the mirror view extended the coverage. More images like this were taken on sols 280, 528, 550, 582 and 594 (Figure 76). On sol 3 the first images from the second camera were taken, and on sol 5 the latch pin was successfully ejected to free the arm. The pin fell just in front of the lander, making a small pit in the dusty surface (Figures 58, 59 and 76).

Surface sampler operations began on sol 8. Pictures were taken to document most sampling and other mission events, and series of before and after images from every surface sampler operation are shown in Figures 59 through 73 to provide complete documentation of the surface activities at this site. These images are grouped by location for easy comparison, and in date order at each location, rather than in purely chronological order. The sampling area was much rockier than typical lunar surfaces examined during the Luna, Surveyor and Apollo missions, in part because the finer components of the surface material can be removed by winds in the thin Martian atmosphere. A fear often expressed during Viking landing site selection was that the surface might be too rocky for the arm to collect a sample, and the area pre-programmed for sampler operations (ICL, the Initial Computer Load) was indeed too rocky. In its place the smooth, fine-grained region called Sandy Flats at the northeast end of the sampling area was chosen for the first sampling. Elsewhere rocks were numerous, but a small hollow (Rocky Flats) offered another possible sampling site. A third small area of fine material at the southwest end of the sampling area was outside the nominal sampling area but was targeted later.

On sol 8 the arm was used to collect its first sample from Sandy Flats at the trench called Biology 1 (Figure 60). The coarse fraction was dumped (purged, in Viking terminology) close to the lander (Figures 61, 62) after the fine fraction was delivered to the biology experiment. The biology analysis continued for 30 sols. The arm then collected another sample at almost the same place for the organic chemistry instrument (GCMS 1). Two attempts were made to collect this sample as the first acquisition appeared to be too small, but an arm stall prevented the second sample from being delivered. No image of the trench was taken after GCMS 1. The instrument team went ahead with the analysis of the small sample on sols 17 (low temperature) and 27 (high temperature), and it proved sufficient for analysis. The second GCMS 1 sample was purged, and then another sample was collected with two scoops from the outer end of the trench for the inorganic chemistry instrument (XRFS 1). This sample of fine material, originally called S1 (Sample 1) and later C-1 (Chryse-1) by the XRFS team, was delivered to XRFS successfully, and any coarse material was purged. All these samples came from essentially the same place. All XRFS samples are described by Clark *et al.* (1982).

On sol 14 a new GCMS sample, GCMS 2, was collected beside the previous trench at Sandy Flats (Figure 60), but not analyzed as the team decided to replace it with a sample from Rocky Flats. Its coarse fraction was held in the scoop and then purged on sol 22. The coarse material fell in a location hidden by the lander body and could not be imaged. As the sampler retracted following the sol 14 dig it had stalled, and diagnostic tests (sol 18) and imaging (sol 20) preceded the decision to purge the sample. On sol 31 another sample, GCMS 3, was taken at Rocky Flats, the rougher area south of the spacecraft (Figure 61). It was partly purged just in front of the trench, and the scattered fragments were imaged. The rest of the sample was analyzed on sols 32, 37 and 43. No organic molecules other than contaminants were detected in these GCMS samples. Rocky Flats was also the source of two sample scoops for the inorganic chemistry instrument, XRFS 2, on sol 34 (Figure 61). After the fine fraction was purged, its coarse fraction (called C-2) was delivered to XRFS and analyzed. Meteorological data were also collected throughout this period and for much of the rest of the mission.

On sol 36 attention passed back to Sandy Flats, where the Biology 2 sample was taken near the previous sample sites but closer to the lander (Figure 60). The fine material was delivered to the instruments, and the coarse fraction was purged, though little or nothing could be observed at the purge site. Rocky Flats was targeted for XRFS 3 on sol 40 (Figure 61). The coarse fractions of two separate scoops were delivered to the instrument as sample C-3, and the fine material and any remaining coarse lumps were purged. The sample was not as large as expected. Also on sol 40 an image of a temperature sensor on Footpad 2 was taken using a mirror on the sampler arm (Figure 76).

On sol 41 the sampler was directed back to Sandy Flats for the Physical Properties 1 test (Figure 58). The temperature probe was placed in the soil where it measured a temperature of 269 K (–4° C), 3 K cooler than the air above it. Then a sample was scooped from a shallow trench beside the previous trenches. A magnified image of the backhoe was obtained using a magnifying mirror on the lander body near the colour chart. Fine material adhered to magnets on the backhoe, but much of it was shaken off by vibration of the collector head. The material in the scoop was dumped onto a grid marked on the lander body, forming a conical mound which could be monitored for changes as a way of examining wind transport of sediment on Mars. The pile and its changes over time are illustrated in Figure 75. After this there was a reduction in Lander 1 activity as attention passed to Lander 2, but a final sampler action was performed at the end of the nominal mission on sol 91. A single-scoop sample of fine material was collected at Sandy Flats, near GCMS 2 but closer to the lander (Figure 60). This was Biology 3, and it was delivered for analysis in the biology instruments. In the process some was spilled into the XRFS, forming that experiment's sample C-4. Any coarse material was purged, and the arm was parked for the conjunction period. All activities up to this time are summarized in map and panorama formats in Figure 62.

During the conjunction period as Mars moved behind the Sun, the Viking landers obtained images, continued analyzing samples and collected meteorology data. As the two landers' radio signals passed behind the Sun's disk, they provided information on plasma conditions in its corona. After conjunction the information stored on the landers was transmitted back to Earth, but Lander 1 had suffered the loss of one track of its four track tape recorder, and some of its conjunction data could not be recovered. Conjunction was on 25 November 1976 (MY 12, sol 333), and the lander's first new commands were uplinked on sol 146.

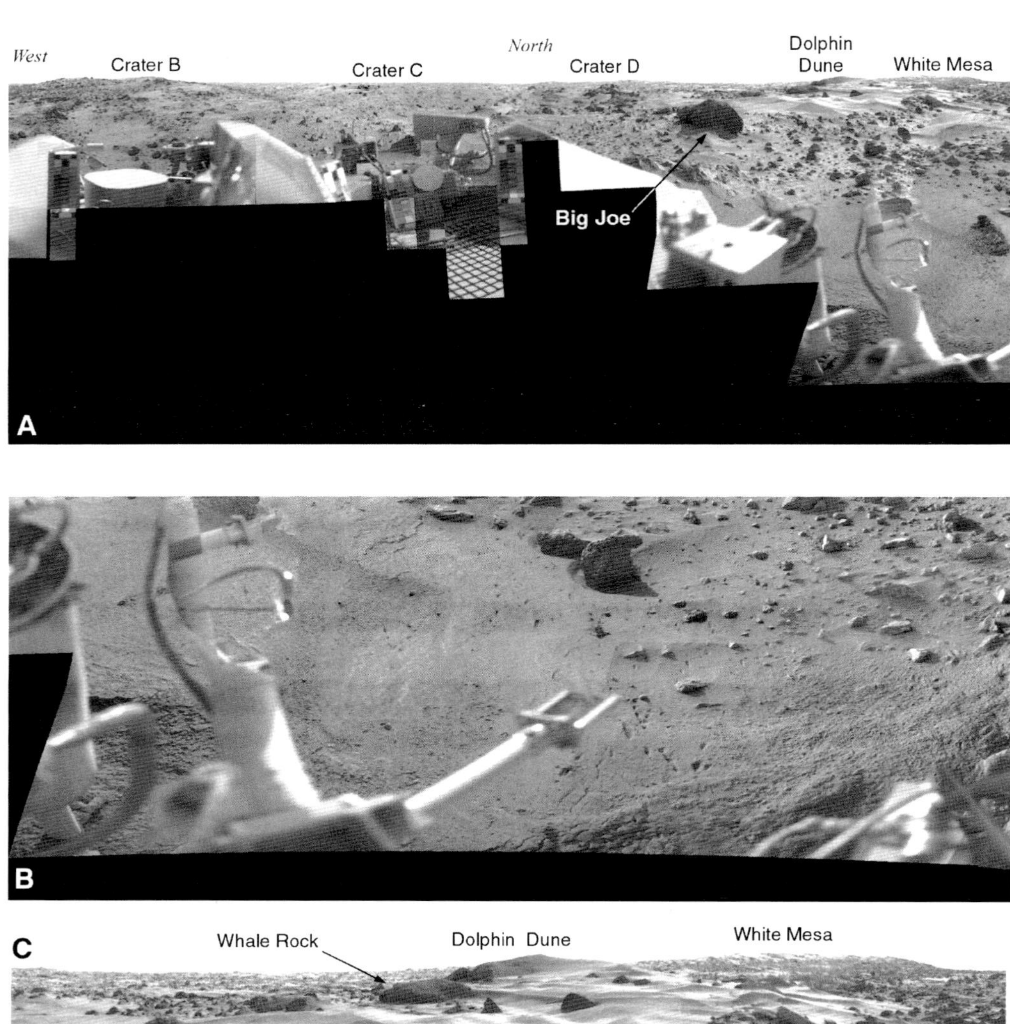

Figure 56 Viking Lander 1 surface panoramas.
Images from both cameras have been combined to minimize the area obstructed by spacecraft components. **A (across both pages):** The full 360° panorama. **B (across both pages):** The sampler arm operating area. **C:** Big Joe rock and the large area of drifts beyond it. **D, E, F:** Horizon craters B, D and A, identified in Figure 55.

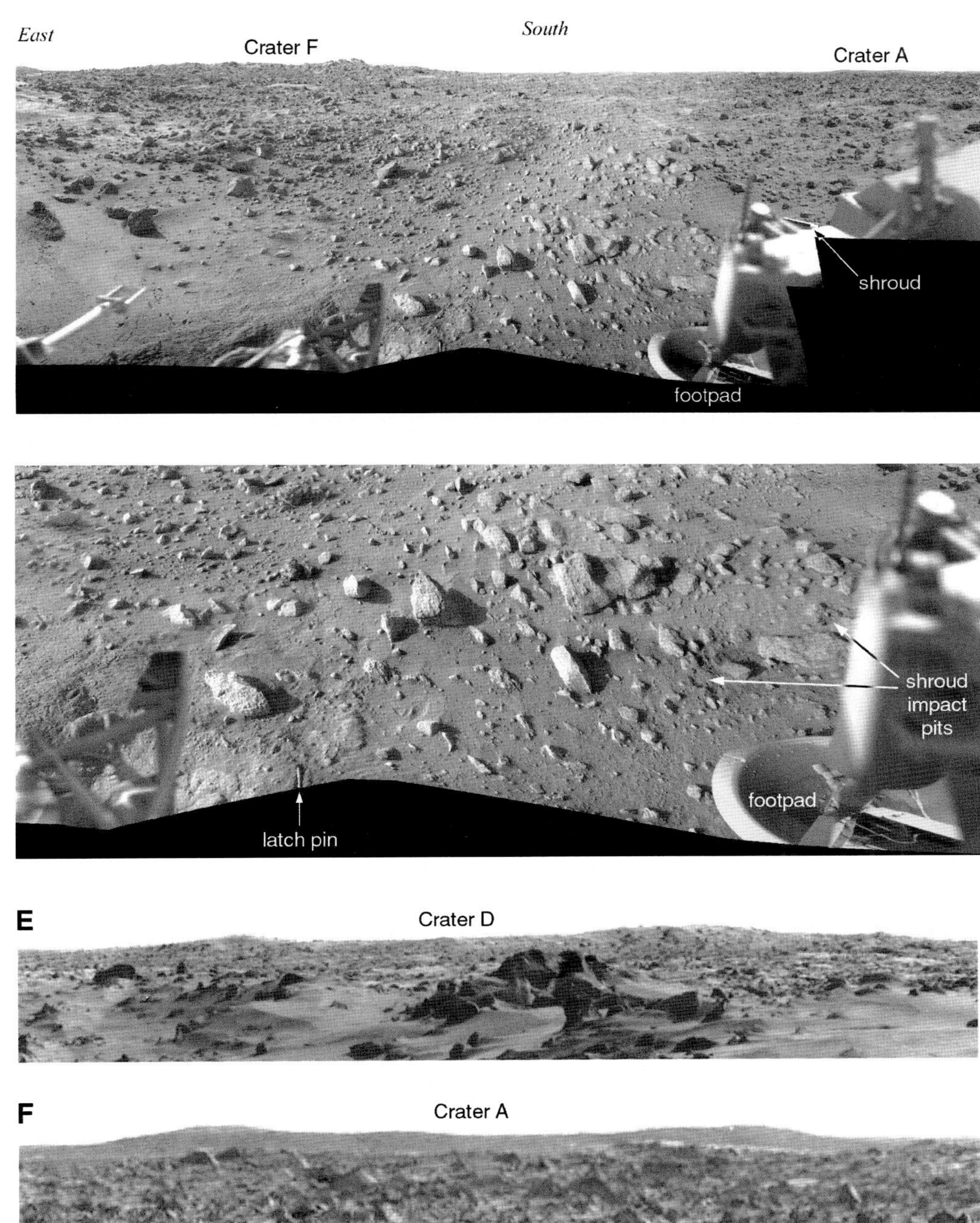

Figure 56 (continued)

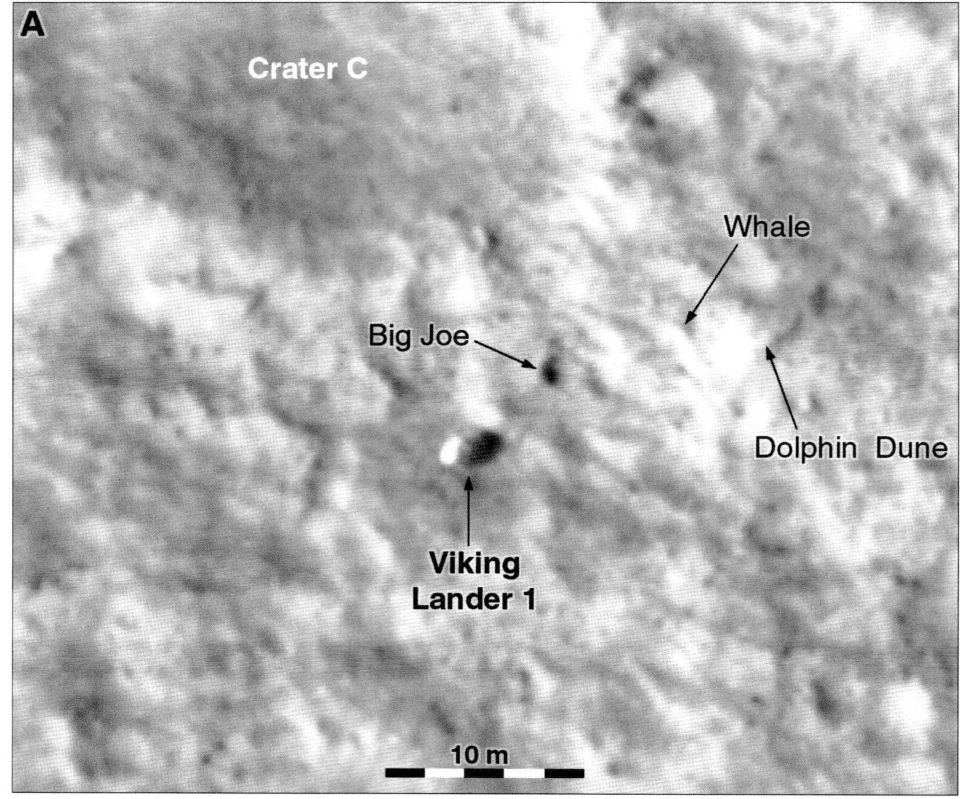

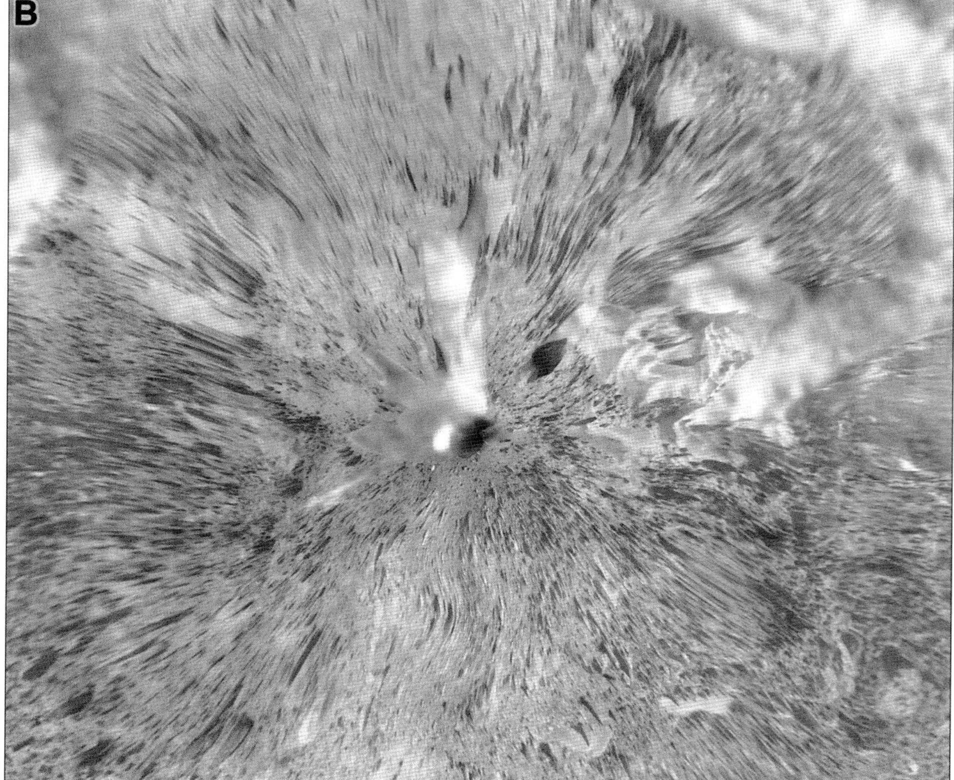

Figure 57 **Viking Lander 1 panorama projected onto HiRISE image PSP_001521_2025.**

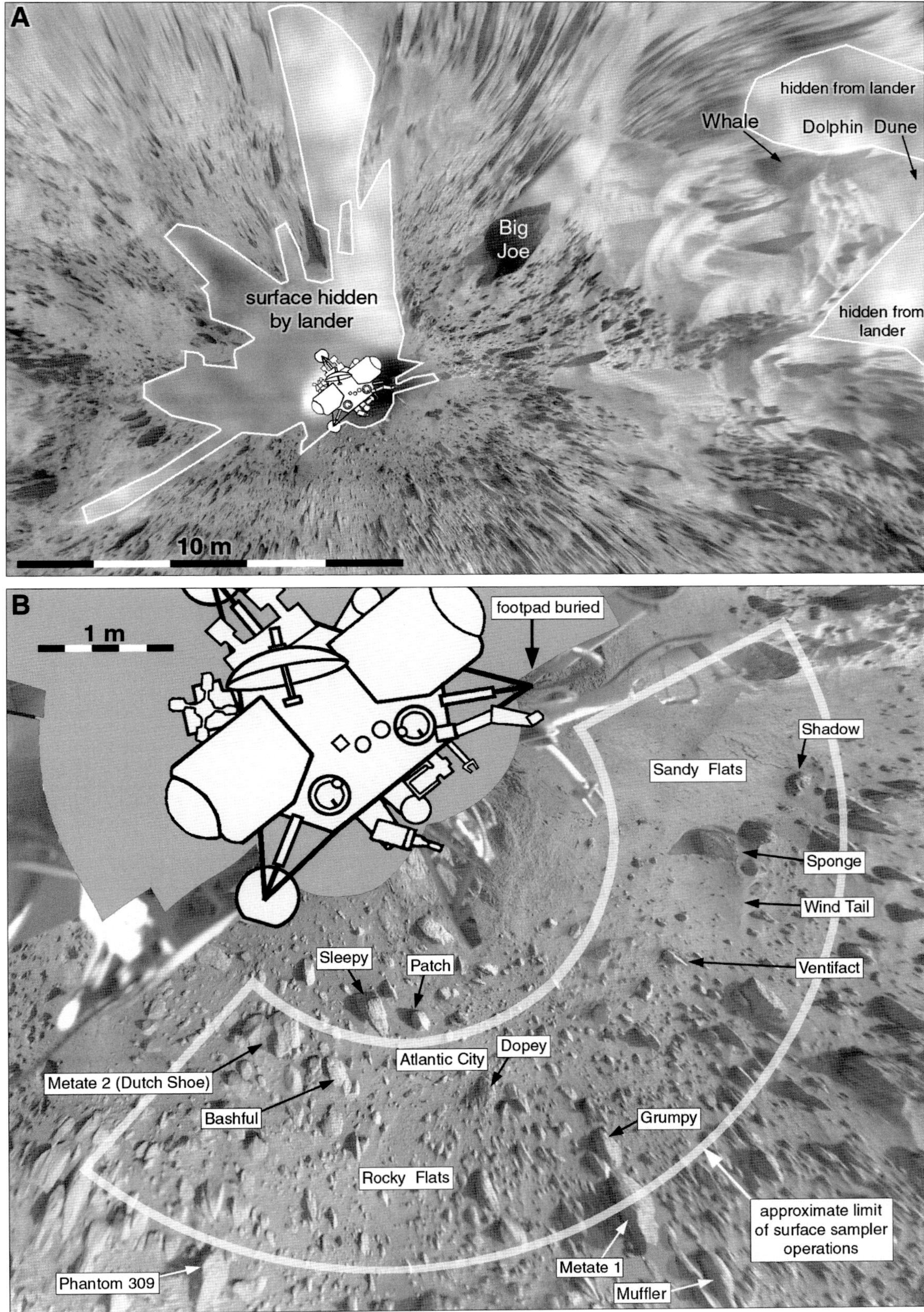

Figure 58 A: Viking Lander 1 vicinity. **B:** Sample field photomap compiled from reprojected panoramas. The surface is shown just before sampling operations began.

Before shroud ejection

Shroud impact pit 1

Before shroud ejection

Shroud and impact pit 2

Before pin ejection

Latch pin impact site

Figure 59 Spacecraft events during preparation for sampler activity.

Table 20. *Viking Lander 1 Nominal Mission Activities*

Sol	Activities
0	Land, biology instrument lid deployed, XRFS calibrated (MY 12, sol 209).
2	Sampler shroud ejected, bounced 1 m from lander footpad, latch pin did not eject to free arm.
3	GCMS instrument lid deployed.
4–5	GCMS atmospheric analyses. Five tests are successful, one ended by anomaly.
5	Sampler latch pin ejected successfully.
8	Biology 1, XRFS 1 samples collected and delivered, GCMS 1 sampling failed.
8	Biology 1 sample analyzed in Pyrolytic, Labelled Release and Gas Exchange instruments.
8–31	XRFS 1 sample analyzed.
14	GCMS 2 sample collected, but not delivered to instrument as sampler stalls.
17–18	GCMS atmospheric analyses, seven tests successfully conducted.
18	Diagnostic sequences on stalled arm.
22	GCMS 2 sample purged, GCMS 1 analysis continues.
24, 27	GCMS atmospheric analyses, ten successful tests on each sol.
27	Biology 1 sample analyzed in Pyrolytic and Labelled Release instruments.
31	GCMS 3 sampling: scoop contents purged, new sample acquired, analyzed on sol 32.
33	GCMS atmospheric analyses. Ten tests successfully completed.
34	XRFS 2 sample taken using two sample strokes, delivered and analyzed on sols 34 to 36.
36	Biology 2 sample taken, then analyzed in the Pyrolytic Release instrument.
37	GCMS 3 sample analysis continued.
39	Biology sample analyzed in Labelled Release instrument.
40	XRFS 3 sample collected using two sample strokes, then analyzed over sols 40 to 60.
41–42	GCMS atmospheric analyses, one test on each sol, both successful.
41	Physical Properties 1 test, magnetic properties, soil dump onto lander grid.
43	GCMS 3 sample analysis continued.
52–102	GCMS atmospheric analyses on sols 52, 62, 72, 82, 92 and 102.
66–103	XRFS performs intermittent analysis of residue from purged third sample.
91	Biology 3 sample collected and delivered to Pyrolytic Release instrument.
115	End of Nominal Mission (15 November 1976) (MY 12, sol 323).
125	Conjunction (25 November 1976) (MY 12, sol 333).

The sampler arm was not used again until sol 177, well after conjunction. An air temperature profile was measured by holding the sampler arm temperature sensor at various heights above the ground. Then, on that and the three following sols, digs were made at a site called Atlantic City, on the edge of Rocky Flats near the earlier samples XRFS 2 and XRFS 3 (Figure 63). This was trench XRFS 4, and the four scoops comprised sample C-5. Coarse material was collected each time and delivered to the instrument, and any fine material was purged near Sponge (Figure 63). The last of these digs was on sol 180, or 21 January 1977 (MY 12, sol 389). The GCMS instrument began to experience problems with its ion pump on sol 190.

Back at Sandy Flats on sol 202 a new task was undertaken, the digging of a very large and deep trench called Deep Hole 1. First, the arm was moved to press the backhoe down on the surface of the drift, a contact called Backhoe Touchdown 1. Six arm retractions one after the other pulled the backhoe through the drift repeatedly to dig a trench, and then the process was repeated with six more retractions. This made a trench much larger than

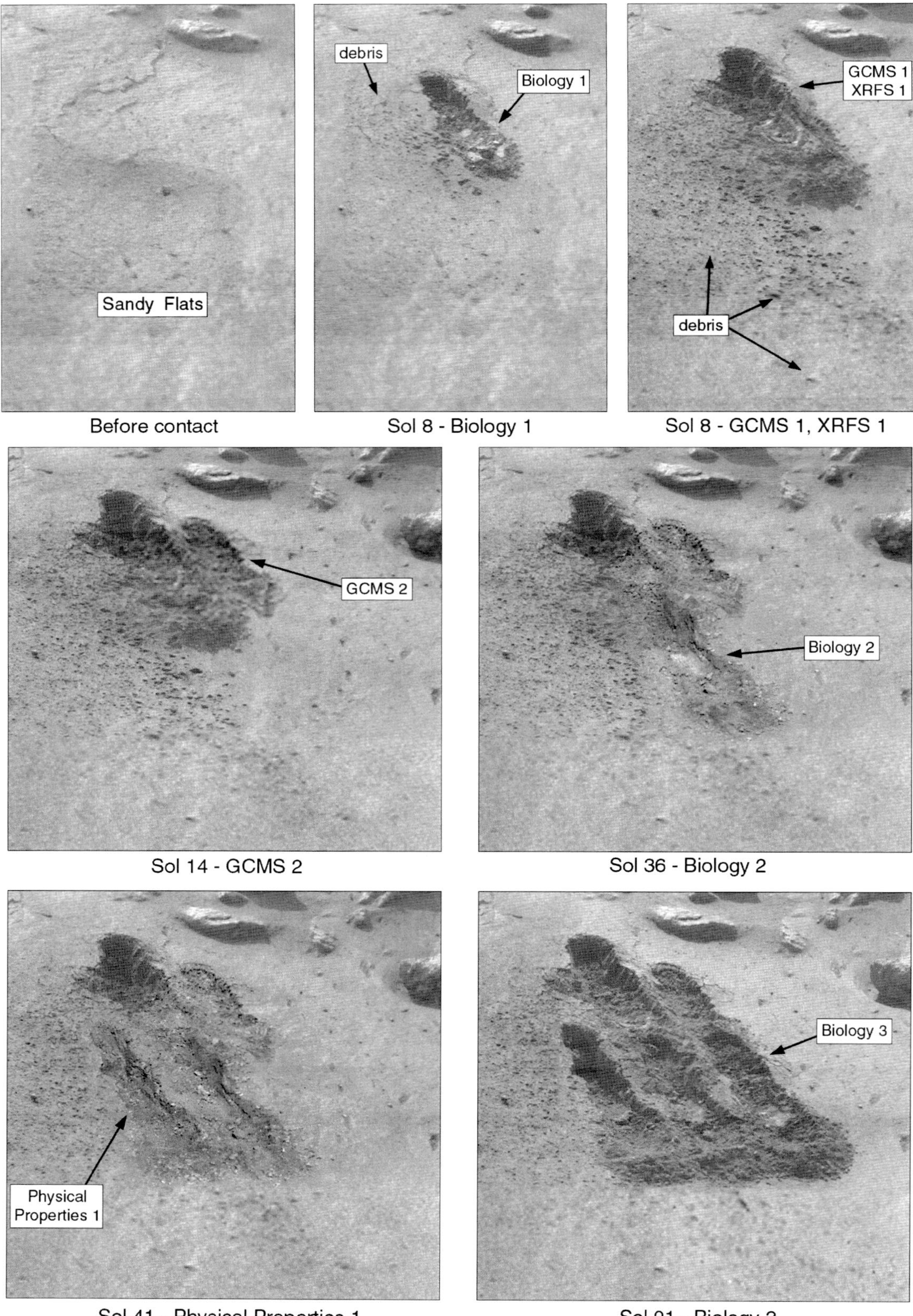

Figure 60 Viking 1 activities in Sandy Flats during the Primary Mission.

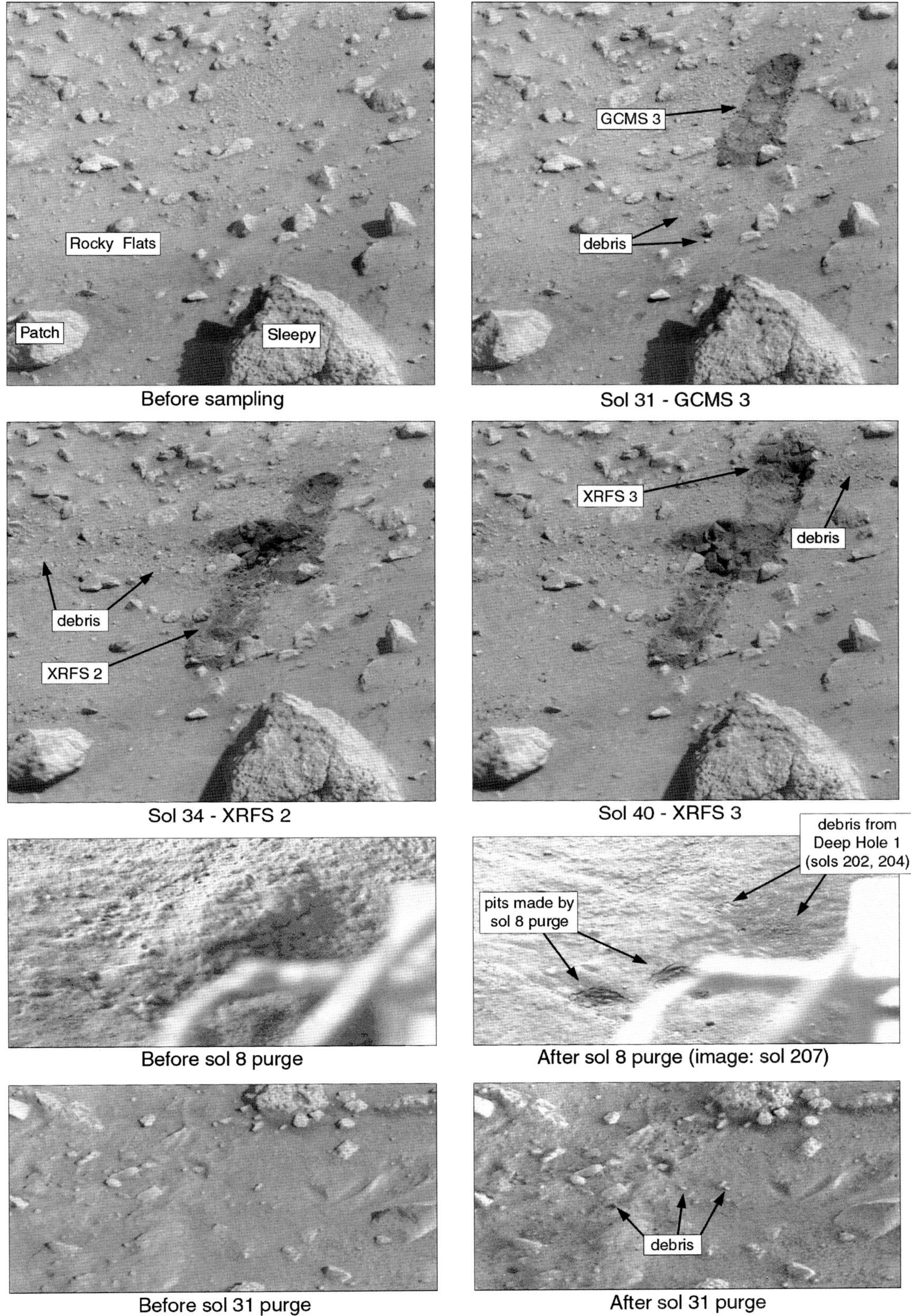

Figure 61 Viking 1 activities in Rocky Flats and at two purge sites during the Primary Mission.

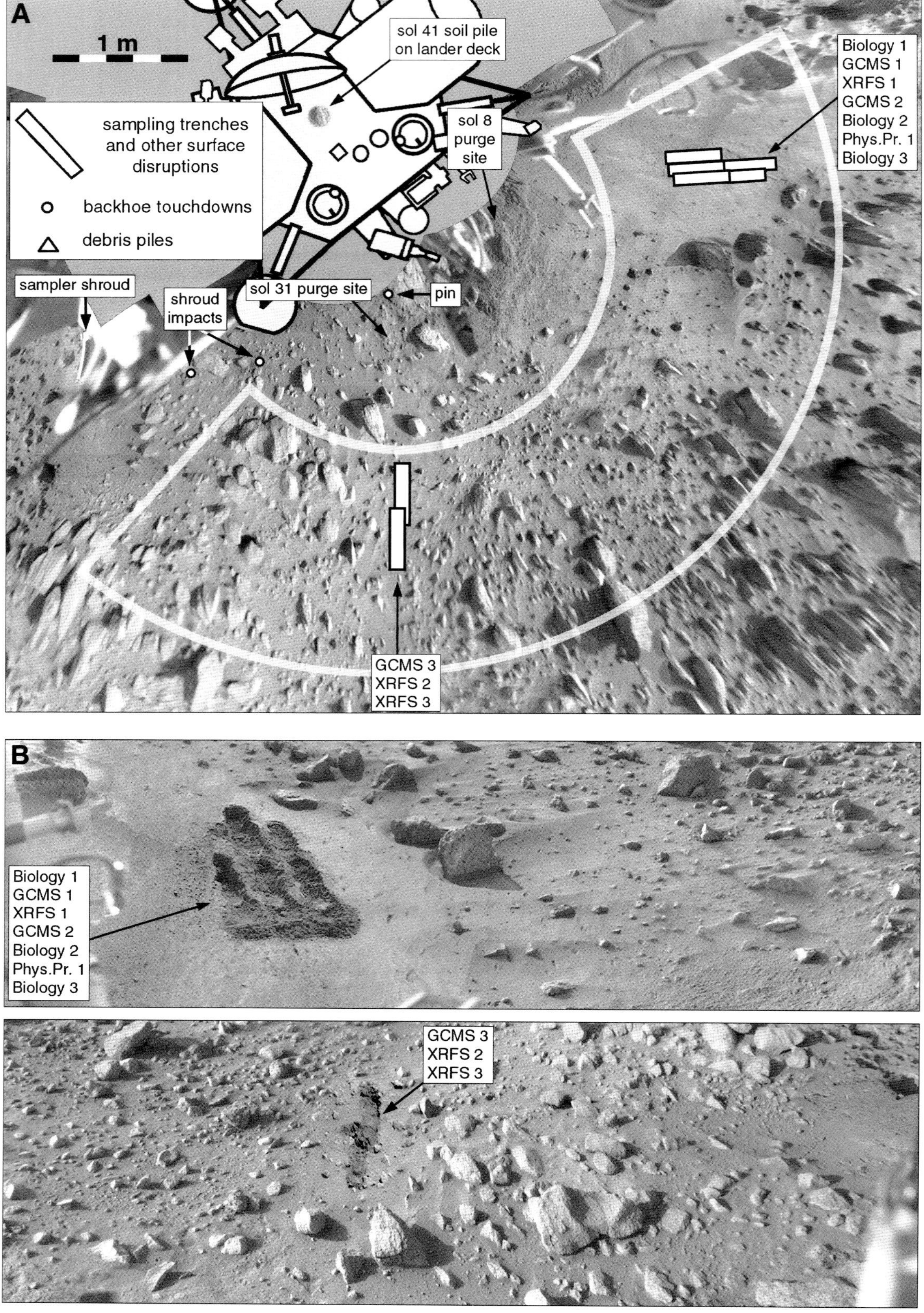

Figure 62 Viking 1 activities up to sol 100, the end of the nominal mission.
A: Map of the sample field. **B:** Panoramic view of the sample field. Figure 58B shows the area before activities began.

any made previously, though it was not well imaged on this sol. Following this digging operation, the arm was moved to place its mirror where it could view the temperature sensor on Footpad 2 late on sol 202, and an image of it was taken on sol 203 (Figure 76). Probably at that time some debris fell from the scoop near the sol 8 purge site (Figure 61). On sol 204 the digging was repeated, again with two sets of six retractions, to enlarge the trench (Figure 64). A 14-cm rock was excavated during this process. The excavation was extended again on sol 218 with two more sets of six retractions closer to the lander, and finally Deep Hole 1 was completed with two more sets of six arm strokes on sol 219 (Figure 64). This big trench was now about 22 cm deep, 30 cm wide and 160 cm long. Meanwhile a dust storm had developed (Pollack *et al.*, 1979), detected first by Viking Orbiter 2 on sol 205, then by Lander 1 on sol 208 and finally on sol 210 by Lander 2 as the storm spread around the planet. These dates are all stated in Lander 1 sols. The storm dissipated slowly over the next 30 sols.

Deep Hole 1 provided access to deeper material, and on sol 229 a single scoop sample, C-6, was collected from the XRFS 5 trench inside Deep Hole 1. It was intended to provide fine material, but there was not enough for a satisfactory analysis. Any coarse fragments were purged into the sol 34 trench (XRFS 2), but images showed no indication that anything had fallen there. On sol 230 the GCMS pump failed, ending that experiment apart from the use of its comminutor for measuring grain hardness. The lander's cameras detected a slump in the drift at the base of Big Joe rock on sol 240 (Figure 74). Another sample of deep material, Biology 4, was taken from Deep Hole 1 on sol 250, but not documented with images. Fine material was delivered to the instrument, and again any coarse fragments were dropped into the XRFS 2 trench, but none were seen in images. Later that same sol, another single-scoop XRFS 6 sample, called C-6 as the previous C-6 was inadequate, was taken from Deep Hole 1. Once again, fine material was delivered to the instrument, and coarse material was dropped in XRFS 2, but none was seen. On sol 252 the Gas Exchange part of the biology instrument failed.

During the northern hemisphere winter on Mars, corresponding to April to September 1977 on Earth, Viking Lander 2 was placed in a low-activity mission phase (Survival Automatic Mission, SAM) to maximize its chances of survival. Viking Lander 1 took advantage of the reduced VL2 operations with a busy work schedule. On sol 270 it tried to perform a magnetic properties experiment. The backhoe was wiped against two brushes to remove dust from its magnets, then pushed into the regolith six times to see what adhered to the magnets. This ground contact, called Backhoe Touchdown 2, was made in Sandy Flats just to the right (south) of Deep Hole 1. The experiment gave poor results because the brushing had not adequately cleaned the magnets, so it was repeated on sol 357.

On three occasions the arm positioned its sampler head to occult the Sun so the camera could observe the brightness of the sky near the solar disk. The first time on sol 280 the arm position was not correct, so another attempt was made on sol 321. Also on sol 280 two more pictures were taken under engine 2 using the mirror on the sample arm to show the region under the spacecraft scoured by engine exhaust during landing, and then the arm was moved to Rocky Flats for a physical properties study. A sample was scooped up from the Physical Properties 2 trench (Figure 63), the fine fraction was purged, and any remaining larger fragments were purged over the drift area adjacent to GCMS 2 to see what effect they had. They were dropped from a height of 85 cm, and fragments up to 1 cm across penetrated less than 0.5 cm. Immediately following that purge, Backhoe Touchdown 3 was commanded in Sandy Flats (Figure 64), erasing signs of the purged material. The purged fragments were only seen in one image of the touchdown event just before retraction.

On sols 285 and 286 sample C-7 was collected from the shallow trench XRFS 7 by skimming the collector along the surface once on each sol. This sample of near-surface material was taken from Wind Tail, a north-south–oriented linear dust deposit in a sheltered area south of Sponge rock (Figure 63), a location Clark *et al.* (1982) called Jonesville. The lowest temperature yet recorded by Lander 1, 177 K, was measured on sol 291, only 15 sols before the winter solstice. Next, on sol 296, the arm made Backhoe Touchdown 4 near Bashful rock in Rocky Flats and then returned to the area it had first worked in at Sandy Flats. It scooped up some drift material at a trench called Physical Properties 3 and dumped it nearby to build a conical pile of soil, called Conical Pile 1 (Figure 63). The pile deposited on the lander on sol 41 had not been given a name, but it and Conical Pile 1 and several others made later would be monitored for wind erosion effects throughout the mission. A second sample was not delivered to the pile. Sol 300 was 24 May 1977 (MY 12, sol 509). Events during the period from conjunction to sol 300 are illustrated in Figures 63 and 64 and mapped in Figure 65. The labelled release and gas exchange experiments ended during this mission phase, whereas the pyrolytic release experiment, though ended, was still used to monitor the radiation background from the RTG power sources. The last labelled release analysis ended on sol 306.

The next operation was an attempt to roll or move a rock on sol 302. The arm touched the rock Grumpy and rolled it away from the lander before sliding over the rock and digging a trench about 30 cm beyond it (Figure 66). On sols 311 and 312, two single scoops collected fine material from Rocky Flats for the XRFS instrument, making up sample C-8 from the XRFS 8 trench. The coarse material left on each day was purged over the sol 34 trench, and some was observed to fall on the left edge of the trench (Figure 66). On sol 312 (MY 12, sol 521), a second major dust storm was observed at this site (Pollack *et al.*, 1979). On the next sol it was also detected by Viking Lander 2. The XRFS 8 purge was followed on sol 324 by the construction of a pair of conical piles. Material from the Physical Properties 4 trench in Sandy Flats was dumped to create Conical Pile 2 on top of Metate 1 rock, at the outer edge of the sample field. Then the process was repeated with material dug from the Physical Properties 5 trench in Sandy Flats (Figure 67) and used to create Conical Pile 3 on a patch of drift material between rocks (Figure 66). The name Metate, used for two rocks at this site, refers to hollowed stones traditionally used for grinding grains in the Americas and elsewhere. These rocks both had hollow tops. Metate 2 somewhat resembled a clog (Figure 68) and so was also referred to as Dutch Shoe (Olivier de Goursac, personal communication, 9 September 2011).

The measurement of sky brightness near the sun, first attempted on sol 280, was repeated on sol 321. This time the levels of atmospheric dust during a major storm were so high that the images were useless, so a third attempt was scheduled for sol 376. Colour images of the sky during the dust storm were taken on sol 325, and 60 images of the sky were obtained for a photometric study on sol 328. Atmospheric studies like these occurred throughout the mission, in addition to routine monitoring of the solar disk on almost every active sol.

Several types of material were present at the Viking 1 site, most obviously the fine-grained drifts of Sandy Flats and the coarser fragments found in Rocky Flats. A third type of material visible in images and sometimes disturbed by the arm operations appeared as thin platy sheets, fragile enough to break during digging but coherent enough to hold together when disturbed, or to resist wind erosion. These were interpreted as fine-grained drift-like material, cemented by some process which might include the precipitation of salts from drying brines. By analogy with terrestrial desert materials, these were called 'duricrust' in mission reports. On sols 336, 337 and 338, attempts were made to sample duricrust from a trench called XRFS 9 at the far right-hand end of the arm workspace. The arm skimmed the surface three times, once each day, each stroke a little farther to the right (Figure 68). Fine materials were purged, leaving any clods or lumps of duricrust to be delivered to the XRFS instrument, but none were delivered. It appears that they disintegrated during collection or delivery.

Soil temperature measurements were obtained on sols 343 and 344 as the arm was pushed into drift material in the Physical Properties 6 trench at Sandy Flats on sol 343 (Figure 67) and left with the temperature probe buried in the drift for a full day-night cycle. Temperatures varied from 187.5 K early in the morning to 215.5 K in the mid afternoon. The arm may have struck a buried rock as it moved into place. Later on sol 344, material picked up in the scoop from Physical Properties 6 was used to build Conical Pile 4 near Ventifact rock (Figures 67, 74). Physical studies continued on sol 350 at the trench Physical Properties 7, just to the left of the old XRFS 2 trench in Rocky Flats (Figure 66). The intention was to collect coarse material from Rocky Flats and grind it in the GCMS comminutor. This would give an estimate of grain hardness, but it appeared that insufficient material was delivered to the comminutor. The residue in the scoop was dropped from a height of 82 cm over the drift area, but no fragments or surface pits made by fragments were seen.

The magnetic experiment attempted on sol 270 was repeated about 90 sols later on sol 357, this time using three brushing strokes to clean the magnets, but the

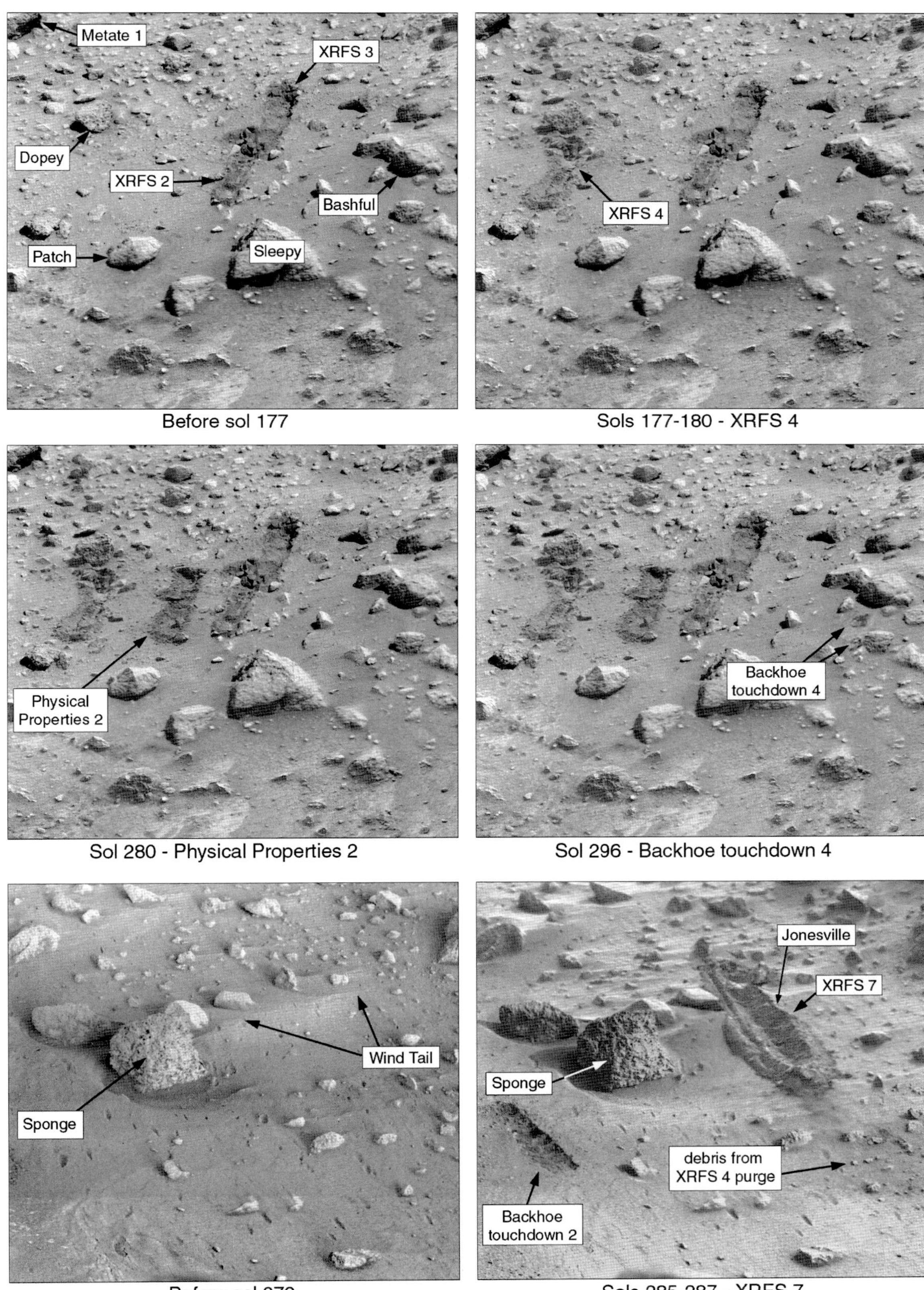

Figure 63 Viking 1 activities in Rocky Flats between sols 101 and 300.

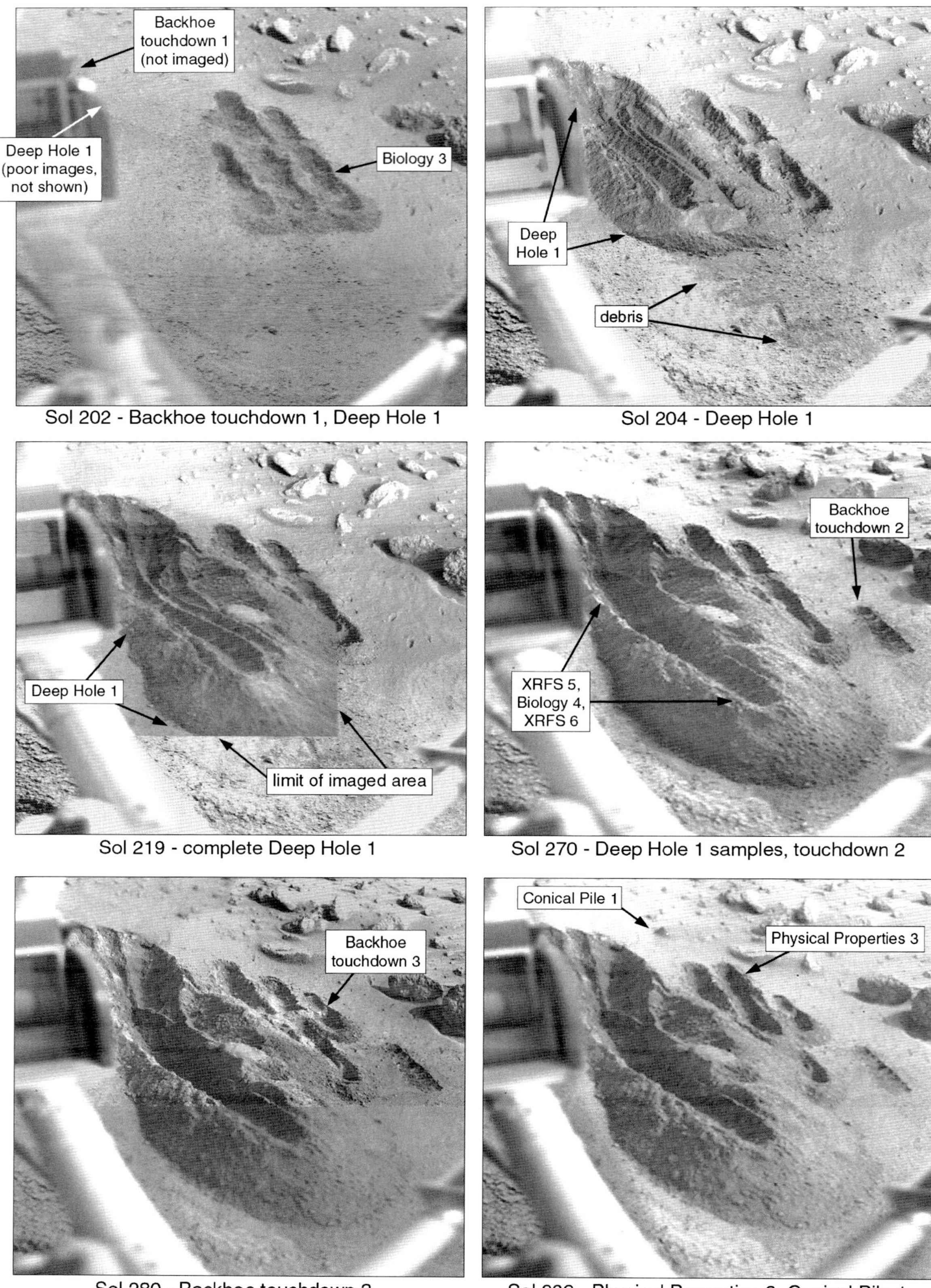

Figure 64 Viking 1 activities in Sandy Flats between sols 101 and 300.

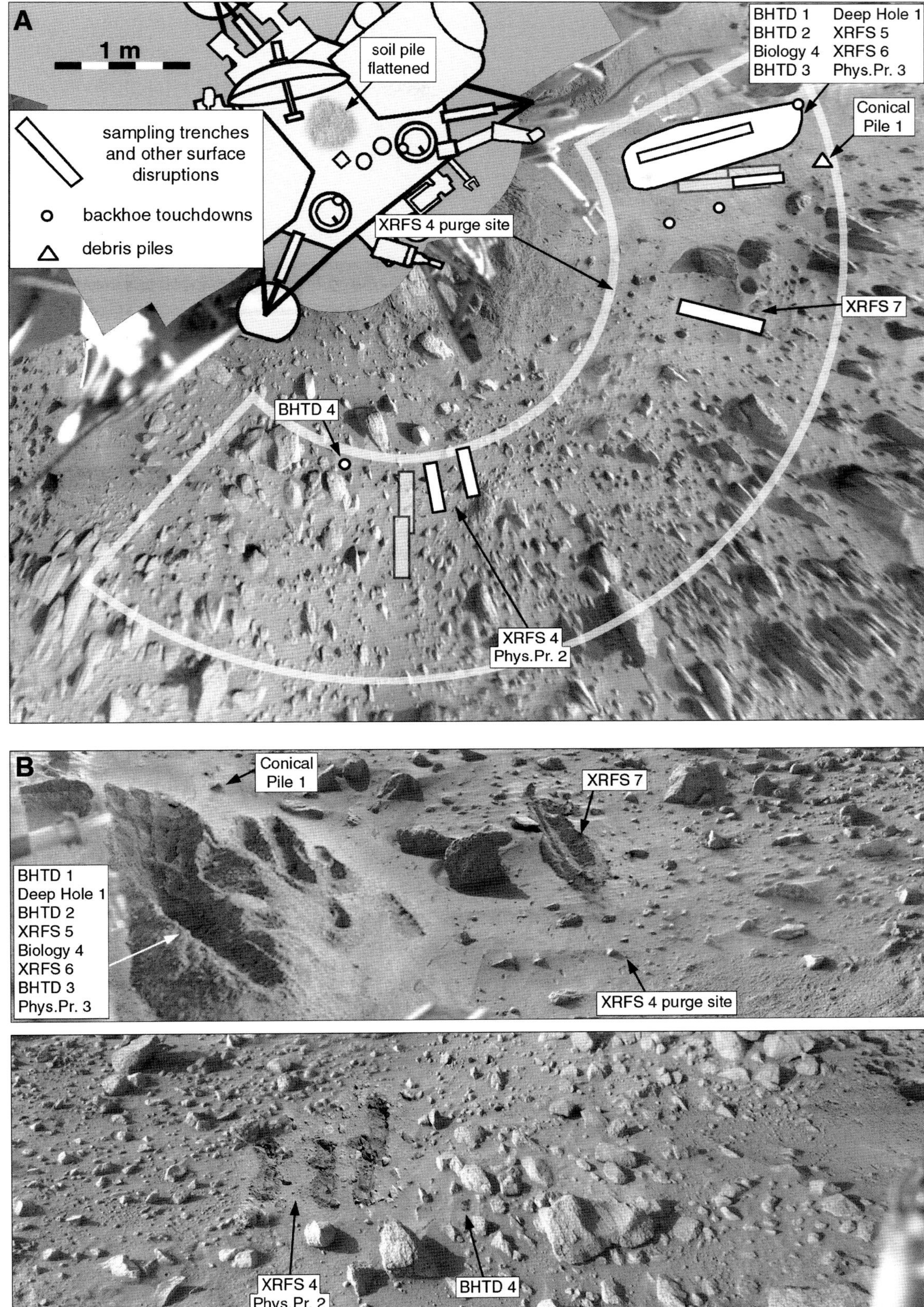

Figure 65 Viking 1 trenching activities after 300 sols.
A: Map of the sample field. **B:** Panoramic view of the sample field. Only events occurring between sols 101 and 300 are labelled.

backhoe snagged on the brush, stalling the process. About 150 sols later on sol 520, the procedure was successfully completed. On sol 369 the arm was used to roll Bashful rock. It passed over the rock, making a small trench beyond Bashful, and rolled the rock towards the lander as it was retracted (Figure 68). The movement, which was partly towards the left, exposed a mound of soil previously hidden on the right side of the rock. Three attempts to strike and chip the rock Metate 2 (Dutch Shoe) were made on sol 376 to test its hardness. No damage to the rock was observed, but the contact rotated the elongated rock counterclockwise (Figure 68). Also on that sol, the sky brightness experiment started on sol 280 was successfully concluded. Sampling resumed on sol 378 with single-scoop sample C-9 from the XRFS 10 trench near Bashful rock. The rock was moved as the arm retracted. The intention was to collect a sample from underneath the rock's original location, but the trench was in the wrong place, behind the rock's original position. The coarse fraction was purged, but nothing was seen. Sol 400 was on 4 September 1977, or MY 12, sol 609.

Phobos transited the Sun's disk, as seen from the landing site, on sols 415, 419 and 423. The lander measured the drop in light levels during these partial eclipses. The timing of the eclipses would help locate the lander on the surface. The last event coincided with the passage of a local dust storm over the landing site, but no obvious surface changes were observed. On sols 430 and 431, another attempt was made to sample duricrust with four scoops from the far right end of the sampling area, an area Clark *et al.* (1982) referred to as Angie's Place. This was XRFS 11 (Figure 68), and the coarse fraction of the sample was delivered to XRFS as intended. The fine fraction was purged near Footpad 3. On sol 441 an attempt was made to scrape another rock, this time Sponge rock, but it rolled towards the lander without showing any sign of damage. All rocks tested in this way were found to be hard, not chipped or scratched by the arm. Events during the period from sol 301 to 450 are illustrated in Figures 66 through 68 and mapped in Figure 69.

A new deep excavation in Rocky Flats was commenced on sol 456. This was Deep Hole 2, and the first stage in its construction involved four sets of six backhoe retractions, resulting in a trench 8 cm deep, 32 cm wide and 92 cm long (Figure 70). Then on sol 468, a single-scoop sample (C-11) was dug from trench XRFS 12 within Deep Hole 2 and delivered to the XRFS instrument. The trench was extended again on sol 486 with four more sets of six backhoe retractions. The new trench was 11 cm deep at its deepest point, 30 cm wide and 110 cm long. This exposed some blocky material, and an attempt to collect some was made later on the same sol. The trench created within the larger hole was XRFS 13, but no sample from it was delivered to the instrument. A repeat on sol 502 at XRFS 14, essentially the same place as XRFS 13 (Figure 70), was successful in collecting two scoops of blocky material (sample C-12) and delivering it to the XRFS. Sol 500, just prior to XRFS 14, was on 16 December 1977 (MY 13, sol 40).

On sol 520 the arm brushed its backhoe magnets clean and then touched the surface at the Backhoe Touchdown 5 location near the edge of Sandy Flats (Figure 71). The magnets were inspected to see whether anything adhered to them, finally completing the magnet experiment attempted on sol 270. Then a mirror on the arm was used to obtain a picture of the Footpad 2 temperature sensor, revealing that it was buried in the drift material (Figure 76C). Earlier images had not shown the sensor itself. A week later on sol 527 a sample was collected from XRFS 15, a location originally under Bashful rock (Figure 72). If anything was picked up, it was not delivered to the instrument. Bashful and two other rocks were moved in the process. Then the arm attempted to perform Backhoe Touchdown 6, but the collector head struck a rock or other surface feature, and the backhoe itself did not actually reach the ground. At the end of sol 527, the arm was left in a position where its mirror would permit imaging under the lander on sol 528, showing an area under the engine which had been scoured by exhaust. Four more images like this were taken on sol 550, using small arm movements to reposition the mirror so the images would form a mosaic of the otherwise hidden area (Figure 76). They showed the scoured area and a pit dug by the exhaust during landing. Later on that sol, the arm performed Backhoe Touchdown 7 in the Deep Hole 2 tailings (Figure 71).

Operations continued in the Deep Hole 2 tailings on sols 558 to 561, four sols in which 16 arm strokes, four per sol, were used to dig a trench, XRFS 16 (Figure 70). The intention was to collect a sample, C-13, by purging fine material each time and accumulating coarse material

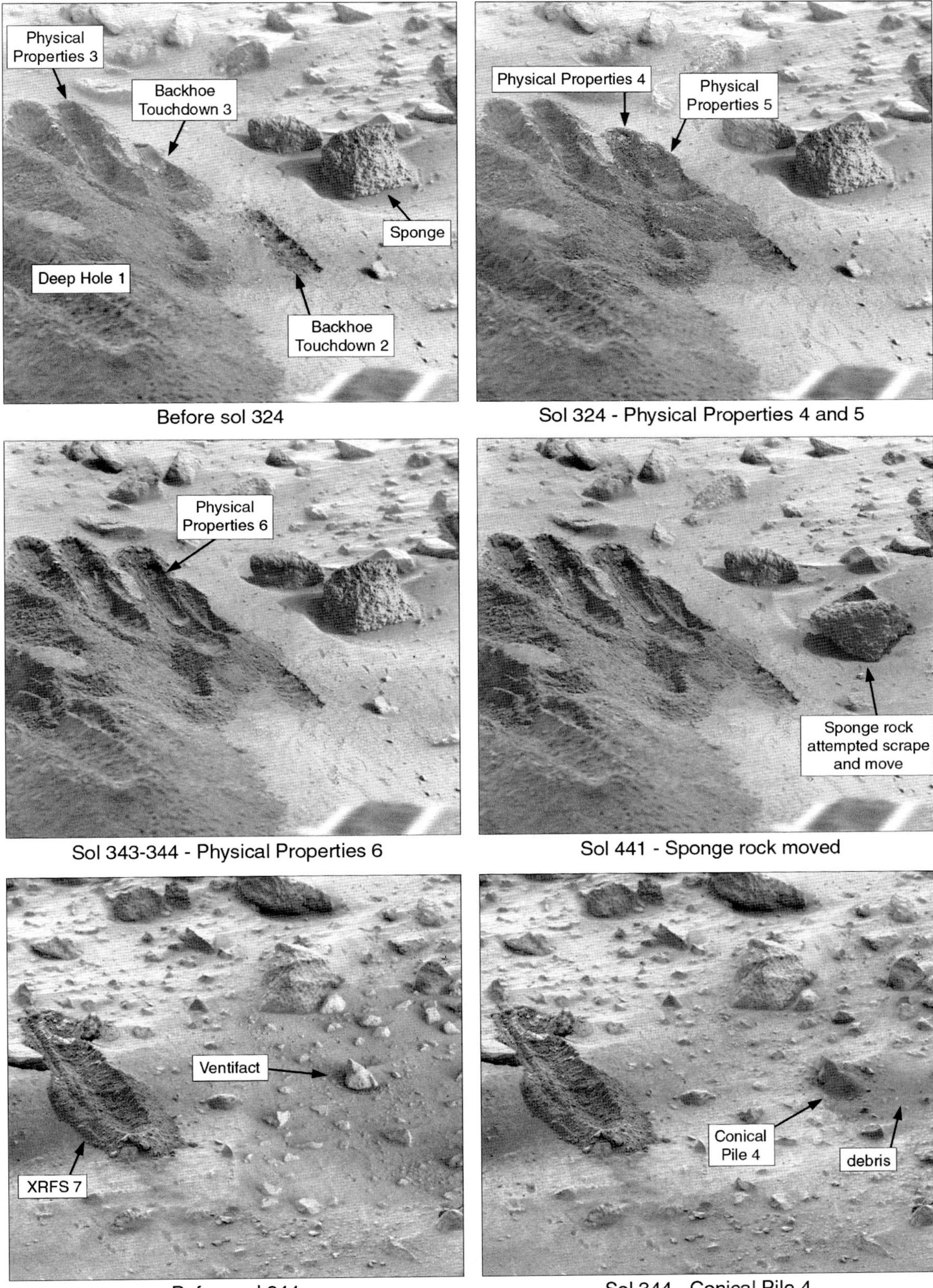

Figure 66 Viking 1 activities in Sandy Flats between sols 301 and 450.

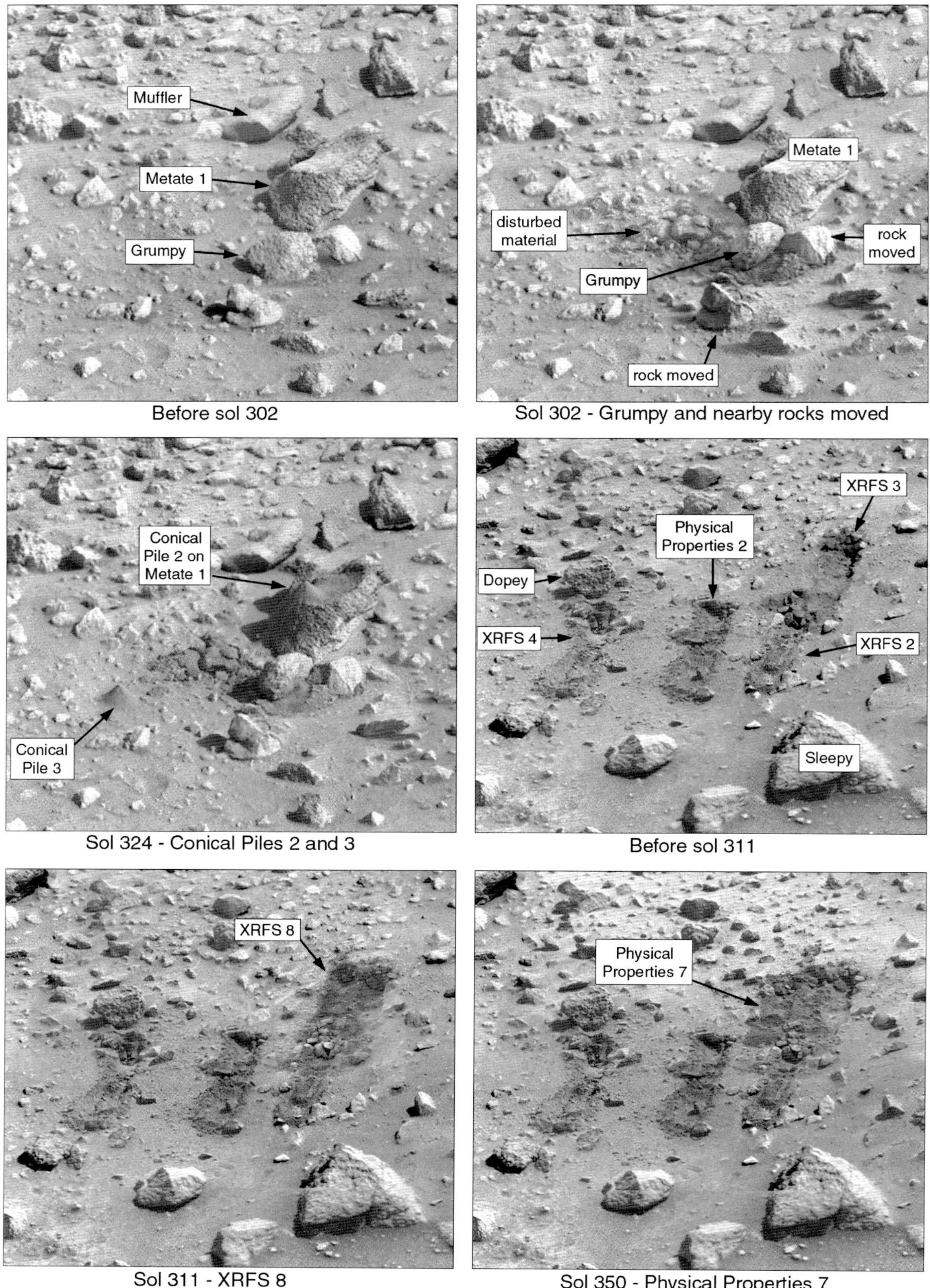

Figure 67 Viking 1 activities in Rocky Flats between sols 301 and 450.

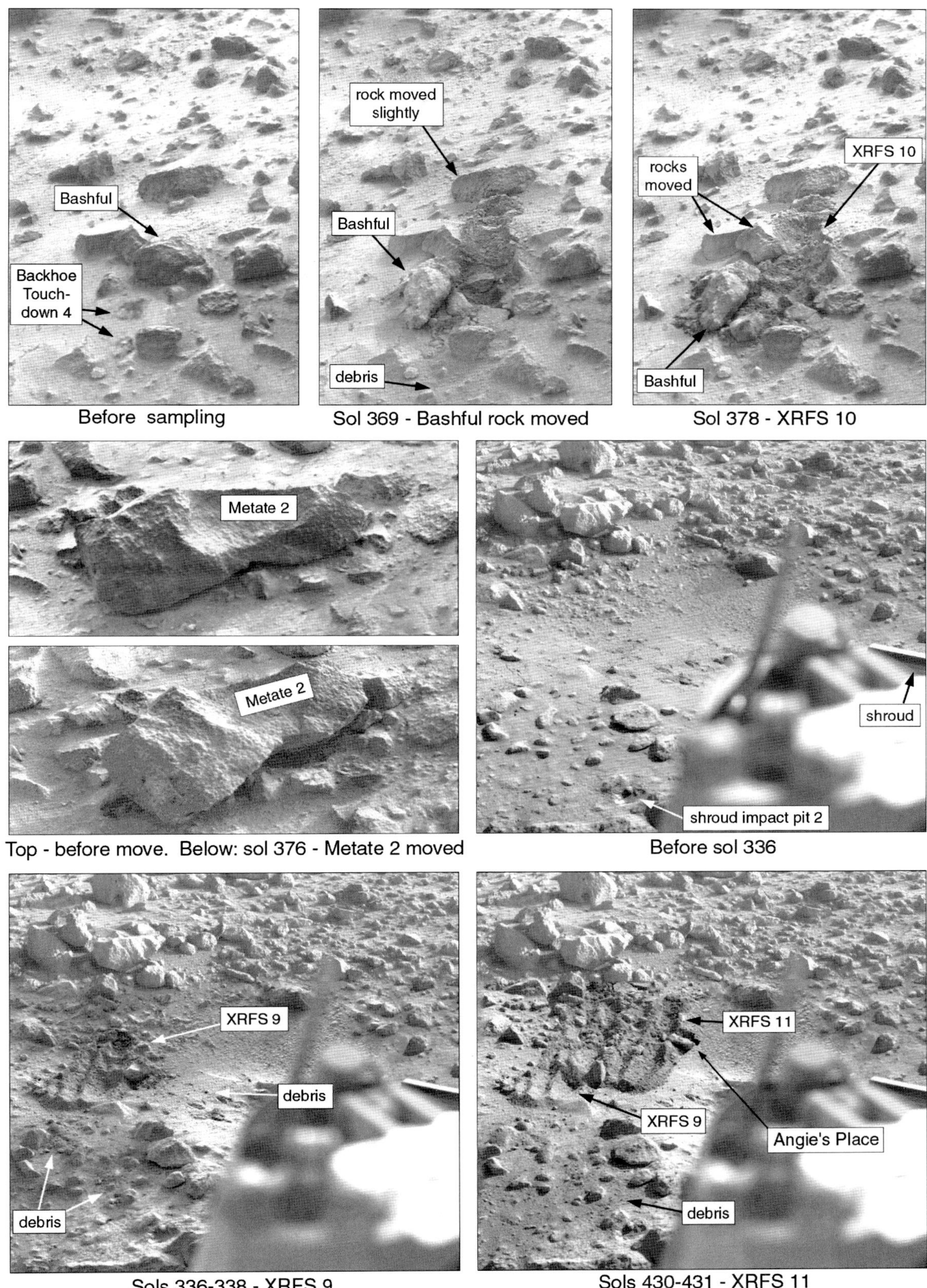

Before sampling

Sol 369 - Bashful rock moved

Sol 378 - XRFS 10

Top - before move. Below: sol 376 - Metate 2 moved

Before sol 336

Sols 336-338 - XRFS 9

Sols 430-431 - XRFS 11

Figure 68 Viking 1 activities in Rocky Flats between sols 301 and 450 (continued).
The rock Metate 2 was also called Dutch Shoe.

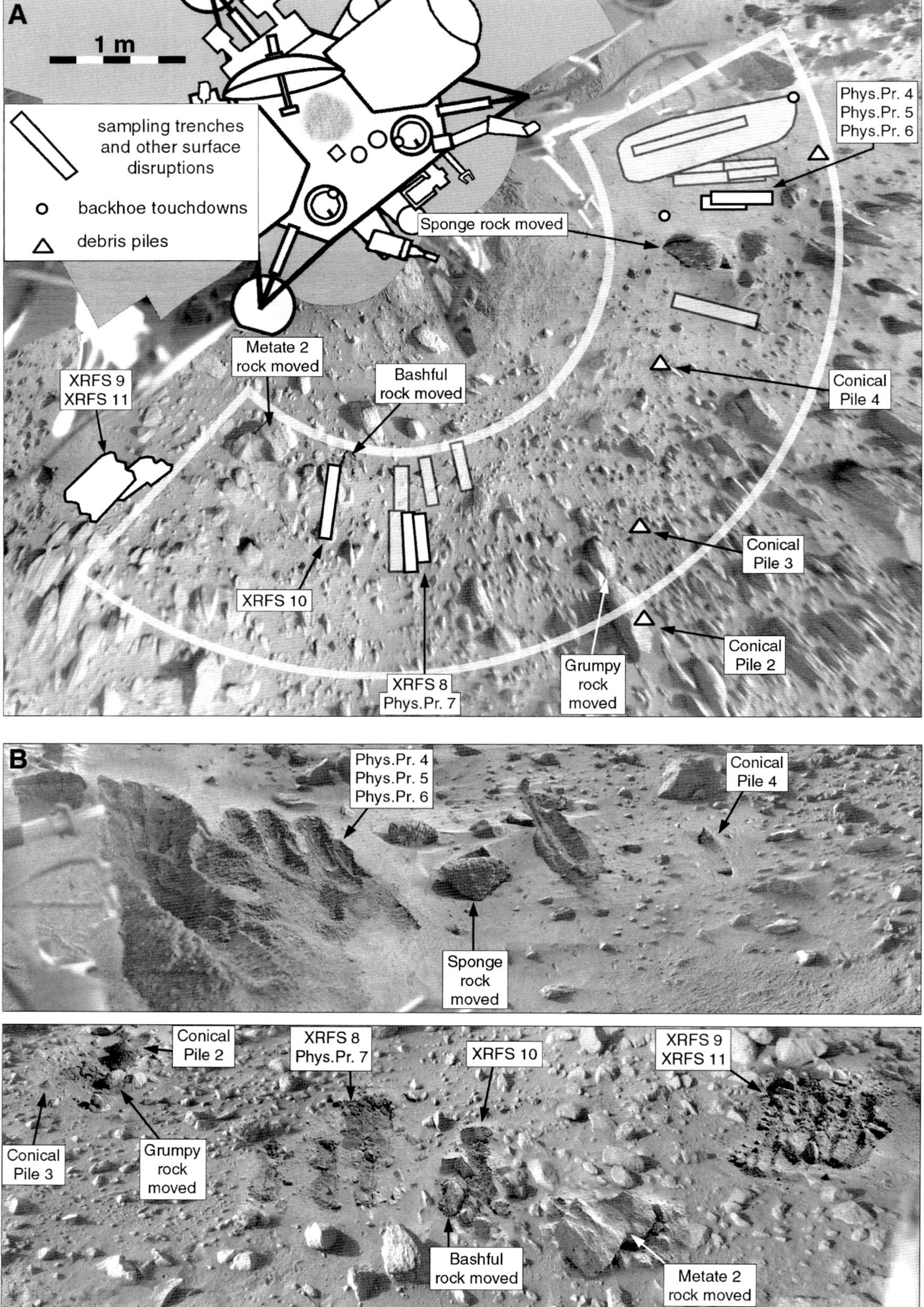

Figure 69 Viking 1 trenching activities after 450 sols.
A: Map of the sample field. **B:** Panoramic view of the sample field. Only events occurring between sols 301 and 450 are labelled.

for XRFS. Nothing was successfully delivered, presumably because there was little or no coarse component to the tailings. At the same time, the final attempts to free the seismometer were made, still without success. On sol 581 the process was repeated, this time with all 16 arm strokes in one sol (Figure 70, XRFS 17). The arm probably dug through the tailings into the underlying original surface, so samples would be a mixture of both types of material. Fine materials were purged and dispersed, but enough coarse fragments were collected to fill the XRFS chamber, forming sample C-13 (because the previous attempt at C-13 failed). At the end of sol 581, the arm was positioned for more imaging under the lander, and one image was taken on sol 582.

Another attempt was made on sol 586 to collect samples from the Deep Hole 2 tailings area (XRFS 18, Figure 70), this time digging deeper to gather material from the original surface under the tailings. Again 16 strokes were made, and fines were dispersed each time, but nothing could be delivered to XRFS as the sample delivery funnel was blocked by large fragments from XRFS 17. This sequence was in fact an accidental repeat of the sol 581 sequence. Afterwards, the arm mirror was positioned to take another image under the lander. The picture was taken on sol 594, becoming the last frame in the mosaic underneath Engine 2 (Figure 76). Sol 600 was on 28 March 1977 (MY 13, sol 140).

On sol 608 a sample was dug from Rocky Flats at a site called Physical Properties 8 (Figure 72) and delivered to GCMS for comminution. Physical Properties 9, on sol 612, was a bearing test in Sandy Flats (Figure 71). The arm's collector head was pushed down into the drift material to observe the degree of surface deformation with a known load. Also on that sol, Physical Properties 10 repeated that process at a site in Rocky Flats (Figure 72), and the temperature sensor on the arm was raised to different levels to construct a vertical profile of air temperatures. Sampler arm operations concluded on sol 639, beginning with two surface contacts to measure soil properties, Backhoe Touchdowns 8 and 9 in Rocky Flats. Then a sample was dug up at Physical Properties 11 in Rocky Flats and used to create Conical Pile 5, about 3 cm high, between two rocks (Figure 71). After that the arm was parked and never used again.

Events during the period after sol 450 are illustrated in Figures 70 through 72 and mapped in Figure 73.

After the end of arm operations, the lander continued monitoring weather and looking for changes in the surroundings. The first such change had already occurred, and it encouraged the search for additional changes. On sol 183, an image of a sloping drift surface near Big Joe (Figure 74) revealed a slump scar that had not been present on sol 74 when the area had last been observed. This drift and others were observed repeatedly in the hope of seeing another event like this, but only one was ever seen. It occurred between sols 767 and 771 near the large rock Whale (Figure 74). Both events appeared to involve a thin surface crust breaking and sliding down the sloping surface. Many areas show other small changes as thin layers of fine material were removed or deposited by the wind (Jones *et al.*, 1979; Guinness *et al.*, 1982; Arvidson *et al.*, 1983).

Several conical soil piles were created so that wind erosion could be monitored during the mission, and several of them were observed to change (Arvidson *et al.*, 1983). Conical piles 2, 3 and 5 had changed very little, if at all, by sol 1601 but were largely or completely removed by sol 2068. Figure 75 shows these changes, incorporating a higher resolution image from sol 2209 which covers only piles 3 and 5. Conical Pile 4 was constructed on sol 344 and changed little until sol 1765, when part of the pile was seen to have been stripped away by the wind to expose part of the underlying rock (Figure 74). Other changes occurred on the mound of soil at the edge of Deep Hole created by the XRFS 18 excavation (Figure 72). Sometime between sols 1313 and 1757, two soil clods or small stones were dislodged from the mound, one falling into XRFS 18 before sol 1720, the other rolling down onto a rock beside the Physical Properties 11 trench after sol 1720. If all those changes occurred at about the same time, the events may have coincided with a period of atmospheric pressure fluctuations and dusty skies in the mid 1700s range of sols (Arvidson *et al.*, 1983). Another soil pile, not given a name or number, was placed on a gridded surface on top of the lander on sol 41 and observed repeatedly (Figure 75). It changed fairly rapidly, collapsing into a broad low patch of soil between sols 76 and 207, perhaps aided by vibrations caused by spacecraft operations. After that time, wind made small changes to its distribution. Similar small changes, not illustrated here, were observed in soil thrown onto a footpad during landing (Jones *et al.*, 1979).

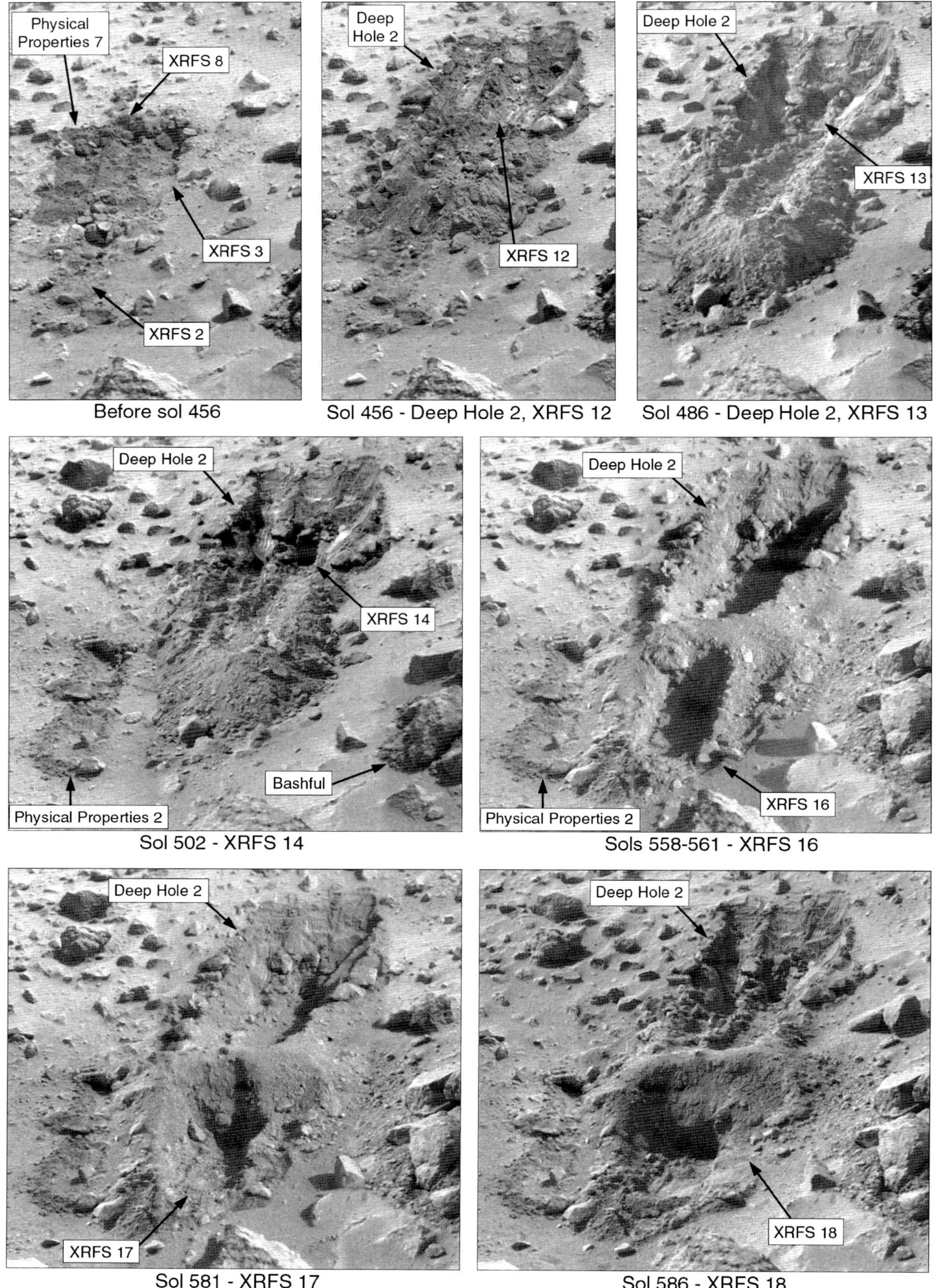

Figure 70 Viking 1 activities at Deep Hole 2 between sol 451 and the end of sampler operations.

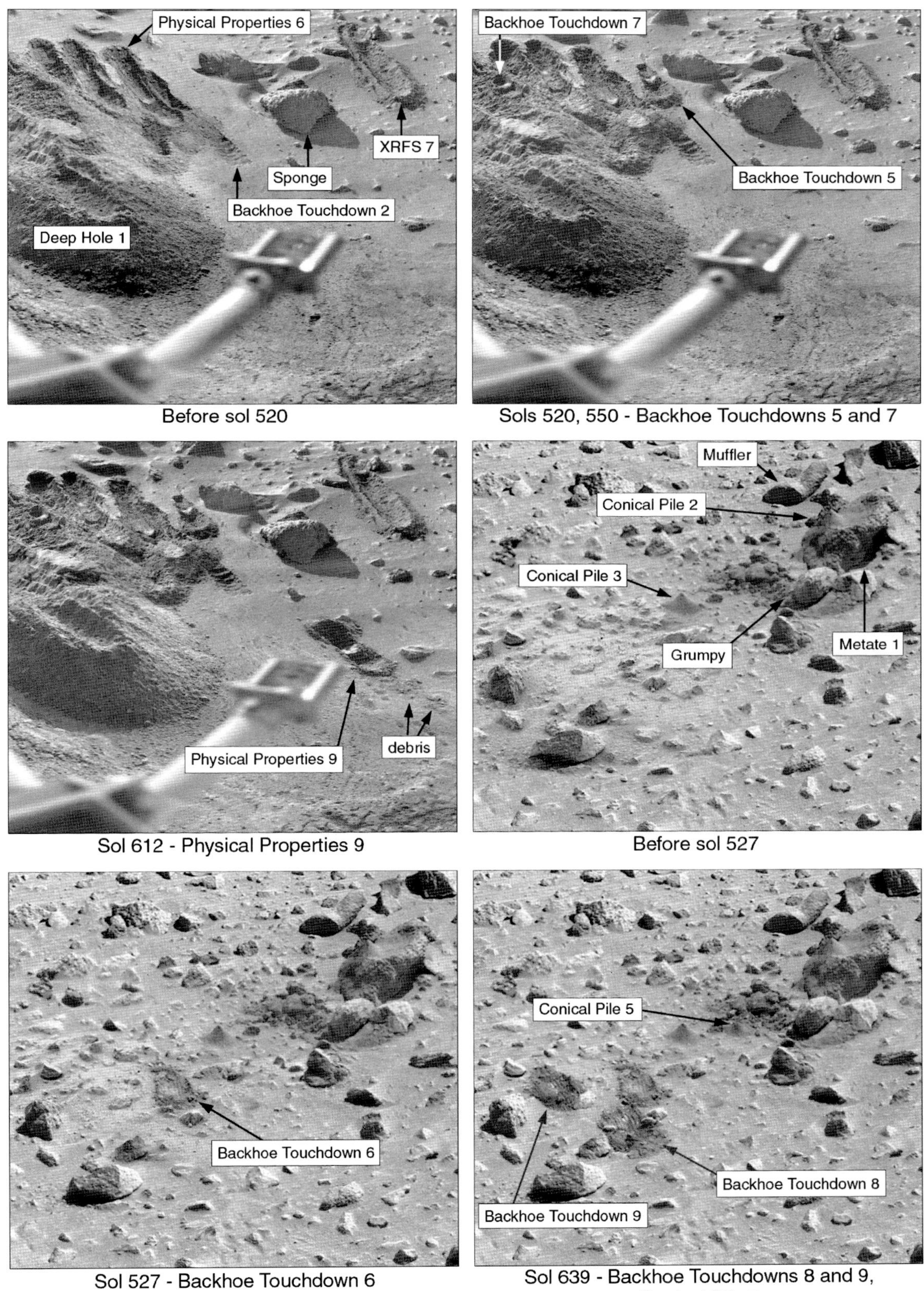

Figure 71 Viking 1 activities between sol 451 and the end of sampler operations.

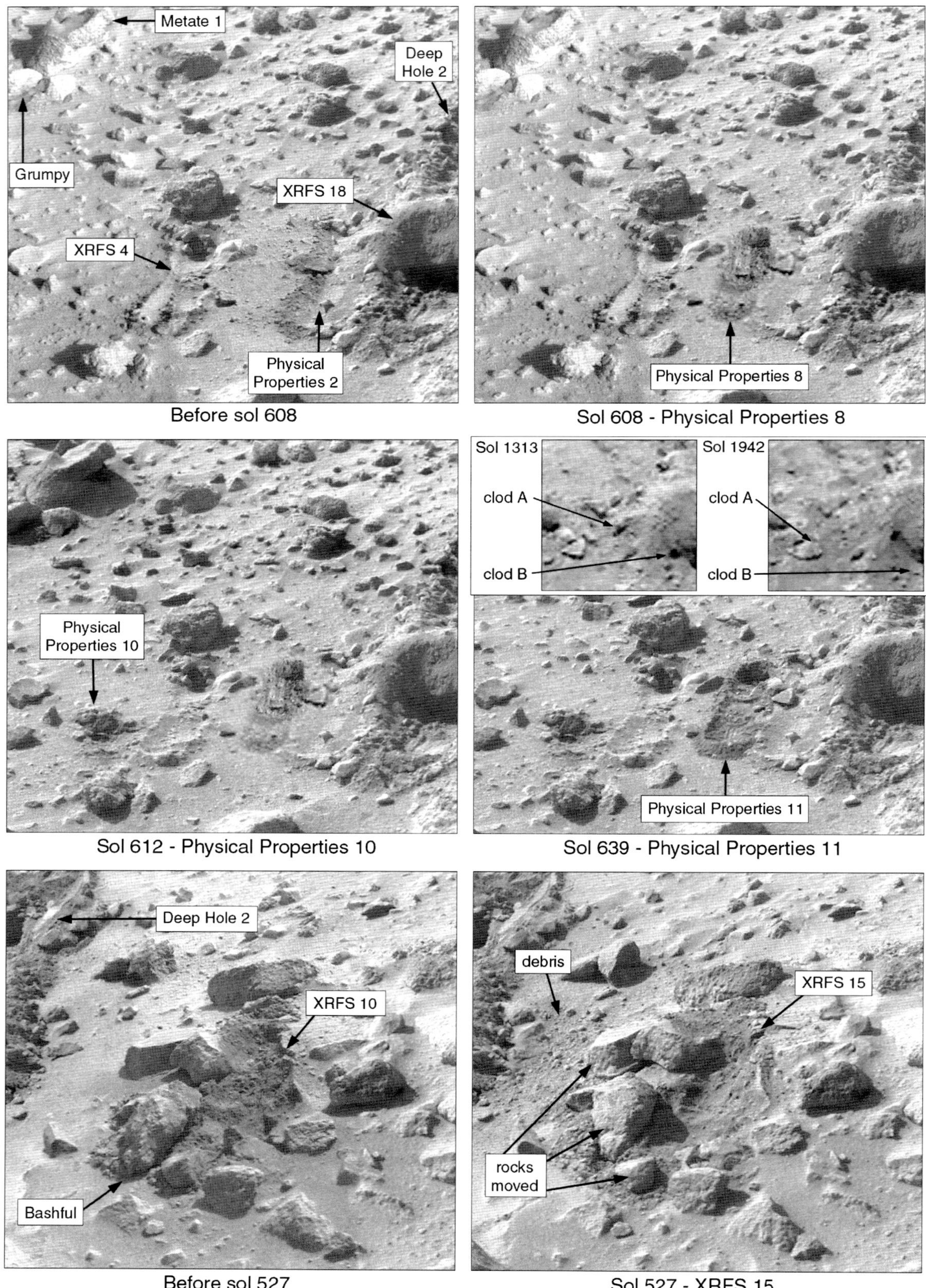

Figure 72 Viking 1 activities in Rocky Flats between sol 451 and the end of sampler operations. Surface changes later in the mission are shown in an inset with the sol 639 image.

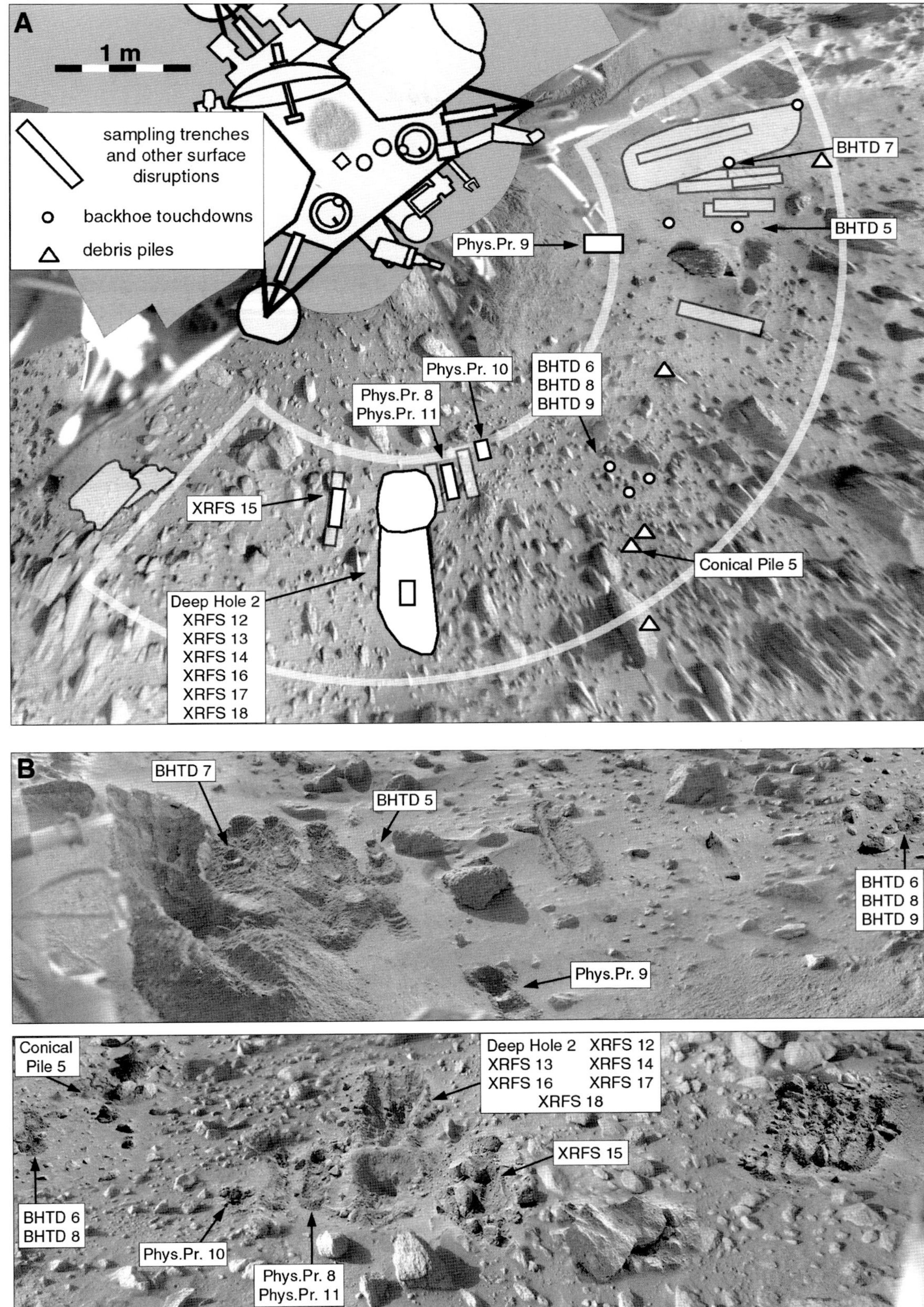

Figure 73 Viking 1 trenching activities up to the end of Surface Sampler operations.
A: Map of the sample field. **B:** Panoramic view of the sample field. Only events occurring after sol 450 are labelled.

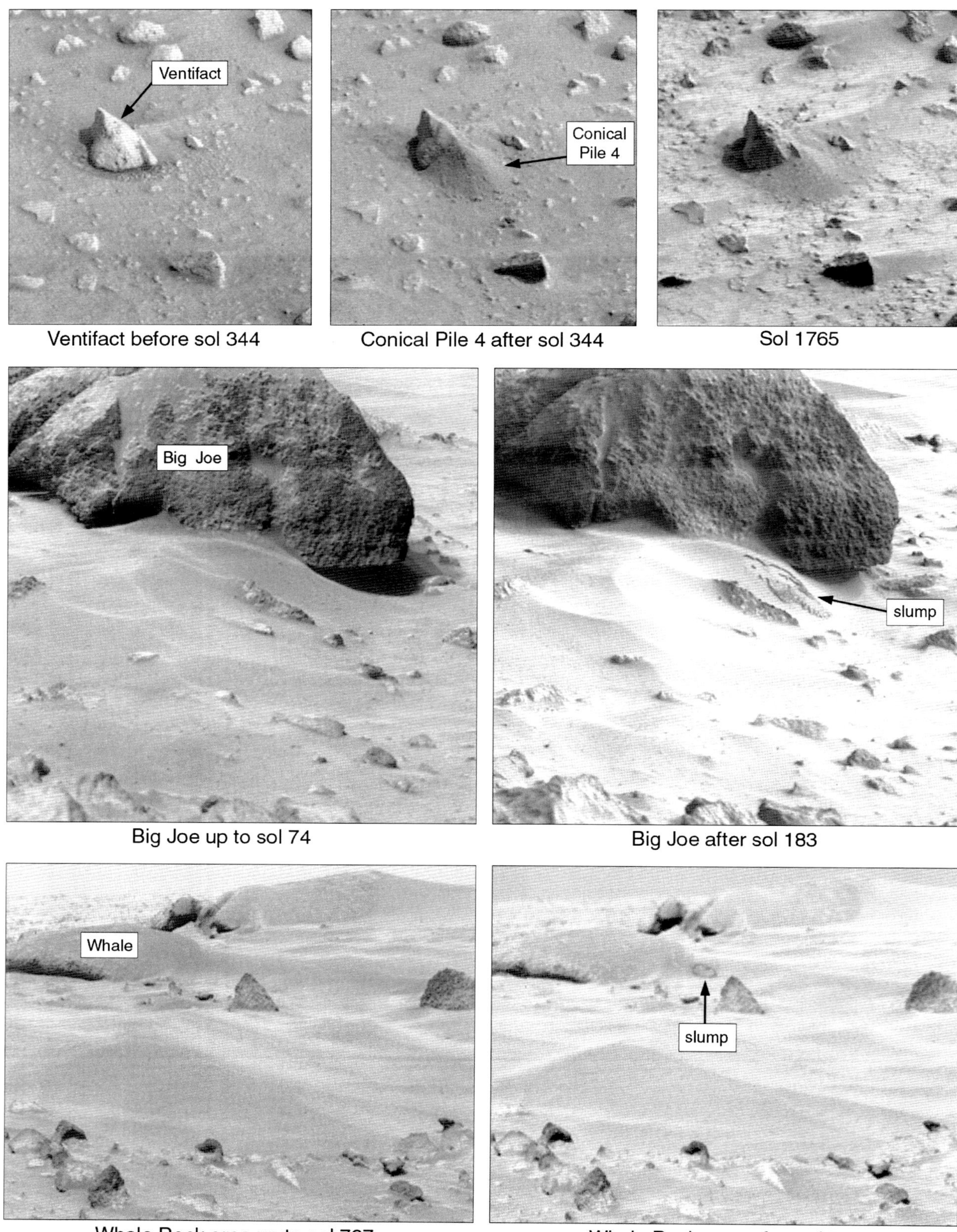

Figure 74 Changes observed at the Viking 1 landing site.

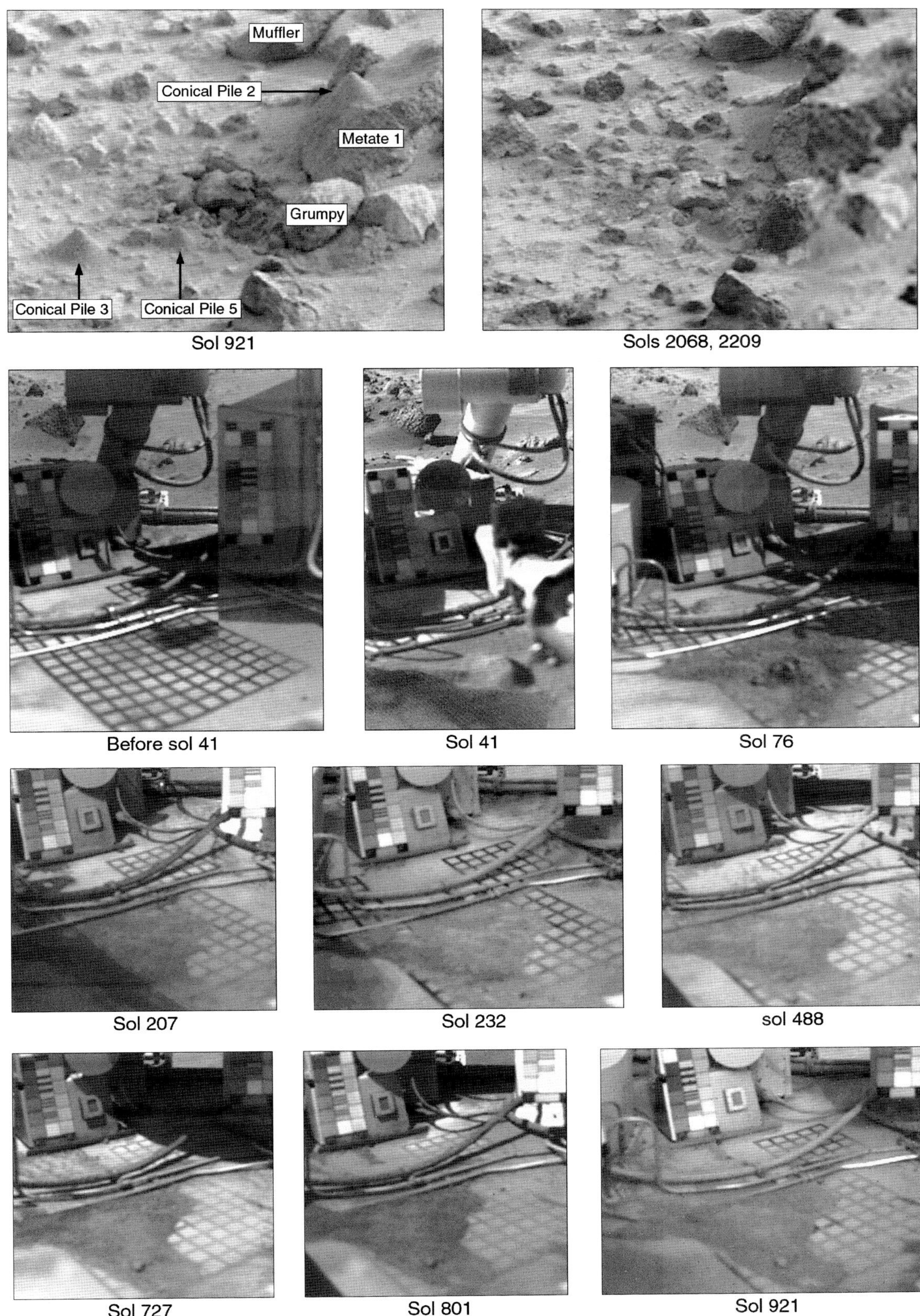

Figure 75 Changes observed near Conical Piles 2, 3 and 5 (top) and on the Viking Lander 1 spacecraft deck. Colour scale targets and the base of the high-gain antenna are visible beyond the gridded deck area.

Table 21. *Viking Lander 1 Extended Mission Activities*

Sol	Activities
146	First commands of the extended mission received (MY 12, sol 355)
177–180	XRFS 4 samples collected, coarse regolith fraction, four strokes
202	Backhoe Touchdown 1, in drift materials for surface properties study
204–219	Deep Hole 1 excavation in drift materials, dug on sols 202, 204, 218, 219
209	First evidence of dust storm observed by lander (MY 12, sol 418)
229	XRFS 5 sample collected from Deep Hole 1, purged to form rock pile
240	Slump in drift observed in drift near Big Joe rock, the first natural surface change
250	Biology 4, XRFS 6 samples collected, both from Deep Hole 1
252	Gas Exchange helium supply depleted, ending operations
270	Backhoe Touchdown 2 in drift materials for surface properties study
280	Physical Properties 2 sample dug and purged, backhoe touchdown 3 on drift
285–286	XRFS 7 sample collected from wind tail of Sponge rock
290	Lowest winter temperature recorded, 177 K (MY 12, sol 499)
296	Backhoe Touchdown 4, physical properties 3, purged to form conical pile 1
302	Grumpy rock rolled successfully
306	Labelled Release experiment ends (MY 12, sol 515)
311–312	XRFS 8 sample from blocky material, purged to form rock pile, second major dust storm seen
321	Highest wind speed measured, with maximum velocity 27.5 m/s
324	Physical Properties 4, drift material, used to make conical pile 2 on Metate 1 rock
	Physical Properties 5, drift material, used to make conical pile 3 near Metate 1
336–338	XRFS 9 sample attempt, three sample strokes to collect duricrust, not successful
343–344	Physical Properties 6, diurnal temperatures measured in drift material
344	Conical Pile 4 constructed near Ventifact rock (MY 12, sol 553)
350	Physical Properties 7 to comminutor, coarse material purged over drift but nothing seen
357	Sampler arm problem, 'no-go' during magnet cleaning prevents sample collections
369	Bashful rock pulled towards lander by sampler
376	Metate 2 rock pushed in attempt to chip it, no damage to rock observed (MY 12, sol 583)
378	XRFS 10 sample collected from behind small rock
419	Phobos shadow transit observed by orbiter and lander and by lander alone on sols 416, 423
430, 431	XRFS 11 sample collected from duricrust, four sample strokes
441	Sampler used to scrape Sponge rock as a test of hardness, rock moved but not damaged
456	Deep Hole 2 excavated in blocky material, XRFS 12 sample attempt fails (MY 12, sol 665)
486	Deep Hole 2 enlarged, XRFS 13 sample collected
502	XRFS 14 sample collected from Deep Hole 2, deeper than XRFS 13 (MY 13, sol 42)
520	Backhoe Touchdown 5 on drift material
527	XRFS 15 sample collected from behind small rock, backhoe touchdown 6 on blocky material
550	Backhoe Touchdown 7 on Deep Hole 1 tailings
558–561	XRFS 16 samples collected from Deep Hole 2 tailings, 16 strokes, delivery not successful
581	XRFS 17 samples from Deep Hole 2 tailings and surface below, 16 strokes, successful
586	XRFS 18 samples collected from surface below Deep Hole 2 tailings, 16 strokes, successful
608	Physical Properties 8 sample taken for comminution test (MY 13, sol 148)
612	Physical Properties 9, bearing test on drift material, physical properties 10 on blocky material
639	Backhoe Touchdowns 8, 9 on blocky material, physical properties 11 sample taken from blocky material and purged to make Conical Pile 5, then surface sampler parked, operations end
656	Continuation mission begins, automated monitoring for surface and atmospheric changes
668	Landing anniversary (one Mars year), 7 June 1978 (MY 13, sol 209)
2245	End of mission, 13 November 1982, bad command ends communication (MY 15, sol 448)

Between them, these surface changes hinted at the rate of surface modification on Mars. Though each event was small, if similar changes occurred roughly annually for a million years, they would substantially alter the appearance of the landscape. Much of the change probably occurs during infrequent but strong wind gusts, perhaps during dust storms, with little change between these events. When a dust storm passed over Viking Lander 1 on sol 423 (28 September 1977, MY 12, sol 632), observations during the event showed no surface changes or particle motion. Jones *et al.* (1979) suggested that 'the large global dust storms, which appear from orbiter images to be maelstroms of destruction, may cause no more erosion, even in their source regions, than we have observed at the lander sites'.

Figure 76 shows an area roughly 1 m square immediately in front of and beneath the lander. A mosaic of the small images seen in the arm on the mirror is placed approximately in the correct location under the lander. Fractured, platy material appears to have been exposed as the landing engine exhaust blew a dusty covering away from it. The shadow of a spacecraft component falls near the middle of the mosaic.

Viking Lander 1 operated for 2245 Sols until 13 November 1982 (MY 15, sol 448), when a bad command from Earth inadvertently ended the mission. The last communication from Viking 1 had been two days earlier. The lander was beginning to show signs of incipient battery failure similar to those seen on Viking Lander 2 before its mission ended in April 1980. The mission no longer had a regular flight team since automated operations began and people moved to other projects, so a group of engineers was assembled at JPL to address the problem. They designed and transmitted instructions to the spacecraft, which ceased operating immediately afterwards. The new instructions probably overwrote essential software which allowed the antenna to track Earth in the Martian sky (E. Strickland, personal communication, 22 September 2006). Attempts to contact the lander continued until 21 May 1983 when the mission was formally ended. If that problem had not occurred, the lander was programmed to continue operations until sol 4835 (24 February 1990), if its power supply lasted that long. The design lifetime had been only 120 sols.

Viking 2 Site Certification

As with the site certification for Viking 1, this section is based on the minutes of the Landing Site Staff (LSS), kindly provided by Norman Crabill. The B-1 site for Viking 2 was originally chosen for its low elevation, its position near the maximum southern extent of the 'polar hood' of haze over the northern pole during winter where water vapour might be found, and because it would lie under the same orbiter ground track as A-1. One orbiter could serve as a communication relay for both landers, freeing the second to explore other regions. The first Viking 1 orbiter images of B-1 seemed very bad to Masursky, who proposed extending the imaging in a long strip from B-1 to the northeast, up to about 57° N. He called this the 'northeast noodle' and hoped that terrain hazards might be mantled (covered) somewhere along that imaging strip. Jim Martin cut the 'noodle' short, stopping imaging at 50° N, prompting Masursky to rename it the 'northeast rigatoni', a shorter variety of noodle. This was the B1-NE region mentioned at LSS 12.

The Viking 1 site was chosen at the 22nd meeting of the Landing Site Staff (LSS 22) on 12 July 1976. At the next meeting, LSS 23 on 14 July, the latitude of the second orbiter's periapsis was again discussed. The estimated elevations of the C sites (Table 17) were considered again, as were expected winds and the ground tracks of the orbiters for either periapsis latitude. Press representatives and photographers sat in on this and several later meetings for background material for their stories about the upcoming landings.

Site geology was the subject of LSS 24 on 16 July. The Viking 1 site was thought to be a basalt plain similar to a lunar mare, with many wrinkle ridges and relatively few craters. The regolith would consist of crater ejecta and sediments from Lunae Planum deposited by several channels west of the site, or aeolian material. If this was correct, how should the B and C sites best complement A? B-1 seemed to consist of polygonally fractured thin lava flows mantled with layers of debris. C sites were cratered uplands with later lava flows and a regolith of ejecta and aeolian material. Geologist Mike Carr preferred C for access to old cratered material, whereas biologists Joshua Lederberg and Harold ('Chuck') Klein preferred B for potential water and organic materials.

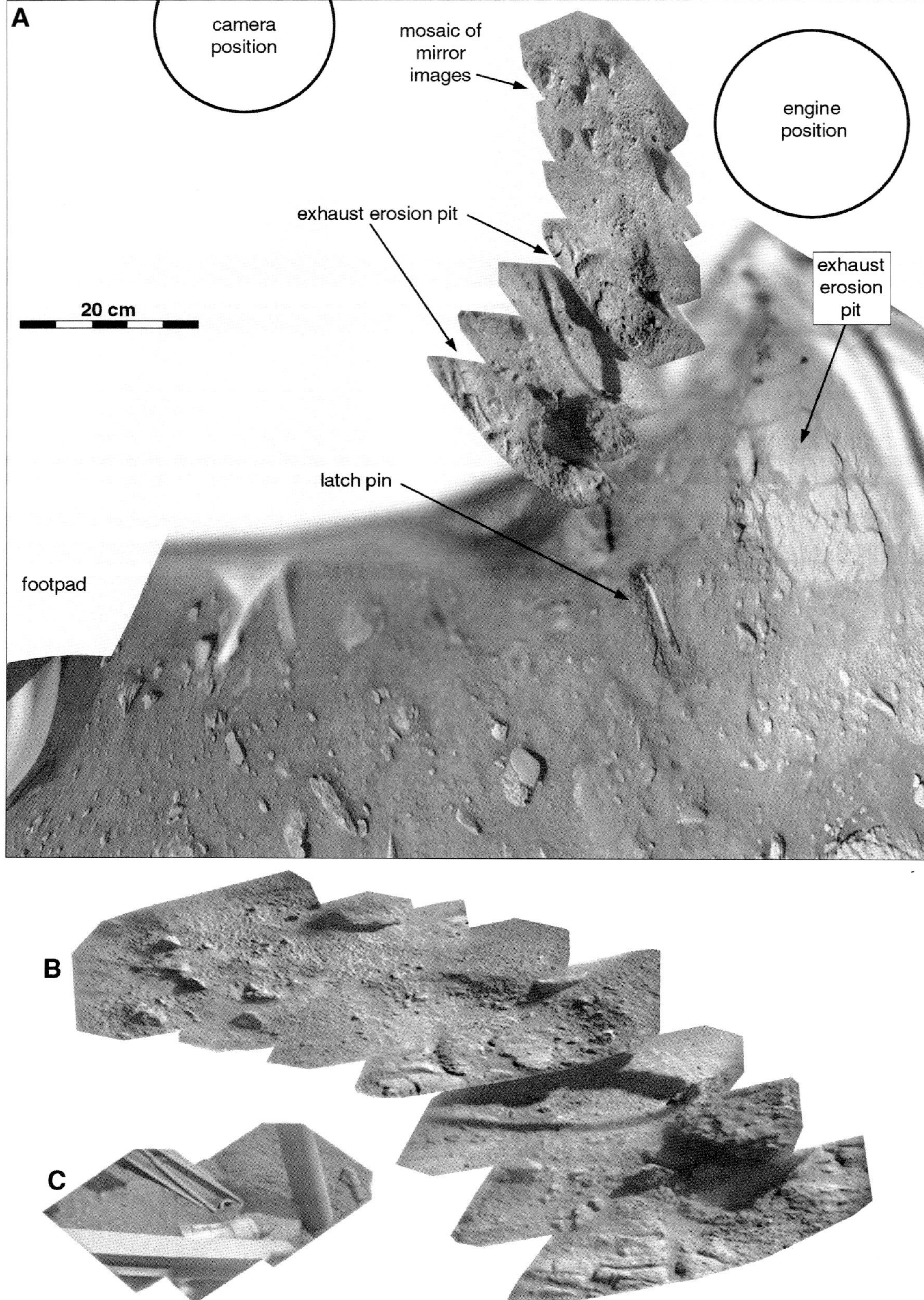

Figure 76 The immediate foreground of Viking Lander 1.
The area mapped is roughly 1 m wide, with north at the top. The mosaic of images taken using the arm-mounted mirror is shown in its approximate location (A), and enlarged (B). Pits dug by the engine exhaust are indicated. Each pit exposes fractured, crusty material. C is another mirror mosaic showing the Footpad 2 temperature sensor.

Table 22. *Viking Site Terrain Hazards Discussed on 17 July 1976*

Site	Ellipse	Location	Hazard*	Notes
A	A-1 WNW	22.5° N, 47.5° W	1–2%	Mare ridge hazard assumed to be 0%
B-1		c. 44° N, 13° W	12–15%	Fractured plains
		c. 43° N, 7° W	13–15%	mantled plains
C-1	α north	c. 5° S, 44° W	15%	99% ellipse 50 km from canyon rim
	α middle		14%	99% ellipse 20 km from canyon rim
	α south		13%	99% ellipse touching canyon rim
	β north	c. 6° S, 43° W	10%	99% ellipse 20 km from canyon rim
	β south		6–8%	99% ellipse touching canyon rim
	γ	c. 5° S, 41° W	11%	99% ellipse 20 km from canyon rim**

*The hazard rating is the probability of landing in terrain mapped as potentially hazardous.

**Position on Figure 77 matches distance from canyon rim rather than centre location.

LSS 25 was held the next day. The B latitude was discussed again at this meeting, so the Viking 2 orbit insertion burn could be planned. The seismologists preferred C sites because they were closer to Viking 1, which would improve the chances of the two seismometers detecting the same signal and being able to locate the source. A hazard assessment of sites discussed at this meeting is given in Table 22, and the sites are illustrated in Figure 77. The number of possible ellipses was increasing, but few seemed very satisfactory.

LSS 26 and 27 on 18 and 19 July dealt with the Viking 1 target coordinates. Viking 2 planning resumed on 20 July 1976, just after the successful landing of Viking 1 and on the 7th anniversary of the Apollo 11 lunar landing, 'First National Space Day' as the LSS 28 minutes called it. The process of B site certification was established. The potential B sites were Cydonia (B-1), Alba Patera (B-2) and Utopia Planitia (B-3). The Alba site was high on the broad dome of this vast shield volcano, now called Alba Mons. The patera, the volcanic caldera (collapse crater) itself, is 200 km south of the ellipse (Figure 45). Further discussion took place at LSS 29 on 21 July. Masursky summarized: A-1 was chosen to be safe and interesting. The B latitude was thought likely to show evidence of water, but lacked radar data and so was more risky. Now B-1 looked hazardous, partly mantled with erosional debris, and C-1 looked interesting, having ancient craters and lava flows and being typical of half of Mars. Ultimately, however, the B latitude was still preferred at this meeting. The final decision was made at LSS 30 on 22 July. The Viking 2 periapsis latitude would be 46° N. The photographic coverage needed for site planning is described in Table 23 and illustrated in Figure 78A. The B-2 site would now be either west or east of Alba Patera (Mons) itself, not on the volcano. The first task would be to select one of the three areas, B-1, B-2 or B-3, for landing, and then the best ellipse in that area would be certified.

Table 23. *Imaging Coverage Needed for Viking 2 site Selection, 22 July 1976*

	B-1	B-2	B-3
Central latitude	47° N	45° N	45° N
Latitude range	44°–51° N	41°–50° N	40°–50° N
Longitude range	350°–30° W	90°–120° W	200°–265° W

LSS 31 on 23 July had to deal with an unexpected question. Viking Lander 1's surface sampler, the robotic arm, had failed to unlatch from its flight position. If it proved impossible to free it and no samples could be acquired and analyzed, would that constitute a failure for the Viking 1 mission? If so, should Viking Lander 2 continue to the B site, or as mission rules had suggested in the event of a failed landing, go to the very safe C site? A vote was held and B was still the preferred location. The Viking Lander 1 arm did deploy successfully soon afterwards. Discussions of site certification procedures and imaging requirements continued through LSS 32 on 24 July. At LSS 33 on 27 July, B-1 was reported to be too rough and complex to be satisfactory. An ellipse, not specified but not the later epsilon site, could be fitted into the orbit 35A imaging area, but it was too small to be useful. It lay near 42° N somewhere between 4° W and 15° W, in an imaging sequence taken on 25 July which included the first view of the infamous 'face on Mars' later popularized as a supposedly artificial structure. The mottled terrain northeast of B-1, 'pustular' in appearance, was hilly and partly eroded, but it could not support a landing. The next three LSS meetings on 29 July, 1 and 3 August were taken up with data and planning issues while the B sites were studied further.

On 4 August at LSS 37, all three B sites were discussed. B-1 consisted of mottled and mantled plains. B-2 near Alba Patera showed volcanic rocks and fractured plains with an impact-derived regolith and some aeolian cover. B-3 consisted of volcanic plains with a thick aeolian mantle. Masursky reported that it was not possible to locate a 99% ellipse (130 by 50 km) in B-1, but a smaller 50% ellipse (52 by 20 km) could be fitted in a few places. This was not ideal, but could it be justified based on the experience with Viking 1? He showed four possible sites (Table 24, Figure 78B) and pointed out that they might not do any better at B-2 or B-3. Don Wise suggested moving the B-2 site 30° east (presumably from the original B-2 location) to the northern part of Tempe Terra (Figure 78A), with a landing site on the flat top of one of the layers seen in Mariner 9 images of the area. The layers were expected to contain ice, which might account for the observed erosion of the layers, and could be relatively free of blocks. This area of layered material is near 45° N, 60° W, west of the crater Sytinskaya.

A new site in B-1 was identified as a result of detailed terrain analysis and presented at LSS 38 on 11 August. This analysis predicted the probability of landing in terrain other than smooth plains, so the areas with the lowest values would be preferred. The new location was at 44.5° N, 5.2° W (Figure 78B). At LSS 39 on 12 August, the organic chemistry group stated their preferences for the second landing site, in order of priority: first, a sedimentary area; second, evidence of fluvial processes; third, aeolian deposition with a northerly source. Their general comment was 'prefer less rocks than present A-site to increase sampleability', which, according to the minutes, 'provoked general laughter'. Everybody hoped for fewer rocks, as the Viking 1 site looked more hazardous at the surface than it had seemed from orbit. In fact much of the area accessible to the Viking 1 surface sampler initially appeared too rocky to be easily sampled, though this turned out not to be a serious limitation. The biologists wanted to land in an area with morning fogs, seen in the approach images and at the very least suggesting the presence of water in some form. Attention also shifted to B-2, including the original ellipse at 44° N, 110° W (Figure 45). At the next meeting, LSS 40 on 13 August, Gentry Lee referred to another promising area in the B-2E area east of the original B-2 site. It was not specified precisely, but it had been seen in Orbiter 2 images from orbit 004 (Figure 78A).

So Alba had promising sites, but it had other drawbacks. At LSS 41 on 16 August, Masursky described the current situation. B-1 was 'semi-catastrophic' and could not be used. B-2 was the best area seen to date, but it had no evidence of water and bad infrared data indicative of a rocky surface. At the next meeting, LSS 42 on 17 August, Lee reviewed the B-2 region and concluded that its smooth areas were not smooth enough and must be rejected. The site certification team was exhausted and could not keep multiple sites in consideration. The bold but controversial decision was now taken to choose B-3. There were no useful images for that site yet, but B-1 and B-2 were both bad. Two ellipses called B-2E and B-2 Late had now been rejected.

The contingency plan if B-3 failed to provide an ellipse would be the small plateau in B-1 imaged on orbit 52A (12 August), which was the only really satisfactory 50% ellipse in B-1. That site (Figure 78B) was now called B-1 epsilon, at 45.5° N, 352.5° W, and although the 50% area on the plateau looked safe, the area around it out to the 99% ellipse was bad. Ninety-nine percent ellipses would be used for hazard evaluation of new sites, but they would be shrunk so that the new 99% ellipse would correspond to the old 68% ellipse. The 99% ellipse used for Viking 1 was 130 by 50 km, with a 50% ellipse size of 52 by 20 km. Now, for Viking 2, the 99% ellipse would be 100 by 37 km, and the 50% ellipse only 40 by 15 km. Despite what was said about not keeping multiple site studies going, several ellipses in B-2 and B-3 continued to be studied over the next few days.

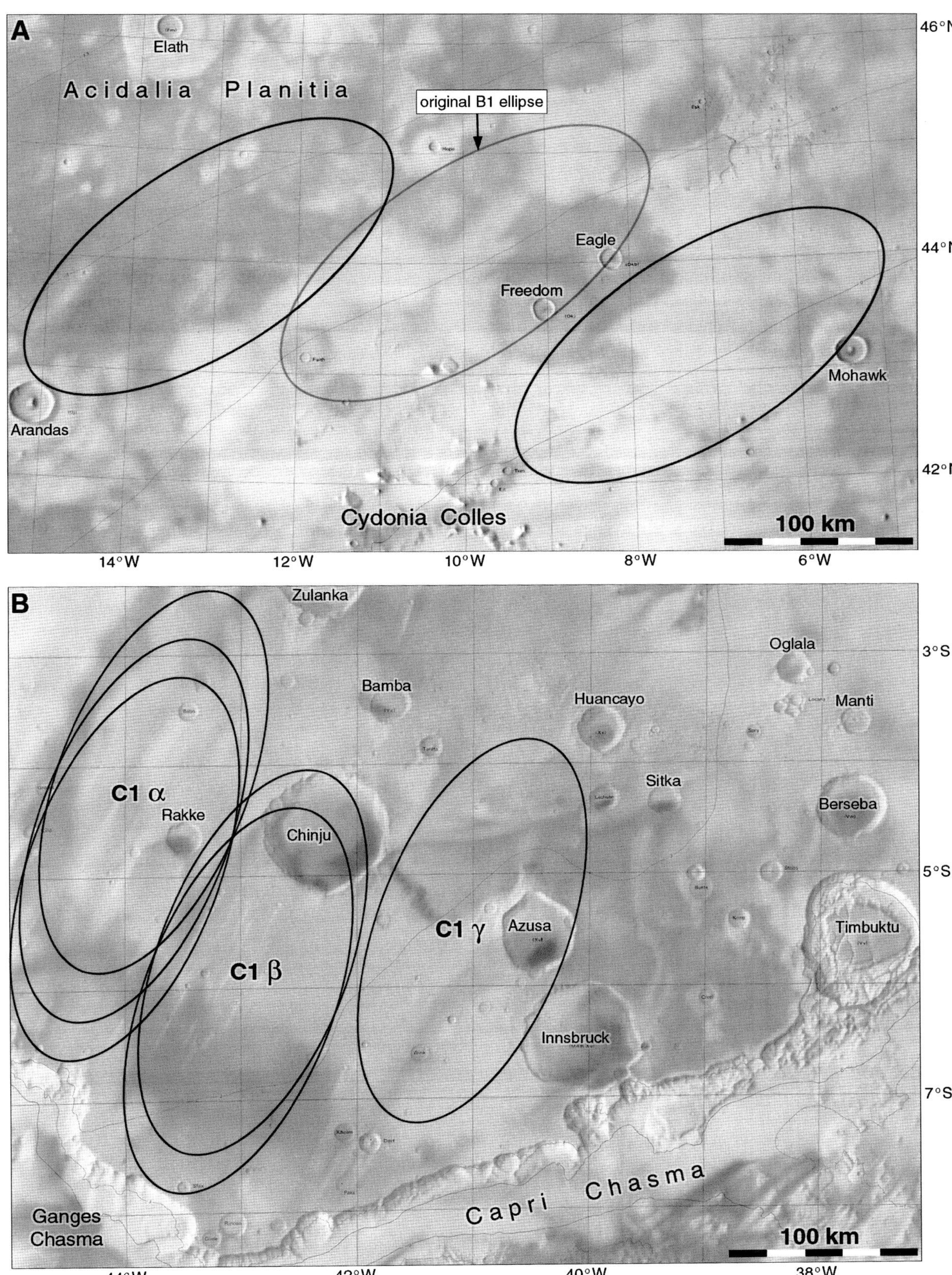

Figure 77 B and C sites considered for Viking 2 on 17 July 1976, from Table 22.
The base maps are USGS maps I-988 (77A) and I-1046 (77B).

Table 24. *Possible Ellipse Locations in B-1, 4 August 1976*

Site	Location	Images	Site	Location	Images
Alpha	46.8° N, 357.2° W	36A55, 57	Gamma	44.3° N, 11.5° W	09A48, 50
Beta	43.7° N, 357.2° W	39A24, 26	Delta	51.0° N, 353.7° W	26A69, 70

One reason B-2 W had looked promising was that images suggested that large sand dunes might mantle the surface, covering hazardous rocks and surface irregularities. Henry Moore now spotted a dune-like texture covering parts of B-3 in the vicinity of the large crater Mie. The dunes appeared to cover Mie's ejecta and parts of the crater itself. This might be the long-awaited safe site.

At LSS 43 on 18 August, B-3 planning was the main subject. Mosaics of new images from Viking 2 were delayed by a lack of plots showing expected image locations, but preliminary geological maps were discussed. An area called 'Better B-3', extending from 50° to 55° N, 160° to 190° W, was looking interesting. At the next meeting, LSS 44 on 19 August, a map of B-3 was presented showing the geology as then understood. The northeast corner of B-3 was covered with sandy material, forming dunes in places. The rest of the site had been stripped of sand, revealing fractured plains. A region just west of the 100-km crater Mie consisted of crater ejecta covered with sandy deposits, the dunes identified by Henry Moore.

This sandy region was now the preferred landing site as any rocks or other hazards were likely to be buried. Orbital images containing possible sites at B-2 were also identified at the meeting. Four of them (Viking Orbiter 2 images 8B16, 8B18, 8B25 and 8B27) correspond to ellipses in Figure 79A around 45° N, 153° E. A fifth (image 8B08) was at 45° N, 140° W near the ellipse B-2 W V in Figure 79A. It had sand dunes, high water vapour, and the best thermal inertia data but contained crater ejecta and appeared rough in images.

Site planning made significant progress by the time of LSS 45 on 20 August. Carl Sagan showed a map of Soviet data from Mars 3 and Mars 5, which used passive microwave systems to estimate the electrical properties and density of surface materials. The data suggested that the surface materials at northern latitudes were not unusual. Areas covered by these datasets are shown in association with those missions in Figures 24 and 36. Next, five possible landing sites were shown and described in B-3 and two in B-2 (Table 25, Figure 79). B-3 was seen to have no smooth mare-like surfaces, and its geology was not yet understood. Masursky preferred B-2 to B-3. Hugh Kieffer suggested looking outside existing B-3 image coverage, perhaps north of 50° N near 252° W. He also thought B-2 was better.

LSS 46 on 21 August made the decision between B-2 and B-3. A new B-2 ellipse was added, but at B-3 only the eastern sites were considered, giving five ellipses. Two additional B-2 sites were proposed by Kieffer (Table 26). A hazard map of B-3 East was prepared by Larry Crumpler and Paul Spudis. Finally the group voted, preferring B-2 over B-3, but Jim Martin overruled them, deciding on B-3 because there was so little to choose between them and B-2 would require a delay of about a week to obtain more data. The recommended site now was at 48.0° N, 226.0° W, from the minutes, though the table reproduced here as Table 26 gives a longitude of 228.0° W. Masursky's written recommendation was for 48.0° ± 1.5° N (areographic), or 47.7° ± 1.5° N (areocentric), and 226.0° ± 2.0° W, reflecting uncertainties on the order of 100 km. The 47th LSS meeting, held on 23 August, repeated the 48° N, 226° W position and discussed the final details needed for planning the landing. Updated coordinates would be estimated from newly processed images before the next meeting so that a final recommendation could be made.

The final planned LSS meeting, LSS 48, was held on 30 August. Masursky summarized the work to date. The Viking 2 site search had made use of 1056 orbital images, compared with about 800 for Viking 1. The target coordinates shown on an annotated image in the LSS minutes were 47.89° N, 225.86° W. The weather looked good, apart from a troubling Viking Orbiter 2 camera artifact. Electrical transients introduced strips of parallel lines

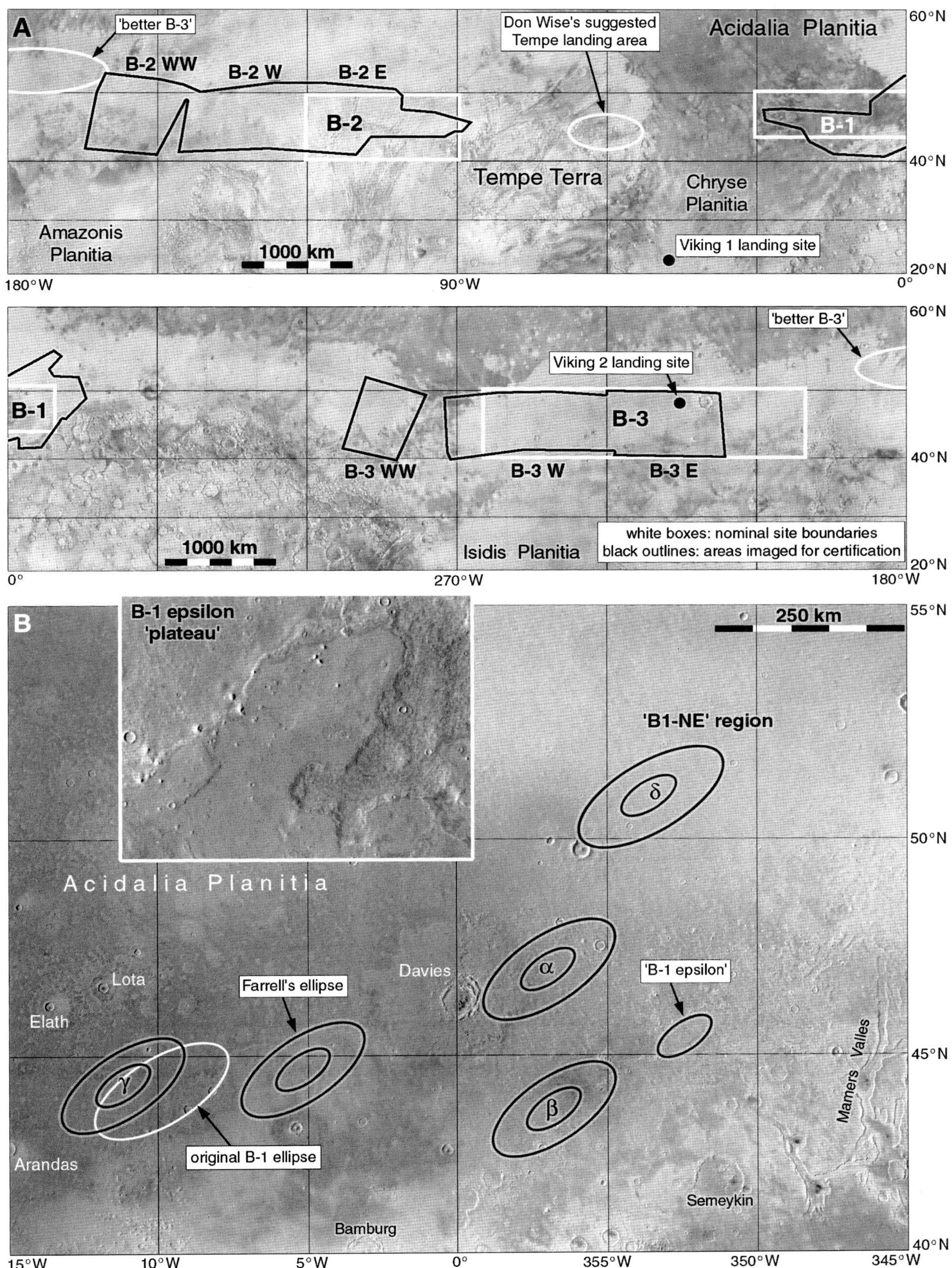

Figure 78 A: Location of B site candidate areas and certification images. **B:** Possible ellipse locations in B-1, from Table 24 and the text. The base maps are USGS map I-961 (78A) and part of the global Viking mosaic (78B), and the inset illustration is part of Viking image 673B40. The crater Davies on the 0° meridian commemorates veteran planetary geodesist Merton E. Davies.

Table 25. *Viking Sites from LSS 45 Meeting, 20 August 1976*

Site	Location	Notes
B-2 West α	43.9° N, 154.4° W	Similar to ellipse B-2 West I in Table 24.
B-2 West β	47.3° N, 149.5° W	Coordinates from text. A map presented at the meeting and by Masursky and Crabill (1976) shows this ellipse at about 47.5° N, 152° W.
B-3 East α	47.0° N, 224.9° W	Textured sand deposits estimated 25 m thick, covering smaller craters.
B-3 East β	47.5° N, 228.1° W	Fractured terrain partly mantled by sand.
B-3 West α	48.0° N, 255.5° W	Clouds partly obscure fine details.
B-3 West β	47.0° N, 248.5° W	Clouds partly obscure fine details. Minimum thermal inertia site.
B-3 West γ	40.0° N, 267.0° W	Debris cover less evident, but site extends outside image area.

Table 26. *Viking Sites from LSS 46 Meeting, 21 August 1976*

	H. Masursky sites in B-2 West and B-3 East		H. Kieffer sites from infrared data	
Site	Ellipse	Location	Ellipse	Location
B-2 West	I	44.1° N, 154.9° W	IV	42° N, 150° W
	II	47.3° N, 156.6° W	V	47° N, 135° W
	III	43.5° N, 153.0° W		
B-3 East	α	47.2° N, 224.9° W		
	β	48.0° N, 228.0° W		

across parts of many images, a problem called 'microphonics'. Doyle Vogt's weather report concluded that 'the landing will occur through heavy microphonics'.

A special meeting of LSS was held on 12 October as a 'post-mortem' to consider what changes in site selection procedures would be made if a future landing had to be planned or what data the Viking Orbiters might obtain to help with future planning. Two fictional missions were considered, another Viking landing in January 1977 after the upcoming conjunction and a separate Viking 3 mission in 1981. NASA was considering a Viking 3 mission for 1981 at the time. Several suggestions were made. First, the landing ellipse could be smaller, perhaps about 40 by 20 km for a landing from orbit. Second, future sites should be made in warmer locations or where water was more likely to be found. Mutch preferred a polar site, and Kieffer suggested landing on a polar cap. Seymour Hess favoured a polar site or a location with morning fog. Third, high-resolution images, up to 1 m/pixel, would be needed for planning future rover operations, and that would also require spacecraft with a higher data transmission rate. For a hypothetical 1977 landing, a site between 200 and 600 km from Viking 2 would favour seismic studies by enabling simultaneous detections of smaller signals. Viking 1's seismometer had not deployed. A 1981 landing should consider deploying the seismometer away from the lander to reduce wind and equipment vibrations. Penetrators would be better than soft landers for a seismic network. Descent imaging would be useful for site location and context. Gerald (Gerry) Soffen, the Viking Project Scientist, said at this meeting that he was 'amazed we really got two landers down safely'. Hal Masursky wanted to lower the orbiters to take images of the landing sites with the highest possible resolution. This was done for the Viking 1 site but not for Viking 2.

9 September 1975: **Viking 2 (United States)**

Viking 2 was launched at 18:39 UT from Cape Canaveral, and after a 333-day cruise, it entered a 1496 by 35 800 km orbit about Mars with an inclination of 55.2°

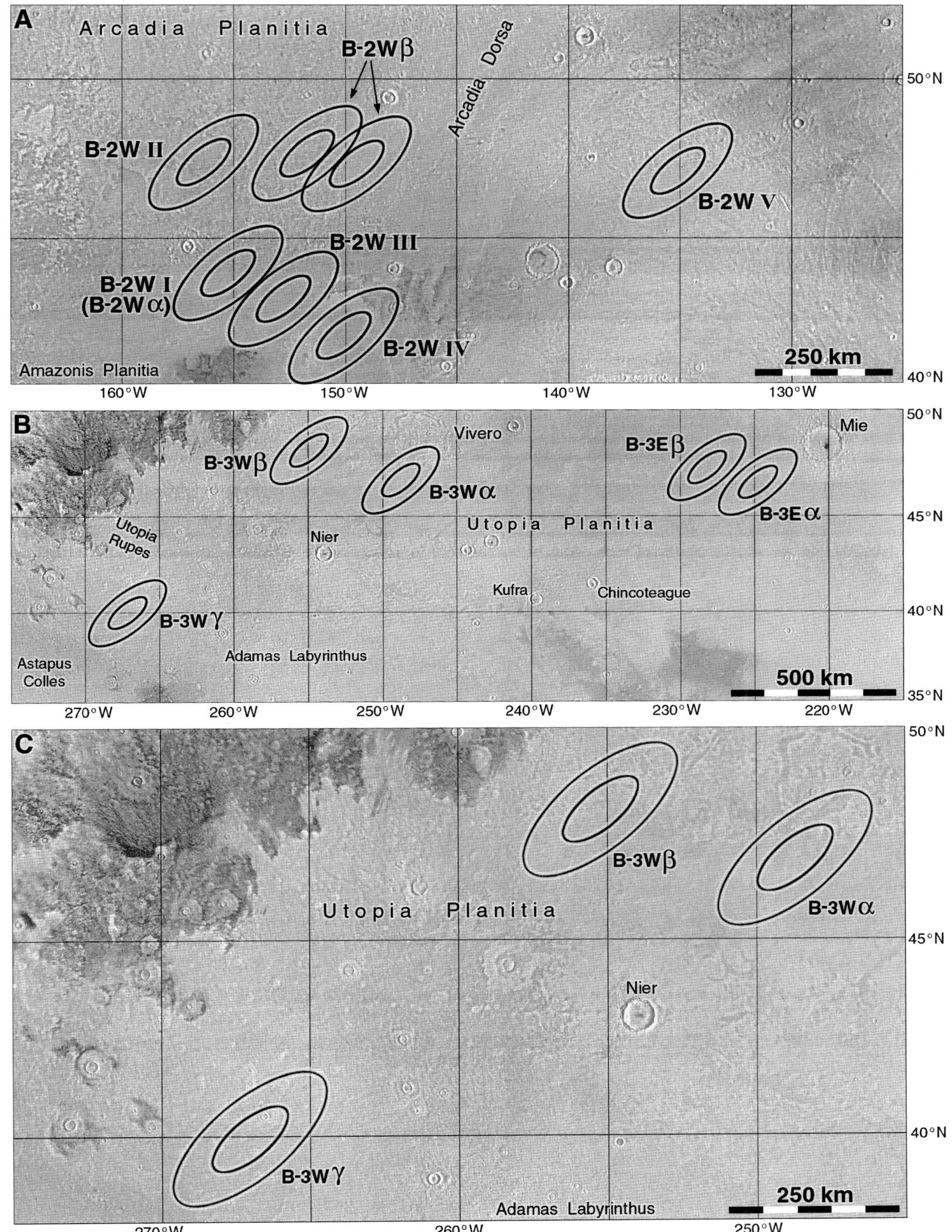

Figure 79 **A:** Candidate Viking 2 ellipses in the B-2 area. Site B-2Wβ has two possible ellipses (Table 23). **B:** Candidate Viking 2 ellipses in B-3. **C:** The B-3 West ellipses at higher resolution. The map bases are parts of the global Viking mosaic.

and a period of 27.4 hours on 7 August 1976 (MY 12, sol 226). The first task was to complete the landing site certification already begun by Viking Orbiter 1. Viking Orbiter 2 activities are summarized in Table 27 based on JPL (1978) and Snyder (1979), and Viking Lander 2 activities in Tables 30 through 32. The mission resulted in the second successful landing and surface operations on Mars, and, between the two orbiters, with the acquisition of global imaging. Viking Orbiter 2 also imaged both satellites. Its first Phobos encounter on 17 September 1976 (MY 12, sol 266) revealed the distinctive pattern of grooves on that satellite. Its Deimos close encounters in October 1977 provided the first high-resolution images of that small moon.

Between the two Viking Orbiters, the entire surface of Mars was covered with images greatly superior to those from Mariner 9. This global coverage was enabled by the long operational lives of the orbiters, but in the early planning stages, global mapping could not be guaranteed. Assuming coverage might be limited, the Viking Project designed a set of high-priority observations to be added to the primary goal of landing site certification as circumstances permitted (Viking Project, 1972). These consisted of 81 targets for high-resolution imaging, mainly for geological analysis, and 25 more for low-resolution (high-altitude) imaging of clouds, wind streaks, variable features and regions in which dust storms usually begin. Other targets were defined for the infrared and atmospheric water mapping instruments. These targets are listed in Tables 28 and 29, and the high-resolution imaging targets are illustrated in Figure 81.

Figure 82 presents some examples of Viking Orbiter 2 images. Figures 82A, 82B and 82C are approach images showing a crescent phase. Figure 82D shows the Viking 2 landing site and was the best available image of it before Mars Global Surveyor. Figure 82E is part of Deuteronilus Mensae at 40° N, 336° W. The feature in Figure 82F is a mountain at the southern edge of Argyre at 56° S, 39° W, and Figure 82G shows the edge of the south polar cap at 60° S, 200° W in Terra Cimmeria. Viking Orbiter 2 also obtained bistatic radar data at near-polar latitudes (Simpson and Tyler, 1981).

Figures 83 and 84 show the combined coverage from both orbiters at several resolutions. This is far from uniform because the longitude of periapsis varied with respect to solar illumination during the Martian year. High-resolution imaging would be obtained when periapsis occurred at longitudes with good lighting. At other times medium-resolution mapping of broader regions, or global monitoring for clouds, dust, variable features or polar cap changes, would take precedence. Viking Orbiter 2 returned 15 582 images to Earth. Viking Orbiter 2 was shut down when its attitude control fuel was exhausted on 25 July 1978 after 706 orbits, and it is expected to burn up in about 2025. The Mars Exploration Rover Spirit may have imaged Viking Orbiter 2 (JPL, 2004) before dawn on 7 March 2004 (MY 27, sol 2; Spirit's sol 63).

The lander was released at 19:40 UT on 3 September 1976 (MY 12, sol 253) and immediately experienced unexpected motions and a power loss which threatened the spacecraft's survival. The onboard computer detected the problem and switched to its backup systems, stabilizing the lander and enabling it to enter the atmosphere safely. The problem was attributed to the explosive bolts used to separate the lander, and to avoid further possible problems, a biological barrier installed between the orbiter and lander was not separated as intended. It restricted the fields of view of the orbiter's remote sensing instruments, so it was eventually released without incident on 3 March 1978.

At 12:58 UT on Earth (9:49 a.m. local time on Mars), Viking Lander 2 settled safely on the ground at Utopia (Figures 85 to 88), 44 sols after Viking Lander 1. The first images showed a surface covered with rocks but no sign of the expected mantle of sand dunes. Later, Viking Lander 2 imaged its entire surroundings (Figure 89). The landing site was very rocky, with fewer dust drifts than at the first Viking site. The horizon was extremely flat (Figure 89), making identification of the site very difficult. Compounding the problem was the low resolution and extremely low contrast of the orbital images. The Viking 2 site could not be located precisely during the mission, but later high-resolution orbital images permitted a more reliable result, and the HiRISE camera on Mars Reconnaissance Orbiter eventually located the lander. Figure 90 shows the surface panorama projected onto a HiRISE image.

Analysis of radio transmissions after landing allowed Mayo *et al.* (1977) to estimate the lander coordinates as 47.670° N, 225.59° W, accurate to within about 120 m (0.002°) in latitude and 12 km (0.2°) in longitude.

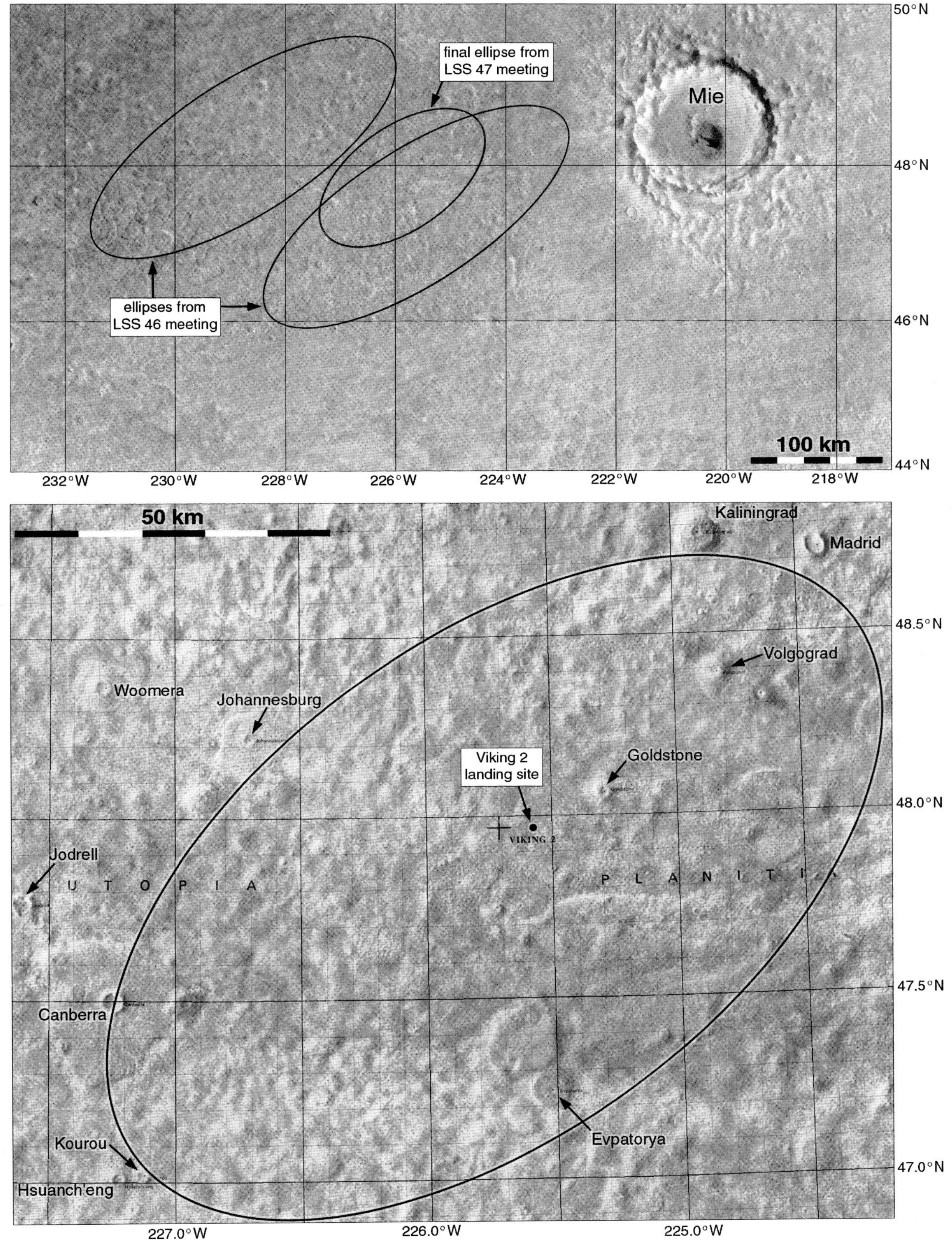

Figure 80 A: Viking 2 landing site ellipses in area B-2 East. **B:** The final landing ellipse. The base maps are USGS maps I-1061 (top) and I-1060 (bottom). The actual landing site is shown correctly relative to surface features, not the grid. Craters near the landing site were named after tracking stations, including the Deep Space Network sites on Earth.

Attempts to locate this point on orbital images gave varied results, and all locations discussed here are plotted on Figure 85B. The method of locating the lander by identifying horizon features was rendered difficult by the flat topography, but a hill and several ridges east of the lander were suggested by Mutch *et al.* (1977) to be parts of the rim or outer ejecta lobes of the large crater Mie (Figure 80A). They placed the lander about 15 km WSW of a pedestal crater (a crater on a hill or plateau) called Goldstone. USGS maps (Figures 80B, 85A) place it near that location but a little to the northwest. An inset map on a set of USGS panoramic images of the Viking 2 site (*e.g.* USGS, 1984) placed it about 10 km northeast of the previous location, only 8 km from Goldstone.

Stooke (1997) addressed the problem of limited horizon detail by using vertical exaggeration of subtle horizon details to enhance their visibility. The largest hill east of the lander was assumed to be Goldstone, and a low hill north of the lander, apparently not described before, gave a direction to a second feature. The two vectors from Goldstone and this 'northern hill' would intersect at the lander location if the northern hill could be identified in orbital images. Stooke described three possible matches for that hill in Viking images, leading to three possible landing sites (Figure 85B). He preferred the site closest to Goldstone as it gave the best match to the USGS (1984) location. A similar approach led Parker and Kirk (1999) to a site further from Goldstone, just north of Stooke's third site. Oberst *et al.* (2000) performed a new control network analysis to try to improve the location from the tracking coordinates, finding a site slightly farther south than that of Mutch *et al.* (1977). Stooke (2004) revised his analysis when Mars Global Surveyor images and topography became available. The new orbital data showed the candidate northern hill more clearly than before, suggesting that the middle of his three locations matched topography better than the earlier preferred site. He identified a specific feature as a candidate for Viking Lander 2 itself, which proved to be incorrect. Soon after that, Malin Space Science Systems (2005a) announced that they had found the lander near Stooke's new location. When HiRISE images from Mars Reconnaissance Orbiter became available, the Malin feature was seen to be the Viking 2 backshell, not the lander (Parker *et al.*, 2007). The lander was found nearby at a location that had not been suggested by anybody (Figure 86).

Viking Lander 2 used its arm to dig trenches and collect samples, whose analytical results were very similar to those from Viking 1. The lander completed its primary mission on 15 November 1976 but continued to operate until 11 April 1980. It observed the surface repeatedly to search for changes but did not find obvious changes like those seen in Chryse. Viking Lander 2 was designated the Gerald A. Soffen Memorial Station in 2001. Soffen, the Viking Project Scientist, had died on 22 November 2000.

Viking Lander 2 surface activities are described in Tables 30, 31 and 32 and illustrated in Figures 92 through 105. Figures 96, 99, 102 and 105 show the sample area as it evolved during the mission. The summaries of activities presented here are based on Soffen (1976), Shorthill *et al.* (1976), Snyder (1979) and especially Moore *et al.* (1987). The Viking Mission Status Bulletins released by JPL during the mission were also consulted, with the last of the set (JPL, 1978) being particularly useful.

On sol 1 the sampler arm shroud was ejected successfully. It hit a rock near Footpad 3, moving it very slightly, and bounced onto the soil about 60 cm farther away, before coming to rest about 110 cm beyond the rock (Figure 92). The arm latch pin was also ejected successfully but was not seen in images. Then on sol 8 the arm was used for the first time to collect the Biology 1 sample in an area of crusty and cloddy material called Beta (Figure 92). The fine fraction was delivered to the Biology instrument, and the coarser material, if any, was to be dropped over the XRFS instrument funnel. The first attempt to deliver the XRFS sample caused a computer failure, but the problem was corrected, and a second attempt followed on sol 13. Nothing was deposited into the funnel, so if there were any larger objects in the sampled material they broke up when excavated. The XRFS problem delayed the first GCMC acquisition.

The next sampling was on sol 21, when the arm performed Backhoe Touchdown 1 on the crusty or cloddy material at the Bonneville Salt Flats area in front of the rock Bonneville, then retracted, extended and retracted again to break up the crusty surface. Then the arm scooped up a sample, GCMS 1, from the same area for the GCMS instrument (Figure 93). The fine fraction was delivered to the instrument and the coarse fraction was purged. Objects as large as 27 mm across were seen at the purge site near the future Physical Properties 1 trench (Figure 95). On

Table 27. *Viking Orbiter 2 Orbital Events*

Date	Orbit	Events
7 August 1976	0	Mars orbit insertion
9 August 1976	2	Adjust orbit for site certification; walk periapsis westward
14 August 1976	6	Increase orbital period and periapsis walk rate
25 August 1976	16	Reduce walk rate to place periapsis over landing site
27 August 1976	18	Begin synchronous orbit over landing site (B3) (dates over B1, B2?)
3 September 1976	25	Viking Lander 2 separation and landing
17 September 1976	39	Close Flyby of Phobos, north polar imaging
30 September 1976	51	Increase orbital inclination to 75°, begin westward periapsis walk
8 November 1976	85	Last data transmission of Primary Mission
25 November 1976	101	Mars at conjunction (15 November, end of the Primary Mission)
15 December 1976	119	First commands after conjunction
20 December 1976	123	Lower periapsis to 800 km, increase inclination to 80°, periapsis crosses morning terminator into night side
14 January 1977	146	Earth occultations start, continue daily for 134 days
28 January 1977	159	Sun occultations start, continue daily for 172 days
16 February 1977	176	First observation of major dust storm from orbit
2 March 1977	189	Return to synchronous orbit over Viking Lander 2
18 April 1977	235	Adjust orbital period so 13 orbits take 12 sols
30 May 1977	279	End of the series of Earth occultations, northern winter sostice
17 July 1977	330	End of the series of solar occultations
9 September 1977	388	Periapsis crosses evening terminator into daylight
25 September 1977	404	Adjust orbital period to allow close approach to Deimos
9 October 1977	418	Adjust orbit to be synchronous with Deimos
10 October 1977	418	First of three close Deimos encounters
15 October 1977	423	Closest Deimos encounter (33 km altitude)
20 October 1977	428	Last Deimos close encounter
23 October 1977	432	Adjust orbital period to 24.0 hours, lower periapsis to 300 km
1 November 1977	440	Gas leak in attitude control system
2 November 1977	441	One attitude control system shut down, northern vernal equinox
8 November 1977	447	First bistatic radar observations
19 November 1977	458	First of a series of 176 Earth occultations
15 January 1978	515	Valve opened to share helium gas between attitude control systems
30 January 1978	531	First of 26 long solar occultations near apapsis
12 February 1978	544	Longest solar occultation of the mission, 4.5 hours
24 February 1978	556	Last of the series of 26 solar occultations
3 March 1978	563	Lander bioshield ejected to improve instrument view of planet
24 March 1978	583	Major propellant leak forces shutdown of one attitude control system
29 March 1978	589	Orbiter placed in roll-drift mode to prolong lifetime
12 May 1978	633	Last in the series of Earth occultations
17 June 1978	668	Orbiter uses Jupiter for attitude lock for ten orbits of observations
25 June 1978	676	One Mars year in orbit
26 June 1978	678	Orbiter returned to roll-drift mode
19 July 1978	700	Orbiter uses Procyon for attitude lock for last science observations
25 July 1978	706	Periapsis crosses equator, spacecraft powered down as attitude control fails

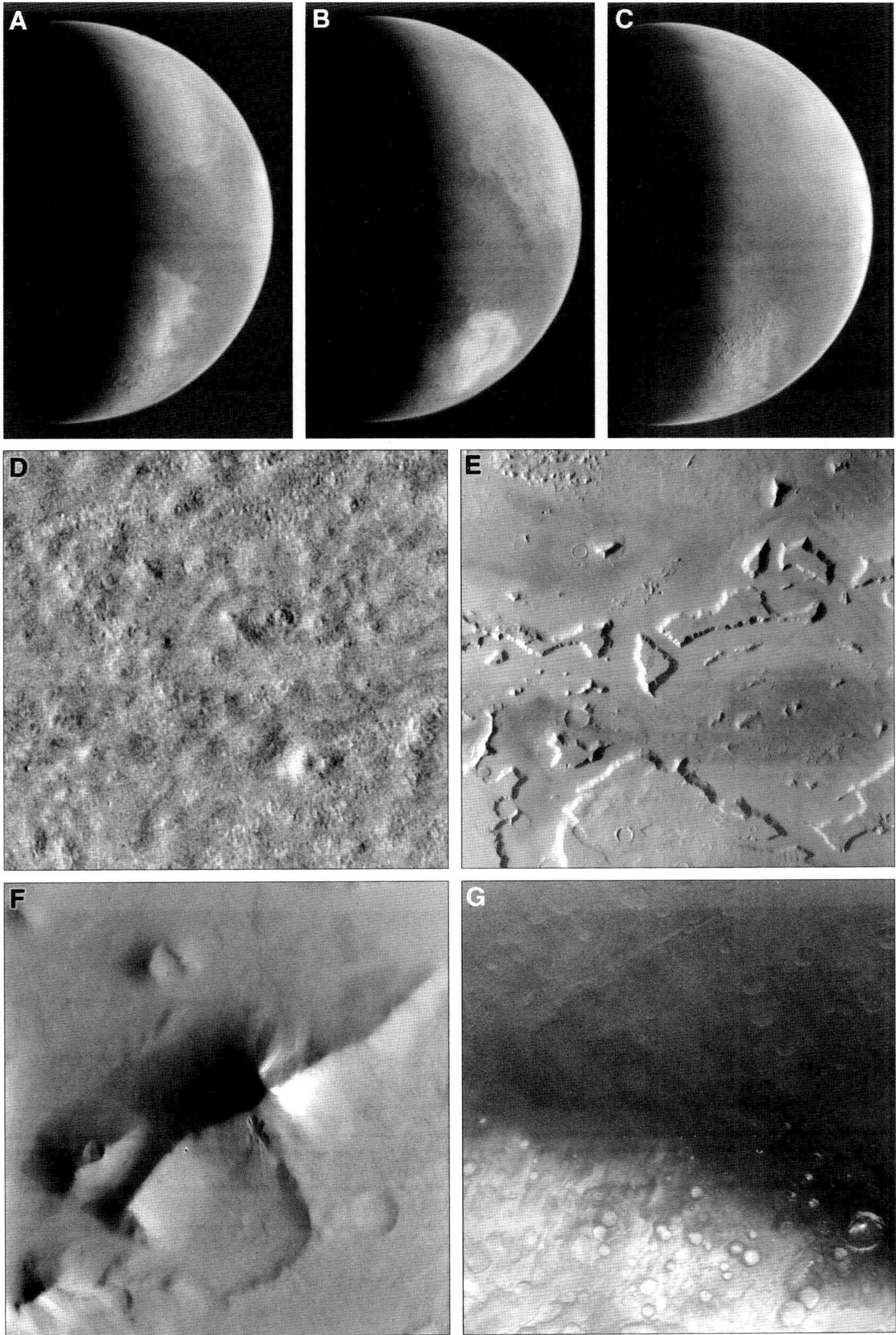

Figure 82 Viking Orbiter 2 images.
A: f218d12. **B:** f218d20. **C:** f218d24. **D:** 009B15. **E:** 675B64. **F:** 569B21. **G:** 167B25. The prominent hill at lower right in 82D is Goldstone.

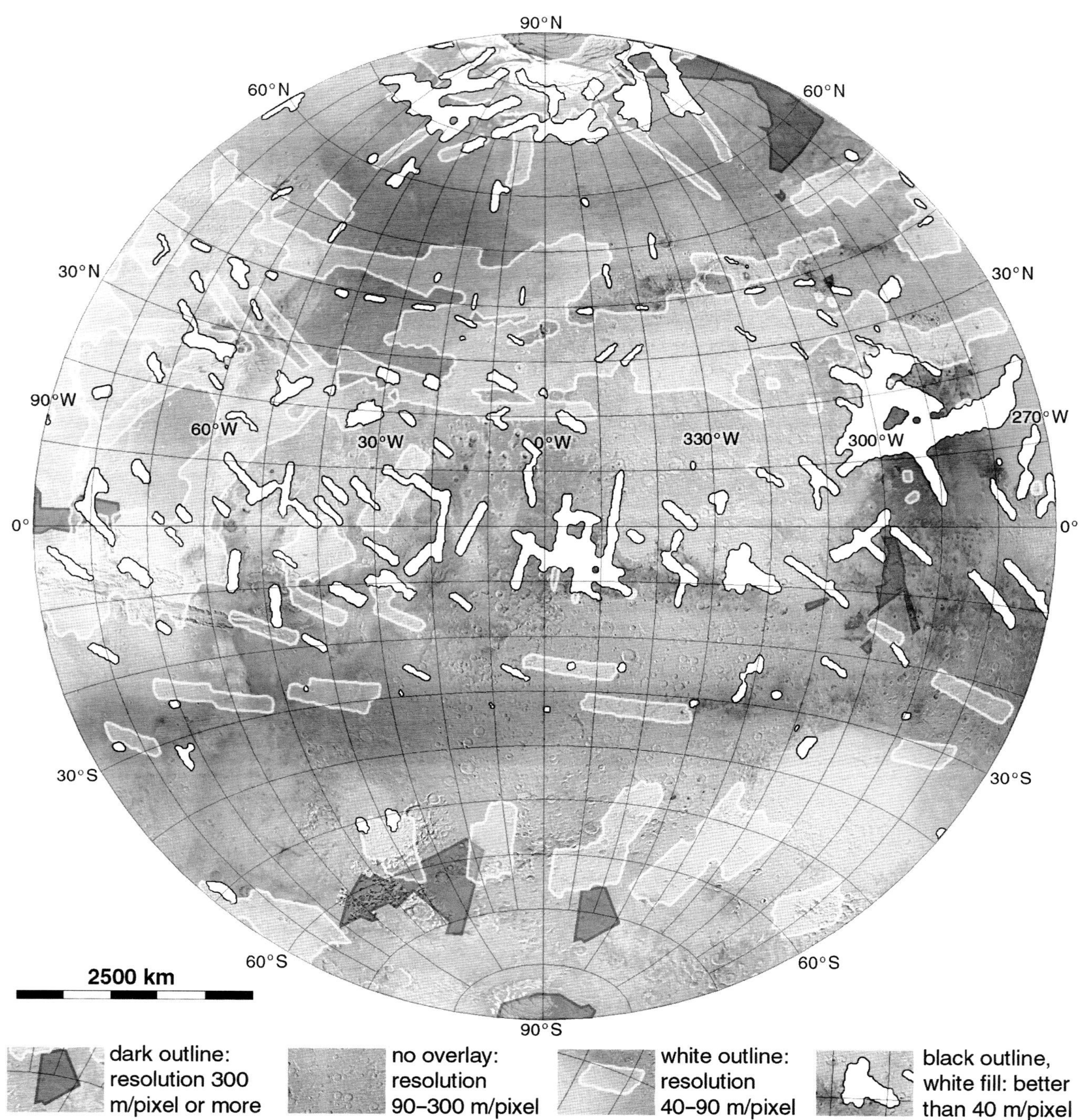

Figure 83 Viking Orbiter image coverage and resolution, 0° longitude hemisphere.
Data from both Viking orbiters are combined in this map. The resolution ranges and areas covered are derived from maps on the Arizona State University website (themis.asu.edu/maps) providing online access to Viking images. Some small errors are known to exist in the source materials.

Figure 84 (opposite page): **Viking Orbiter image coverage, 180° longitude hemisphere.**
The images below the map (Viking Orbiter 1 images 488a16, left, and 746a46, right) show the craters Airy and Airy-0, the origin of the Martian longitude system.

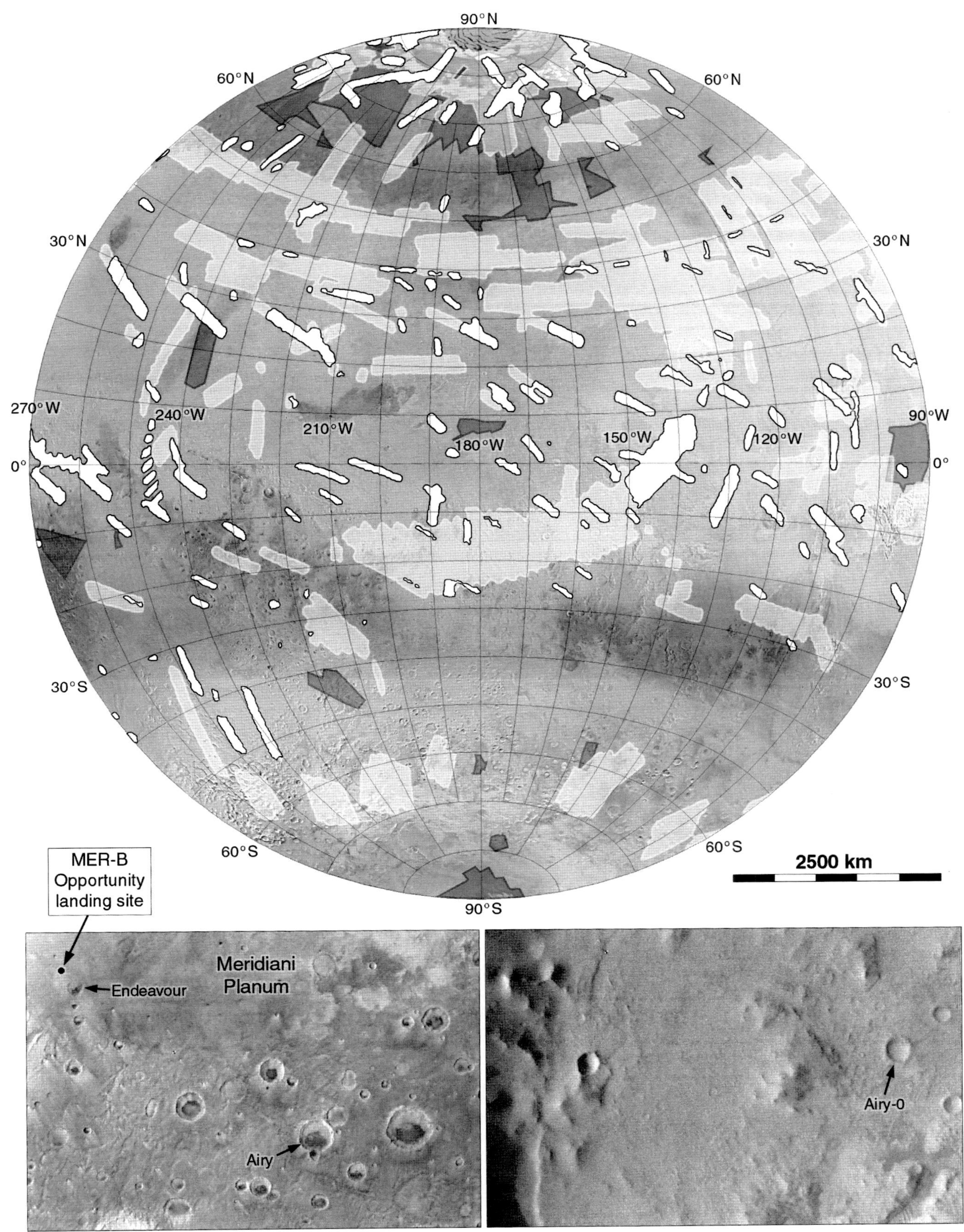
90°N
60°N
60°N
30°N
30°N
270°W
240°W
210°W
180°W
150°W
120°W
90°W
0°
0°
30°S
30°S
60°S
60°S
90°S
2500 km
MER-B
Opportunity
landing site
Meridiani
Planum
Endeavour
Airy
Airy-0

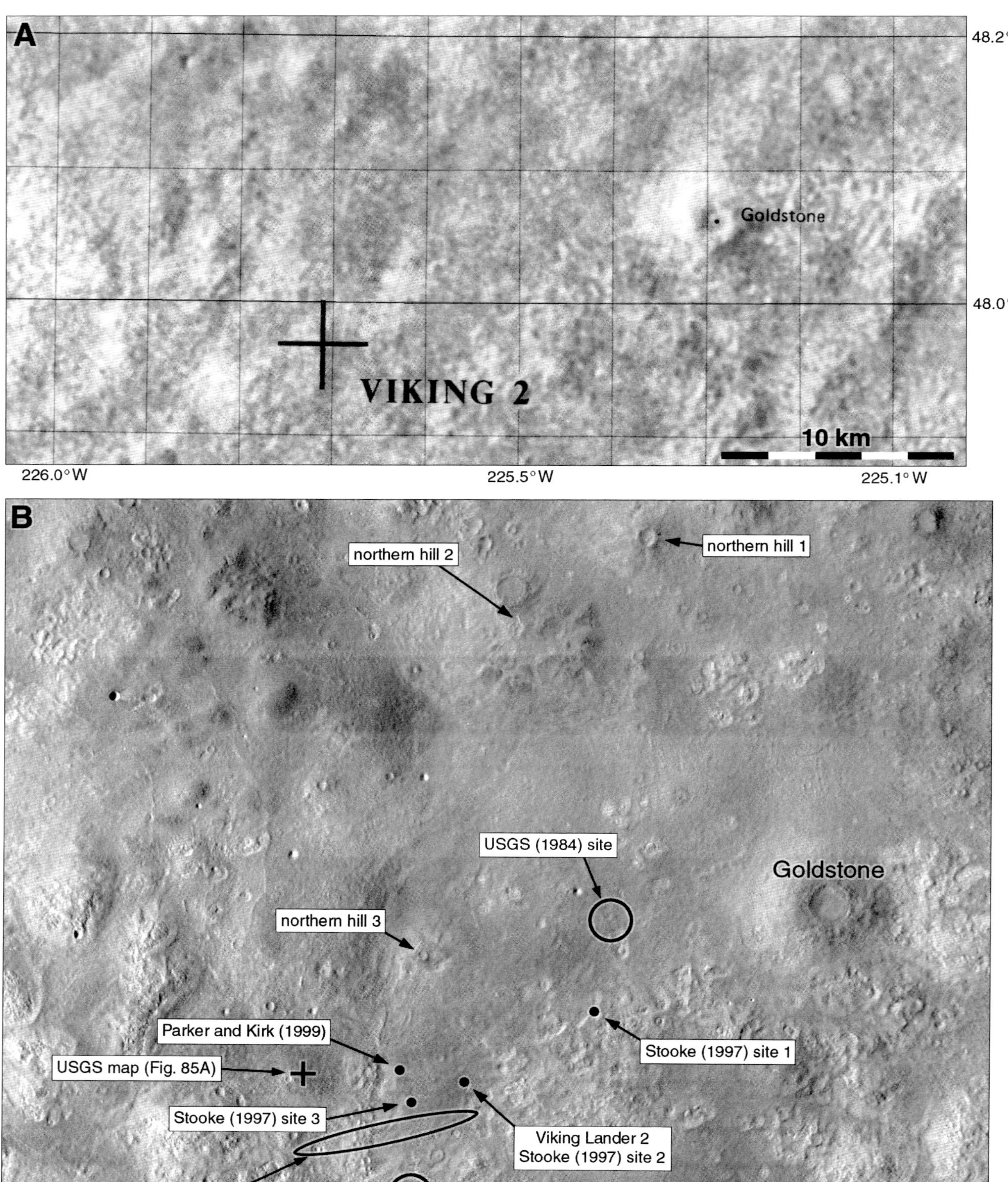

Figure 85 Attempts to locate Viking Lander 2.
A: USGS map I-1060 showing an early estimate of the Viking Lander 2 location. **B:** Mars Odyssey THEMIS visible image mosaic showing several suggested locations. The three Stooke (1997) sites are derived from the three candidates for the northern hill.

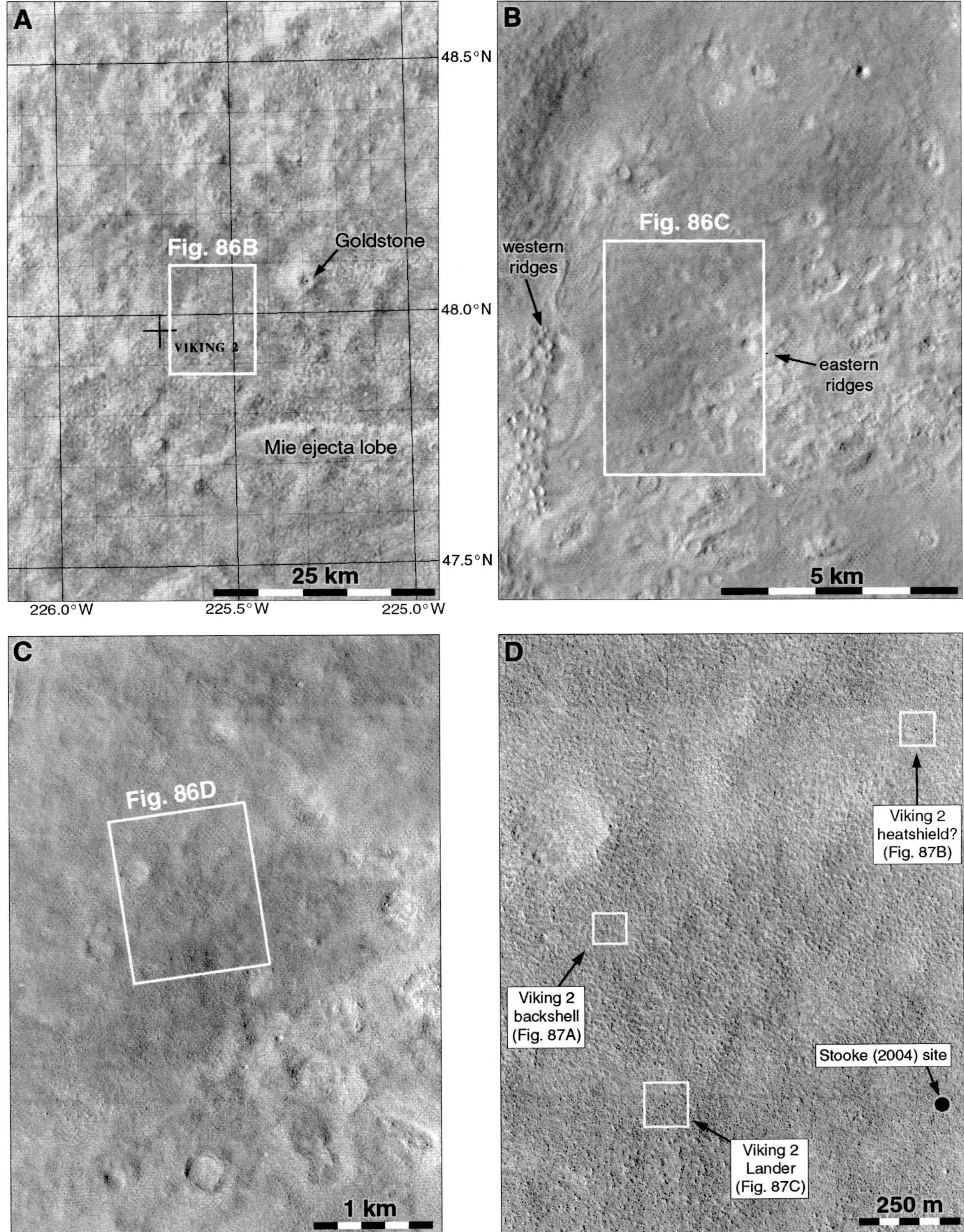

Figure 86 Viking 2 landing site.
A: Detail of US Geological Survey map I-1060 showing an early estimate of the landing site location. **B:** Mars Odyssey THEMIS visible image mosaic. **C and D:** Mars Reconnaissance Orbiter HiRISE image PSP_001501_2280.

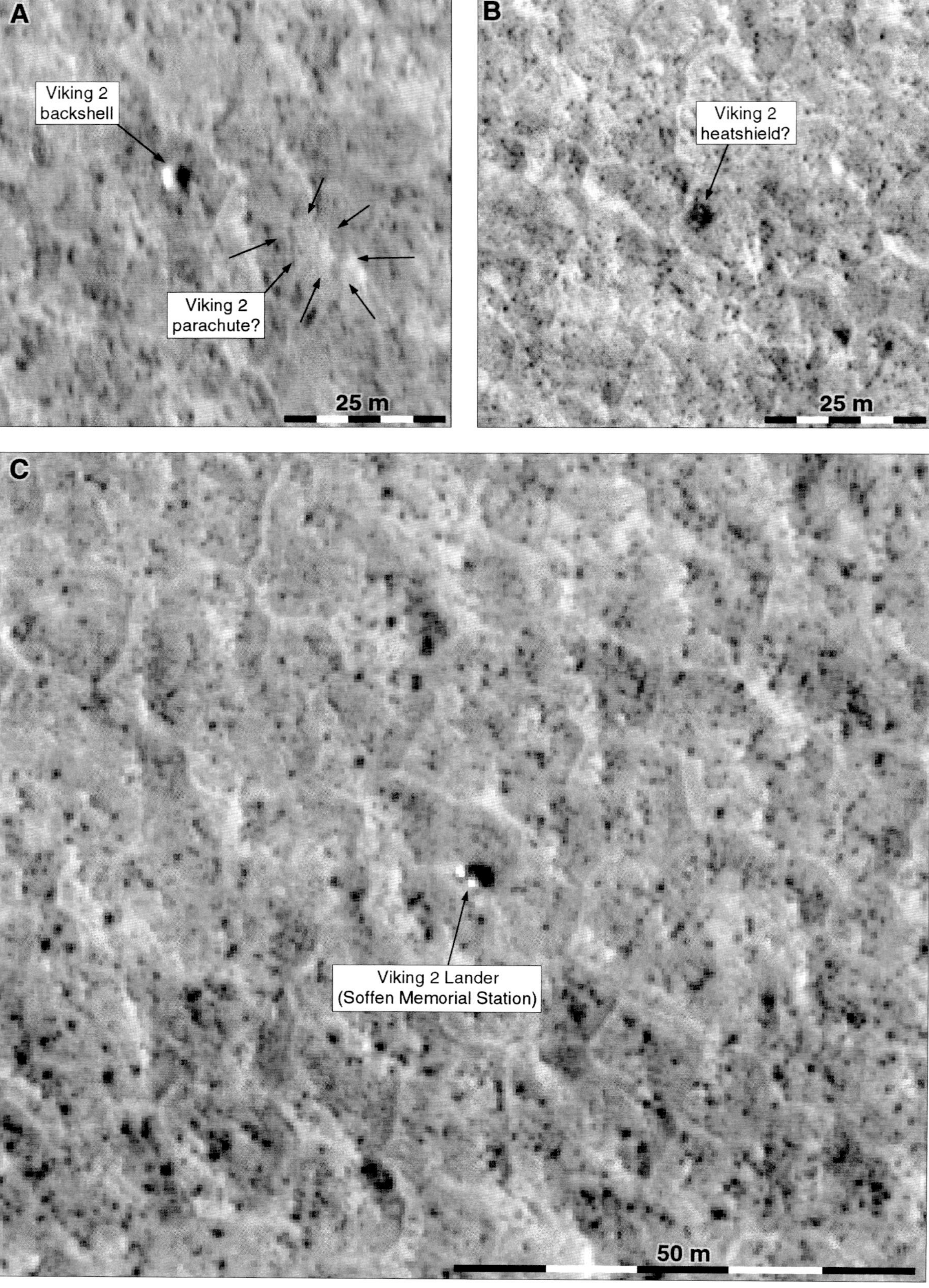

Figure 87 Viking 2 hardware on the surface of Mars, seen in HiRISE image PSP_001501_2280. The heatshield and parachute identifications are tentative.

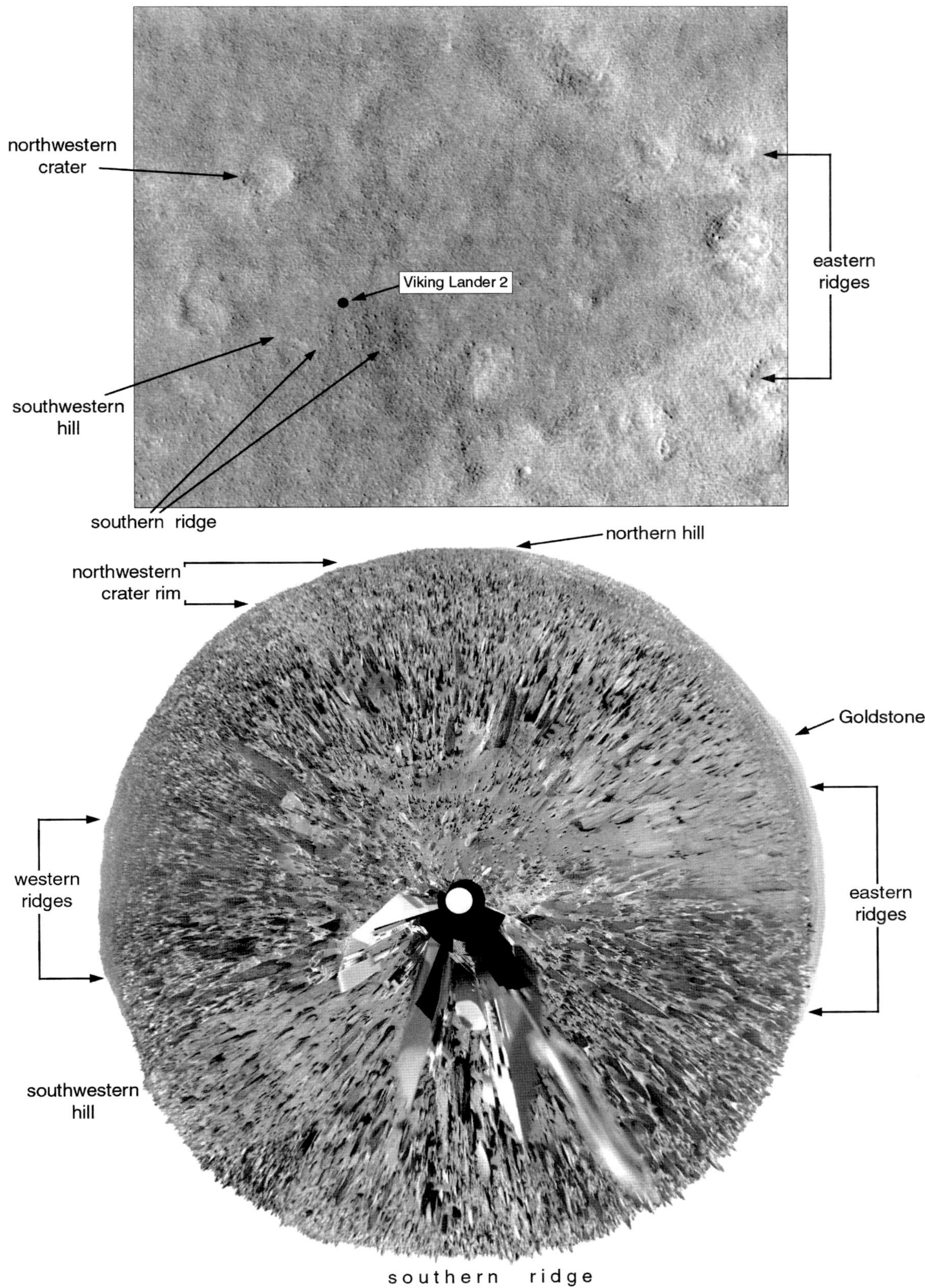

Figure 88 Comparison of HiRISE image PSP_001501_2280 and Viking 2 panoramas reprojected into a circular view with exaggerated horizon relief.

Table 28. *Prime Targets for Viking Low-Altitude Imaging*

Site	Location	Notes	Site	Location	Notes
1	14° N, 180° W	Amazonis irregular crater	42	37° N, 37° W	Northern plains
2	10° S, 175° W	Cratered terrain boundary	43	22° S, 27° W	Mare Erythreum
3	8° N, 174° W	Amazonis hills	44	30° N, 17° W	Cratered terrain boundary
4	40° N, 174° W	Propontis hills	45	57° N, 9° W	Frost-rimmed crater
5	34° S, 170° W	Cimmerium channels	46	3° N, 353° W	Edom cratered terrain
6	16° S, 156° W	Memnonia upland plains	47	3° S, 340° W	Edom channels
7	22° S, 152° W	Memninia escarpment	48	c. 37° N, 330° W	Ismenius channels, stereo
8	7° S, 150° W	Memnonia channel	49	25° N, 327° W	Arabia basin, colour
9	54° N, 147° W	Castorius Lacus crater	50	3° N, 322° W	Sabaeus Sinus craters
10	28° S, 144° W	Memnonia dark marking	51	9° S, 321° W	Sabaeus Sinus craters
11	3° N, 144° W	Amazonis furrowed ridge	52	19° S, 318° W	Mare Serpentis channels
12	23° N, 143° W	Amazonis grooved area	53	14° N, 314° W	Aeria channels
13	32° N, 143° W	Amazonis grooved area	54	c. 33° N, 300° W	Cratered terrain boundary
14	38° N, 137° W	Cratered terrain boundary	55	13° S, 303° W	Iapygia, stereo
15	17° N, 134° W	Nix Olympica, stereo	56	35° S, 300° W	Hellas stereo swath
16	11° N, 125° W	Fractured plain boundary	57	27° S, 295° W	Trinacria cratered area
17	3° N, 121° W	Volcanic crater	58	10° S, 294° W	Libya escarpment
18	27° S, 123° W	'Comet' (wind) tails	59	25° N, 284° W	Plateau edge
19	9° S, 120° W	South Spot, stereo, colour	60	4° N, 273° W	Libya edge, stereo
20	1° N, 114° W	Middle Spot, stereo	61	31° S, 268° W	Trinacria channels, colour
21	10° N, 105° W	North Spot, stereo	62	15° S, 260° W	Channelled crater rim
22	c. 42° N, 110° W	Arcadia volcanic complex	63	25° S, 255° W	Dandelion (Tyrrhena Pat.)
23	35° S, 100° W	Claritas fractures, colour	64	5° S, 250° W	River channel
24	24° N, 97° W	Mariotis fractured dome	65	33° S, 245° W	Southern plains
25	27° N, 93° W	Mariotis volcano, stereo	66	10° N, 243° W	Plains boundary
26	20° S, 92° W	Syria channeled plains	67	22° S, 235° W	Channelled terrain, colour
27	13° N, 90° W	Cratered dome	68	0°, 235° W	Broken plains, stereo
28	33° S, 88° W	Fractured uplands	69	20° N, 237° W	Valleys*
29	26° S, 81° W	Ridged plains	70	12° N, 214° W	Eunostos fresh crater**
30	15° S, 10–108° W	Grand Canyon, stereo, col.	71	27° N, 221° W	Rilles, stereo
31	c. 11° N, 68° W	Plains boundary, stereo	72	6° S, 217° W	Broken uplands
32	3° N, 75° W	Canyon	73	24° N, 215° W	Shield volcano, stereo
33	30° N, 72° W	Tempe cratered terrain	74	19° N, 216° W	Shield volcano (Albor)
34	45° N, 72° W	Tanais fractured area	75	31° N, 210° W	Volcanic dome (Hecates)
35	15° N, 55° W	Cratered terrain boundary	76	23° S, 212° W	Channelled terrain
36	c. 50° S, 45° W	Argyre, stereo	77	14° N, 198° W	Cerberus plumes, colour
37	62° N, 48° W	Dark crater	78	7° N, 192° W	Plumes (wind streaks)
38	27° S, 42° W	Red River (Nirgal Vallis)	79	20° S, 185° W	River channel (Ma'adim)
39	1° N, 44° W	Chaos, colour	80	86° S, 45° W	Water ice lake, colour
40	5° N, 37° W	Chaos	81	80° N, 210° W	Olympia polar cap remnant, stereo
41	18° N, 37° W	Rivermouth (Viking A-1)			

Notes: *No description in source. **214° W in source table, 219° W on source map.

Table 29. *Other Viking Orbital Data Targets*

29a. Viking high-altitude imaging targets

Site	Name	Location	Notes
1	South Spot	8° S, 123° W	Very repetitious white clouds – Arsia Mons
2	Arcadia	43° N, 114° W	spawning area for moving clouds
3	Nix Olympica NW	20° N, 135° W	Frequent white clouds in local summer ('July')
4	North Spot NW	15° N, 110° W	Frequent white clouds in local summer – Pavonis
5	Elysium	23° N, 213° W	Frequent white clouds
6	Albor	Not specified	Not specified, but also frequent white clouds
7	Libya	5° N, 266° W	High regional frequency of white clouds
8	North Hellas	37° S, 300° W	Transient yellow markings (dust storms)
9	Mare Serpentis	30° S, 320° W	Frequent origin of yellow storms
10	West Hellas	45° S, 305° W	Region of seasonal brightening
11	Utopia	Not specified	Not specified
12	Xanthe	23° N, 44° W	Active light area, low elevation, compare with 14
13	Aeria	13° N, 305° W	Active light area, compare with 17
14	Nilokeras canyon	27° N, 54° W	Dark area with unusual high contrast
15	Claritas	30° S, 100° W	Active light area
16	Solis Lacus	24° S, 85° W	Wind streaks behind craters, variable markings
17	Syrtis Major	12° N, 284° W	Light and dark streaks
18	Zephyria	8° N, 191° W	Crater with very long, dark wind streak
19	Hesperia	22° S, 241° W	Craters with light wind streaks
20	Marsissippi Delta	10° S, 37° W	Low-elevation channel, water candidate
21	Red River	29° S, 39° W	Meandering channel, water candidate
22	Pyrrhae Regio	17° S, 27° W	Very low elevation, water candidate
23	Chryse Canyon	15° N, 34° W	Low-elevation northern channel, water candidate
24	North cap boundary	82° N	Any longitude, boundary changes slowly
25	South polar clouds	35° S–60° S	Any longitude, polar cloud hood varies

29b. MAWD observation targets, medium and low altitude

Name (priority)	Location	Name (priority)	Location
N polar cap edge (1)	55° N, any longitude	Nepenthes-Thoth (2)	25° N, 250° W
Amazonis (1)	32° N, 160° W	Memnonia (2)	20° S, 150° W
Nix Olympica (1)	30° N, 140° W	Zephyria (2)	5° S, 195° W
Nix Olympica (1)	18° N, 134° W	Margaritifer Sinus (2)	0° N, 20° W
Nix Olympica (1)	20° N, 121° W	Deucalionis Regio (3)	15° S, 10° W
Chryse (1)	36° N, 32° W	Hellas (3)	35° S, 298° W
Elysium (2)	24° N, 216° W	Hellas (3)	44° S, 265° W
Aetheria (2)	41° N, 232° W	Aurorae Sinus (3)	5° S, 55° W

Table 29. *(cont.)*

29c. IRTM Special Interest Areas

Description	Location	Description	Location
Landing Site A	Not yet known	Roughened area	10° N, 127° W
Landing Site B	Not yet known	River junction	24° N, 61° W
Hellas	40° S, 300° W	Rough	2° N, 37° W
Canyonlands Delta	25° N, 40° W	Crater tails	13° N, 283° W
Nix Olympica south flank	26° N, 133° W	Crater tails	8° N, 192° W
North Polar Cap	> 85° N	Canyonlands	12° S, 66° W
West Nix Olympica plains	28° N, 160° W	Jumble	8° S, 33° W
South Tharsis	30° S, 120° W	Mound	23° S, 254° W
Syrtis Major Center	10° N, 290° W	IRR hot areas	27° S, 76° W
Iapygia	13° S, 305° W	IRR hot areas	27° S, 263° W
South Spot Caldera	9° S, 120° W	IRR hot areas	46° S, 115° W
Plateau bordering meander	19° N, 235° W	IRR hot areas	8° S, 86° W
South Polar Cap	70° S, 90° W	IRR hot areas	30° S, 290° W
South Cap, dark valley	85° S, 10° W	IRR cold areas	44° S, 213° W
Cerberus caldera	20° N, 210° W	IRR cold areas	0° N, 110° W
Parallel fractures	25° N, 107° W	IRR cold areas	32° S, 75° W
Elysium plain	23° N, 213° W	Note: IRR refers to the Mariner 9 infrared radiometer.	

Table 30. *Viking Lander 2 Nominal Mission Activities*

Sol	Activities
0	Viking 2 lands, first images taken, biology lid deployed, XRFS calibrated (MY 12, sol 253)
1	Sampler shroud ejected
3	GCMS lid deployed, and nine atmospheric analyses made on sols 3, 4, 5, 8 and 9
8	XRFS sample attempt, but arm failed to deliver sample to instrument
8	Biology 1 sample delivered to Pyrolytic, Labelled Release, and Gas Exchange instruments
10	Sampler head diagnostics
13	XRFS 1 sample delivered successfully, but did not contain desired rocks
14–17	XRFS sample analysis attempted made on all four sols, but no useful material in instrument
15, 16	GCMS atmospheric analyses, one on each sol
21	Backhoe Touchdown 1, GCMS 1 sample collected, then analyzed over sols 24 to 37
28	Biology 2 sample collected for Pyrolytic and Labelled Release analyses
29	XRFS 1 sample collected successfully, and analyzed over sols 29 to 42
30	Attempted rock nudge, but arm contacted ground and ICL rock was not touched or moved
34, 37	GCMS 2 sample – 'Mr. Badger' rock moved on both days, soil under rock sampled
40	GCMS 2 sample delivered and analyzed over sols 40 to 47
45	Biology double rock nudge, 'Notch' rock moved slightly
46, 47	XRFS 2 sample collected, two attempts per day, partially successful but not enough rocks
47–52	XRFS 2 sample analysis attempted, but sample is insufficient
51	Notch Rock pushed, Biology 3 sample collected beneath rock, delivered to Biology instruments
52–61	GCMS atmospheric analyses, four analyses conducted on sols 52, 53, 57 and 61
56, 57	Physical Properties 1, soil temperature data, imaging under lander
57, 58	XRFS 3 sample collected, two attempts per day, partially successful, but not enough rocks
57–60	XRFS 3 sample analysis attempted, but sample insufficient
80	Conjunction, possible seismic event (25 November 1976) (MY 12, sol 333)

Sol 28 the Biology 2 sample was dug from a trench overlapping Biology 1 in the crusty and cloddy material (Figure 93). The fine component was delivered to the Biology instrument, and any remaining coarse material was purged, but this time nothing was seen.

After the sampling on sol 8 failed to deliver anything to XRFS, a new attempt was made on sol 29. The XRFS 1 trench was dug just to the right of GCMS 1 in the crusty and cloddy material at Bonneville Salt Flats (Figure 93). The arm lifted Bonneville rock as it extended during trenching. This sample was called U-1 (Utopia-1) by the XRFS team. One scoop of fine material was delivered to the instrument, and coarse material, if any, was purged. On sol 30, a second scoop was collected at the same site to add to U-1. Coarser material was again purged near the sol 21 purge site, including several small objects. Then, later that day, the arm attempted to move the rock ICL. (Initial Computer Load, a rock at the location where a pre-programmed dig would have taken place if commands could not be received from Earth after landing. The name was pronounced 'ickle'.) The arm intentionally contacted the surface in front of the rock, leaving a mark (Figure 93), and was then raised to contact and move the rock. It either missed the rock or hit it but failed to move it.

The failed rock push on sol 30 was repeated with a new rock, Mr Badger, on sol 34. This time the rock was moved about 10 cm away from the lander and rotated slightly, and a small trench was dug in front of the rock (Figure 94). The same rock was pushed a little more on sol 37, moving it about 12 cm farther away from the lander. Then the GCMS 2 sample was dug from the surface where the rock had been to obtain material which had been buried for a long time. First the surface was cleaned by dragging the backhoe over it to remove surface material disturbed by the rock push activities, and then the sample was collected. This was repeated twice. The first time, a sample was delivered to GCMS; the second was purged on sol 40, but nothing was observed. Another rock push was attempted on sol 45, this time with the rock Bonneville in the Bonneville Salt Flats area. The arm contacted the surface and then was elevated and extended to move the rock. Bonneville was rotated away from the lander, but as the arm retracted the rock rolled back over the trench (Figure 94). A small rock fragment was seen to fall into the trench. Later that day another rock, Notch, was moved in the same way. This rock rotated and was moved away from the lander (Figure 94). On the next sol the arm was used to collect a sample (XRFS 2) from a trench at the Beta site in the crusty and cloddy material (Figure 93). This was an attempt to collect rock fragments. None were obtained, and apparently all the objects observed in images were clods which just broke up when disturbed. Two sampling strokes were made, and each was purged of fine material, but nothing was seen. Two more strokes were made on sol 47, and purged, and again nothing was seen.

Notch rock was pushed again on sol 51, rotated and moved away from the lander. Any material in the scoop was then purged, forming a small pile between Notch and the lander (Figure 94). Then the Biology 3 sample was taken from under the rock's original position, and its fine material was delivered to the Biology instrument. The soil was exposed for no more than 30 minutes before sample collection at the request of the biologists. The coarse fraction was purged, and some fragments 2 mm across or larger were observed to have fallen on the purge site. The regular wind patterns observed until now suddenly changed for several days beginning on sol 51, as northern hemisphere winter approached. With conjunction and the end of the nominal mission not far away, attention now moved to the physical properties investigations. On sol 56 a shallow trench, Physical Properties 1, was dug in the Alpha area near the centre of the sample field (Figure 95). The trench was shallow partly because of the slope of the surface in this location and partly because the arm encountered and moved a rock buried just under the surface. The temperature sensor was operated here, finding soil temperatures of about 272 K. Then images of the scoop were taken using a magnifying mirror on the lander body to examine its collector head, and three images were obtained using another mirror to look at footpad 2 from a viewpoint the camera alone could not provide. The footpad carried a temperature sensor, but it was not clear whether it was measuring the temperature of the soil or the air. These images did not show the sensor, but images taken later in the extended mission showed it to be buried.

On sol 57, three images were taken using the mirror to examine an area under the front of the lander. A pit or crater eroded by the descent engine exhaust was seen (Figure 106). Later on that sol, the Physical Properties 1

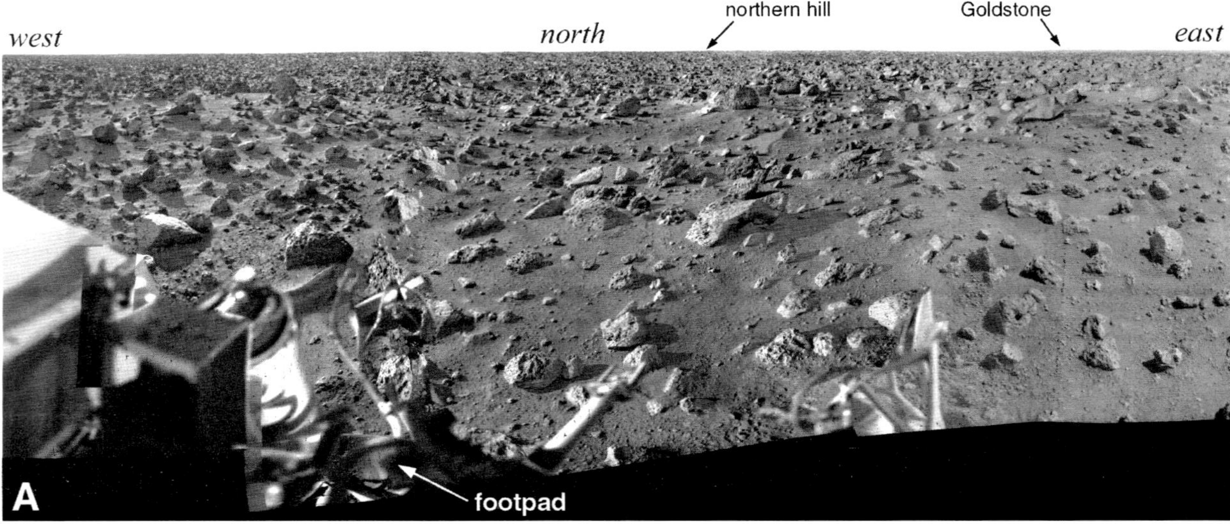

Figure 89 Viking Lander 2 surface panoramas.
A (across both pages): The full panorama. **B (across both pages):** The sample area. **C:** The northern hill, shown with ten times vertical exaggeration. **D:** Goldstone and the eastern ridges, with five times vertical exaggeration. Goldstone is the prominent hill on the horizon. The darker hills in the middle distance have not been identified.

Figure 89 (continued)

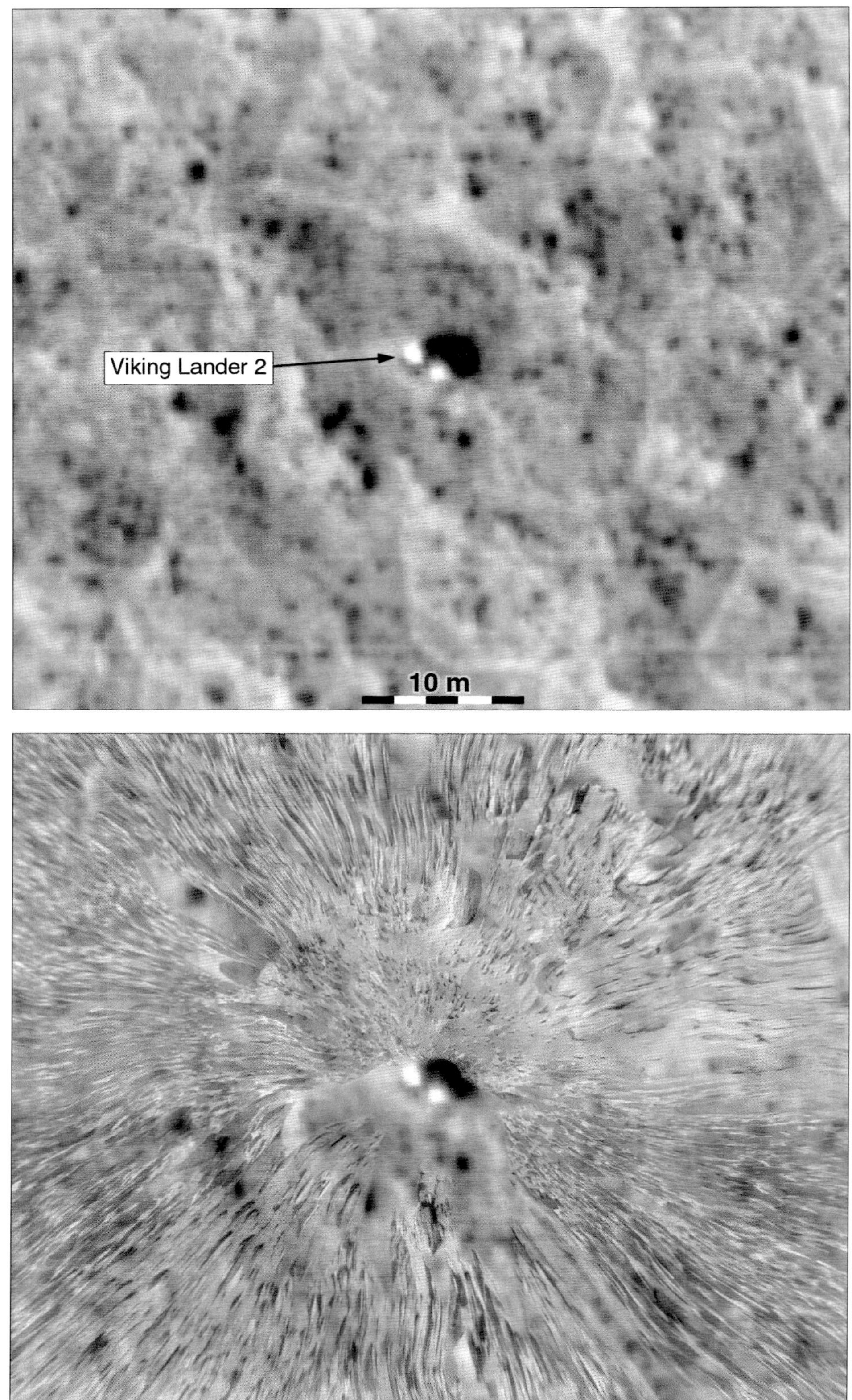

Figure 90 Viking 2 landing site in HiRISE image PSP_001501_2280 (top), and overlaid with a reprojected panorama (bottom).

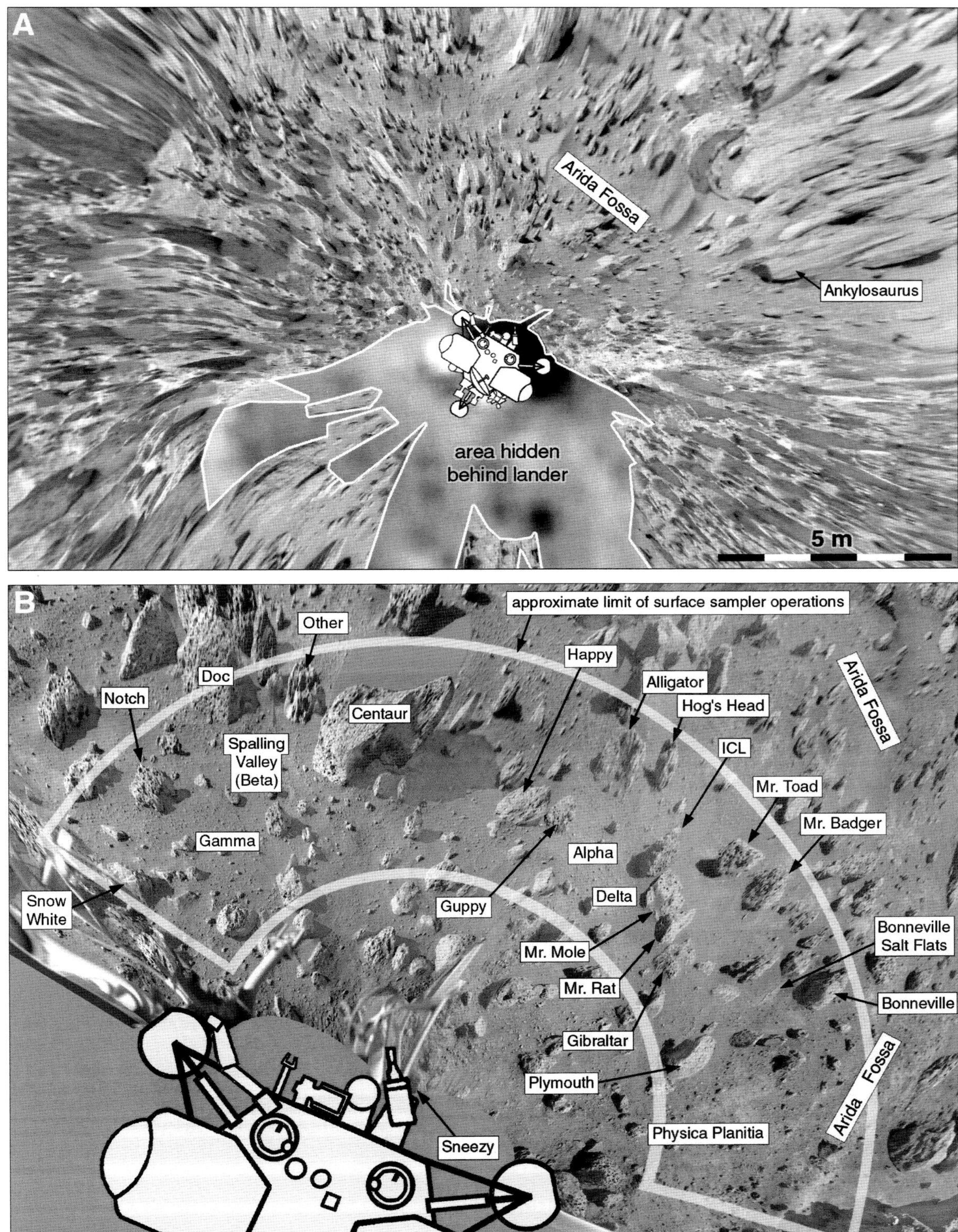

Figure 91 A: Viking Lander 2 vicinity. **B:** Sample field photomap compiled from reprojected panoramas. The surface is shown just before sampling operations began.

Figure 92 Viking Lander 2 shroud ejection and first sample trench.

sample was purged and a few dropped objects were seen, and finally, sample XRFS 3 was dug from the Alpha area using two digging strokes (Figure 95). Two more strokes on sol 58 added to that sample, extending the left side of the sol 57 trench. The surface was broken into clods, but very little coarse material was delivered to XRFS, too little to be analyzed. Fine material was purged after each of the four digs on these two sols. At the end of sol 58, the arm was parked for conjunction and not moved again for two months. Nominal mission events are illustrated in Figures 92 through 95 and mapped in Figure 96.

Recorded meteorological and seismic data transmitted after conjunction included a possible seismic event early in the morning of sol 80. Wind data were not collected continuously and did not cover this period. Most signals from the seismometer were caused by concurrent wind gusts, but the early morning was usually still, so the interpretation as a seismic event is plausible but not certain. No other such events were detected during the Viking mission, suggesting a very low rate of seismic activity.

Sampler activities began again on sol 131 with the collection of a single-scoop sample (U-2) from trench XRFS 4 at the original location of Notch rock. Enough fine material was transferred to the XRFS instrument for a successful analysis, and the coarse fraction was dropped on the 'rock pile', an area adjacent to the Biology 3 trench (Figure 97). No images were taken to observe any fragments. Similarly, on sol 145 the Biology 4 sample was taken from the Beta area (Figure 98) and the fine component delivered to the Biology instrument. Again the coarse fraction was dropped on the rock pile, but no images were taken. Later that same sol, another attempt was made to collect rocks from trench XRFS 5 in Spalling Valley. This rock hunt involved four digging strokes in the crusty/cloddy material, comprising two separate trenches, with the left trench superimposed on the right one. The fine component was purged and dispersed each time, and the coarse material was added to the rock pile. Images showed several new fragments on the pile.

On sol 154 the arm was positioned so an image of footpad 2 would appear in mirror 1 on the left side of the arm. This was another attempt to observe the temperature sensor, but it was not seen in the image. Then on sol 161, trench XRFS 6 was dug to provide another sample, U-3 (Figure 97). This sample was taken from Spalling Valley, and its collection required two digs, with the left trench being partly filled by debris from the right trench. Fine material was delivered to XRFS each time, and the coarse fraction was dropped on the rock pile. A large dust storm was first observed on sol 165. More material was collected from Spalling Valley on sol 172, from a new trench called XRFS 7. This was another rock hunt, with four digs from which the fines were dispersed and the coarse material, if any, dropped on the rock pile. A few new objects were seen in images. The name 'rock pile' is an exaggeration, but fragments were beginning to accumulate here (Figure 97).

Sampling continued with two attempts to collect fine material for XRFS on sols 185 and 186. One dig each day was made at trench XRFS 8 (Figure 97) under the original position of the rock Badger, which had been moved on sols 34 and 37. Sample U-4 was collected over the two sols and was sufficient to fill the instrument chamber. Each time the coarse fraction was purged over the rock pile, and images showed new additions to the pile, the largest about 1.2 cm across.

An arm problem on sol 195 prevented the collection of Biology and XRFS samples at the same location as XRFS 8 and a planned cleaning of the backhoe magnets. After sol 220, activities were curtailed because winter temperatures dropped below a critical level (–65° C), a period called the Survival Automatic Mission. Limited data were recorded and transmitted as Orbiter 2 passed overhead every twelve days. SAM lasted from sol 220 to sol 373, and on its first sol, temperatures dropped to the condensation point of carbon dioxide, though no CO_2 frost was seen. The last gas exchange analysis ended on sol 230, and the last labelled release analysis on sol 260. A major dust storm was first seen on sol 269 (MY 12, sol 522). Events between sols 100 and 300 are shown in Figures 97 and 98 and mapped in Figure 99. Some events in the images are shown out of chronological order to retain spatial continuity.

After the sampler problem on sol 195, the arm was not used for nearly 200 sols, or about 7 months. A travelling wave tube amplifier on the lander, required for direct transmission to Earth, failed on sol 387, after which all data transmission had to be sent via an orbiter. Activities began again on sol 388 with Backhoe Touchdown 2 and the start of the next trench, XRFS 9 in Physica Planitia (Figure 100). A sample, U-5, was collected here from

Sol 28 - Biology 2

Sols 46, 47 - XRFS 2

Before sampling

Sol 21 - Touchdown 1, GCMS 1

Sol 29 - XRFS 1

Sol 30 - XRFS 1 repeat

Before contact

Sol 30 - touchdown prior to failed rock move (ICL)

Figure 93 Viking Lander 2 activities during the primary mission.

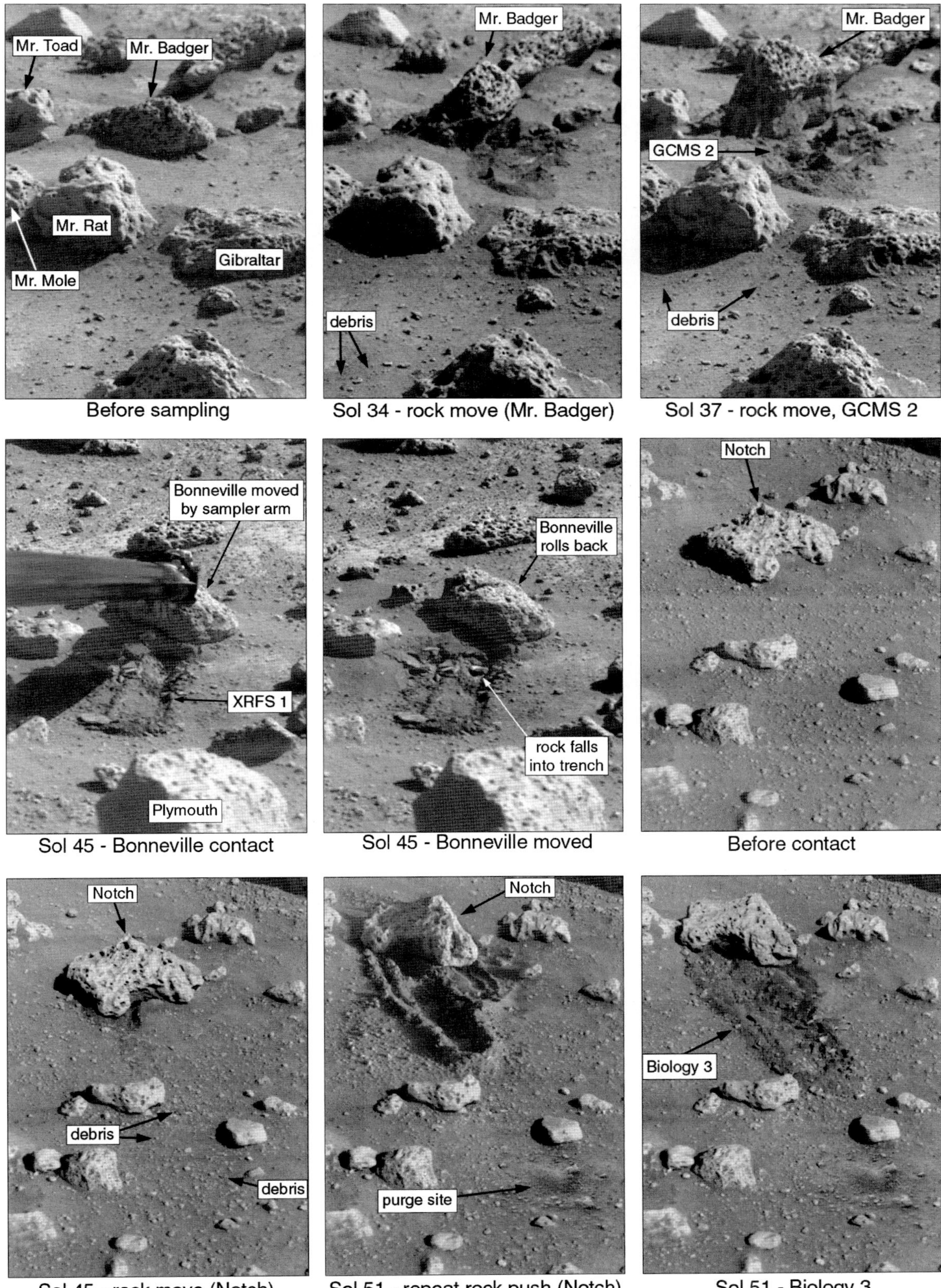

Figure 94 Viking Lander 2 activities during the primary mission (continued).

Figure 95 Viking Lander 2 activities during the primary mission and later purge events.

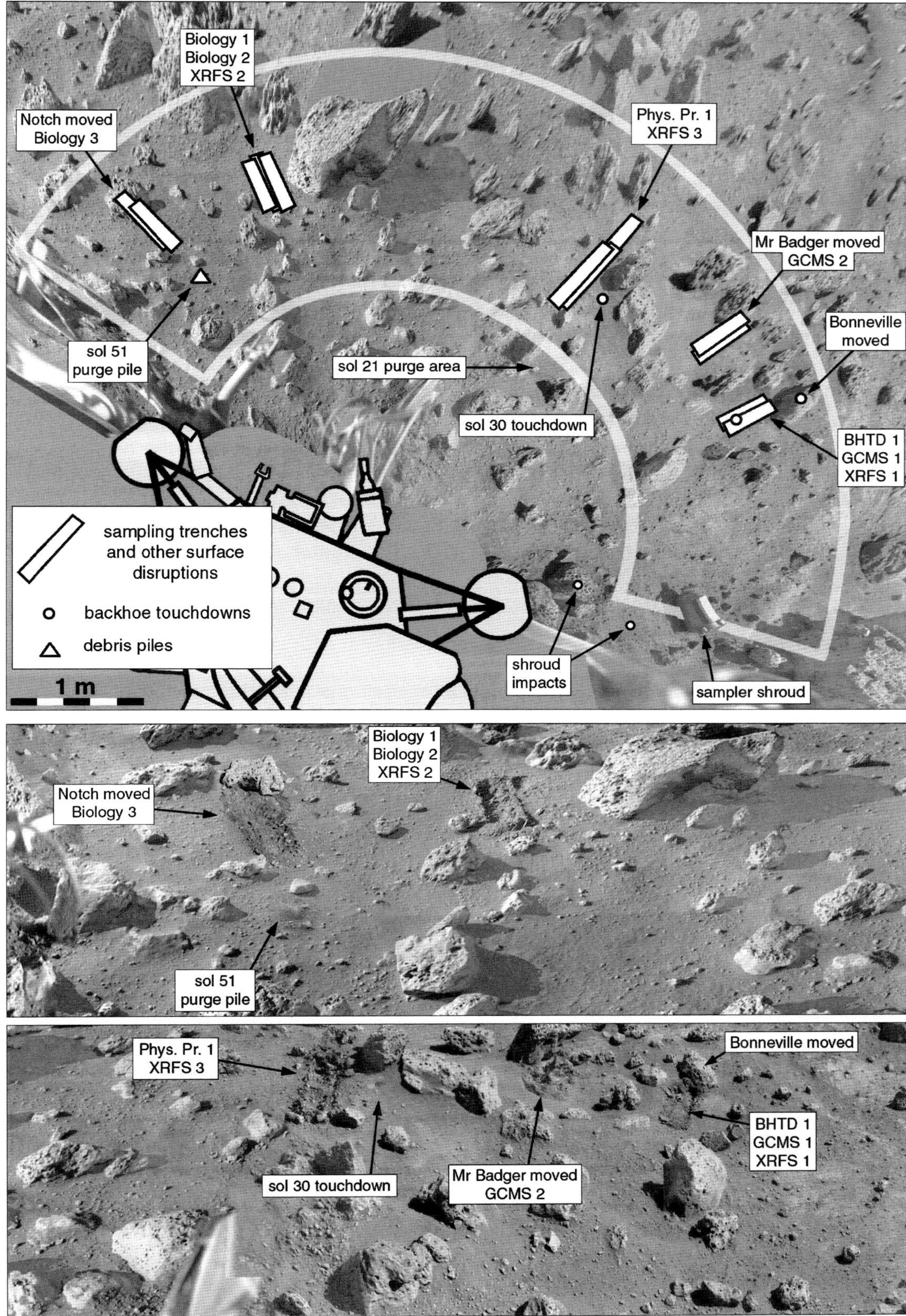

Figure 96 Viking lander 2 primary mission activities.

Table 31. *Viking Lander 2 Extended Mission Activities, Sols 100–300*

Sol	Activities
103	First commands of extended mission transmitted to lander (MY 12, sol 356)
131	XRFS 4 sample collected near Notch, then dumped to make pile of fragments
145	Biology 4 sample collected near Centaur, dumped to make pile of fragments XRFS 5 sample collected, fine material discarded, coarse fraction dumped in pile
153	Pyrolytic Release Experiment switch fails, instrument ends operations
161	XRFS 6 sample collected, fine material delivered to XRFS, coarse fraction dumped in pile
165	First dust storm activity detected by lander (MY 12, sol 418)
172	XRFS 7 sample collected, fine material discarded, coarse fraction dumped in pile
185–186	XRFS 8 sample collected near Mr Badger, dumped to make pile of fragments
195	Arm sequence aborted, preventing Biology and XRFS sample collection and magnet cleaning
208	GCMS part fails, ended organic analysis operations
214	Frost observed on surface for the first time
220	Survival Automatic Mission starts (limited activity during winter) (MY 12, sol 473)
230	Last Gas Exchange analysis before helium supply depleted
232	Atmosphere clearing, dust storm ending, Mars at perihelion on sol 233
252	Arm problem limits reach, extensions beyond 259 cm prohibited after this date
260	Last Labelled Release Experiment analysis ends
269	First effects of second major dust storm observed by lander (MY 12, sol 522)
281	Extensive frost seen, covering almost the entire surface around the lander

two overlapping trenches and successfully delivered to the instrument. Earlier during the SAM, XRFS had tried to gather enough wind-blown dust for analysis, without success. On sol 404 more temperature measurements were made at Physical Properties 2, a site in the middle foreground of the sampling area. The arm was inserted into the soil on sol 405 and removed on the following sol after taking temperature data through the diurnal cycle. The temperatures ranged from 171° K overnight to 218° K in the mid-afternoon. Soil remaining in the scoop after it was retracted on sol 406 was dumped to form Conical Pile 1 (Figure 101), a 5-cm high mound paced on top of the rock Centaur. This and other conical piles would be examined over the remaining lifetime of Viking 2 to look for wind erosion effects.

Viking 2 next began the excavation of a large trench, Deep Hole, to provide access to material which had been buried for a considerable time and protected from the surface environment. On sol 417 the arm made four sets of six consecutive retractions, using the backhoe to move loose material (Figure 98). By the end of the sol the trench was 7 cm deep, 25 cm wide and about 67 cm long, measured from its outer end to the crest of the soil pile at the end closest to the lander. After digging the big trench, the arm made a final digging stroke to collect a sample (U-6) from an ephemeral trench, XRFS 10 inside Deep Hole. The fine fraction was delivered to XRFS for analysis. Then on sol 442 the process was repeated. Four sets of six retractions were made, this time a little closer to the lander to extend the trench. When finished the trench was 12 cm deep, 27 cm wide and 80 cm long. After the digging was completed, two sample scoops were taken from trench XRFS 11 (Figure 98), and the fine part of the sample (U-7) was delivered to the instrument.

On sol 471 the rock Snow White was pushed by the arm, moving 5 cm away from the lander and rotating slightly (Figure 101). The arm did not appear to scratch or mark its surface, giving an indication of Snow White's hardness, but left a mark in the soil in front of the rock. Then on sol 479 the arm touched the surface (Backhoe Touchdown 3) at the site of the next trench. This was Physical Properties 3, and the sampling stroke

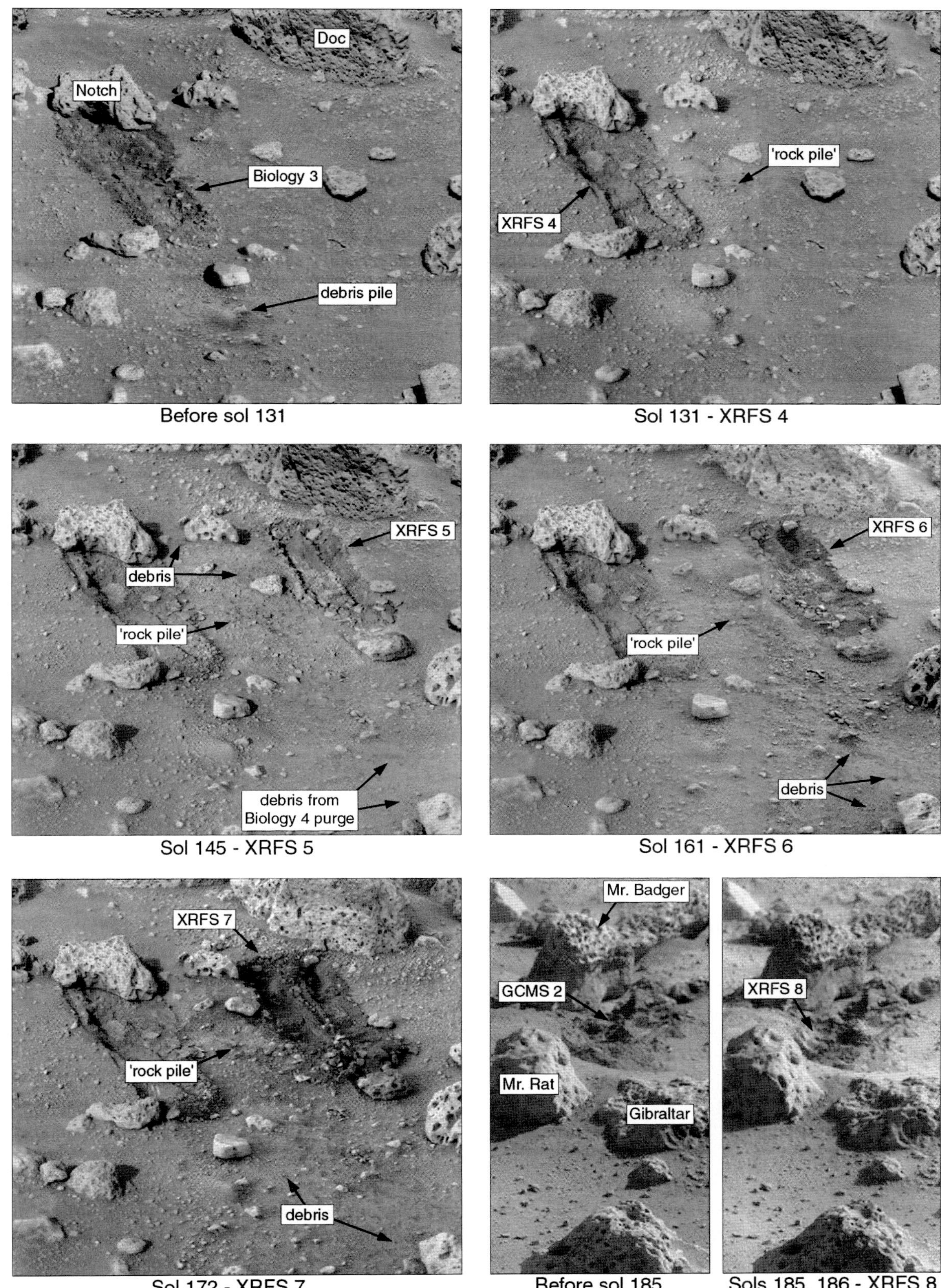

Figure 97 Viking Lander 2 activities between sols 100 and 300.

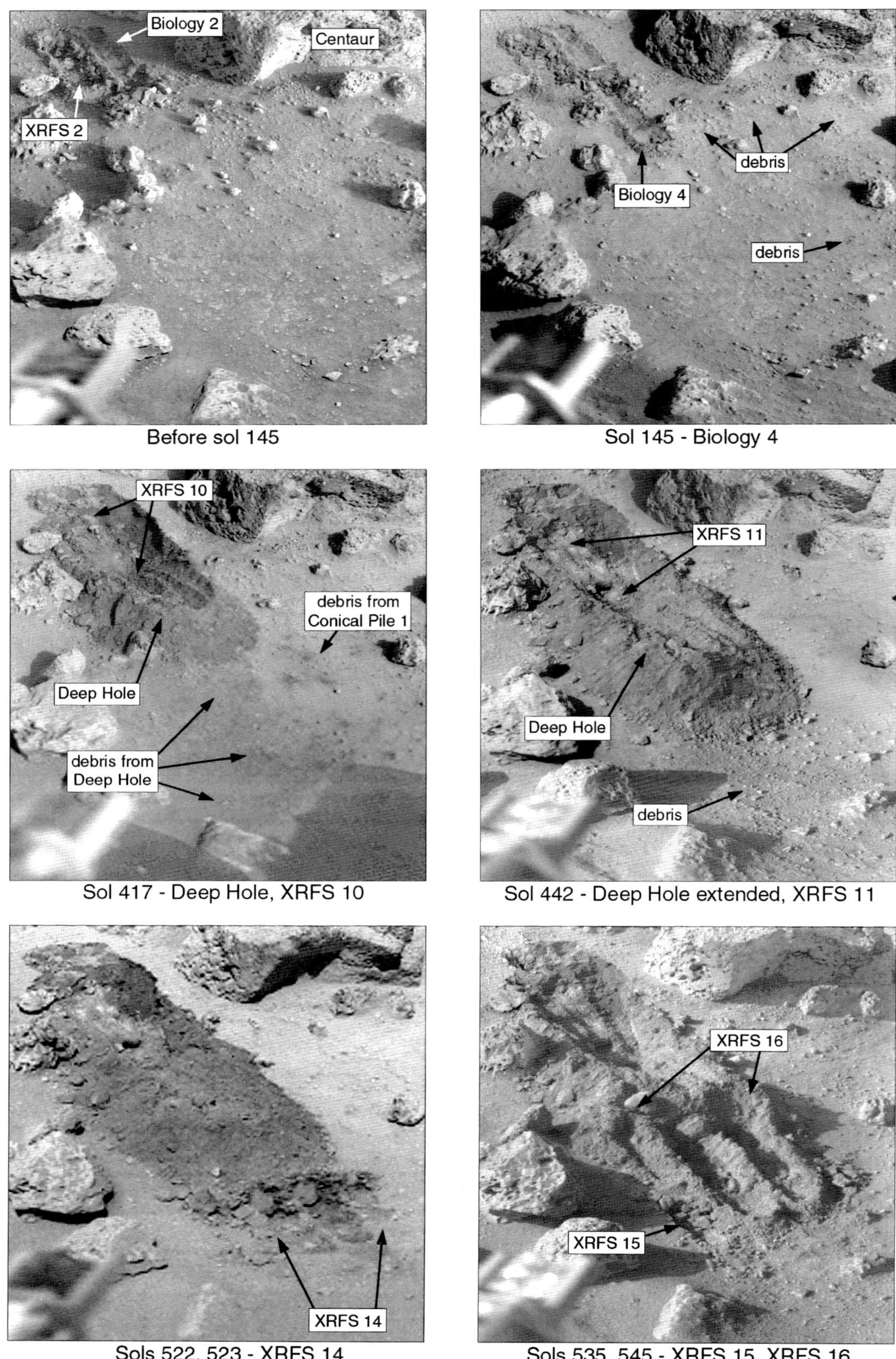

Figure 98 Viking Lander 2 activities near Centaur, including the excavation of Deep Hole.
The last four stages are included here for spatial continuity but occurred in the following time period.

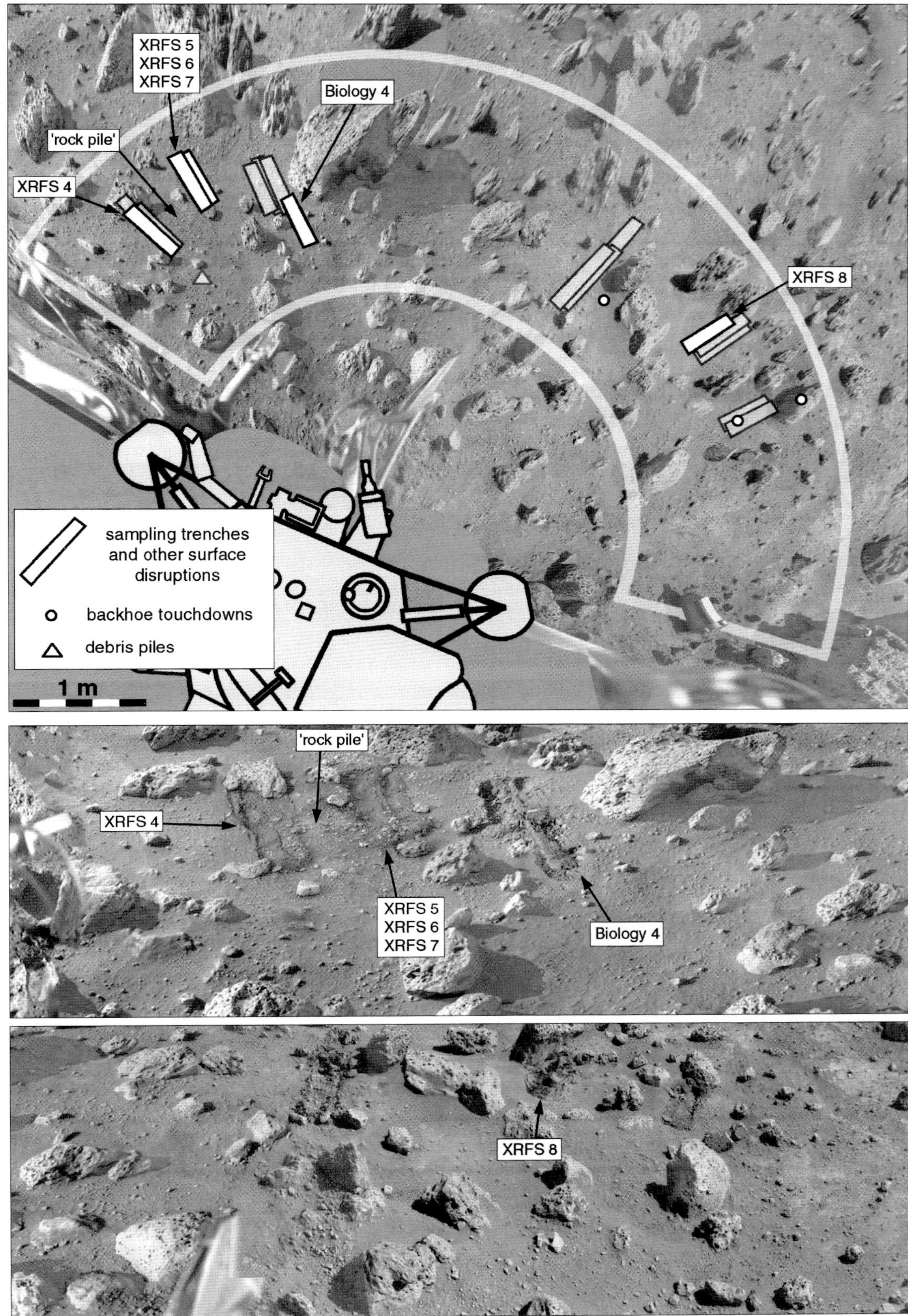

Figure 99 Viking Lander 2 activities, sols 100 to 300.

Table 32. *Viking Lander 2 Extended Mission Activities, Sols 301–1281*

Sol	Activities
310	Gas Exchange experiment, first of five heat of hydration experiments (last on sol 427)
358	Maximum wind gust velocity of the mission recorded, 30 m/s (MY 12, sol 611)
387	Electronics failure renders direct links to Earth impossible, orbiter relay required in future
388	Backhoe Touchdown 2 near shroud, XRFS 9 sample collected after touchdown
405–406	Physical Properties 2 test, diurnal temperature measurements
406	Conical Pile 1 constructed on Centaur
417	Deep Hole excavated near Centaur, XRFS 10 sample collected from Deep Hole
442	Deep Hole extended, XRFS 11 sample collected from Deep Hole
450	Frost no longer present in images after this date as Mars moves into northern spring season
471	Sampler arm touches down in front of Snow White, then moves forward to push rock
479	Backhoe Touchdown 3 near shroud. Physical Properties 3 sample near shroud, fines purged, then coarser part purged
483	XRFS 12 sample attempt, no sample collected
502	Backhoe Touchdown 4 on trench tailings near Badger
506	XRFS 13 sample collected, fine material to comminutor, coarse material dumped
522–523	XRFS 14 sample collected from base of Deep Hole tailings, coarse material delivered
522	Backhoe Touchdown 5 near Notch rock
535	XRFS 15 coarse material sample collected from base of Deep Hole tailings
545	XRFS 16 coarse material sample collected from base of Deep Hole tailings
555	Arm problem, inorganic analysis sample not obtained
559	Physical Properties 4 trench dug with backhoe, attempt to chip Mole rock unsuccessful
562	Digital memory failure, (check) only tape recorder storage remains until sol 664
595	Physical Properties 5 sample taken near shroud, dumped to make Conical Pile 2 nearby Backhoe Touchdown 6 in front of lander Physical Properties 6 sample in front of lander, dumped to make Conical Pile 3 near Centaur Physical Properties 7, bearing test near base of Rat rock
595–596	XRFS 17 sample collected near Bonneville rock
596	Surface sampler parked, assumed to be end of operations, but used again on sol 957
609	Northern summer solstice, seismometer turned off
632	Lander Continuation Automatic Mission (LCAM) starts
664	Memory unexpectedly becomes useable again, operations resumed
669	Landing anniversary, one Mars year on the surface
841	Final LCAM relay, final data relay before solar conjunction
862	Communications resume after conjunction, via Orbiter 1
871	Start of final mission stage (long-term automatic mission). Perihelion on sol 874
957	Final sampler operations: Physical Properties 8, Conical Pile 4, Physical Properties 9
958	Sampler arm parked permanently (MY 13, sol 543)
1050	Real-time imaging and other data relayed by Orbiter 1
1281	Battery failure, end of lander mission, 11 April 1980 (MY 14, sol 197)

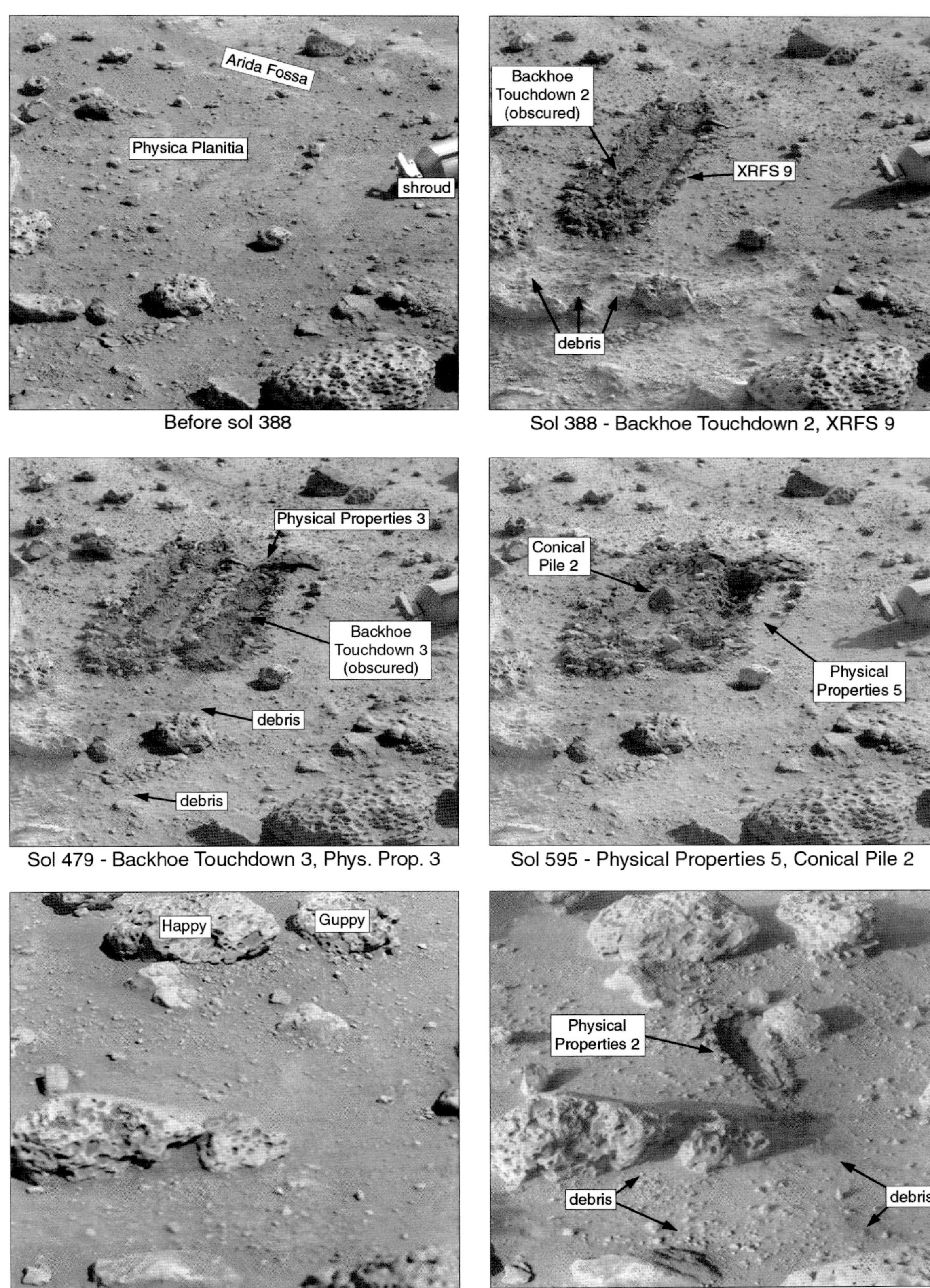

Figure 100 Viking Lander 2 activities, sols 301 to 500.

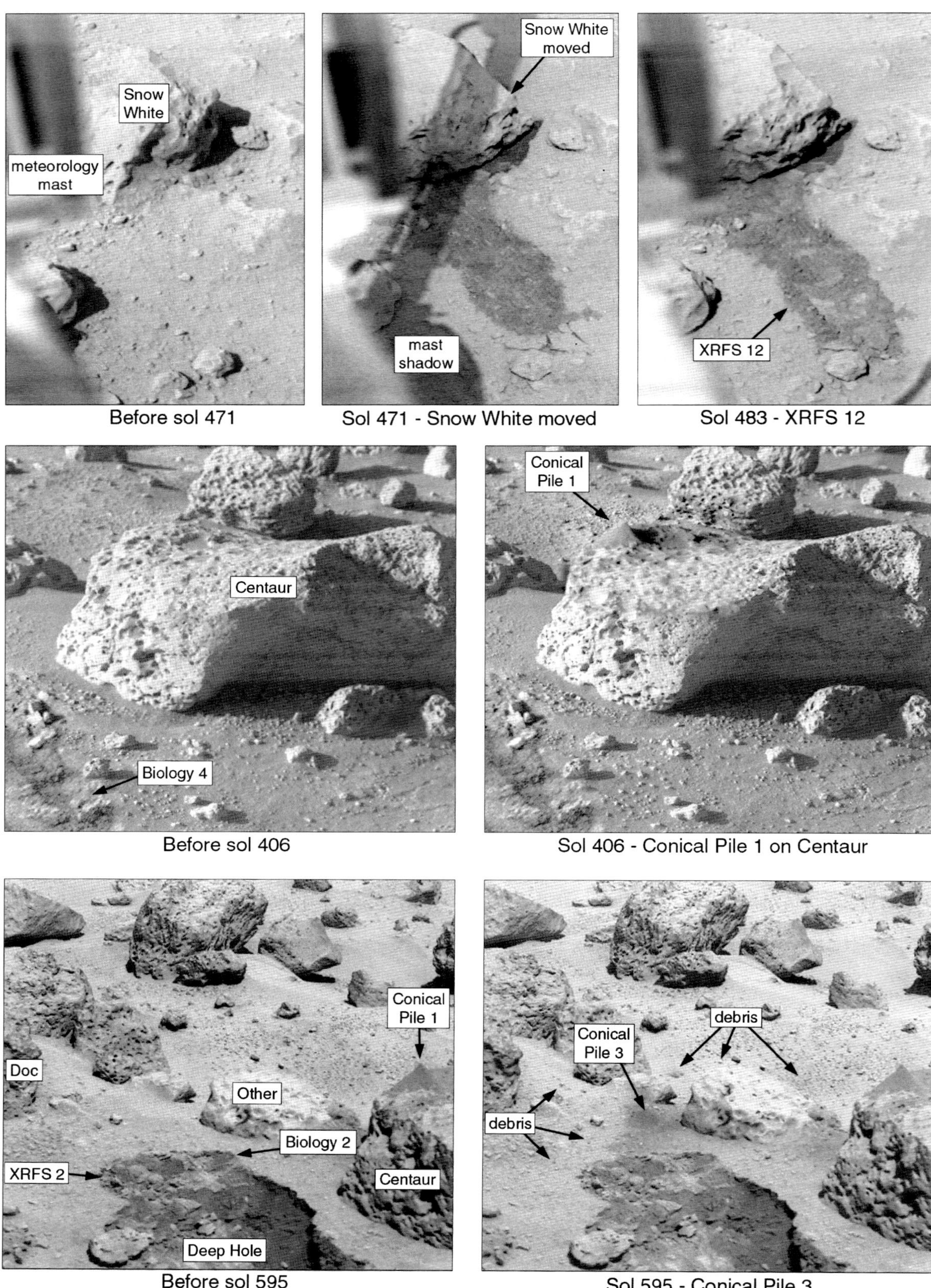

Figure 101 Viking Lander 2 activities after sol 301.

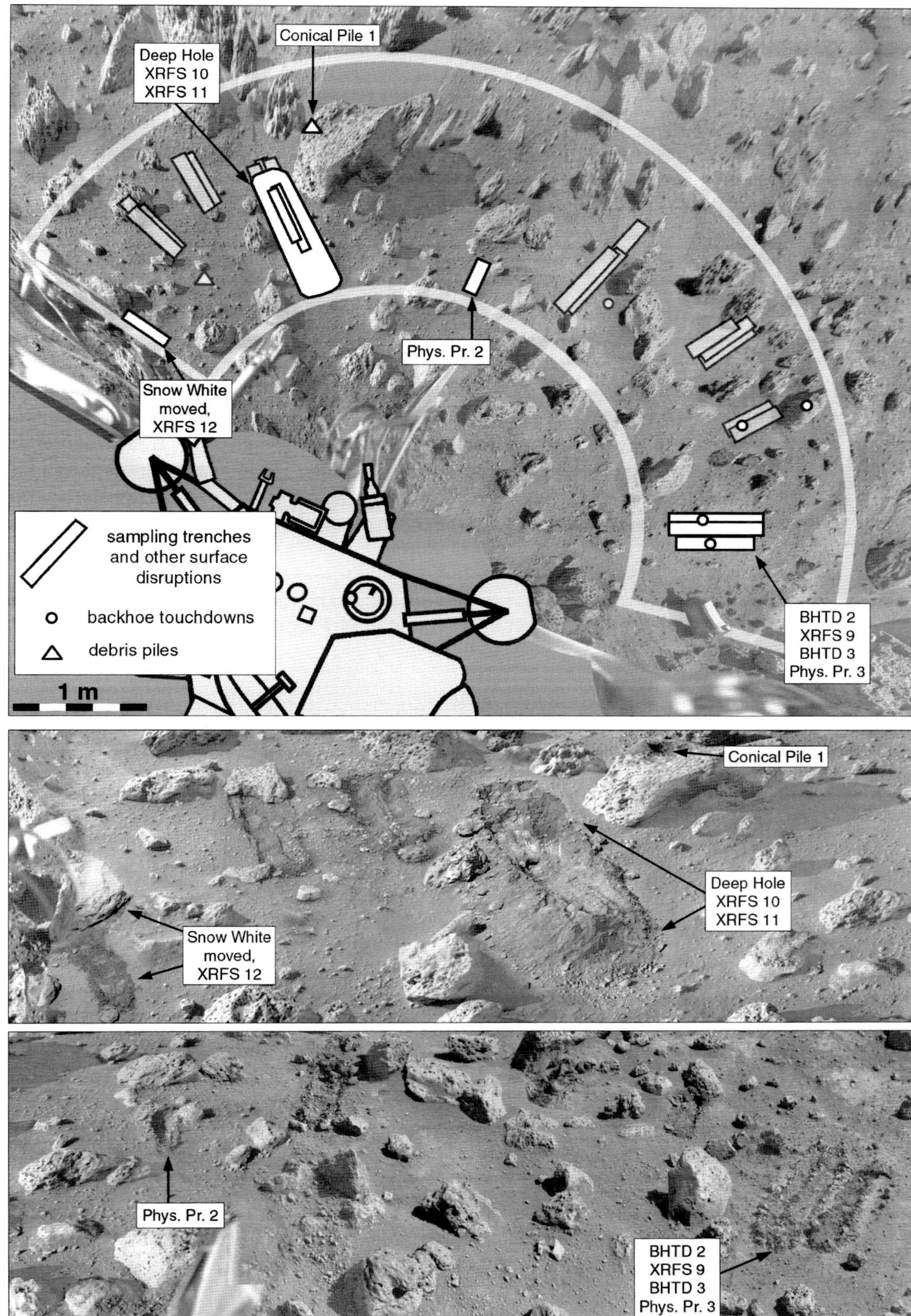

Figure 102 Viking Lander 2 activities, sols 301 to 500.

dug the trench in the crusty/cloddy material by extending from the touchdown point and then retracting again (Figure 100). The sample was lifted up, and its fine component was dispersed. Then any coarse material was purged, and fragments up to 0.8 cm were observed. Later on that sol, images of footpad 2 were finally obtained using mirror 1 on the arm (Figure 106). They showed that the temperature sensors were exposed to the atmosphere, not buried. These sensors on both landers were used during the parachute descent but survived the landing and gave surface data as well. Sampling continued on sol 483 with a dig (XRFS 12) near Snow White rock, which had been moved earlier, to obtain a sample which had been sheltered or protected underneath the rock. The arm did not reach the intended area, probably because it struck a buried rock, so no sample was collected. Events between sols 300 and 500 are illustrated in Figures 100 and 101 and mapped in Figure 102.

A magnetic test was conducted on sol 502, as the arm touched the surface at the Backhoe Touchdown 4 site on disturbed soil in front of Mr Badger (Figure 103). The magnetic properties of the soil were investigated by observing particles adhering to magnets on the backhoe. On sol 506 multiple arm extensions and retractions were made in trench XRFS 13 in Bonneville Salt Flats to obtain a sample for XRFS. The trench showed possible layering in the wall at its outer end, in addition to many clods and platy lumps where the crusty soil was broken up by digging (Figure 104). Fine material was delivered to GCMS for comminution, and the coarse fraction was purged. The grinder in GCMS was operated on sol 522 and recorded the electrical current used as its motor comminuted the sample to give an estimate of mineral grain hardness.

The arm returned to Deep Hole on sol 522 and 523, making 16 attempts over two sols to dig into the ground at the foot of the tailings pile in front of the large trench. This set of digs constituted XRFS 14 (Figure 98), but the arm did not extend as far as it was supposed to, and no useful sample of the desired coarse material for XRFS was collected. After each dig the fine fraction was dispersed. In the middle of the digging sequence on sol 522, the arm was moved to touch the surface near Notch (Backhoe Touchdown 5, Figure 103). The failed sampling was repeated on sol 535 with another 16 digs at the base of the Deep Hole tailings, this time called XRFS 15 (Figure 98). The arm strokes were all a little short of the intended length. Again, fine material was purged after each of the 16 digs. No images were obtained to document these activities.

Then on sol 545 another attempt was made to collect material for XRFS, this one called XRFS 16. Twelve strokes were made, of which 11 were slightly shorter than intended. Images showed several holes cut into the tailings this time (Figure 98), but still no coarse material was collected. Any fine material was dispersed after each dig. On sol 555 images were taken using mirror 2 on the arm to observe the area underneath engine 2 (Figure 106). These revealed three small pits or craters dug in fine-grained material by the engine exhaust, roughly 6 cm wide and 1 cm deep. A rock called Sneezy was also seen in these images. Then a trench called Physical Properties 4 was dug on sol 559 by performing eight backhoe retractions. The trench was about 4 cm deep and its tailings were mostly fine grained, but one 10-cm rock was also excavated. After that, the arm was placed on the rock Mole and pushed downwards. The rock did not move or appear to be scratched or chipped (Figure 103).

The arm was next used on sol 595. First, a sample was dug from Physica Planitia (Physical Properties 5) using two extensions and retractions (Figure 100). The sample was dumped to make Conical Pile 2 in the adjacent XRFS 9 trench. Then the arm was placed on the surface (Backhoe Touchdown 6, Figure 104) and extended and retracted to dig the Physical Properties 6 trench in the Alpha area. This was supposed to provide a measure of surface strength, but the data were lost. A picture of the dirty collector head was taken. Then the soil was dumped out to make Conical Pile 3 among a group of rocks (Figure 101). This was only about 1.5 cm high. Another picture was taken to view the collector head, which still had a film of dust adhering to it. Later that afternoon the arm performed a surface bearing test (Physical Properties 7, Figure 103) in an area called Delta. As it pressed down, the surface around it buckled, allowing estimates of soil properties. The day ended with the start of a two-day operation to collect a final sample for XRFS. This was XRFS 17, in Bonneville Salt Flats. Two digging strokes were made on sol 595 and eight more strokes on sol 596. Clods and lumps

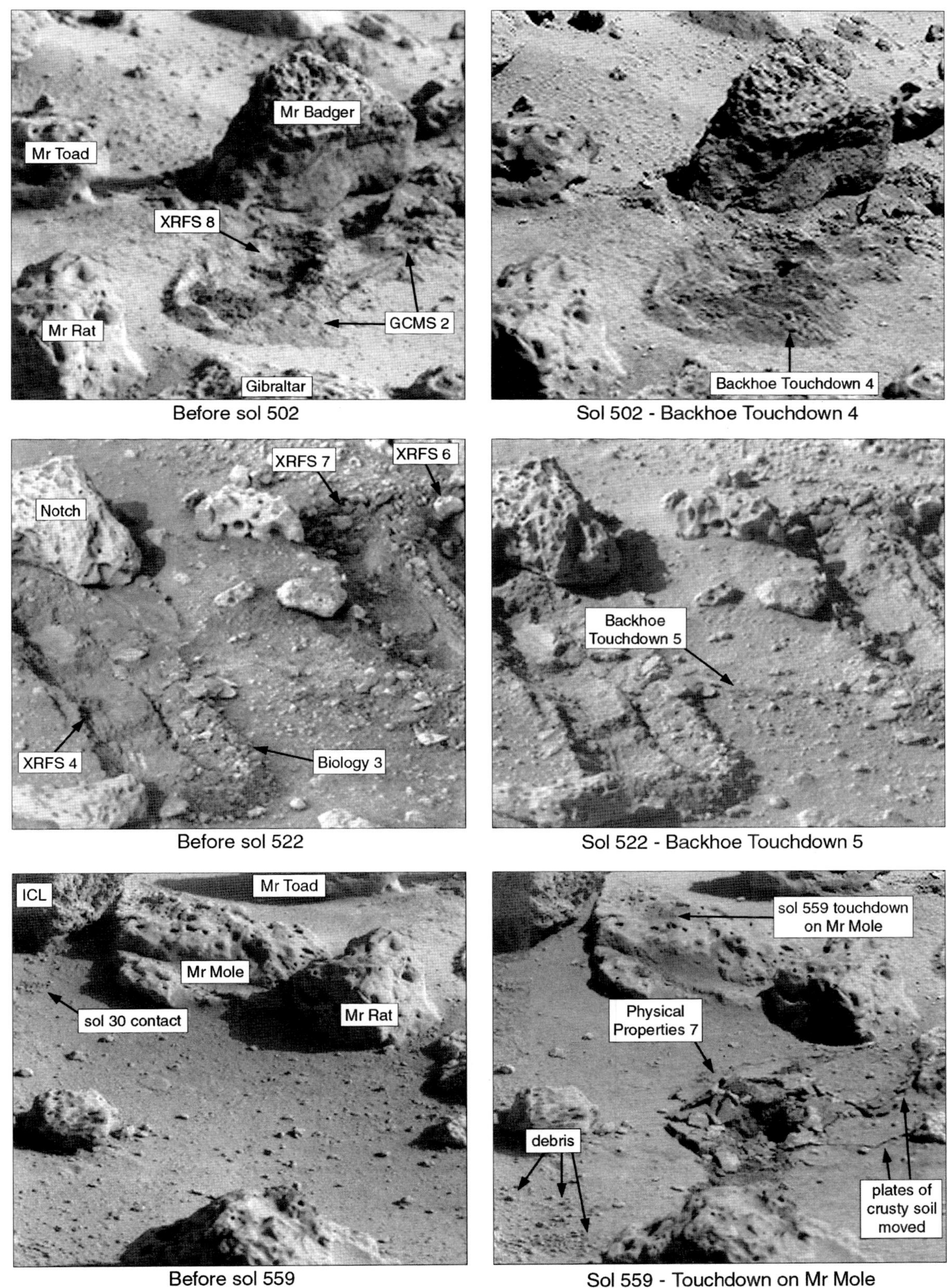

Figure 103 Viking Lander 2 activities after sol 501.

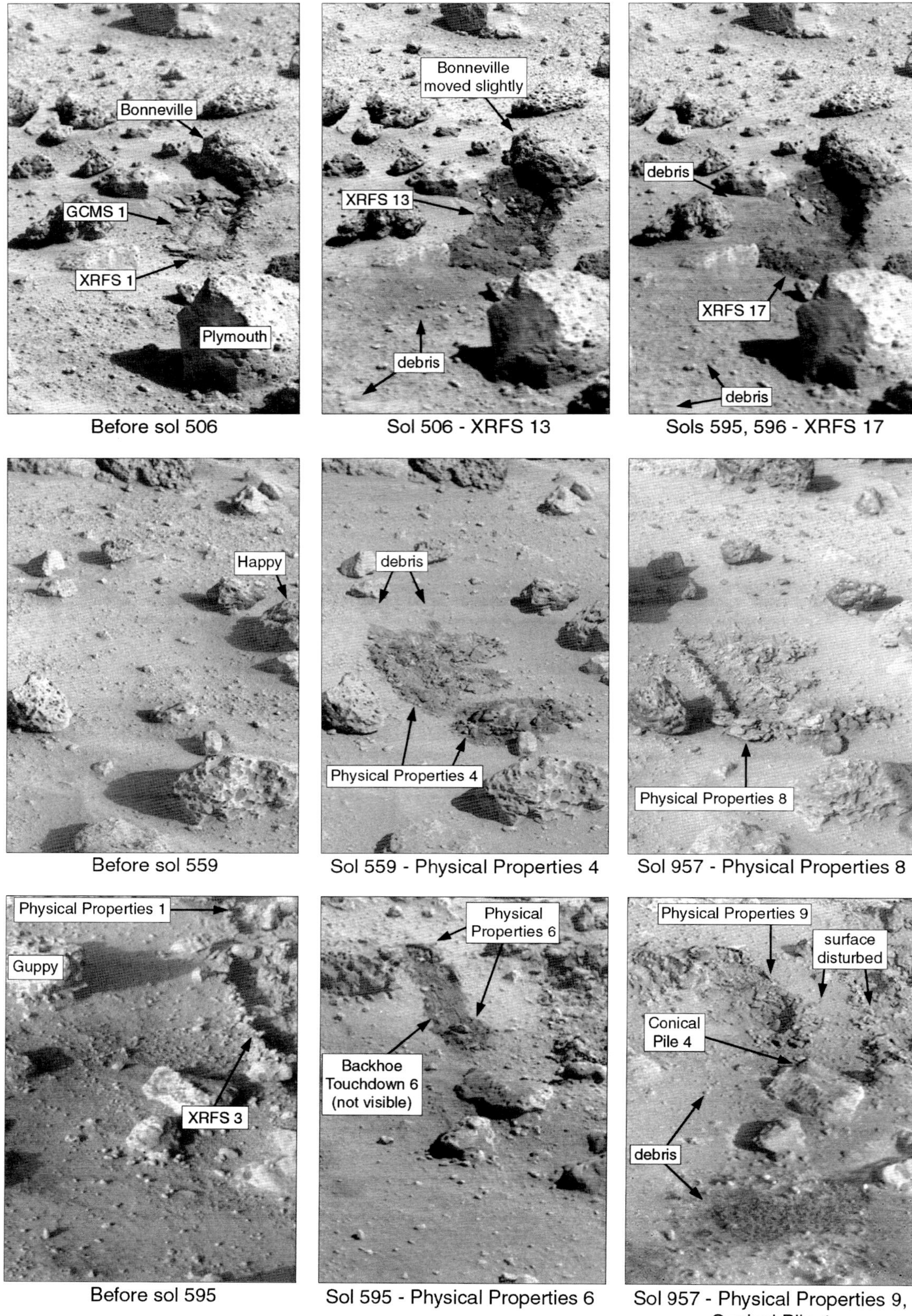

Figure 104 Viking Lander 2 activities after sol 501 (continued).

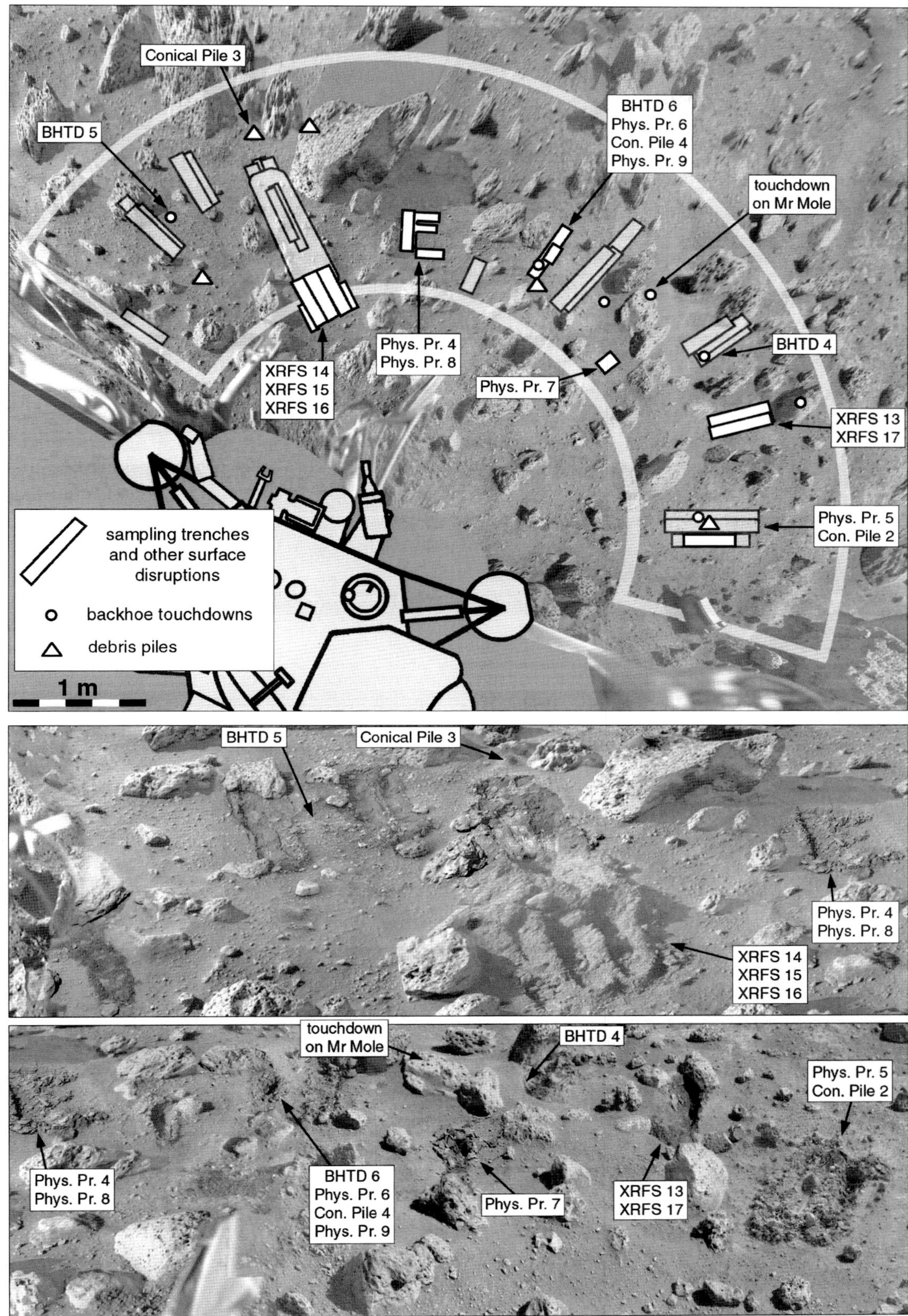

Figure 105 Viking Lander 2 activities, sol 501 to end of mission.

of platy soil from previous operations were moved, and a rock in the tailings pile was pulled forwards into the new trench (Figure 104). Material was removed from underneath Bonneville. This operation obtained a sample (U-8) and delivered it to XRFS. Afterwards the arm was fully retracted and parked. This was thought at the time to be the end of sampler operations, but the arm was to be used again a year later.

The final sampler arm operations occurred on sol 957, during the winter season. Two trenches were dug, but they did not receive names like their predecessors and were not included on mission maps (Moore *et al.*, 1987). Here they are unofficially referred to as Physical Properties 8 and 9. Physical Properties 8 was dug at the Physical Properties 4 site of sol 559 (Figure 104). A scoop of material from this location was used to create a conical pile, also not officially named but referred to here as Conical Pile 4, near Physical Properties 6. Remaining material was purged nearby, showing up clearly against the frosty surface. Physical Properties 9 was a bearing strength test conducted slightly closer to the lander than the Physical Properties 6 trench of sol 595 (Figure 104), just beyond the new conical pile. The arm temperature sensor was then buried in this location for a full diurnal temperature cycle. The soil was partly covered with frost. The sensor was inserted in the midafternoon on sol 957, expected to be the warmest time of day, and measured at a temperature of 182 K. The temperature dropped to 147 K overnight and rose to only 171 K the next afternoon. The arm was parked for the final time after this test, on sol 958 (MY 13, sol 543 or 16 May 1979). Events occurring between sols 501 and the end of the mission are shown in Figures 103 and 104 and are mapped in Figure 105.

Figure 106 shows the area underneath the lander. The main map covers an area 1 m square. A mosaic of images taken with a mirror under the engine is shown in approximately its correct location in the map and enlarged below. Another mosaic of small images of the footpad temperature sensor is shown beside it at a reduced scale. Figure 107 shows frost on the surface during the first and second winters. Surface changes like those seen at the Viking 1 site (Figure 74) were not observed at the Viking 2 site, but fine dust accumulated and buried small pebbles just north of the visible footpad between sols 51 and 447. The activities of the sampler arm moved the lander itself on three occasions, one prior to sol 23, once between sols 51 and 447, and once between sols 502 and 1140.

During the later parts of the mission, Viking Lander 2 could only transmit via the orbiters, and eventually only via Orbiter 1. During the survey mission, these relay opportunities were very rare. Two transmissions of stored and real-time data occurred on 1 and 18 August 1979 (sols 1033 and 1050). The next opportunity on 19 November (sol 1140) revealed a memory failure which caused the loss of much stored meteorology data. On sol 1177 (27 December 1979), stored data and real-time images were obtained, but at the next attempt on sol 1212 (1 February 1980), low-battery voltage interrupted the downlink. Stored data and real-time images were obtained before the shutdown, but the timing required that images be taken very early in the morning, reducing their quality.

More stored data were transmitted to the orbiter on sol 1245 (6 March), but only returned to Earth on 20 March. This revealed that the limited power had now forced the disconnection of the instruments and the tape recorder, so on sol 1280 (10 April) commands were sent to restore power to them. On the next sol data transmission was cut short after only ten minutes, showing that the batteries could no longer hold power, so the spacecraft was shut down on sol 1281, 11 April 1980 (MY 14, sol 197). If it had continued to operate, it would have been silenced only 124 sols later when the last operating orbiter, Viking Orbiter 1, also failed, because direct data transmission to Earth was no longer possible. The design lifetime for the Viking landers was only 120 sols.

1970s: **Pioneer Mars**

Post-Viking missions were already being studied before Viking was launched. Niehoff and Friedlander (1974) described a Pioneer Mars mission to follow Viking as early as 1979, modelled after the Pioneer Venus mission to save money at a time when funding was expected to be very limited. Several versions of the mission were described, including an orbiter with atmospheric, ionospheric and surface remote sensing instruments, and versions with four or more penetrator probes.

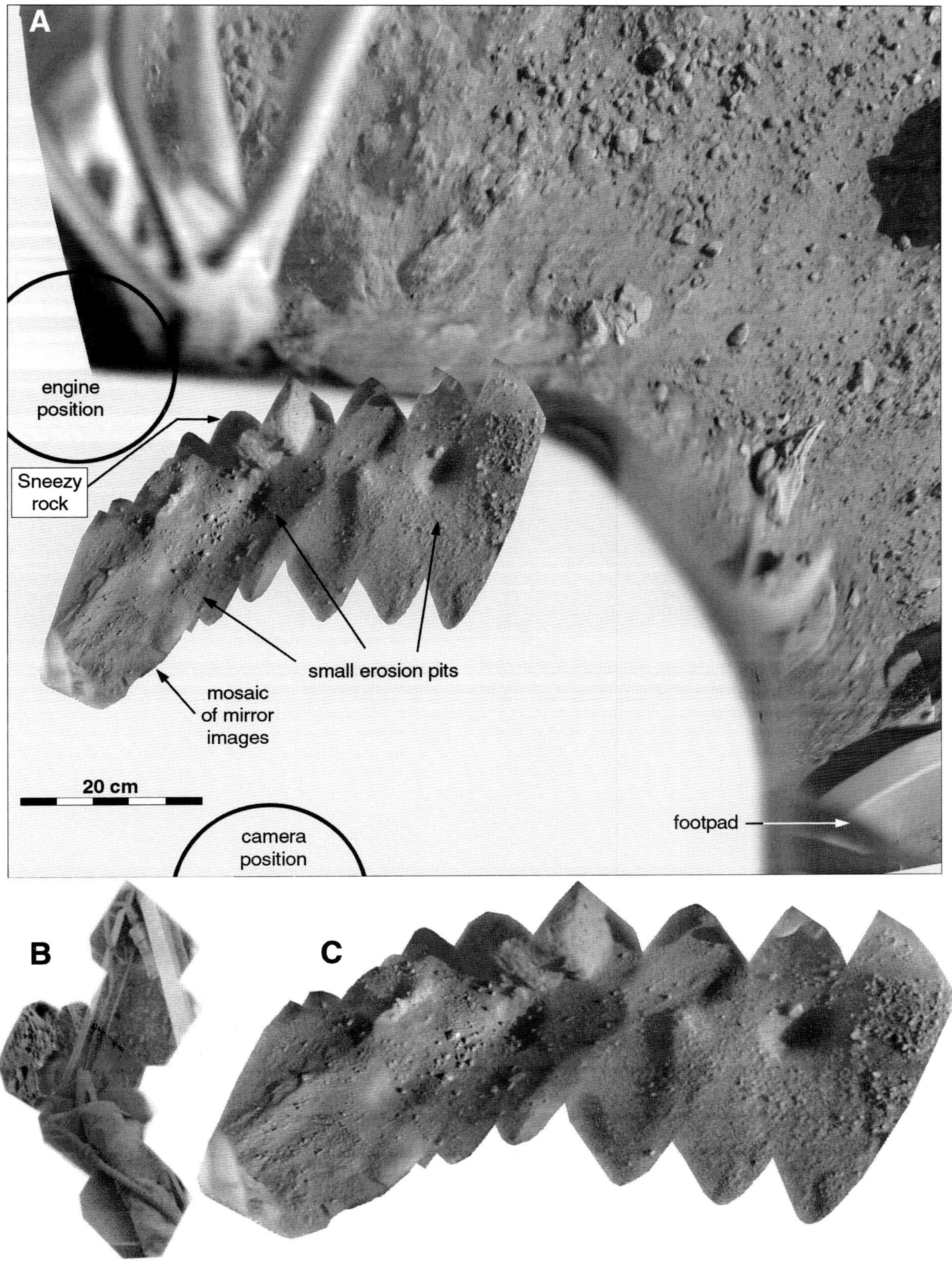

Figure 106 A: The immediate foreground of Viking Lander 2. The area mapped is roughly 1 m wide, with north at the top. **B:** A mosaic of images taken with an arm-mounted mirror to show the footpad temperature sensor. **C:** A mosaic of images taken using another mirror, showing engine erosion under the lander. Its approximate location is shown in (A), where small pits dug by the engine exhaust are indicated.

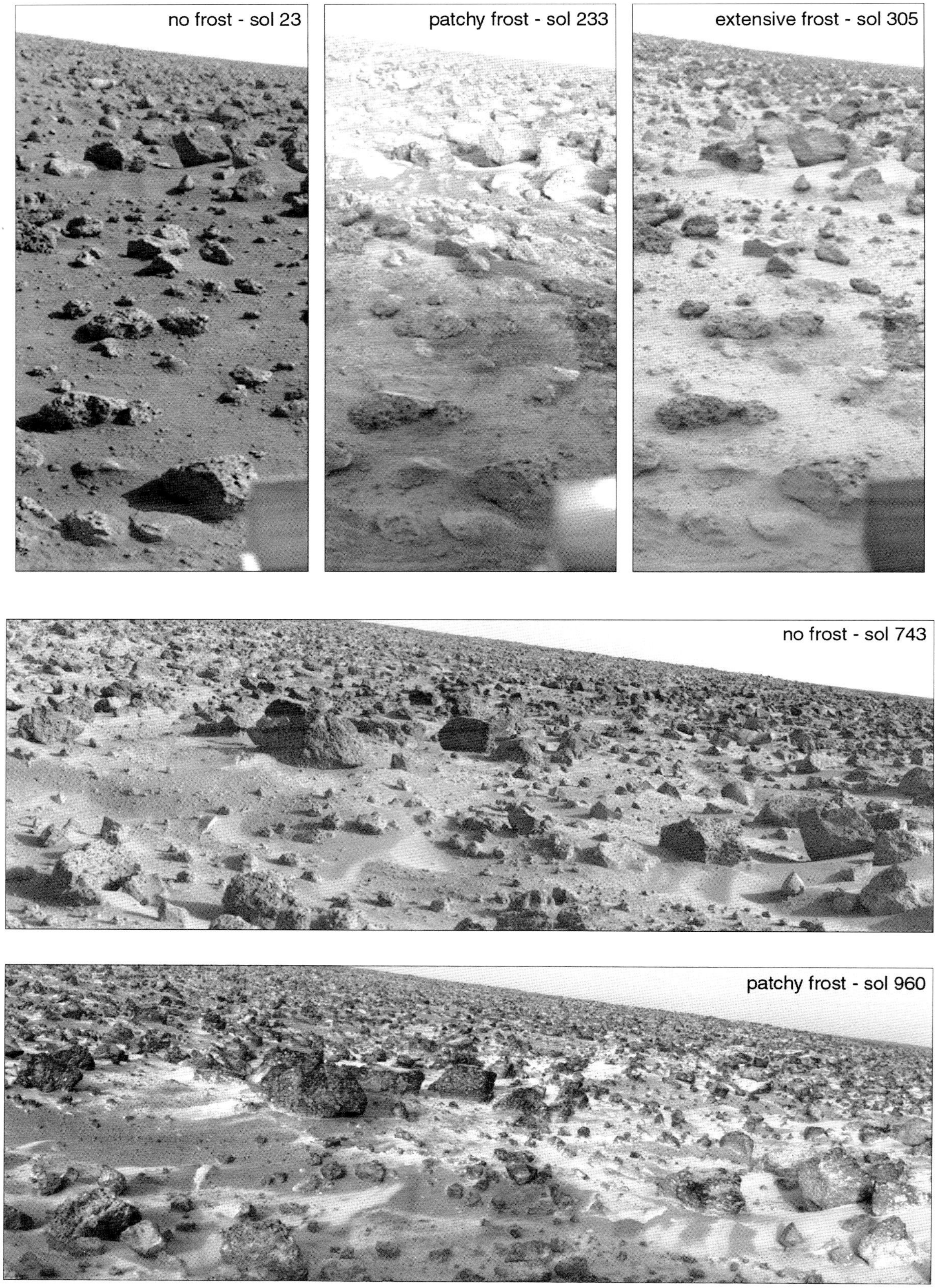
no frost - sol 23
patchy frost - sol 233
extensive frost - sol 305
no frost - sol 743
patchy frost - sol 960

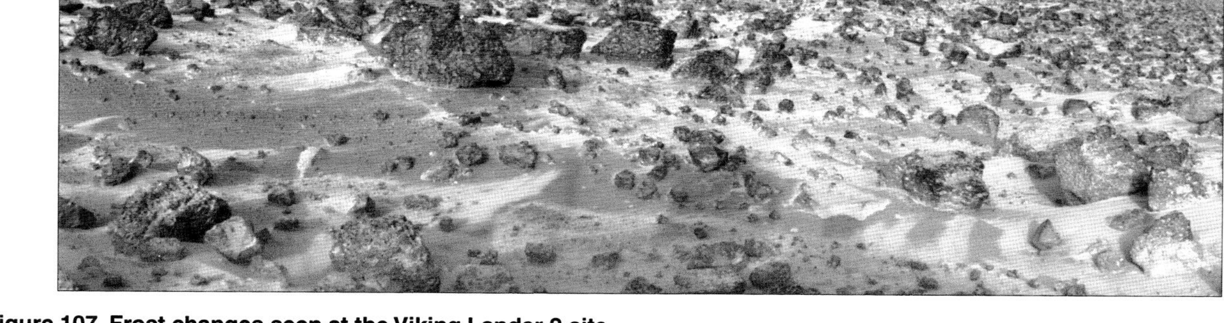

Figure 107 Frost changes seen at the Viking Lander 2 site.
Top: First winter. **Bottom:** Second winter.

Table 33. *Pioneer Mars Penetrator Targets from NASA (1974)*

Penetrator	Deployment time after arrival (days)	Location	Description
Northern mission			
1	20	59° N, 180° W	Northern plains (seismic net)
2	60	70° N, 150° W	Etched plains
3	107	65° N, 108° W	Etched plains (seismic net)
4	130	80° N, 135° W	Northern plains
5	160	90° N, 180° W	Permanent ice, polar cap (seismic net)
6	180	83° N, 0° W	Residual polar cap area
Southern mission			
1	20	86° S, 240° W	Laminated deposits (seismic net)
2	30	70° S, 240° W	Edge of south polar cap
3	50	75° S, 15° W	Etched plains
4	70	75° S, 70° W	Laminated deposits
5	100	53° S, 240° W	Cratered terrain (seismic net)
6	160	53° S, 298° W	Cratered terrain (seismic net)

In one mission design (NASA, 1974), a carrier spacecraft would deploy two sets of six surface penetrators and then enter a highly elliptical orbit to serve as a communication relay, or deploy six penetrators from orbit. The periapsis would be several hundred kilometres during the full Mars year primary mission and would then be raised to more than 1000 km, high enough that the orbit would not decay for several decades. Each probe would partially penetrate the surface on impact and relay data to an orbiter during its periapsis passes. The penetrator payload would consist of an accelerometer to assess surface properties during impact, a seismometer and an alpha-proton spectrometer for composition. Other versions of the penetrator might have included a camera and a meteorology instrument. Three probes would establish a seismic network about 2000 km across, and the other three would study interesting and diverse sites. The penetrator landing ellipse would be about 100 km by 10 km. Six sites around each pole, identified as interesting targets, are listed in Table 33 and illustrated in Figure 108.

Another version of this mission was presented by Friedlander *et al.* (1974) in a report to NASA's Planetary Programs Division. One option was an orbital mission, including a radar altimeter and a gamma-ray spectrometer, operating from an elliptical orbit with a very low periapsis (about 100 km) and a 45° inclination. This would permit upper-atmosphere and ionosphere studies, as well as the surface remote sensing, but would limit the mission lifetime. The second mission option was a polar orbiter with four, or possibly six, penetrators. The orbiter would serve as a communication relay and conduct remote sensing. The penetrators would survive only a week on the surface, but would be deployed individually over a period as long as a Mars year. They would study surface composition and physical characteristics, including water detection. This version of the mission did not include a seismic network.

The landings would all be at medium to high latitudes, and sites were identified in each hemisphere (Table 34, Figure 108). Of the northern sites, N1 and N2 had lower priority and could be omitted in a four-penetrator mission. The long delay after arrival might make the southern mission preferable, and the south also offered more geological diversity. A comparison of the two sets of Pioneer Mars landing sites (Figure 108) shows some similarities between them, especially in the southern hemisphere.

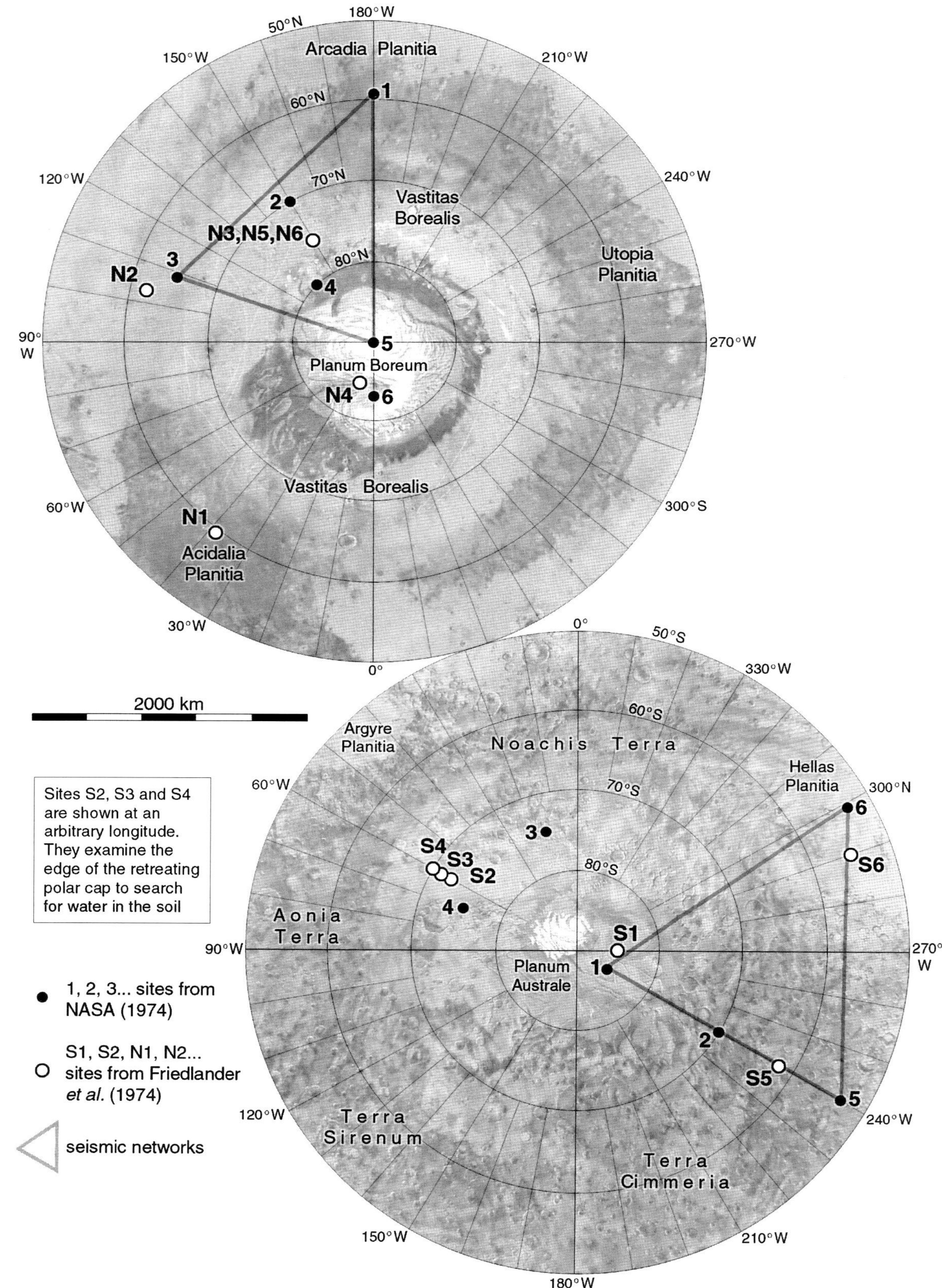

Figure 108 Pioneer Mars penetrator landing sites from NASA (1974) and Friedlander *et al.* (1974). The polar base maps are from Mariner 9 data, modified from the maps in Figures 28 and 29.

Table 34. *Pioneer Mars Penetrator Targets (Friedlander et al., 1974)*

Penetrator	Deployment, months after arrival	Location	Description
Northern polar mission			
N1	0.1	59° N, 38° W	Moderately cratered dark plain, soil composition
N2	1.4	62° N, 103° W	Volcanic plains, mountain – test for basalt composition
N3	3	76° N, 150° W	Etched plains, composition and water content
N4	8	85° N, 16° W	Layered deposit, examine soil stratigraphy
N5	16	76° N, 150° W	Repeat N3, now covered with polar cap, test for water
N6	17	76° N, 150° W	Repeat N3 as polar cap retreats
Southern midlatitude mission			
S1	0.5	85° S, 270° W	Polar cap, 'Australis Chasma', water or carbon dioxide?
S2	4.1	72° S	Study edge of shrinking south polar cap, seek water, discriminate between water ice and carbon dioxide. Longitudes were not specified, but are assumed to be similar to isolate the latitude variable.
S3	4.2	71° S	
S4	4.3	70° S	
S5	8	62° S, 240° W	Cratered uplands, low albedo, composition, grain size
S6	13	55° S, 290° W	Floor of Hellas, high albedo, composition, grain size

1979: **Viking Rovers**

Another post-Viking plan was for a mobile Viking spacecraft capable of traverses of tens of kilometres and the return of samples to Earth as well as in situ analyses (Masursky *et al.*, 1974). The ability to land within a small ellipse close to the pre-planned target would be essential for a hypothetical 1979 mission. The USGS designed four examples of possible traverses in representative areas covered by high-resolution Mariner 9 B frames, which were the only data suitable for analysis at the time of this study. Geologic mapping suggested that between five and ten different types of material could be studied in each location. The sites and traverses are listed in Table 35 and illustrated in Figures 109 and 110. Their locations are shown in Figure 111. At each stop indicated by dots along a traverse, images and geochemical data would be obtained, and at some stops material would be collected for return to the lander for more thorough analysis. These were the first Mars rover traverse plans, though many would follow before the first rovers arrived at the planet.

Table 35. *USGS Viking Rover Candidate Landing Sites, 1979 Launch*

Location	Name	Goals
32° N, 211° W	Cebrenia	Crater ejecta, plains materials, volcanic shield (may instead be a large pingo, an ice-cored hill). Three traverses (A, B, C) and one alternative traverse (D).
6° 54' N, 292° 45' W	Syrtis Major	Dark plains, wind streaks, craters. Two traverses (A, B) and one alternative (C).
6° S, 150° W	Mangala Vallis	Fluvial channel, interchannel materials, craters. Two traverses (B, C), one alternative (A).
82° S, 83° W	South Pole	Layered materials, etched terrain. Two traverses (A, C), one alternative (B).

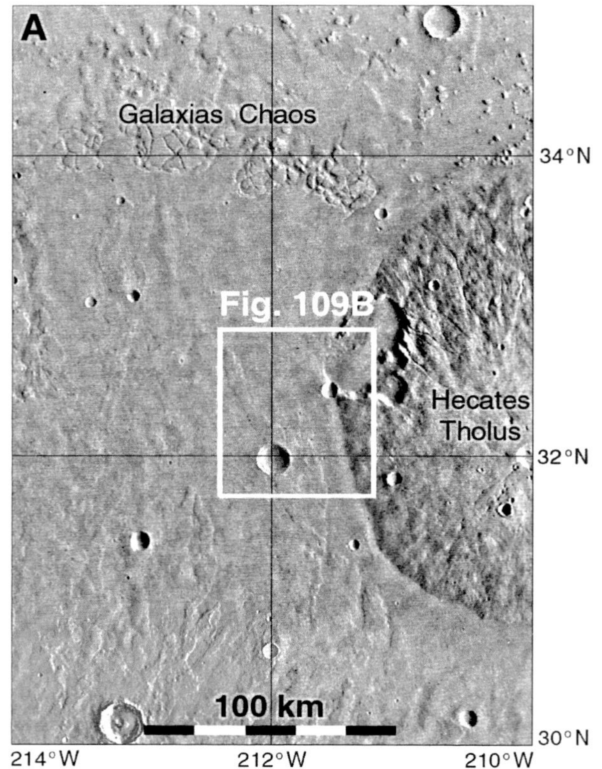

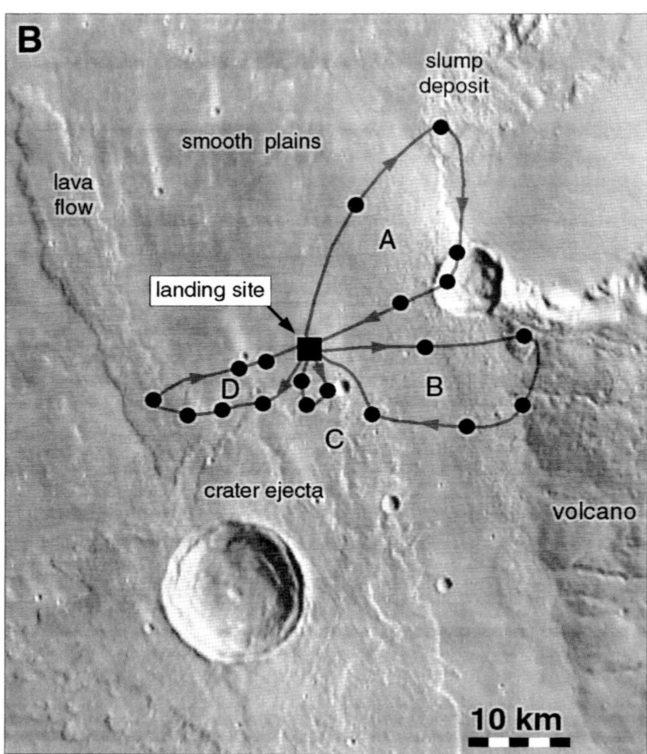

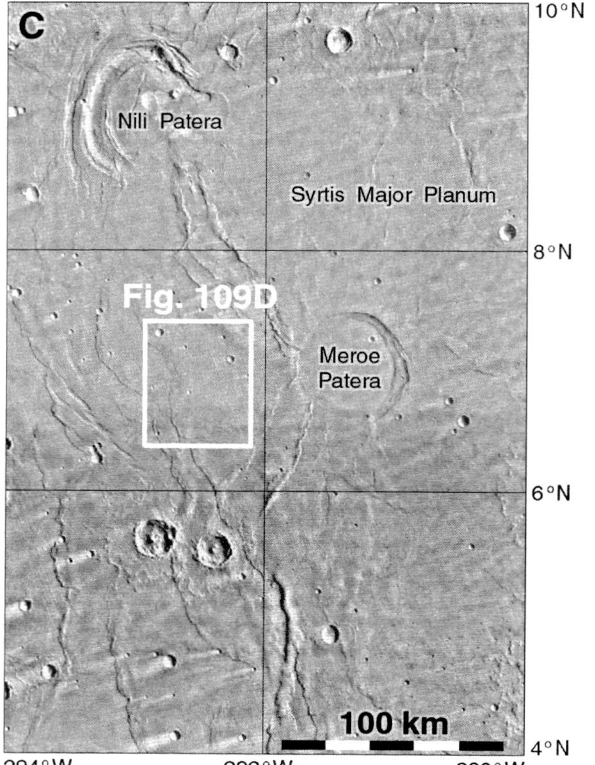

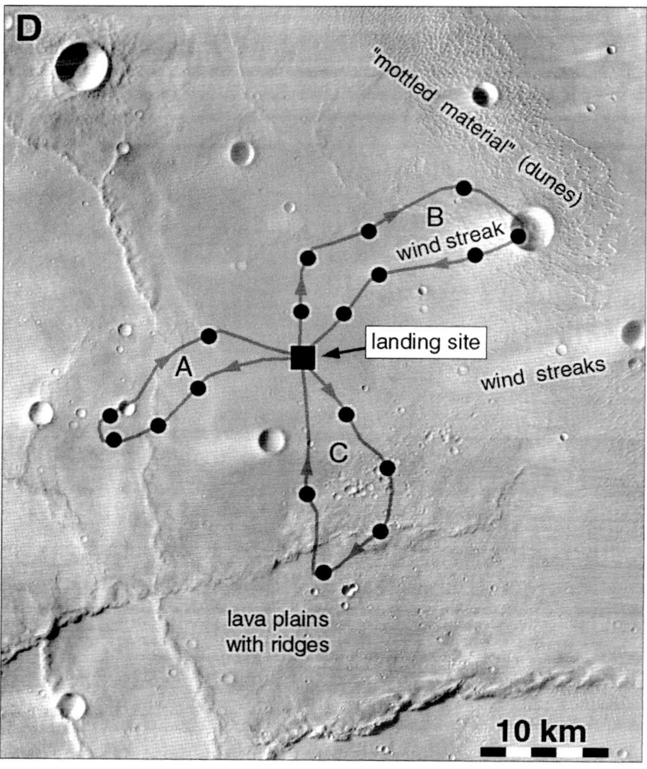

Figure 109 Viking rover sites and traverses from Masursky *et al.* (1974).
A, B: Cebrenia. **C, D:** Syrtis Major. The base maps for (A) and (C) are Viking image mosaics. (B) is a Mars Odyssey THEMIS infrared mosaic with shading inverted. (D) is a Mars Odyssey THEMIS visible image mosaic with wind streaks modified to match Mariner 9 source data used in the original mapping.

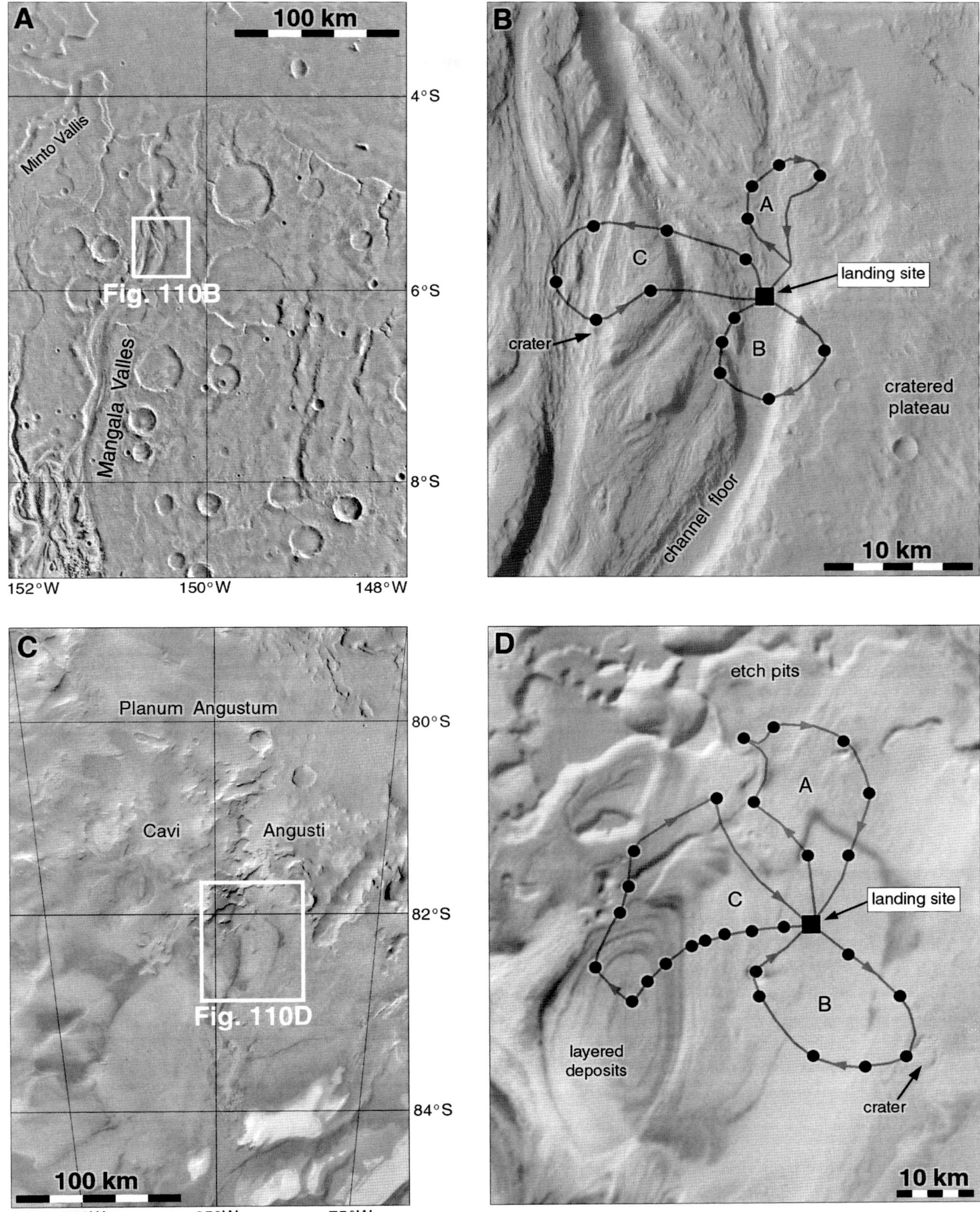

Figure 110 Viking rover sites and traverses from Masursky *et al.* (1974).

A, B: Mangala Vallis. **C, D:** South Pole. The base maps for (A) and (C) are Viking image mosaics. (B) is a Mars Odyssey THEMIS visible image mosaic, and (D) is a THEMIS infrared mosaic with shading inverted to approximate albedo. Large squares are landing sites; small dots are rover science stops.

Another possible rover mission (Moore, 1970) would launch in October or November 1979. The voyage to Mars would take about 270 days with a trajectory correction about ten days after launch and arrival at the planet in August 1980. Two days after entering orbit, the spacecraft would adjust its path to prepare for landing between 30° N and 30° S. Three days later the orbiter would eject a bioshield cover and deploy the lander, which would then fire its thrusters to begin the descent.

The rover would operate for about an Earth year, from August 1980 to August 1981. The vehicle, designed at JPL, had three compartments, each with its own pair of wheels. The front compartment or 'science bay' would carry a Viking-style sampling arm with a magnetic properties experiment attached, a second arm for more difficult sampling, biology experiments, a mass spectrometer for composition analyses, a meteorology instrument and a seismometer. Typically the rover would move 50 to 100 meters at a time, then stop, image its surroundings, perform a science experiment, transmit its data to Earth, and await new commands. JPL assumed that science sites would be about 14 km apart, and estimated that early in its mission the rover would travel about 300 m per day, enabling it to traverse the distance between two science sites in 47 days. Distance traversed would increase as controllers gained confidence in their remote driving capabilities, so that in one Earth year the rover might traverse up to 500 km.

Another study (Martin Marietta, 1977b) of a mobile Viking lander referred to as Viking 3 mentioned a canyon as a possible landing site, probably referring to Candor. That rover would use an image-based automated hazard detection system to achieve a safe landing. The date of that report distinguishes this 'Viking 3' from the earlier Viking Rover study, but it suggested a launch in 1981 rather than 1979. Darnell and Wessel (1974) described a different concept in which a Viking lander could deploy a rover, potentially at a site on the edge of the south polar cap. Rovers with ranges of 100 m, 200 m and 1000 m were considered, with the short-range rover tethered to the lander and the long-range rover able to deploy explosive charges for the lander seismometer. All rovers would collect samples for analysis on the lander.

1980s: **Mars 1984 Rover/Penetrator Mission**

A Mars Science Working Group (MSWG) chaired by Thomas Mutch was established by NASA to develop a science strategy for a future mission (Mars Science Working Group, 1977). It met four times in 1977. The plan assumed two Space Shuttle launches in December 1983 or January 1984, each carrying a spacecraft consisting of an orbiter, a lander with rover, and three penetrators, set to arrive at Mars in September or October 1984. The penetrators would be deployed just before arrival, but the rovers would wait in orbit until the dust storm season was over. Highly elliptical initial orbits would permit magnetospheric studies. After the rovers landed, the orbiters would enter circular orbits, one near-polar at 500 km altitude for global mapping and communication with the penetrators, the other 1000 km high with about a 30° inclination for rover communications. As the rovers might each deploy an instrument station with a seismometer, there could be ten simultaneously operating landed components.

As each orbiter neared the planet, it would deploy three penetrators which would fall on a circle around the centre of the planet's disk as seen from the approach direction. After deployment the orbiters would be deflected off the approach path to enter orbit. The six penetrators, carrying seismometers and soil and atmospheric analysis equipment, would form a global array. Three would be placed about 500 km apart in an area likely to be seismically active, such as Tharsis. The other three would be spaced about 5000 km apart to give global coverage. Two additional and more sophisticated seismometers would be deployed by the rovers in areas partly shielded from the wind. Latitudes between 50° N and 87° S would be accessible, and the landing ellipses were 200-km-diameter circles. Slopes would have to be less than 45° at the impact point. Site selection was reported in a Penetrator Site Studies document preserved in Tim Mutch's papers at Brown University. One potential array design was described (Table 36), along with four deployment options which include several additional sites (Table 37). Option 1 was the potential array described in Table 36. The penetrator sites in Table 37 were also described in Manning (1977), in which the site selection work was attributed to T. E. Bunch and Ronald Greeley.

Table 36. *Mars 1984 Penetrator Array*

Site	Location	Notes
1. Syria Rise	12° S, 105° W 22° S, 100° W 23° S, 113° W	700-km spacing, one leg of global network, and regional network in active area
2. Acidalia Planitia	46° N, 5° W	Northern plains
3. Amphitrites Patera	59° S, 297° W	Old volcanic materials
4. South Polar Region	81° S, 190° W	South polar deposits, volatiles

Table 37. *Mars 1984 Penetrator Deployment Options*

Option	Launch date	Penetrator	Site	Location
1	Early	1	Acidalia Planitia	46° N, 5° W
		2	Amphitrites Patera	59° S, 297° W
		3	South Polar Region	81° S, 190° W
	Late	1–3	Syria array	15° S, 105° W
2	Early	1–3	Syria array	15° S, 105° W
	Late	1	Cratered Highlands	30° N, 10° W
		2	Hellas Rim	29° S, 303° W
		3	South Polar Cratered Terrain	68° S, 160° W
3	Early	1	Canyonlands	10° S, 73° W
		2	Acidalia Planitia	46° N, 5° W
		3	South Polar Region	81° S, 190° W
	Late	1–3	Syria array	15° S, 105° W
4	Early	1–2	Syria array	15° S, 105° W
		3	South Polar Region	81° S, 190° W
	Late	1	Arsia Mons Plains SW	8° S, 130° W
		2	Cratered Uplands	25° N, 15° W

The rover landing ellipses were roughly 50 by 80 km across. Five landing sites were studied using Viking data, in addition to the four sites previously considered by USGS for the Viking Rover (Figures 109 and 110). Only two sites were identified in the MSWG report, Capri and Candor (Table 38, Figures 111 and 112). The other sites were identified in Working Group documents among Tim Mutch's papers in the archives at Brown University.

The rover landing ellipses in these documents were 65 by 40 km across. The polar orbiter would be able to deploy its rover from orbit at latitudes between 30° N and 50° N (this range could vary, depending on the launch date), whereas the low-inclination orbit could deliver a rover to latitudes between 20° S and 20° N.

Six rover landing sites were identified in a Rover Site Studies report prepared for the Working Group (Table 38a, Figure 111). Most derived from work done earlier for Viking or the Viking rover study, including proposals to land near Viking 1 and visit it or to explore the abandoned A-1 site with its complex geology. In a memorandum dated 9 May 1977, Hal Masursky followed up on discussions at a meeting of the Mars 1984 Mission Study Group held on 1 April. He asked Tim Mutch to request high-resolution stereoscopic Viking imaging coverage of

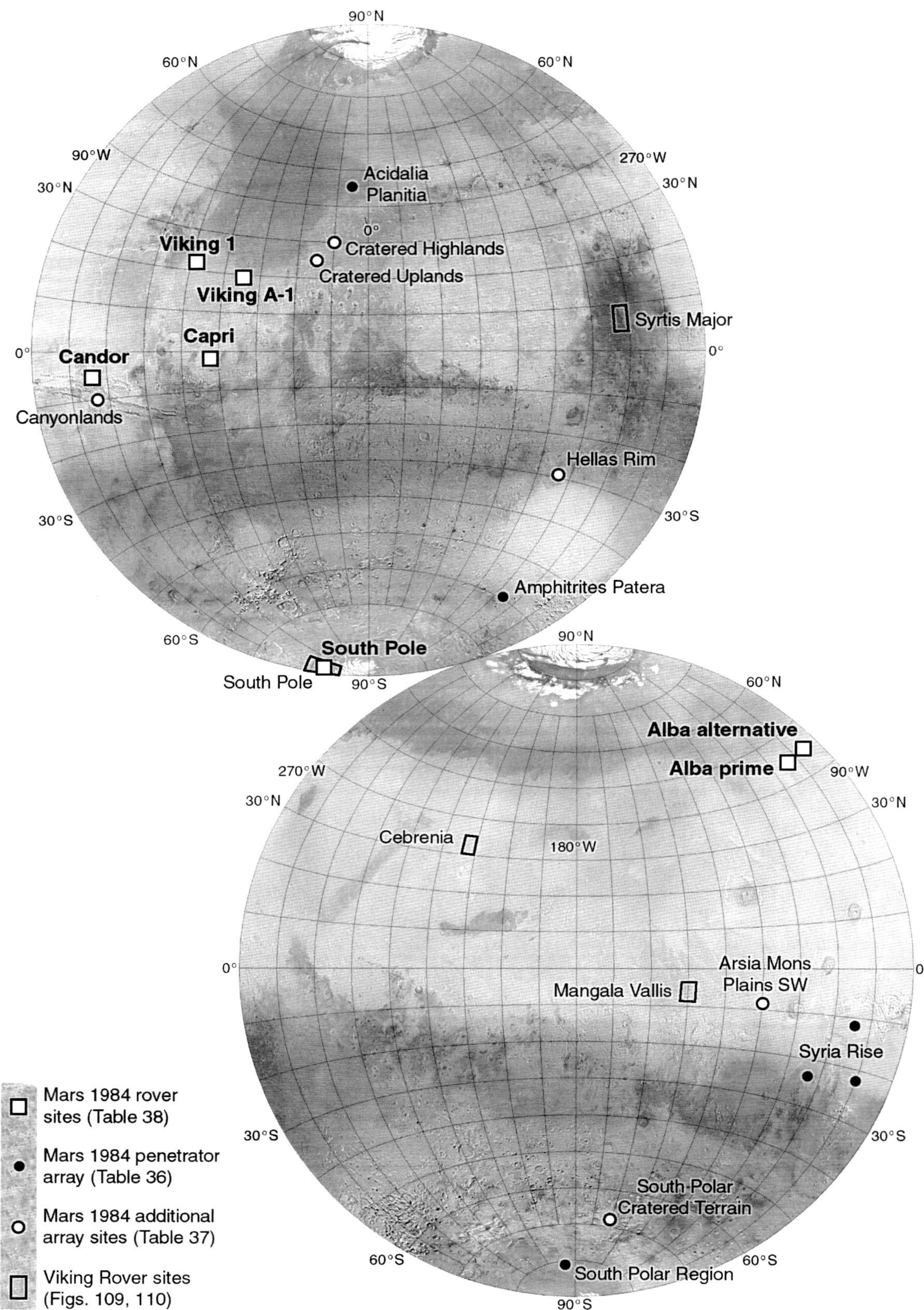

Figure 111 Viking Rover and Mars 1984 sites.

Table 38. *Mars Science Working Group Rover Sites, 1983–1984 Launch*

38a. Sites from Rover Site Studies report

Location	Name	Location	Name
6.3° S, 73.8° W	Candor	19.5° N, 34.0° W	Chryse E (Viking A-1 site)
1.5° S, 41.8° W	Capri (near Viking C-1 site)	22.5° N, 48.0° W	Chryse W (Viking 1 site)
44.4° N, 103.5° W	Alba NE (near Viking B-2)	82.0° S, 83.0° W	South Pole (Viking Rover site)

38b. Sites from Masursky memo to Mutch, 9 May 1977

Location	Name	Notes
1.3° S, 42.0° W	Capri Chasma, north border	Also stated as 1.0° S, 41.8° W
6.25° S, 73.75° W	Candor Chasma	Also stated as 5.7° S, 74.9° W
44.3° N, 104.0° W	Alba Patera southeast	Prime Alba site
44.0° N, 96.0° W		Alternative Alba site

38c. Preferred sites from NASA (1977), MSWG (1977)

Location	Name	Notes
1.0° S, 41.8° W	Capri	44 stations over 400-km traverse, cratered plateau and large channel
5.7° S, 74.9° W	Candor	55 stations over 300-km traverse, canyon walls and layered deposits

four of these sites, using slightly different coordinates (Table 38b). These, he said, 'were sites for which we have made traverse plans'. He added that 'a backup smoother site near B-1' at Cydonia had also been studied. Eventually the Capri and Candor sites were chosen, and detailed mission plans were prepared (Figure 109). Traverses near the Chryse sites were also prepared, including those in Figure 114.

The Capri site provided access to cratered uplands, crater ejecta and a fluvial channel. Candor was on the floor of the canyon system, with access to thick-layered deposits, canyon wall materials and, at the end of the extended mission, possibly the volcanic plateau surrounding the canyon. Alba had fractured volcanic plains and crater ejecta, but also small channels.

The Mars 1984 rovers had three traverse modes. Mode 1 was for detailed site investigations and involved only short, precise drives as needed for science operations. Mode 2, the 'survey traverse mode', would cover about 400 m per sol and could include some observations along the route. Mode 3, the 'fast traverse mode', could cover as much as 800 m per sol, including travel at night. The goal was to cover about 200 km during one Mars year and up to 200 km more in an extended mission in the second Mars year.

On 13 May 1977, Carl Pilcher, Hal Masursky and Ron Greeley suggested a variation on the role of penetrators in this mission. Two penetrators would be dropped in the lander target ellipse, carrying beacons to help guide the rover to a precision landing. After the landing they would operate with instruments on the lander itself as a local area seismic network.

The Mars 1984 orbiters would carry cameras, spectrometers for surface composition, infrared and microwave radiometers, a magnetometer, a plasma probe, a radar altimeter and communication relay equipment.

The relationship between Mars 1984 and other missions was considered by the Working Group. If Viking Lander 1 survived long enough, it might provide useful meteorological data for a Mars 1984 landing at Chryse, if that site was chosen. Conversely, Mars 1984 might be reconfigured to gather samples for collection by a sample return mission in about 1990.

Mars 1984 was not funded, probably in part because significant opposition to it arose in the science community. Jim Arnold and Mike Duke objected publicly that the final report of the Working Group did not reflect the group discussions, particularly in its assertions that the rovers were the only realistic option, that they were essential for future Mars Sample Return missions, and that simpler missions (orbiters, hard landers) were 'a step backwards'. The report also suggested that only Mars rovers would command broad public interest, whereas missions such as Voyager, Jupiter Orbiter/ Probe (Galileo) and the Lunar Polar Orbiter would not. This mention of Voyager refers to the outer planet spacecraft, not the earliest version of Viking (Table 2), and the suggestion that it would attract little public interest turned out to be the opposite of the truth. Elbert King (University of Houston) wrote to Mutch on 29 August 1977, stating emphatically that Mars 1984 'would only ensure a repeat of the very limited scientific success of Viking – providing mostly only costly clues and ambiguous answers to the important scientific questions'. He argued that only sample return was justified by the cost. This dismal assessment of Viking's scientific worth stems from its failure to detect life, or to definitively rule it out, but overlooks its detailed characterization of surface and atmospheric composition, meteorology and landing site geology, not to mention the mission's orbital data.

That work was followed by a Mars Sample Return Study Effort by MSWG (1980). This included several studies to assess sites and rover mobility needs, briefly summarized here in the same order as the volumes in the report and illustrated in Figures 111, 113 and 114.

A polar landing site study (MSWG, 1980, vol. 1) considered locations for collecting ice or layered terrain (ice and dust) cores rather than rocks, though small rocks might be included in the sample as a bonus. Sites would be in perennial ice or in layered deposits exposed in troughs cutting the ice cap. A rover with a 10-km range would allow sampling of both units. The landing ellipse was 80 by 50 km across, oriented north-south. Two sites were identified, Site A in a large patch of perennial ice at 86.5° or 87° N, 120° W and Site B in an area with ice cut by troughs at about 83.5° N, 100° W (84.5° N, 105° W in the report, probably based on Mariner 9 coordinates). Site A (Figure 113A) would not need a rover as long as it could sample beyond the area affected by its landing, but Site B would benefit from a rover with a 10-km range to examine layered materials in the troughs as well as the surrounding ice. Figure 113A shows the locations of sites A and B. This study made use of an early digital mapping and analysis system, a forerunner of the Geographic Information Systems (GIS) used extensively for later missions, to select an optimal location within Site B based on five possible targets illustrated in Figure 113B. The preferred location in Site B was centred on target point 4, as this would maximize the chance of landing within range of both ice and trough material. The study only considered these five candidate sites and recognized that many more candidates would have to be tested for final site selection.

Another study (MSWG, 1980, vol. 2) identified landing sites at a volcanic target and in Chryse Planitia. The volcanic site in Tharsis or Elysium would provide young volcanic rocks, whereas Chryse would help interpret the Viking 1 results and other sites for which surface data were not available. The six candidate sites for young volcanic materials are identified in Table 39 and shown on Figure 113C.

The preferred site at Arsia Mons West had good imaging coverage, young lava flows with little crater ejecta or wind-deposited sediment to confuse the sampling, and two potential landing ellipses within the site (Figure 113D). One ellipse was oriented parallel to the elongated lava flows, and one perpendicular to them. A rover with 5-km range could collect good rock samples. A 100-km rover could sample ejecta from deeper layers near a crater between the two ellipses. At Chryse the landing ellipse (Figure 113E) was centred on the Viking Lander 1 site, then assumed to be at the Morris *et al.* (1978) location (Figure 52). A landing anywhere in the ellipse would provide rocks from ridges or crater ejecta with only 2 or 3 km rover range.

A third study (MSWG, 1980, vol. 3) considered sites at Apollinaris Patera and near the large impact basin Schiaparelli. The Apollinaris site was said to be at 5° S, 190° W and was shown on an illustration at 6.5° S, 188.5° W (Figure 114A), to the west of the large volcano whose lava flows were the sampling goal. A rover range of only 4 km would provide access to good samples almost anywhere in the ellipse. At Schiaparelli the goal was to collect the oldest highland rocks. An ellipse said to be at 8° S, 336° W and mapped at 7.25° S, southeast of the basin rim, should provide good sampling opportunities with a rover range of up to 25 km (Figure 114B).

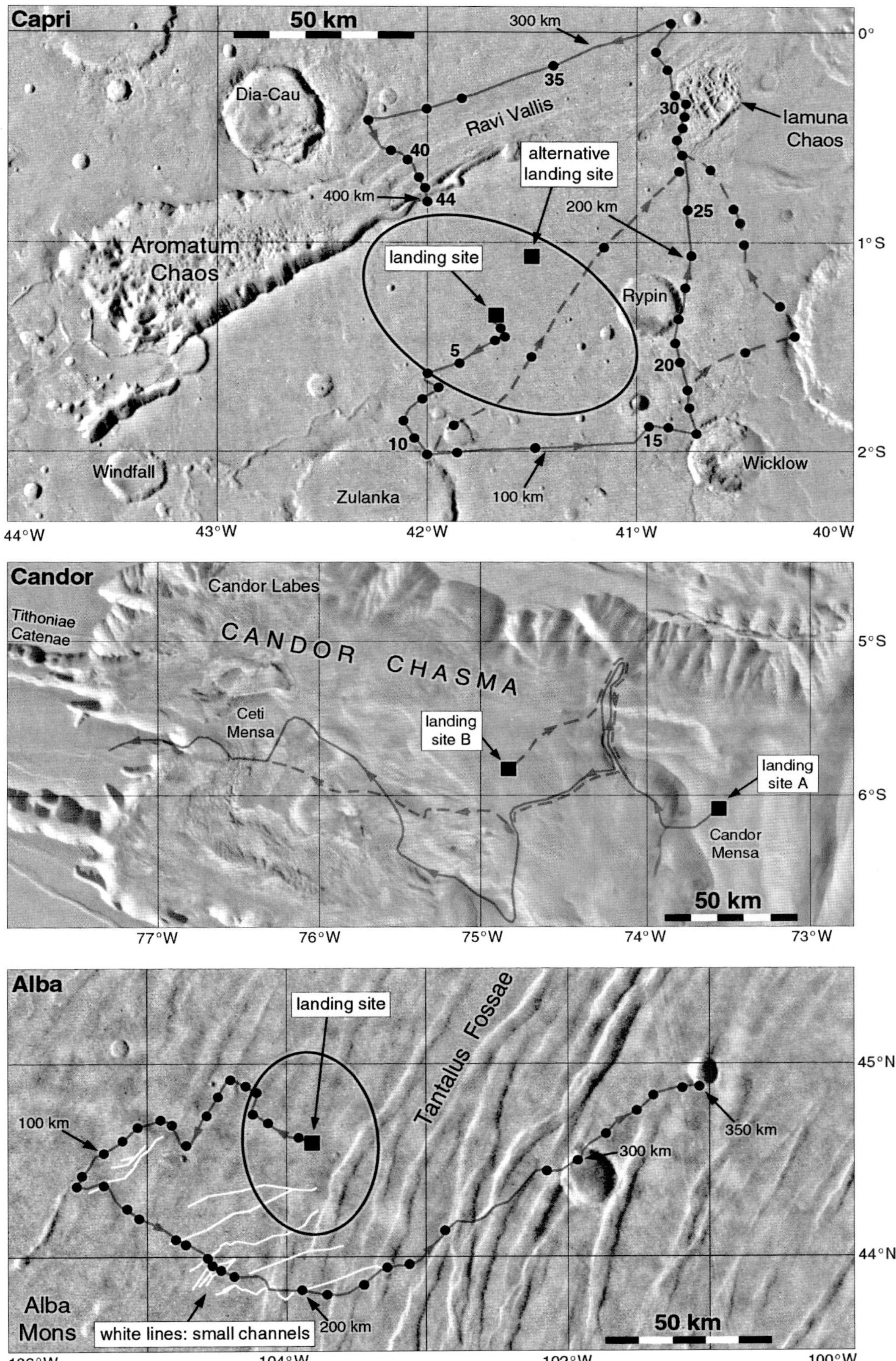

Figure 112 Mars 1984 Mission Rover traverses at Capri, Candor and Alba.

Table 39. *Candidate Young Volcanic Sites (MSWG, 1980, vol. 2)*

Name	Location	Notes
Tharsis North	30° N, 106° W	Flows, mare ridges, may be old, covered with sediment
Alba Patera West	43° N, 114° W	Flows, fractures, channels. Hazardous, rocks may be altered by water, image coverage inadequate
Elysium	25°–30° N, 210°–230° W	Flows and sediments, age uncertain, seems eroded
Tharsis Plains S.	28° S, 135° W	Flows, fractures, highlands mixed together, may be old
Arsia Mons South	25° S, 123° W	Flows, crater ejecta – backup site
Arsia Mons West	8° S, 132.5° W	Flows, little aeolian mantle – preferred site

The last of these site studies, by USGS staff including Harold Masursky (MSWG, 1980, vol. 4), considered sites already discussed in Candor Chasma (Figure 112) as well as new candidates in Hebes Chasma (2° S, 76° W), Iapygia (10° S, 275° W) and Terra Tyrrhena (2° S, 244° W). They are indicated in Figure 111.

A different kind of study by Elbert King (MSWG, 1980, vol. 5) considered where rocks suitable for sampling might be found across the equatorial region. This analysis extended only between 30° N and 30° S. Rocks were important because small soil grains were more likely to be chemically altered. The most important data for this analysis were thermal inertia estimates inferred from Viking Orbiter data. The areas identified as best for sampling rocks are outlined on Figure 114C, though many other areas should also be adequate.

Finally, two studies (MSWG, 1980, vols. 6, 7; Nickle, 1980) considered rovers at the Viking landing sites, directed to sample the range of materials seen by the landers. Figure 114D is a rover route designed by Ray Arvidson and Figure 114E illustrates a 100-m traverse designed by Henry Moore (USGS), both for the Viking Lander 1 site. The rover on the second traverse could collect about 3 kg of rocks, dust and crusty material, as well as an atmospheric sample. Comparison with maps and panoramas suggests the route should be regarded as schematic, but a traverse like this would allow studies of several different rock and soil types.

1987: **Sample Return Planning**

A workshop on Mars Sample Return Science (Drake *et al.*, 1988) was held in Houston from 16 to 18 November 1987. Among other technical papers, four presentations dealt with possible sample return landing sites (Tables 40, 41, 42 and 43).

Harold Masursky discussed four years of work at USGS Flagstaff on a set of ten sample return sites (Masursky *et al.*, 1988a) which offered access to materials of diverse ages and compositions with traverses of no more than 100 km. They are listed in Table 40 and illustrated in Figure 115, and all were included in the later Landing Site Catalog (Table 47, sites 33 to 43, including two at Mangala Valles) with some inconsistency in coordinates. USGS published geologic maps of most of these areas, and on five of them landing sites and rover traverses were shown. A sixth map of Candor Chasma portrayed two landing sites but no rover routes. These sites are identified in Table 40 with their USGS I-map number (I-maps were the Miscellaneous Investigations series maps which then included all USGS planetary maps). Sites without geological maps were plotted on a global map inset, from which their coordinates in Table 40 have been taken. At Apollinaris Patera, no site was illustrated, but a comment on that map noted that the floodplain area west of the Apollinaris volcanic shield would be the best location. This would be similar to the ellipse in Figure 114A. At Elysium the USGS geological map (I-2579, Galaxias) lies farther north than the location shown on the global map, at about 35° N, 217° W, and no site was illustrated on the map. Masursky also mentioned that Soviet scientists were studying seven sites at this time, primarily at the eastern end of Kasei Valles where it opens into Chryse Planitia, and near Uranius Patera. No further details of these sites were provided, but two Soviet sites at these locations are described in Table 44.

Traverse plans and landing sites shown on the USGS maps are illustrated in Figures 116 and 117. At East Mangala Valles traverse 1 would sample the local plains,

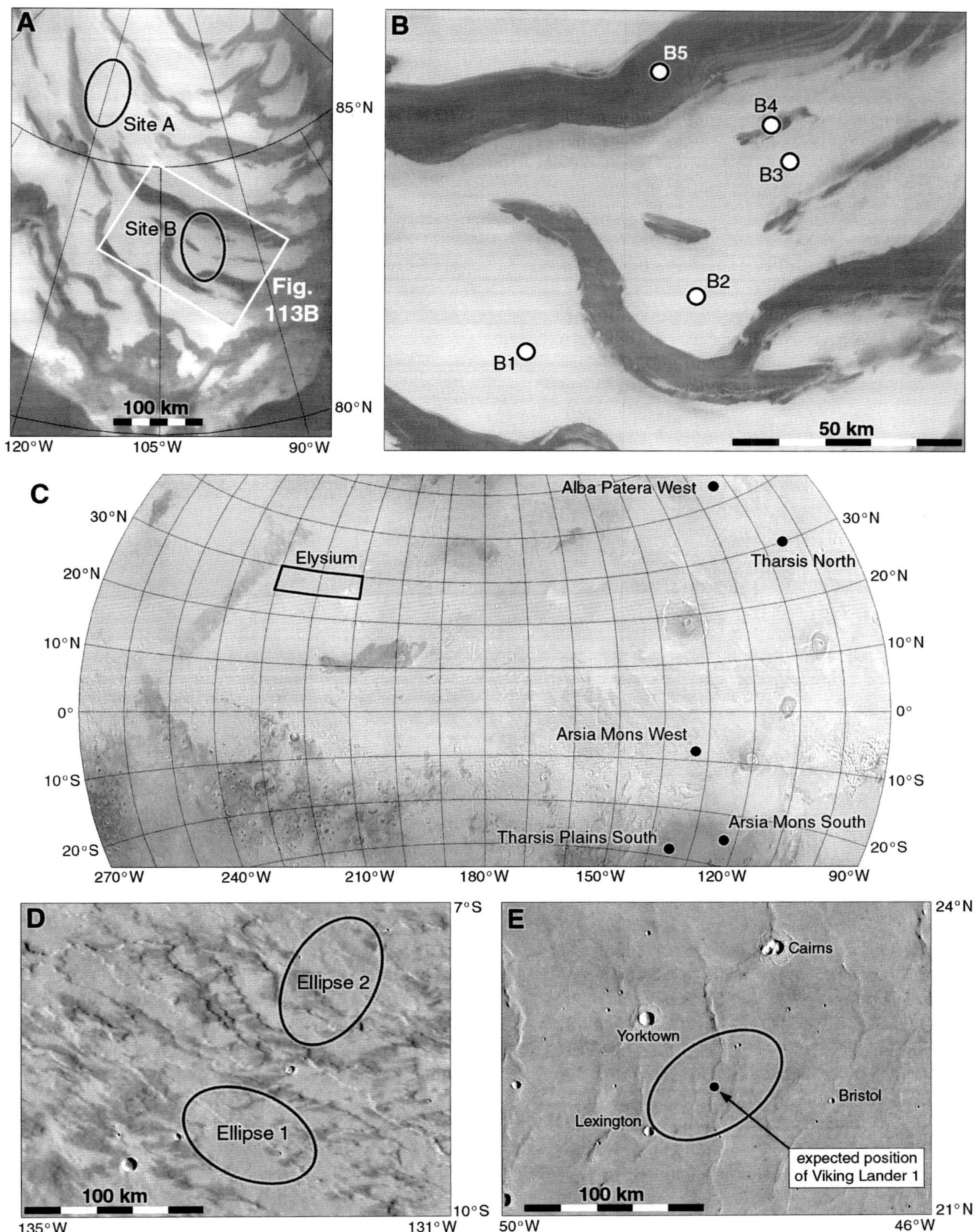

Figure 113 Mars 1984 sites from the Mars Science Working Group (1980) reports.
A, B: North polar sites from MSWG (1980), Vol. 1. **C:** Volcanic sites from Table 39. **D:** Two ellipses in the Arsia Mons West site from Table 39. **E:** Ellipse at the Viking Lander 1 site in Chryse from MSWG (1980), Vol. 2.

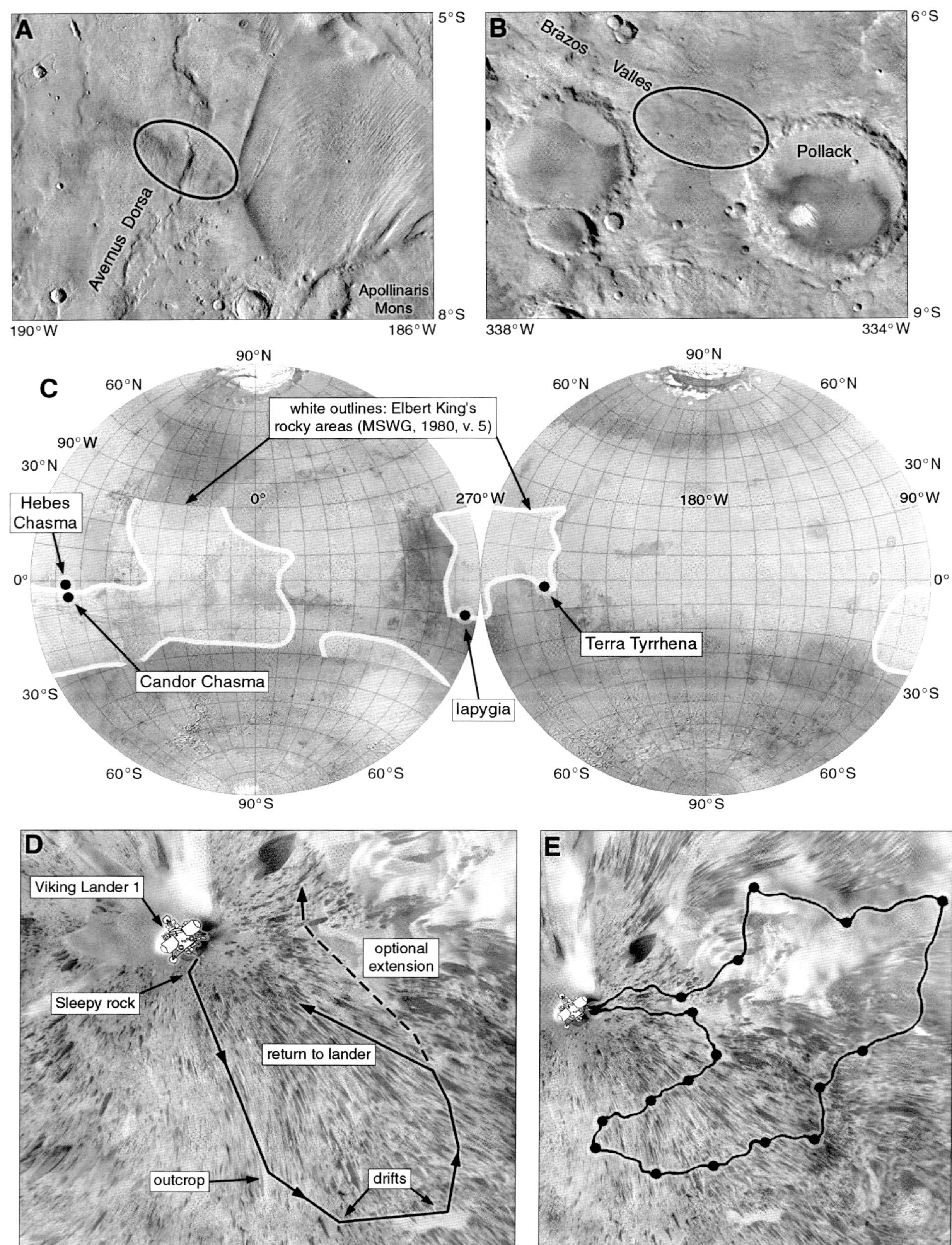

Figure 114 Mars 1984 sites from the Mars Science Working Group (1980) reports.
A, B: Apollinaris and Schiaparelli sites from MSWG (1980), Vol. 3. **C:** Areas where rocks could be sampled, identified by Elbert King (MSWG, 1980, Vol. 5), and additional sites described by Harold Masursky (MSWG, 1980, Vol. 4). **D and E:** Rover sampling routes at the Viking 1 landing site, from MSWG (1980), Vols. 6 and 7, and Nickle (1980).

traverse 2 the highlands and channel deposits, traverse 3 a crater ejecta deposit, and traverse 4 the material of Amazonis Mensa north of the landing site. Each traverse would return its samples to the lander before starting the next drive to guard against later failure. At West Mangala Valles, traverse 1 would sample the plains, traverse 2 the highlands, traverse 3 the material of Eumenides Dorsum, and traverse 4 a mixture of those types of material as well as the Nestus Valles area. At Kasei Valles, the channel floor and wall materials would be sampled.

Several other landing sites with rover traverses were included in the Mars Landing Site Catalog and are illustrated in Figures 138, 139 and 140.

Scott and Tanaka (1988) also described ten sites, chosen to provide samples from several types of material at each location and to span the range of compositions and ages of Martian surface material. These sites are listed in Table 41 and shown in Figure 115. Most of these sites are included in the Landing Site Catalog (Table 47, sites 23 and 44–50). Sites 1 and 2 in Table 41 were the most favoured (Scott, 1988).

Chicarro (1988) identified three regions with many craters exhibiting fluidized ejecta apparently caused by impacts into water- or ice-rich material. All are in ridged plains materials on Mars, which Chicarro observed to have large numbers of these fluidized ejecta craters. He suggested landing 50 km from a 30-km-diameter crater in one of these regions and traversing the ejecta up to the crater rim. Observations would include drilling to sample subsurface ice. For one of the locations, Coprates, he identified a specific landing site and rover traverse. These sites are listed in Table 42 and illustrated in Figures 115 and 117.

Markun (1988) identified 14 locations which might provide specific types of sedimentary material. These are listed in Table 43 and plotted in Figure 118. A report by the European Space Agency (ESA, 1990, Table 6.1) described sample return studies involving rovers that included locations from Tables 40, 41 and 42, but listed some slightly different landing sites (Figure 118, Table 44).

1980s: **Mars Rover Sample Return**

A Mars Rover Sample Return (MRSR) mission was studied between 1987 and 1989. NASA's Mars Exploration Strategy Advisory Group set up a Mars Study Team late in 1986 to examine a sample return mission with significant international collaboration. NASA would provide a lander, rover and communication orbiter, whereas an international partner would provide a lander with Mars ascent vehicle and a spacecraft to carry the sample back to Earth. Launch would be in November 1996, with Mars arrival in September 1997. The orbiter would carry a very high-resolution (1.5 m/pixel) camera for landing site certification. After the dust storm season was over, the partner's lander carrying the ascent vehicle would land first. A radio beacon on that lander would guide the NASA lander to a site nearby.

This study (NASA, 1987) included 11 proposed landing sites (Table 40, counting one at Candor), with one chosen to illustrate a detailed mission scenario. This was the East Mangala Valles site (Figure 116B). The rover would undertake four traverses and stop at 28 sampling sites. Each traverse would begin and end at the sample return lander. The first short traverse provided a contingency sample. It would cover 7 km and collect samples at three sites. The last traverse would be the longest, covering 86 km with seven sampling sites. After each traverse the rover would deliver its samples to the return vehicle's sample container. The total sample mass would be about 5 kg, including a variety of materials and particle sizes. The Mars ascent vehicle would lift the sample container into Mars orbit to rendezvous with the Earth return vehicle. It would leave Mars orbit in August 1998 and arrive at Earth in August 1999, going into orbit where it could be examined initially on the space station as a planetary quarantine measure. After the sample delivery, the rover would continue in an extended mission for two years or more. A second sample return mission might launch late in 1998 and leave Mars for Earth with its sample early in 2001. This second rover's extended mission might still be operating in late 2003.

A series of meetings of the Science Working Group considered mission strategies and landing sites, but the study terminated before specific locations were chosen. The site selection deliberations are briefly summarized here. This study was continued after late 1989 by the Mars Science Working Group.

Possible landing sites were first described at the fifth MRSR Science Working Group meeting held on 11 and 12 January 1988 at JPL. The potential landing sites are listed in Table 45 and illustrated in Figure 119. The site numbers

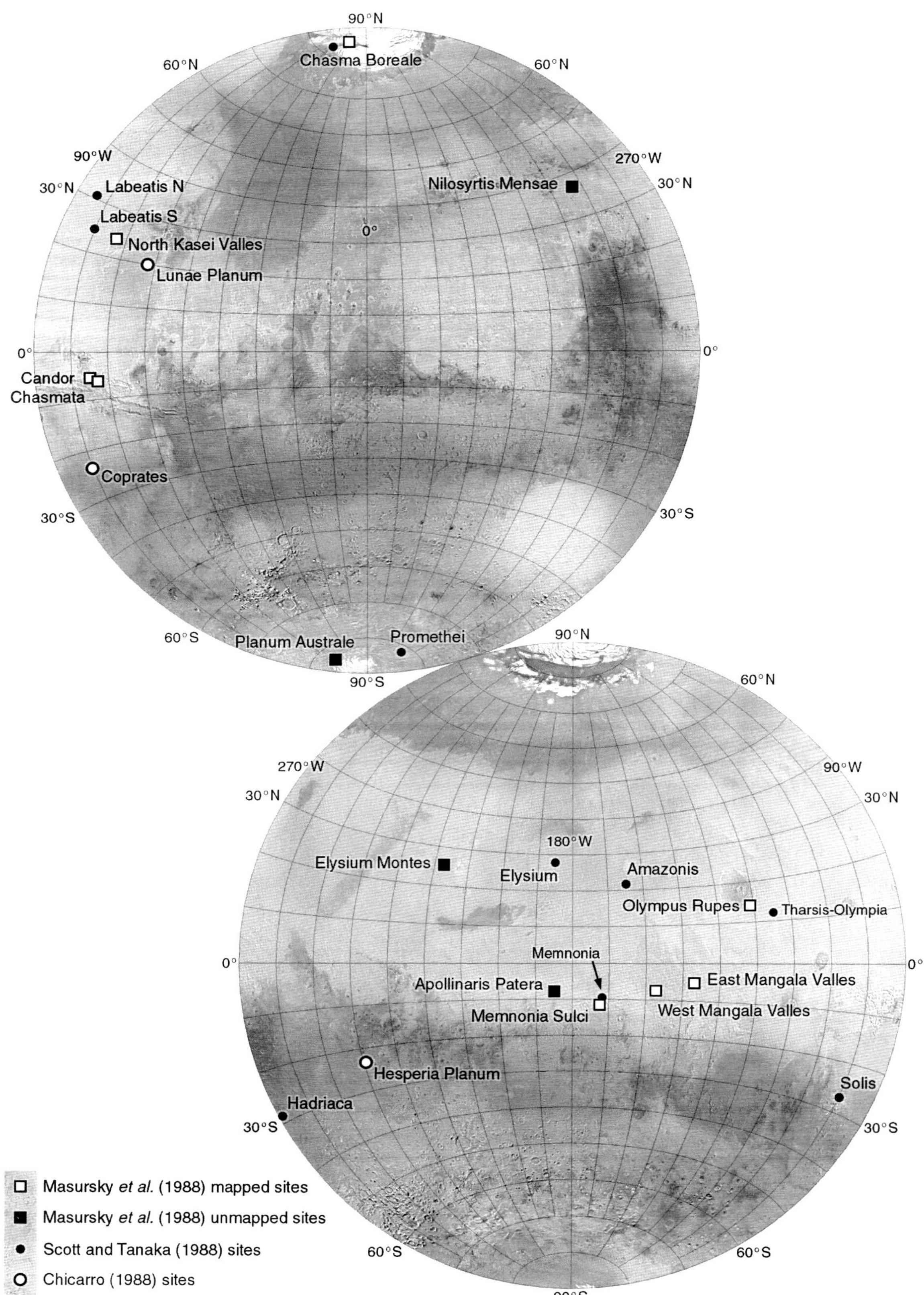

Figure 115 Sample return sites suggested by Masursky *et al.*, Chicarro, and Scott and Tanaka in 1988 (Tables 40, 41, 42).

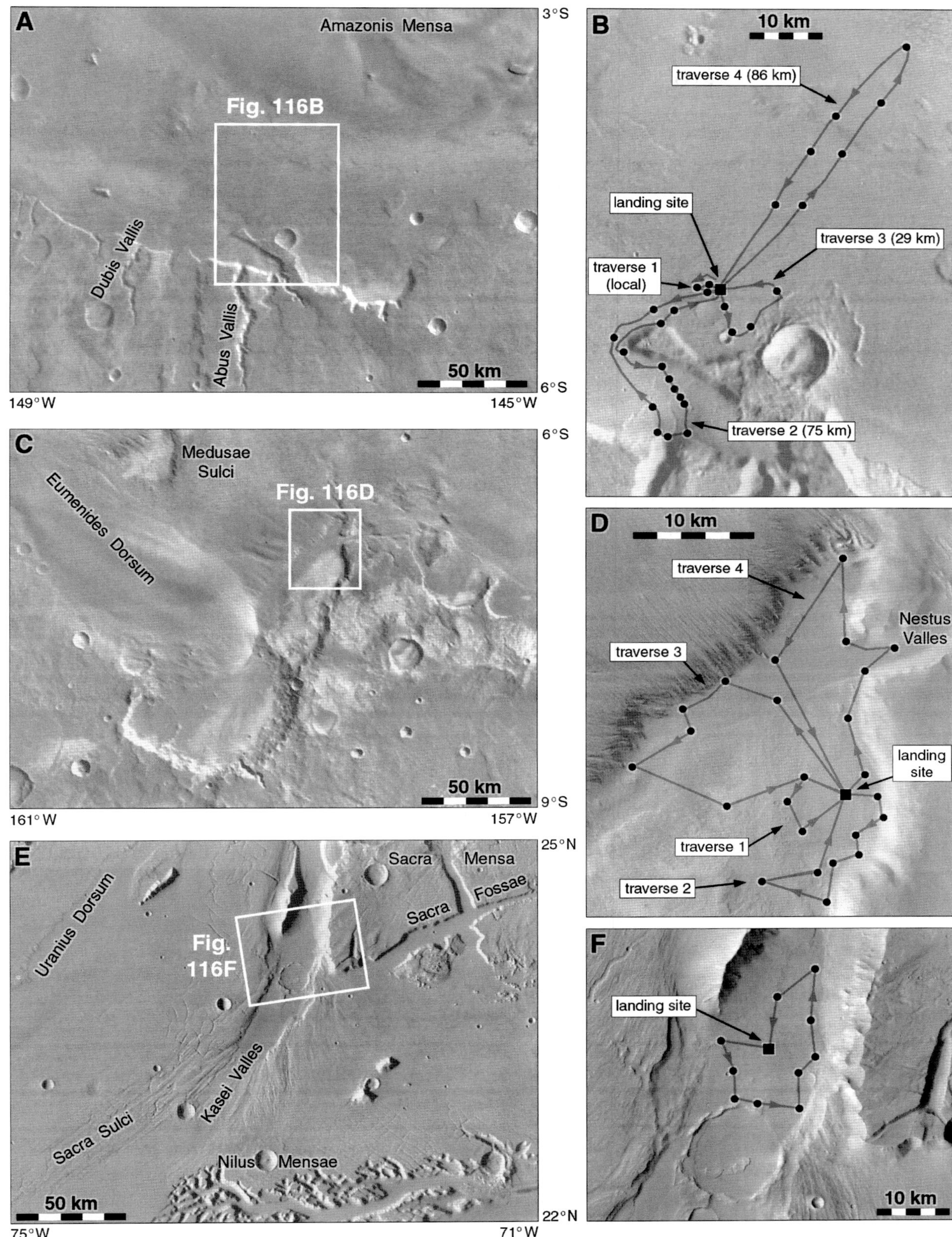

Figure 116 Rover traverses at three sample return sites described by Masursky *et al.* (1988a).
A and B: East Mangala Valles. **C and D:** West Mangala Valles. **E and F:** North Kasei Valles. A variation on Figure 115D is illustrated in Figure 140.

Table 40. *Sample Return Sites Described by Masursky et al. (1988a)*

Name	Location	I-map	Name	Location	I-map
1A. East Mangala Valles	4.7° S, 147.1° W	I-1962	6. Memnonia Sulci	10.9° S, 173.0° W	I-2084
1B. West Mangala Valles	7.2° S, 158.5° W	I-2087	7. Candor Chasmata	5.5° S, 74.5° W	I-2568
2. North Kasei Valles	24.3° N, 72.7° W	I-2107		5.7° S, 71.9° W	
3. Olympus Rupes	14.1° N, 131.0° W	I-2001	8. Apollinaris Patera	8° S, 186° W	I-2351
4. Chasma Boreale	86° N, 45° W	I-2357	9. Elysium Montes	26° N, 217° W	–
5. Planum Australe	83° S, 60° W	–	10. Nilosyrtis Mensae	37° N, 293° W	–

Table 41. *Sample Return Sites Described by Scott and Tanaka (1988)*

Name	Location	Sampled materials
1. Tharsis-Olympia	12° N, 125° W	Olympus flows, fractured flows (Ulysses Fossae), aureole material
2. Chasma Boreale	82° N, 57° W	Polar layered material, grooved plains, crater ejecta
3. Memnonia	10° S, 172° W	Medusae Fossae Formation, ridged plains, highlands material
4. Labeatis North	31° N, 83° W	Older Tharsis flows, ridged plains, faulted highlands
5. Labeatis South	24° N, 80° W	Young and older Tharsis flows, ridged plains
6. Solis	27° S, 100° W	Syria Planum flows, old fractured plains, ancient crust
7. Hadriaca	29° S, 269° W	Volcanic shield material, smooth plains, ridged plains
8. Elysium	27° N, 185° W	Young flows from Elysium Mons, ridged plains, knobby material
9. Amazonis	22° N, 165° W	Young flows of Amazonis Planitia, ridged plains, knobby material
10. Promethei	81° S, 315° W	Polar layered material, basin rim, Dorsa Argentea plains

Table 42. *Sample Return Sites Described by Chicarro (1988)*

Site	Name	Location
1	Lunae Planum	c. 20° N, 62° W
2	Coprates	23° S, 80° W
3	Hesperia Planum	c. 23° S, 249° W

in Table 45 are taken from a list of USGS geologic maps at 1:500000 scale presented at the meeting, because those mapped sites were assumed to be understood well enough to be good candidates for a sample return mission. A minimum mission described at the meeting involved a landing on the Tharsis plains at about 45° N. The lander would take a panoramic image, collect a contingency sample with a robotic arm, and then release a rover which would deploy a seismometer and meteorology package. The rover would then make several short looping traverses out to about 100-m range to retrieve samples for return to Earth. Two other mission plans would include areal sampling (20-km range) and regional sampling (100-km range).

At the eighth MRSR meeting, held on 31 August and 1 September 1988 at JPL, the same sites identified in January 1988 were still being considered, but by the tenth meeting on 30 November 1988 at JPL, different mission scenarios were examined. One included a twin lander mission beginning in 1998 with a rover launch. It would land on 1 July 2000, and over 32 months it would drive between 30 and 300 km while collecting 100 samples. It would rendezvous with a sample return lander, launched in 2001 and landing in March 2002. This lander would obtain a regolith core as a contingency sample. The rover would transfer samples to the sample return lander, which would leave Mars on 1 March 2003 to return to Earth.

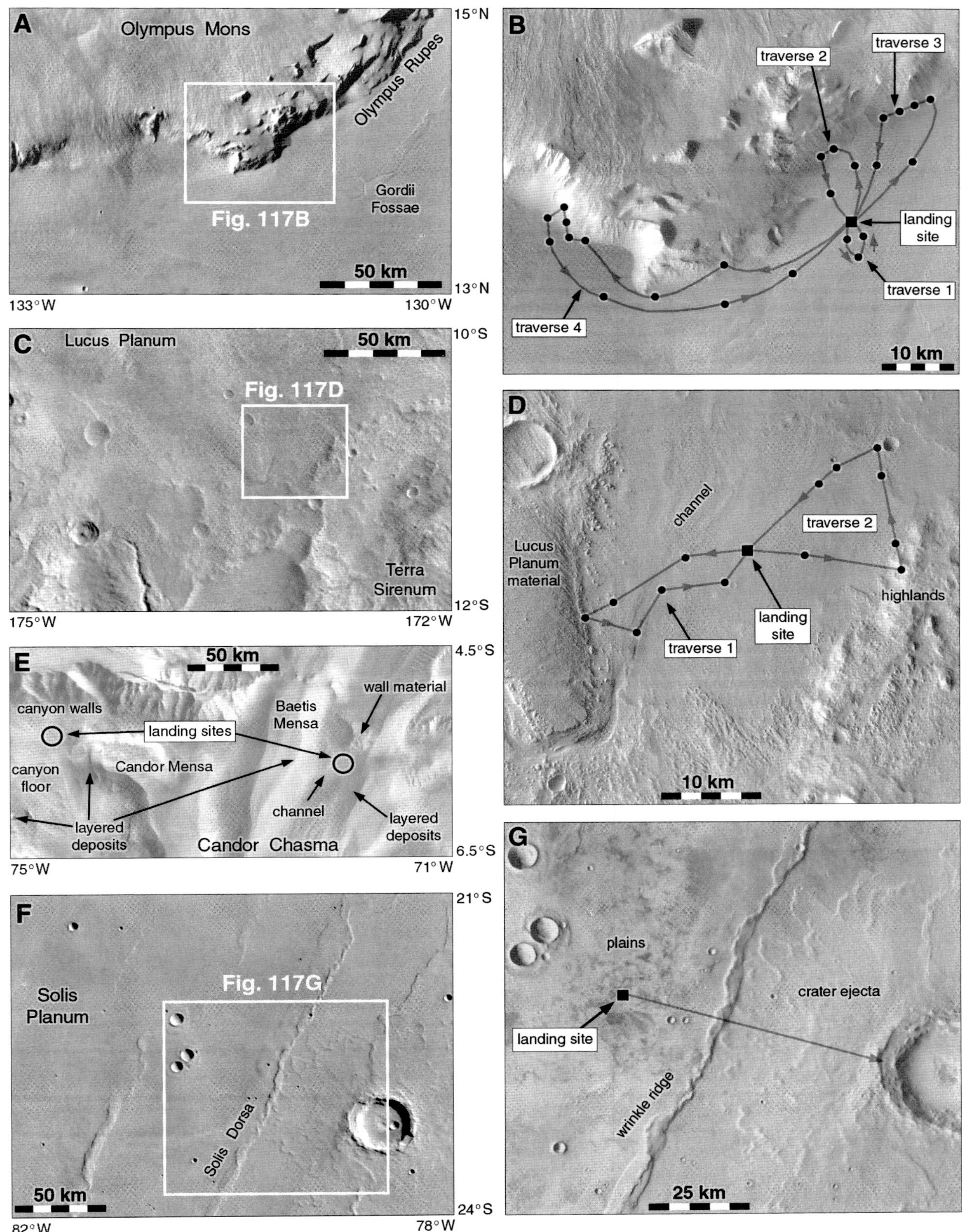

Figure 117 Sample return sites described by Masursky *et al.* (1988a) and Chicarro (1988).
A and B: Olympus Rupes. **C and D:** Memnonia. E: landing sites without specified traverses in Candor Chasma.
F and G: Coprates region (Chicarro, 1988).

Table 43. *Sample Return Sites Described by Markun (1988)*

Rock type	Setting	Location
Conglomerate, glacial	Etched terrain	76° S, 74° W
	Fretted Terrain	38° N, 65° W
	Patterned plains	33° N, 91° W
Conglomerate, fluvial	Channels, Chryse Planitia	23° N, 33° W
Breccia, debris fans	Canyon walls, Ganges Chasma	8° S, 46° W
	Canyon walls, Capri Chasma	12° S, 46° W
Sandstones, aeolian	Old cratered terrain	47° S, 160° W
Sandstones, fluvial	Channels, Chryse Planitia	23° N, 33° W
	Old cratered terrain	48° S, 98° W
Arkose	Graben/horst, Ceraunius Fossae	24° N, 97° W
Shale, lacustrine	Channels, Chryse Planitia	23° N, 33° W
Siltstone, loess	Polar areas	80° S, 270° W to 350° W
		85° N, 330° W through 0° to 90° W
Evaporites	Channels, Chryse Planitia	23° N, 33° W

Table 44. *Sample Return Sites from European Space Agency (1990)*

Name	Location	Notes	Source
Mangala Valles	3° S, 150° W	Highlands, volcanism and a large channel; 50-km traverse.	Masursky, H. (USGS)
Planum Australe	84° S, 65° W	Ice cap, dust, climate studies; 50-km traverse.	
Nilosyrtis Mensae	37° N, 290° W	Highlands, plains, dichotomy boundary; 100-km traverse.	
Memnonia Sulci	10° S, 172° W	Dating of multiple rock types; 100-km traverse.	
Olympus Rupes	17° N, 139° W	Young volcanic plains and flows; 50-km traverse.	
Apollinaris Patera	9° S, 183° W	Volcanism and channels; 100-km traverse.	
Tharsis-Olympus	12° N, 125° W	Older crust, volcanism, aureole material; 50-km traverse.	Scott, D. H.
Chasma Boreale	82° N, 57° W	Ice cap, layered materials; 50-km traverse.	
Coprates	23° S, 80° W	Ridged plains, tectonics, water or ice; 50-km traverse.	Chicarro, A. (CNES)
Hesperia Planum	29° S, 242° W	Fluidized crater ejecta. Organics in ice? 50-km traverse.	
Valles Marineris		Tectonism, landslides, flooding, layered deposits.	Lucchita, B.
Hellas basin		Basin ejecta, weathering, stratigraphic marker.	King, E. A.
Kasei Valles	23° N, 74° W	Channels, volcanism, sediments; 50-km traverse.	USSR
Uranius Patera	28° N, 95° W	Volcanic plains and shield; 100-km traverse.	
Highlands		Age dating of a variety of materials.	Neukum, G.
Cratered Terrain			
Tharsis plains			
Canyon system		Layered sediments, possible early biological evolution.	Klein, H. P.

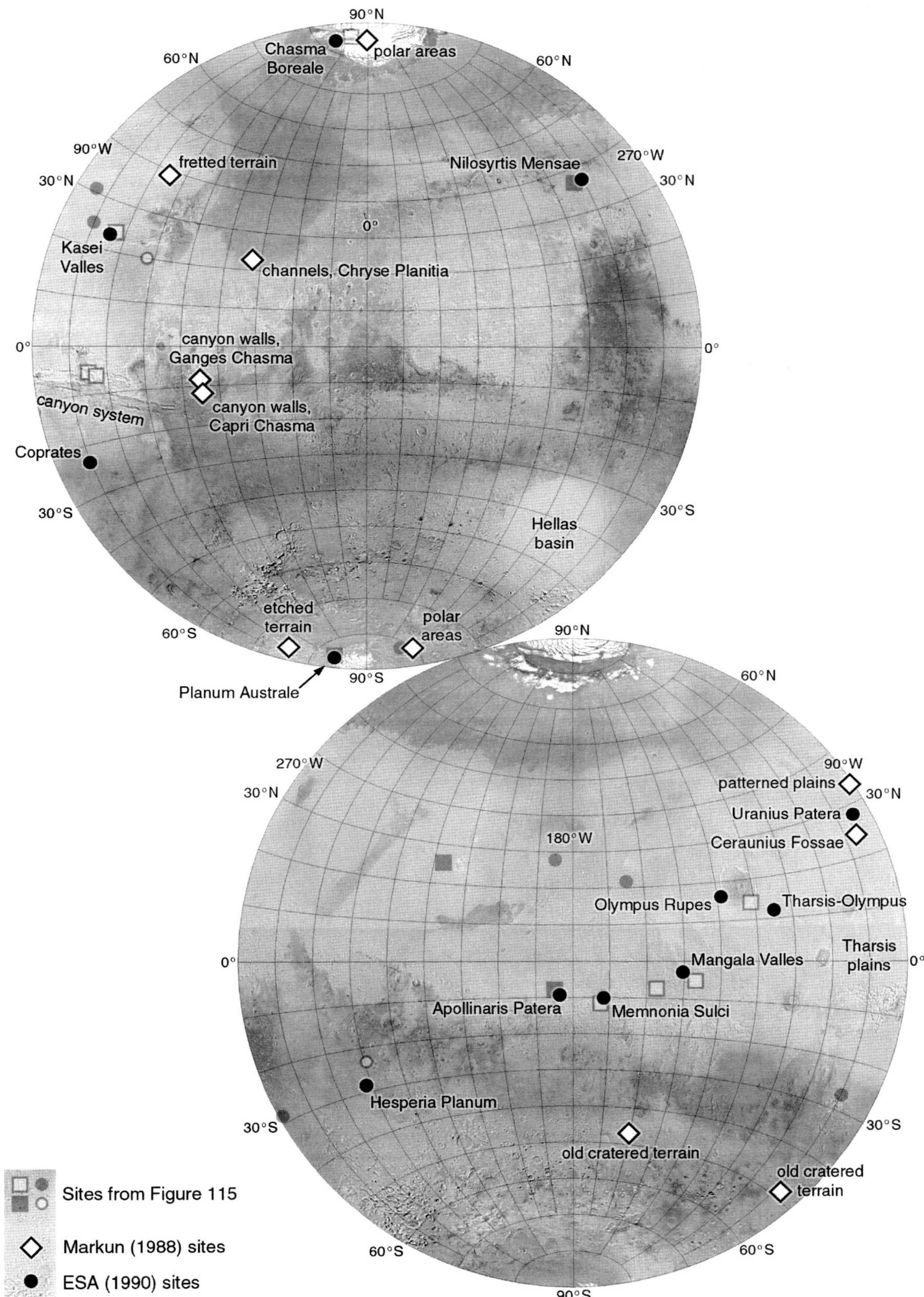

Figure 118 Sample return sites from Markun (1988) and ESA (1990).
Sites from Figure 115 are also shown to illustrate revised locations for several sites in the ESA list (Table 44).

Table 45. *Sample Return Sites Considered at the Fifth MRSR Meeting, 1988*

Site	Location	Notes
1. Alba	c. 35° N, 102° W	Lava petrology, age, tectonism/volcanism relationship
3. Argyre	c. 56° S, 43° W	Basin age, uplifted crust in basin rim, basin fill materials
5. North Elysium	c. 33° N, 213° W	Petrology, ages of volcanics, volatile history and role of water
6. Tyrrhena	c. 21° S, 253° W	Petrology, old volcanics, age and origin of ridged plains
9. Lunae Planum/Upper Maja	c. 17.5° N, 55° W	Old plains, cratered uplands, channel deltas, lake beds upstream of Xanthe Montes
10. Mangala:		
(i) Source	18.5° S, 149.5° W	Outflow channel source at fracture, Mangala Fossae
(ii) Lava bedrock	12° S, 150° W	Rock outside the main channel
(iii) deposits	4° S, 150° W	"Masursky site" at channel mouth (Figure 116B)

A second mission design would include two rovers, landing about 1 July 2000, which could each drive from 400 to 1200 km. They would both rendezvous with a sample return lander, transfer samples to it, and make additional looping traverses to collect more samples near that site. This sample return lander would also collect a regolith core as a contingency sample. A second sample return vehicle landed nearby would serve as a backup, or as the return vehicle for the second rover, and in any case would collect and return a second regolith core. The sample return landers would land about 1 March 2002 and leave Mars about 1 March 2003.

The 11th meeting was held on 23 and 24 February 1989 at JSC. Numerous sites were described. First, 27 sites were listed (Table 46, Figure 119), most corresponding to sites in the Landing Site Catalog (Table 47) but with slightly varying coordinates. The list given here was a preliminary version, which was revised before it was included in the Catalog. Then several specific sites were described. Seventeen were said to have been considered, but they were not identified. Site 5 (Table 46, Figure 125) in Iapygia, near 9° S, 280° W, was a place where valleys enter a crater which might have been an ancient lake bed. A small fresh impact crater would give access to material from deeper strata (Figure 120B). Another site at Western Daedalia Planum (Figure 138C) would be suitable for either a single lander with short rover looping traverses to collect samples or a twin lander mission with a rover traverse of up to 500 km to bring samples to the second lander for return to Earth. Other sites described at the meeting were at Chronium Planum (60° S, 213° W) and at Gusev crater (16° S, 183.5° W). The Chronium mission was designed for three rover loops of 4, 26 and 70 km, visiting 16 sampling stations and collecting 56 rock samples and six soil samples and digging three trenches. The Gusev mission would be one 500-km traverse with 21 sites, 84 rock samples, ten soil samples and four trenches. It would extend from the cratered uplands into Ma'adim Vallis, across Gusev crater and onto the volcanic deposits of Apollinaris Patera. These routes were not illustrated.

The Mars Rover Sample Return deliberations also resulted in a preliminary mission study (Rea *et al.*, 1988). This was prompted by a report by retired astronaut Sally Ride (the first female US astronaut) which recommended that NASA strive for global preeminence in robotic planetary exploration. A major theme in this exploration program would be a Mars sample return mission to be completed before 2001. The mission study was undertaken at JPL and drew on previous studies by JPL, JSC and the Mars Study Team (NASA, 1987).

This version of MRSR would obtain samples to support studies of the planet's composition, evolution, volatiles and biology. Each landing site would provide access to one of the planet's major geologic units, and other materials of differing ages and origins, in a reasonably well-understood context. Samples would be collected by a rover, with a backup sampling system on the lander. The rovers would use scoops to collect soil, rakes for pebbles, drills for rock samples and a container for an atmospheric sample. Remote sensing instruments on the rover would allow the science team to select interesting samples. The

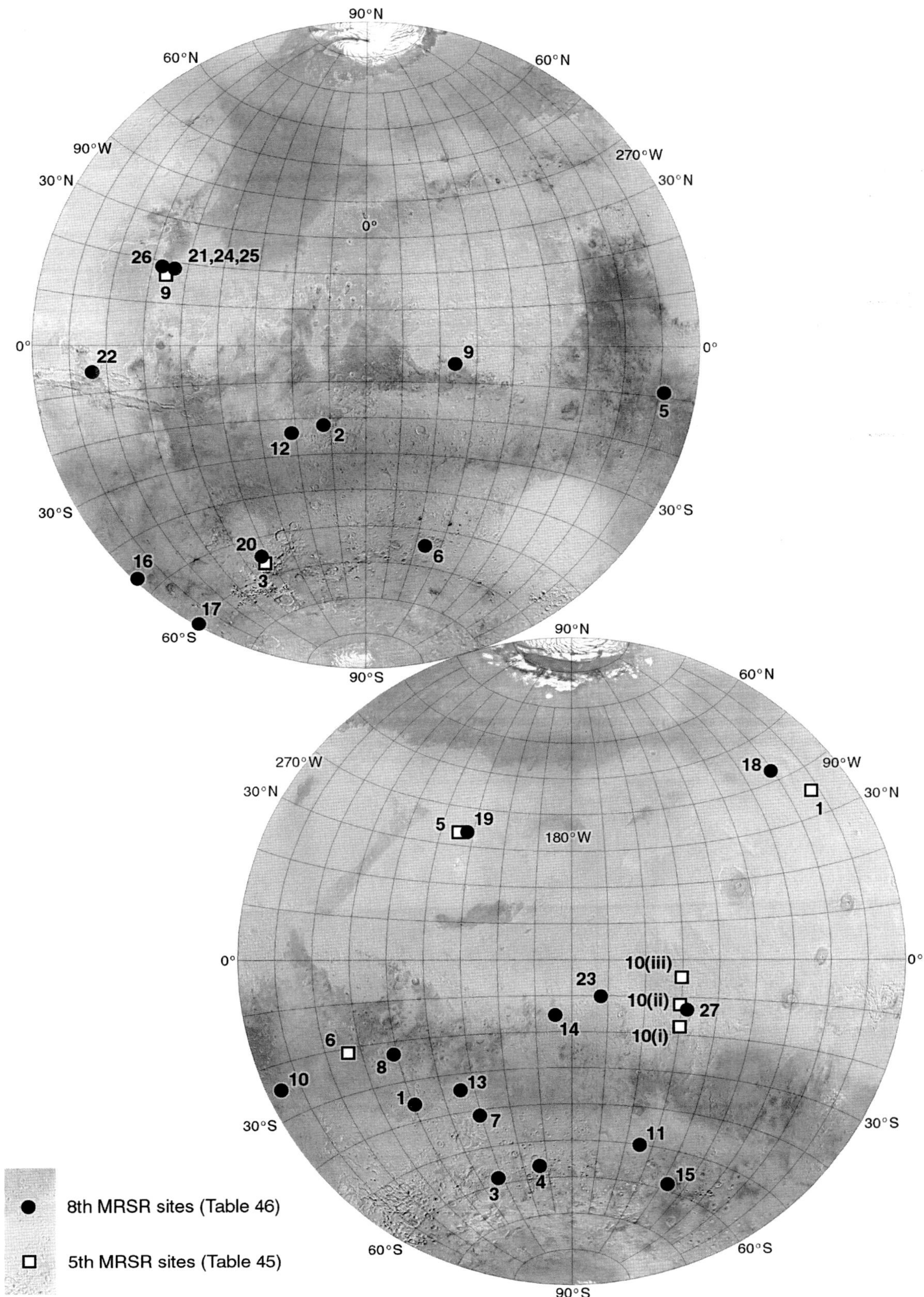

Figure 119 Sample return sites suggested at the fifth and 11th Mars Rover Sample Return (MRSR) meetings.

Table 46. *Sites from the 11th MRSR Science Working Group Meeting, 1989*

Site	Location	Site	Location	Site	Location
1	36° S, 229° W	10	25° S, 266° W	19	33° N, 212° W
2	22° S, 12° W	11	50° S, 154° W	20	55° S, 43° W
3	58° S, 212° W	12	23° S, 21° W	21	18.95° N, 53.50° W
4	57° S, 193° W	13	34° S, 214° W	22	5.5° S, 74.5° S
5	9° S, 279° W	14	15° S, 184° W	23	10° S, 172° W
6	55° S, 336° W	15	58° S, 137° W	24	17.9° N, 53.8° W
7	42° S, 210° W	16	47° S, 89° W	25	17.65° N, 54.2–54.3° W
8	23° S, 231° W	17	60° S, 89° W	26	18.2° N, 57° W
9	5° S, 336° W	18	42° N, 110° W	27	13.8° S, 148.1° W

Notes: (1) An alternate site 27 is at 13.7° S, 148.7° W. (2) Most of these sites are in the Landing Site Catalog (Table 47), with the same numbers given here. Sites 13 and 18 are near Catalog sites, 14 is Catalog site 112, and sites 15, 16 and 17 are not in the Catalog. Site 5 is Figure 120B.

rover could also deploy several seismic or weather stations to create a local network of instruments. Apart from the rover, the mission components would be an orbiter for imaging and communications, an ascent vehicle and an orbiting sample return vehicle. These could be used in four different types of mission, designated Local D, Areal B, Areal D, and Areal B-Heavy. All would launch in 1996 or 1998 and return samples to Earth in 2001.

The Local D reference mission would land a short range (100 m) rover on the broad shield volcano Alba Patera. The first of two launches would carry the rover and ascent vehicle to Mars, where it would aerobrake into orbit and land when ready. The two orbiter components would be launched together on a second rocket, separate, and each brake into orbit separately using rockets. The sample would be lifted to orbit, rendezvous with the return orbiter, and eventually use aerobraking at Earth to enter orbit there. Aerobraking into orbit like this is also called aerocapture.

The Areal B reference mission would place a long-range (20 to 40 km) rover in Mangala Vallis. In this variant, the ascent vehicle and return vehicle would be launched first and would aerobrake into Mars orbit, after which the ascent stage vehicle would land. The second launcher would carry the communications orbiter and rover together, using rockets to enter Mars orbit. Then the rover would separate and land near the ascent stage. After retrieving the sample in Mars orbit and carrying it back to Earth, the return vehicle would brake into orbit using rockets.

The Areal D reference mission would begin in 1996 with the launch of the two orbiters. They would then separate and brake into Mars orbit using rockets. The communications orbiter would carry cameras to help with landing site selection. Then in 1998 the rover and ascent stage would be launched. They would aerobrake into orbit and land together at a site chosen from the orbital data. No site was suggested in the report. The return vehicle would again use rockets to brake into Earth orbit. The Areal B-Heavy reference mission would land a heavy rover in Candor Chasma. Aerocapture would be used at Mars for all mission components, and back at Earth the Mars sample capsule would not enter orbit, but descend directly to Earth. These modifications would save propellant mass to accommodate the larger rover.

Additional studies like these are scattered throughout meeting abstracts of the period. One further example is a study reported by Masursky *et al.* (1988b) of a rover mission to the edge of the north polar cap (Figure 120C). Numerous 1-m-long cores would be taken over a traverse across layered terrain and correlated to form a 'complete section of the polar deposits'. The same report also included a traverse at Mangala Valles (Figure 140, Site 038) which differed from a previous traverse plan at that site (Figure 116D).

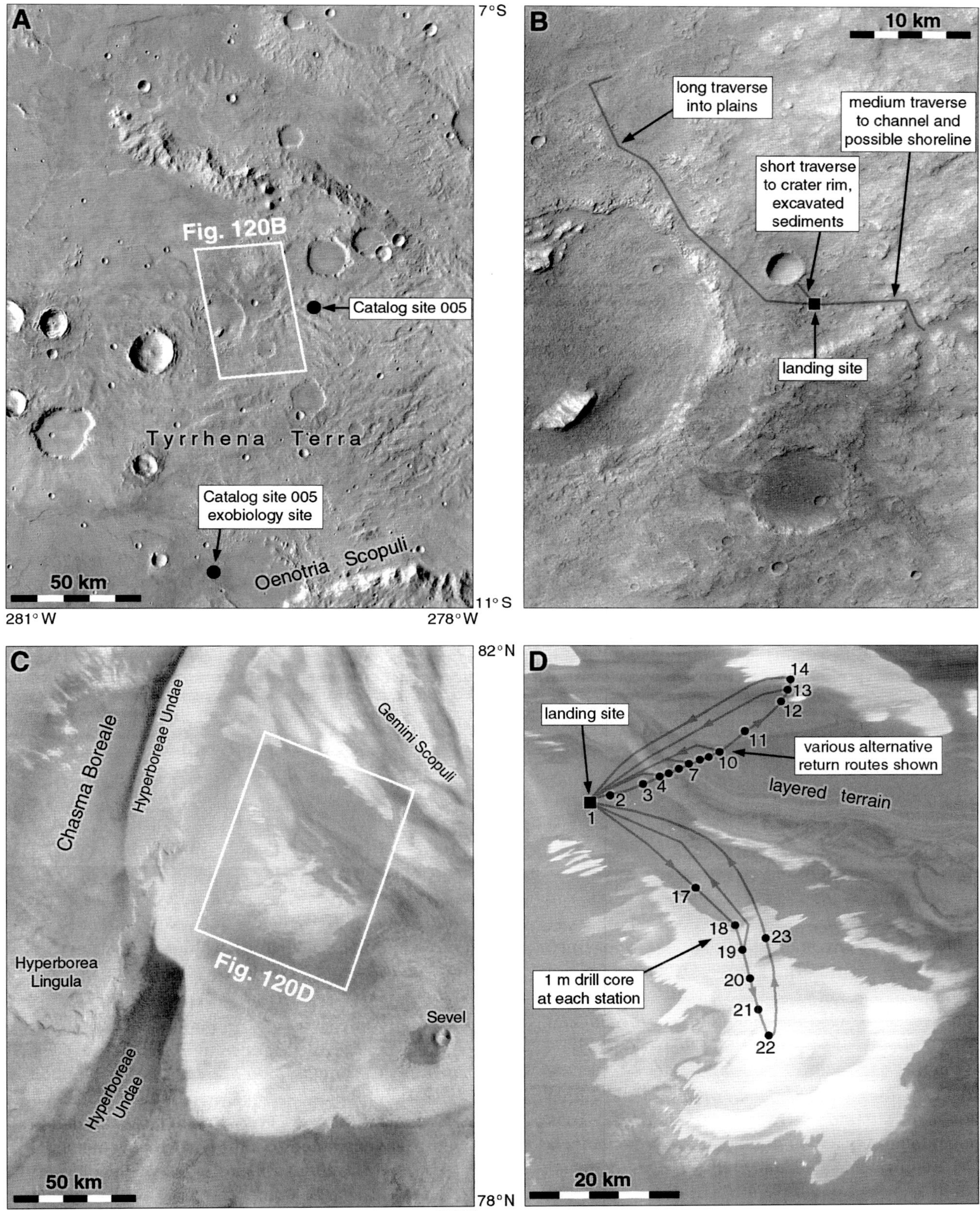

Figure 120 Mars Rover Sample Return sites.
A and B: A site in Iapygia described at the 11th Mars Rover Sample Return Science Working Group Meeting (Table 46, Site 5). (B) includes Viking images 754A08 and 754A10. **C and D:** A north polar site with rover traverses and core drill stations from Masursky *et al.* (1988b). D includes Viking images 058B34 and 058B36.

7 July 1988: **Phobos 1 (Soviet Union)**

Previous Soviet probes to Venus and Mars had been variations on the same basic structure. The twin Phobos spacecraft were the first to use a completely new design, also used later for Mars 96. Phobos 1 and 2 were designed to orbit Mars, rendezvous with Phobos and deposit landers on the surface of the moon. The 6220-kg Phobos 1 was launched from Baikonur at 17:38 UT and placed on a trajectory which would deliver it to Mars. A trajectory correction was made on 16 July, and the spacecraft made observations of Earth's bow shock (where the magnetic field meets the solar wind) on 8 July and of the Sun and the interplanetary environment. Solar x-ray and ultraviolet images were taken on 26 August. The last transmission from Phobos 1 was on 29 August 1988, but on 31 August an erroneous command was transmitted to the spacecraft, shutting down its attitude control system. Phobos 1 began tumbling and turned its solar panels away from the Sun. Its batteries lost power and could not be recharged, ending the mission. It would have arrived at Mars on 25 January 1989 (MY 18, sol 647) (Zakharov, 1988), but as it was unable to enter orbit, it flew past the planet at an altitude of approximately 1000 km.

The Phobos orbiter carried solar x-ray and ultraviolet sensors, several particles and fields experiments, a neutron spectrometer, a low-frequency radar sounding instrument for studying the subsurface of Phobos, and cameras and an infrared scanning instrument for observing Mars and Phobos. The Phobos lander carried spectrometers to analyze the surface of Phobos, a seismometer to probe the moon's internal structure, a penetrator to measure subsurface temperatures and an accelerometer to measure physical properties of the regolith. A second small hopping lander, intended to be carried on Phobos 1, had to be omitted but was carried on Phobos 2. The landing site on Phobos would have been near the equator in the region opposite Mars, near the sites shown in Figure 200.

12 July 1988: **Phobos 2 (Soviet Union)**

Phobos 2 was identical in design and mission goals to Phobos 1 except for some equipment, notably the addition of the small hopping lander omitted from Phobos 1. The spherical hopper, oriented and propelled by protruding rods, carried an x-ray spectrometer, a magnetometer and an accelerometer. Phobos 2 was launched from Baikonur at 17:02 UT UT, entered a heliocentric transfer orbit, made observations of the solar wind during cruise, and arrived at Mars on 29 January 1989 (MY 18, sol 651), entering an equatorial orbit where it operated successfully for two months. The initial orbit ranged from 870 to about 81 000 km above the planet, inclined 1° to the equator, with a 72-hour period. On 12 February periapsis was raised, giving an orbit of 6400 km by 81 200 km and a period of 86.5 hours. Then on 18 February, the apoapsis was dropped to give a nearly circular orbit at a height of 6145 km and an 8-hour period, approaching that of Phobos itself.

The spacecraft imaged Phobos when in range on 21 and 28 February and made long scans of the equatorial regions of Mars (ASCONT, 1998) with its infrared instrument (Termoscan) on 11 February, 1 and 26 March. The imaging spectrometer observed areas of the planet on 8, 11, 14, 21, 22, 27 and 28 February and 1, 7, 12, 13, 14, 21and 26 March. A camera and spectrometer (VSK) mainly designed to observe Phobos also imaged parts of Mars, usually when the little moon was crossing the planet's disk. Figures 121 and 122 show some of the images of Mars and the areas observed by these instruments. The Termoscan images are composites of the visible and thermal channels processed by P. Stooke from images kindly provided by J. Rodionova. In Figure 121E the contrast in the thermal part of the image is reversed to correspond better with albedo in the visible channel. Phobos observations are shown in Figure 200. The infrared observations of Mars suggested the presence of hydrated materials in the regolith (Bibring *et al.*, 1989). Proton and ion data indicated that the planet's atmosphere was being eroded and swept away by the solar wind.

On 7, 15 and 21 March, the orbit was adjusted to bring it closer to Phobos on each pass, and on 25 March, images and infrared observations of Phobos were obtained. Then on 27 March 1989 (MY 19, sol 37), during one of the last planned approaches to Phobos before the landings, communications were lost and the mission ended. On its previous approaches Phobos 2 had turned away from Earth to make observations and then turned back to transmit its data. On the last approach, it failed to re-establish communications after collecting data. A failure in the computer command system was the probable cause.

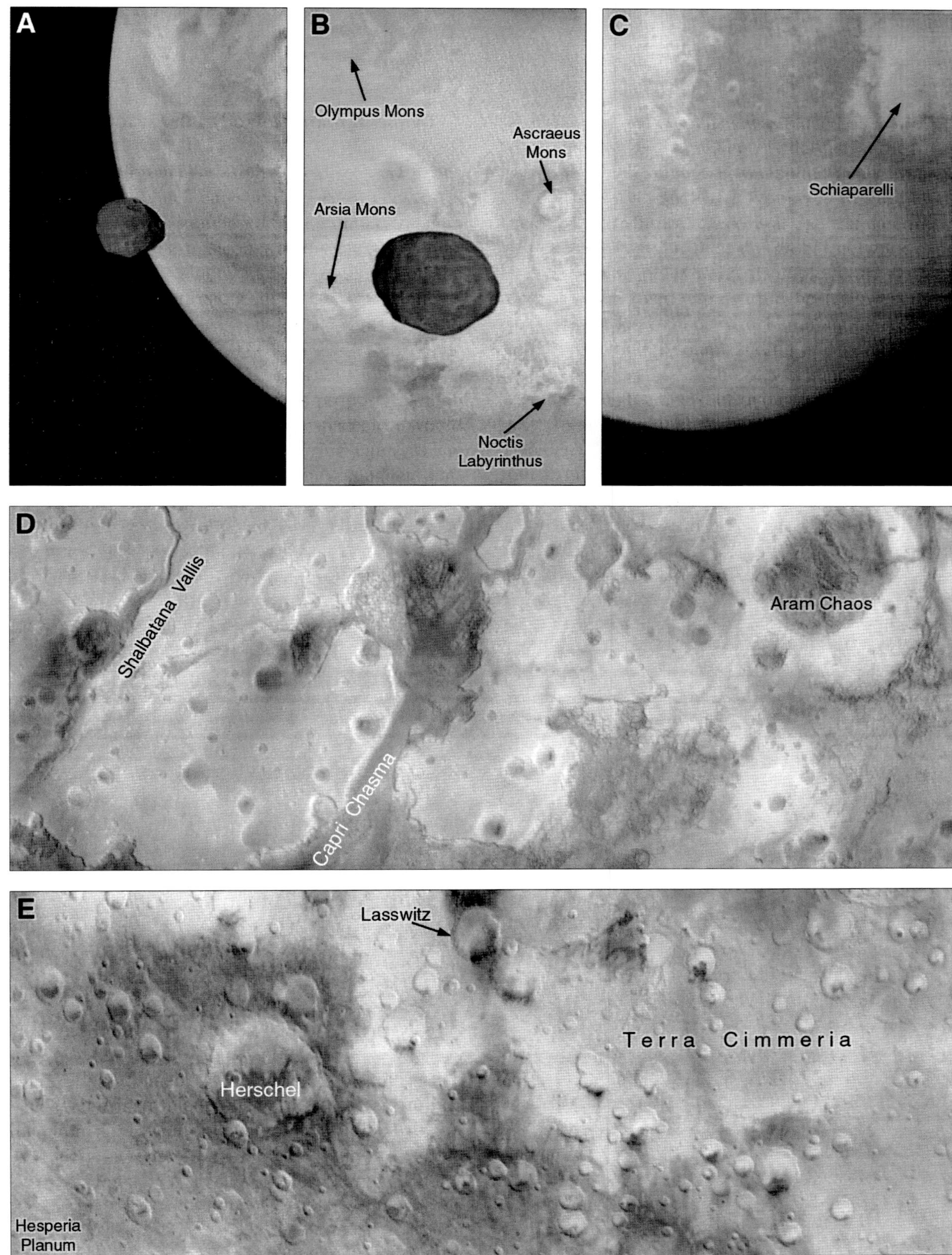

Figure 121 Phobos 2 images.
A, B: Camera and spectrometer (VSK) images of Phobos crossing the disk of Mars over Valles Marineris (A) and Tharsis (B). **C:** VSK image of the Meridiani and Noachis regions. (A), (B) and (C) were provided by T. Stryk. **D and E:** Termoscan composite images of the area south of Chryse (D) and west of the MER-A Spirit landing site (E).

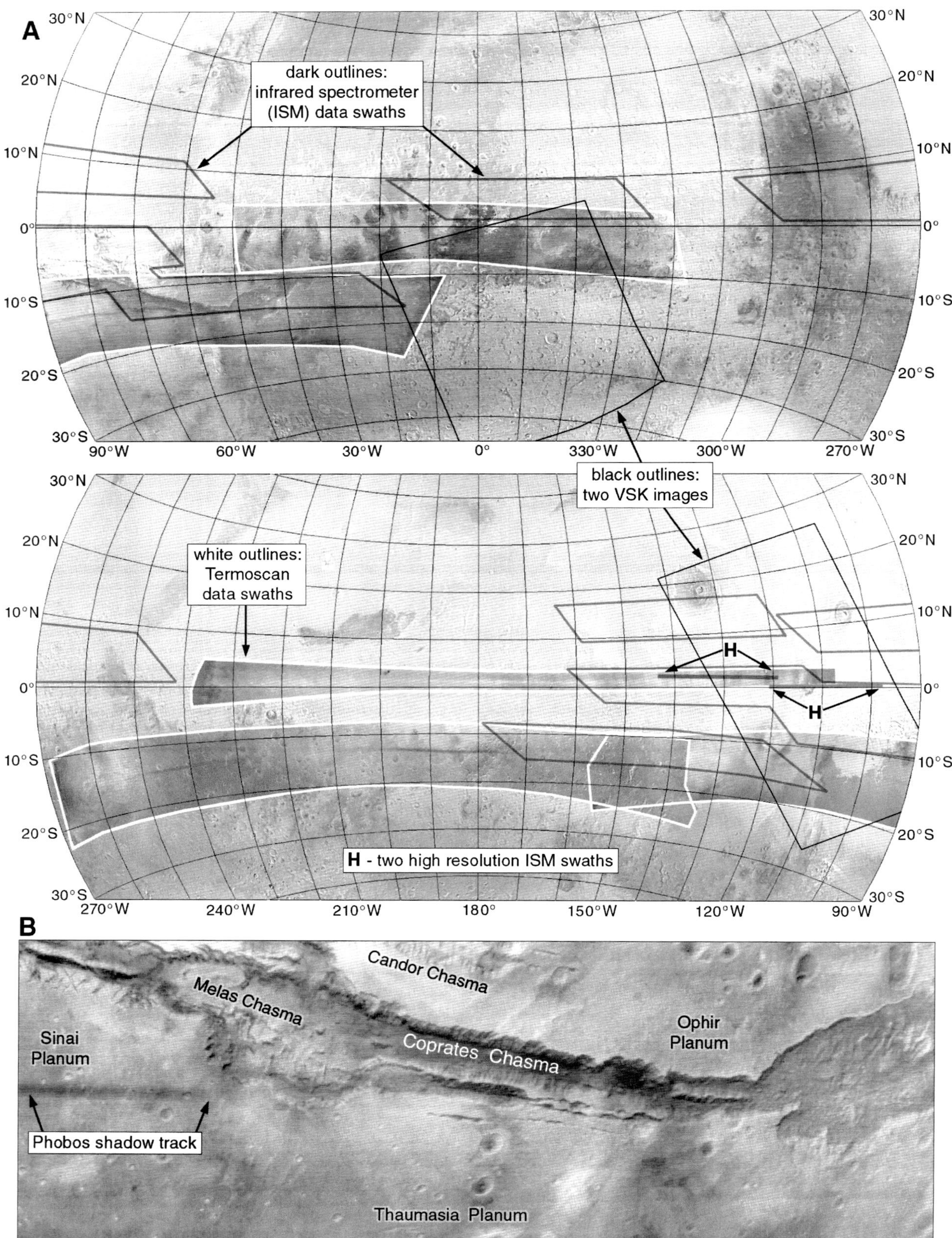

Figure 122 Phobos 2 data coverage map (A) and a Termoscan image of the Valles Marineris (B).
Figures 121D, 121E and 122B are composites of the Termoscan visible and thermal channels processed by P. Stooke from images kindly provided by J. F. Rodionova.

1990s: **Vesta/Mars-Aster (Soviet Union, France)**

This joint Soviet/French mission was discussed in the 1980s for possible launch dates in the 1990s. The Soviet spacecraft would have been based on the Phobos design. The first version of the mission was directed to Venus, deploying landers or balloons like those carried on the Soviet Union's VeGa missions to Venus and Halley's Comet in 1986, and releasing a French-built asteroid flyby and lander spacecraft. The name referred to the most interesting asteroid target, 4 Vesta. In about 1985, a new version renamed Mars-Aster was proposed, which would launch late in 1994. A Russian spacecraft would enter a polar orbit around Mars carrying cameras, an altimeter and spectrometers and dropping a lander, penetrator or balloon. The second Mars-Aster spacecraft, jointly built by Russia and France, would have used one or more gravity assists at Mars to enter the asteroid belt and fly by several asteroids, including Vesta itself, or in another version several comet nuclei. There were several variations of this mission, with different target lists and activities at Mars. The asteroid instruments would include cameras and spectrometers and possibly a laser ranging instrument. Trajectory designs for a 1994 launch included encounters with a variety of asteroids, including Amor-type objects (asteroids whose closest point to the Sun is just outside Earth's orbit), small to medium-sized main belt asteroids, and the large objects 1 Ceres or 4 Vesta, where penetrators would be deployed. Mars-Aster was eventually abandoned due to budgetary and political issues in the 1990s. Nothing is known about possible Mars landing or balloon entry sites for earlier versions of Mars-Aster. Later versions might only have conducted remote sensing.

1990s: **Mars Science Working Group**

Networks of simultaneously operating geophysical and meteorological instruments have often been studied without being approved for development. Some work on this was done by the Mars Science Working Group (MSWG), following the Mars Rover Sample Return studies done between 1987 and 1989. They met for the first time at JPL on 18 and 19 December 1989 to plan the first few missions in a future robotic exploration program. Special emphasis was placed on a Global Network mission, which was poorly defined at the time. The MSWG's studies eventually resulted in the Mars Environmental Survey (MESUR) proposal (Figure 143). The rough plan for future missions at the first meeting began with Mars Observer, then the global network expected in about 1998, followed by a sample return with a local (short-range) rover. This would be followed by a Site Reconnaissance Orbiter with 1-m imaging resolution to select sites for a human landing and several rovers to certify the human site. The network mission at this time was imagined to consist of two carrier spacecraft placed in 200-km orbits inclined 45° and 135° to the equator. Two entry capsules would be deployed on approach, and four from orbit. Landing sites would lie between 45° N and 45° S for the orbit-deployed probes, but the approach-deployed probes could reach from 50° N to 80° S for a 1999 mission. The probes would be penetrators with descent and surface imaging, meteorology, chemistry and seismology instruments. Another mission design might involve six penetrators and 12 to 18 hard landers.

The second MSWG meeting was held at JSC in Houston, Texas, on 26 February 1990. Planning had included a US/Soviet implementation team that met in July 1989, and balloon sites had been discussed in Pasadena the following month. As so many sites were now being proposed, the Mars Landing Site catalog (Table 47) was set up to support this network study and other future missions. Jacques Blamont (Centre National d'Études Spatiales, CNES) described sites for balloon missions, a forerunner of the MESUR balloon proposal illustrated in Figure 144. Their sites were at Acidalia Planitia (Catalog site 81), Arcadia Planitia (Catalog site 82) and Utopia Planitia (Catalog site 83). Jeff Plescia described early work on a human outpost, using an exploration plan set in Mangala Valles (16° S, 149° W). Site study would begin with robotic rovers and then be continued with human explorers doing geological fieldwork, especially in the graben at the source of the valley and in areas of water ponding farther north.

The remaining MSWG meeting results were devoted to MESUR and are described in association with Tables 51 through 54. Other network mission studies from this period are listed in Tables 49, 50, and 55 to 58, and illustrated in Figures 142 through 148A.

1990: **Mars Landing Site Catalog**

Following Viking, studies of a variety of future sample return and network missions were made in the United States, Europe and the Soviet Union, resulting in multiple lists of potential landing sites. In order to gather all this information into one convenient reference, Ronald Greeley (Arizona State University) compiled the Mars Landing Site Catalog. The first edition (Greeley, 1990) included the initial 83 sites in Table 47. The second edition (Greeley and Thomas, 1995) revised some sites and descriptions and added the remaining sites in Table 47. As this was a compilation of reports from various sources, some sites were duplicated (*e.g.* sites 119 and 130) or were very similar to others. Every site in the catalog is illustrated in Figures 123 through 139. Sites associated with a specific mission (*e.g.* Mars 94, MESUR) are also described under the entry for that mission.

The 'top ten' list (actually containing 11 sites) from this catalog is shown in Table 48.

The original catalog contains a description of each site with illustrations. The global distribution of these sites is shown in Figures 123 and 124, and every individual site is portrayed in Figures 125 through 137. The scales of all the small catalog site maps can be estimated from the latitude labels, recalling that 1° of latitude spans 60 km on the surface. Most of the site maps are 3° or 180 km across, and almost all of these maps are Viking image mosaics. An exception is site 31 at the north pole, which is part of Mars Reconnaissance Orbiter Context Camera image T01_000874_2695_XI_89N253W. That image is 30 km across. The unlabelled black dot on each map is the landing site itself, and in some cases an alternative site is indicated with an appropriate label. For site 51, the catalog illustrated a point at 87° W but stated the longitude as 88° W. The 88° W position, adjacent to a small volcanic vent now called West Mareotis Tholus, is a better match to the site's geological description.

Rover traverses at some sites were included in the catalog and are shown here in Figures 138 through 140. Notes on the traverse maps are based on the mission descriptions in the catalog, and the sampling site labels, either letters of numbers, are also from those original maps.

Figure 138 shows two catalog sites with sample return operations. In each case, one option is for a rover and a return vehicle to land far apart, requiring long sample-collection drives before sample delivery. At Olympus Mons (Figures 138A, 138B), the rover collects samples of older fractured lava plains, hills and smooth young plains before reaching the return vehicle at the foot of the giant volcano. The short-range rover would land as close as possible to its return vehicle and make several looping traverses to collect samples from the plains and the foot of the mountain. The Western Daedalia site (Figures 136C, 138D) had similar mission alternatives. Figure 139 includes four different rover traverses at Alba Patera, Hecates Tholus near a previous Viking Rover site, Argyre Planitia near a group of sinuous ridges resembling sub-glacial channel deposits called eskers, and on channel deposits at Maja Valles. The latter site was also considered for Mars Pathfinder (Figure 157), Mars Surveyor 2001 (Figure 180) and Beagle 2 (Figure 191). Figure 140 shows four more sites at Mangala Valles, Chryse and Aeolis. The Chryse rover would land near Viking Lander 1 and visit the old site and the postulated heatshield site just north of it. The Aeolis site is 250 km from the future MER-A Spirit rover site in Gusev crater.

1990s: **Hubble Space Telescope**

Telescopes operating above the atmosphere were suggested as early as 1923 by rocket pioneer Hermann Oberth, and the concept was more fully developed by astronomer Lyman Spitzer, who proposed the spacecraft which eventually became the Hubble Space Telescope (HST). This orbiting observatory, named after the astronomer Edwin Hubble, was built by NASA and ESA, launched on the Space Shuttle flight STS-31 (Discovery) on 24 April 1990 and deployed in orbit on 25 April at an altitude of 600 km. Its 2.4-m mirror and location above Earth's atmosphere gave it unmatched resolution and wavelength range for both deep space and solar system astronomy. The mirror originally suffered from spherical aberration, which was corrected by the addition of a corrective optics system during a servicing mission launched on 2 December 1993 (STS-61, Endeavour). Additional servicing missions were launched on 11 February 1997 (STS-82, Discovery), 19 December 1999 (STS-103, Discovery), 1 March 2002 (STS-109, Columbia) and 11 May 2009 (STS-125, Atlantis). These servicing flights allowed instruments and major spacecraft systems to be repaired or replaced, greatly extending its life and increasing its scientific productivity.

Table 47. *Mars Landing Site Catalog*

Site: Name	Location	Notes
1: Eridania Northwest	36° S, 228° W	RSR: ancient terrain, ridged plains, sediments in crater ejecta, 37° S, 230° W: best access to deposits
2: Parana Valles	22° S, 12° W	RSR: fluvial deposits, backup near NW-flowing channel (11° W)
3: Eridania	58° S, 212° W	RSR: old cratered terrain, younger lavas, ground ice
4: Eridania Southeast	57° S, 195° W	RSR: similar to site 3. 57° S, 197° W: best access to deposits
5: Iapygia	9° S, 279° W	RSR: fluvial deposits. 11° S, 279.5° W: best access to deposits
6: Noachis	55° S, 337° W	RSR: old cratered terrain, younger lavas, ground ice
7: Terra Cimmeria	42° S, 211° W	RSR: similar to site 5. 43.2° S, 208.1° W: best access
8: Mare Tyrrhenum	23° S, 231° W	RSR: similar to site 5. 22.8° S, 230.6° W: best access
9: Sinus Sabaeus	6.5° S, 335.5° W	RSR: fluvial sediments in crater. Not adequately imaged.
10: Terra Tyrrhena	25° S, 266° W	RSR: similar to site 5. 24.8° S, 265.8° W: best access
11: Phaethontis SW	51° S, 153° W	RSR: old cratered terrain, younger lavas, ground ice
12: Samara Valles	23° S, 21° W	RSR: fluvial sediments in two craters. Not adequately imaged
13: Mawrth Vallis	26° N, 21° W	RSR: fluvial deposits of large channel
14: Schiaparelli Crater	5° S, 342° W	RSR: fluvial sediments on large impact basin floor
15: Northeast Arabia	28° N, 293° W	RSR: channel deposits, eroded sedimentary layers
16: Terra Meridiani SE	7° S, 347° W	RSR: smooth plains, fluvial deposits, highland material
17: Kasei Valles North	21° N, 77.5° W	RE: 300 km traverse, deploy instruments, volcanic materials, 21.5° N, 78° W: alternative landing site
18: Alba Patera	47° N, 116° W	RSR: basalts and pyroclastic materials. Land in Alba Fossae
19: Northern Elysium	32.5° N, 212.5° W	RSR: Hecates Tholus, young lava, ground ice, 170-km route
20: Argyre Planitia	55.5° S, 42° W	RSR: lake sediments, ridges, basin massifs
21: Maja Valles	18.95° N, 53.50° W	RSR: fluvial deposits, ejecta, ancient material, 100-km route
22: Candor Mensa	5.5° S, 74.5° W	RSR: thick-layered deposits. 6.05° N, 73.75° W: best access
23: Memnonia	11° S, 173° W	RSR: volcanics, ancient crust, channel deposits, ejecta. 75 km
24: Maja Valles Mouth	17.9° N, 53.8° W	RSR: eroded highlands, fluvial deposits, 50-km route
25: Maja Valles Floodplain	17.65° N, 54.25° W	RSR: fluvial and lake deposits, safer than 24, 50-km route
26: Upper Maja	18.2° N, 57° W	RSR: lake sediments over basaltic ridged plains, 50-km route
27: Upper Mangala Valles	13.8° S, 148.1° W	RSR: lake or fluvial deposits, old cratered plains, 100-km route, 13.7° S, 148.7° W: alternative site
28: Olympus Mons SE(alternative site:	14° N, 131° W13° N, 125° W)	RSR: young lavas, talus of large scarp, 100-km rover route RSR: young lavas, fractured plains, 500-km rover route
29: Daedalia Planum W(alternative site:	19° S, 144° W18.5° S, 147° W)	RSR: young lavas, crater central peak, 100-km rover route RSR: young lavas, highland materials, 500-km rover route
30: North Polar Cap	80°–85° N	RSR: polar ice and layered deposits, drill core, 20-km route

Note: RSR: rover/sample return mission.

Table 47. *(cont.)*

Site: Name	Location	Notes
31: North Pole	90° N	RSR: ice samples, 2-m core, layering history, meteorology. If possible, long rover traverse off cap
32: Dao Vallis	33° S, 266° W	BR: channel deposits, Hadriaca Patera volcanics, 300-km rover, 1000-km balloon route down-channel to Hellas
33: Chasma Boreale	80.8° N, 44° W	RSR: ice cap, layered deposits, northern plains
34: Planum Australe	82.5° S, 60.0° W	RSR: carbon dioxide ice cap, layered deposits, southern plains
35: Memnonia Sulci	9.7° S, 174.2° W	RSR: ancient crust, lava flows, pyroclastic deposits, plains
36: Olympus Rupes SE	13.8° N, 131.2° W	RSR: scarp and plains materials, lavas of diverse ages
37: Kasei Valles	15.1° N, 75.8° W	RSR: young and older volcanic flows, channel deposits
38: Mangala Valles W	7.2° S, 158.6° W	RSR: ancient cratered terrain, diverse flows and pyroclastics
39: Mangala Valles E	4.7° S, 147.5° W	RSR: ancient cratered terrain, diverse flows and pyroclastics
40: Elysium Mons	24.3° N, 214.8° W	RSR: lava flows of two distinct ages
41: Apollinaris Patera	7.5° S, 187.2° W	RSR: lava flows of two distinct ages
42: Nilosyrtis Mensae	35.5° N, 302.5° W	RSR: ancient cratered terrain, younger lava flows
43: Candor Chasma I	10.5° S, 74.5° W	RSR: layered material in canyon walls
44: Tharsis – Olympus	12° N, 125° W	RSR: plains flows, basal Olympus aureole, fractured plains
45: Chasma Boreale	82° N, 57° W	RSR: polar layered material, grooved/polygonal material
46: Labeatis North	31° N, 83° W	RSR: younger flows, older ridged plains, fractured plains
47: Labeatis South	24° N, 80° W	RSR: two types of Tharsis flow, ridged plains material
48: Solis	27° S, 100° W	RSR: ancient faulted terrain, cratered plains, fractured plains.
49: Hadriaca	29° S, 269° W	RSR: Hadriaca shield, plains units, Hellas rim materials
50: Elysium	27° N, 185° W	RSR: Elysium flows, ridged plains, erosional remnants
51: Mareotis Volcanic Field	36° N, 88° W	RSR: volcanic materials in old faulted terrain
52: Tempe Volcano	39° N, 76° W	RSR: volcanic materials in old faulted terrain
53: Ceraunius Fossae Volcanic Field I	20° N, 112° W	RSR: volcanic flows, fractured plains
54: Candor	5.5° S, 74.5° W	RSR: iron deposits at fluvial source locations, 50-km route
55: Cydonia N–Acidalia SE	40° N, 10° W	RSR: plains, fractured terrain, highland remnants, volcanics, ground ice interactions, 100-km rover route
56: Tharsis-Olympus	12.5° N, 125.5° W	RSR: young basalts, fractured plains, aureole material, 30 km
57: Labeatis	25–30° N, 81–83° W	RSR: young basalts, fractured plains, 30-km rover route
58: Chryse Planitia–Viking 1	22.52° N, 47.97° W	RSR: ridged plains, Viking aeroshell, Viking lander parts, ridge material, crater ejecta, 100-km rover route
59: Amenthes SE	4° N, 247° W	PEN: US/Russian site: highlands composition, seismic and meteorology data, 150-km target circle
60: Isidis North	16°–36° N, 267–282° W	BAL: US/Russian site: Isidis rim, plains units, meteorology
61: Arabia	30° N, 327° W	PEN: US/Russian site: old volcanics and plains composition, seismic and meteorology data, 150 by 250 km target ellipse.
62: Hadriaca	36° S, 271° W	PEN: US/Russian site: Hadriaca Patera material composition, seismic and meteorology data, 150-km target circle
63: Arabia South	15° N, 333° W	PEN: US/Russian site: highlands composition, seismic and meteorology data, 150 by 250 km target ellipse

Table 47. *(cont.)*

Site: Name	Location	Notes
64: Isidis Planitia	10° N, 275° W	PEN: US/Russian site: Isidis plains composition, seismic and meteorology data, 150 by 250 km target ellipse
65: Utopia	45° N, 251° W	PEN: US/Russian site: plains sediment composition, ground ice, seismic and meteorology data, 150 by 250 km target ellipse
66: Elysium Southwest	17° N, 224° W	PEN: US/Russian site: volcanic plains composition, seismic and meteorology data, 150 by 250 km target ellipse
67: Elysium South	4–7° S, 205–232° W	BAL: US/Russian site: plains, highlands, possible volcanic materials, fluvial materials, meteorology
68: Medusae Fossae	2° S, 159° W	PEN: US/Russian site: volcanic plains, thick mantling deposits, seismic and meteorology data, 150 by 250 km target ellipse
69: Olympus South	11° N, 137° W	PEN: US/Russian site: volcanic plains composition, seismic and meteorology data, 150 by 250 km target ellipse
70: Alba West	45° N, 126° W	PEN: US/Russian site: volcanic plains, ground ice, seismic and meteorology data, 150 by 250 km target ellipse
71: Kasei South–Echus Chasma	10° N, 80° W	PEN: US/Russian site: young volcanic plains, fluvial sediments, seismic and meteorology data, 150- by 250-km target ellipse
72: Candor Chasma II	6° S, 73° W	PEN: US/Russian site: thick-layered sediments composition, seismic and meteorology data, 150- by 250-km target ellipse
73: Lunae Planum East	16° N, 63° W	PEN: US/Russian site: ridged plains composition, seismic and meteorology data, 150- by 250-km target ellipse.
74: Capri Chasma	14° S, 46° W	PEN: US/Russian site: thick-layered sediments, canyon wall materials, seismic and meteorology data, 150- by 250-km ellipse
75: Meridiani SW – Margaritifer Terra	12° S, 7° W	PEN: US/Russian site: cratered highlands and sedimentary mantle, seismic and meteorology data, 150- by 250-km ellipse
76: Tempe	25–52° N, 30–60° W	BAL: US/Russian site: highlands, ground ice, fluvial materials
77: Ares	10° N–13° S, 0–25° W	BAL: US/Russian site: highlands, channels, chaos, plains(specific sites at 2° N, 16° W or 3° N, 19° W, Figure 131)
78: Tempe Fossae	40° N, 76.5° W	RSR: ground ice, resource assessment, highland composition
79: Aeolis Southeast	15.5° S, 188.5° W	RSR: volcanics, plains, channel deposits, ejecta, 105-km route
80: Dao Vallis – Hadriaca Patera	31° S, 263° W	BR: plains, volcanic shield, highlands, channels, 300-km rover, 1000-km balloon route down-channel to Hellas
81: Acidalia Planitia	40–60° N, 10–40° W	BR: French two balloon mission, northern plains, ground ice, highlands, 200-km rovers, 1500-km balloon routes
82: Arcadia Planitia	40–60° N, 150–180° W	BR: French two-balloon mission, plains, ice, 200-km rovers, 1500-km balloon routes (rover site: 38–40° N, 160–164° W)
83: Utopia Planitia	40–60° N, 240–270° W	BR: French two-balloon mission, northern plains, ground ice, channel deposits, 200-km rovers, 1500-km balloon routes
84: Acidalia	50° N, 26° W	RSR: French mission, volcanic plains, fluvial deposits, highlands, 200-km rovers
85: Arcadia	40° N, 173° W	RSR: French mission, cratered plains, young volcanic deposits

Notes: BAL: balloon; **BR**: balloon plus small rover; **PEN**: penetrator; **RSR**: rover/sample return mission.

Table 47. *(cont.)*

Site: Name	Location	Notes
86: Seasonal North Polar Ice Cap Traverse	79.3° N, 210° W	RSR: land at edge of ice in summer, deploy instruments, drive south at least 850 km towards Viking 2, meteorology
87: South Amazonia	6° N, 148.5° W	PEN: Mars 94 site, thick mantling deposit of uncertain origin
88: West Alba Patera	46° N, 128.5° W	PEN: Mars 94 site, Mars 96 site, older volcanic materials
89: Terra Meridiani S	13° S, 4° W	PEN: Mars 96 site, old crust and smooth plains
90: North Tempe	53° N, 85° W	PEN: Mars 96 site, older volcanics, possible ground ice
91: Cydonia	52° N, 357° W	PEN: Mars 96 site, older plains, hills, ground ice
92: West Arabia	23° N, 336° W	PEN: Mars 96 site, ancient crust, volcanic and fluvial material
93: South Amazonia	11° N, 151° W	SS: Mars 96 site, thick mantling deposit of uncertain origin
94: North Amazonia	30° N, 156° W	SS: Mars 96 site, young volcanic materials
95: North Tempe	55° N, 80° W	SS: Mars 96 site, younger volcanics, possible ground ice
96: Deuteronilus Mensae W	45° N, 348° W	SS: Mars 96 site, older volcanic material, periglacial deposits
97: Ladon Vallis	20° S, 31° W	RSR: near valley mouth – fluvial or lake deposits, ancient crust
98: Marte Vallis	22.5° N, 174.5° W	RSR: young lavas and fluvial deposits, plains, ancient crust
99: Utopia Planitia–Viking Lander 2	47.96° N, 225.77° W	RSR: ejecta of Mie crater, plains material of uncertain origin, retrieve parts from Viking Lander 2 to study degradation
100: Margaritifer Sinus SE	24° S, 12° W	RSR: lake deposits, ancient highland materials
101: Valles Marineris I	6° S, 58° W	MES: ridged plains composition, meteorology, geophysics
102: Valles Marineris II	5° S, 54° W	MES: highland crust composition, meteorology, geophysics
103: Chryse Planitia – Viking Lander 1	23° N, 48° W	MES: fluvial plains composition, meteorology, geophysics
104: Olympus Mons	13° N, 130° W	MES: young lava composition, meteorology, geophysics
105: Valles Marineris III	2° S, 54° W	MES: Lunae Planum composition, meteorology, geophysics
106: Hadriaca Patera	32° S, 268° W	MES: volcanic shield composition, meteorology, geophysics
107: Northwest Hellas	40° S, 310° W	MES: rough plains composition, meteorology, geophysics
108: Argyre Planitia	37° S, 44° W	MES: highland crater composition, meteorology, geophysics
109: South Pole	86° S, 315° W	MES: polar layers composition, meteorology, geophysics
110: Sirenum Terra	45° S, 185° W	MES: ancient crust composition, meteorology, geophysics
111: Northern Plains	60° N, 50° W	MES: northern plains composition, meteorology, geophysics
112: Gusev crater	15° S, 185° W	MES: crater lake bed composition, meteorology, geophysics
113: Syrtis Major	5° N, 295° W	MES: volcanic plains composition, meteorology, geophysics
114: North Arabia	38° N, 309° W	MES: ridged plains composition, meteorology, geophysics
115: North Polar Region	82° N, 55° W	MES: polar layers composition, meteorology, geophysics
116: Aonia Terrae	66° S, 66° W	MES: ridged plains composition, meteorology, geophysics
117: Mangala Valles (PM)	10° S, 151° W	PEN: Mars 94 backup site
118: West Diacria Patera	35° N, 137° W	PEN: Mars 94 backup site
119: South Amazonia (S1)	3° N, 160° W	SS: Mars 94 site, thick mantling deposit of uncertain origin

Notes: BAL: balloon; **BR**: balloon plus small rover; **MES**: MESUR mission lander; **MN**: Marsnet site; **PEN**: penetrator; **RSR**: rover/sample return mission; **SS**: small station (Mars 94/96 lander).

Table 47. *(cont.)*

Site: Name	Location	Notes
120: North Amazonia (S2)	40° N, 149.5° W	SS: Mars 94 site, young volcanic materials
121: South Amazonia (S1')	10° N, 154° W	SS: Mars 94 backup site, similar to site 119
122: North Amazonia (S2')	42° N, 142° W	SS: Mars 94 backup site, similar to site 120
123: South Amazonia (P1')	3° N, 153° W	PEN: Mars 94 site, similar to site 119
124: West Alba Patera (P3')	46° N, 133° W	PEN: Mars 94 site, older volcanic materials
125: Mangala Valles (PM')	9.5° S, 155° W	PEN: Mars 94 backup site, highlands, channel deposits
126: South Amazonia (S1")	9.5° N, 163° W	SS: Mars 94 site, thick mantling deposit of uncertain origin
127: North Amazonia (S2")	40° N, 153° W	SS: Mars 94 site, younger volcanics, mantling deposits
128: South Amazonia (S1"')	3° N, 159.5° W	SS: Mars 94 backup site, younger volcanics, mantling deposit
129: North Amazonia (S2"')	39° N, 147.5° W	SS: Mars 94 backup site, young volcanic materials
130: South Amazonia (P1")	3° N, 160° W	PEN: Mars 94 site, younger volcanics, mantling deposits
131: Mangala Valles (PH)	11.5° S, 162.5° W	PEN: Mars 94 site, highlands, channel deposits
132: South Amazonia (P4)	20° N, 156° W	PEN: Mars 94 backup site, young plains, ridges
133: South Amazonia (S1')	11° N, 169.5° W	SS: Mars 94 site, younger volcanics, mantling deposits
134: North Amazonia (S2')	40° N, 160.5° W	SS: Mars 94 site, young volcanic plains
135: South Amazonia (S1")	5° N, 166.5° W	SS: Mars 94 backup site, younger volcanics, mantling deposit
136: North Amazonia (S2")	40° N, 154° W	SS: Mars 94 backup site, young volcanic plains
137: Mangala Valles	6.3° S, 149.5° W	EXO: lake and delta deposits
138: Aeolis NE (Gusev)	15.5° S, 184.5° W	EXO: fluvial and delta deposits in crater near channel mouth
139: Oxia Palus NW	22.1° N, 36.7° W	EXO: possible ground ice, periglacial features
140: Iapygia Northwest	7.3° S, 305° W	EXO: crater ejecta sampling buried fluvial deposits
141: Hebes Chasma	1.5° S, 76.5° W	EXO: layered lake deposits forming mesa in canyon
142: Mangala Valles	8.5° S, 159.5° W	EXO: lake and delta deposits
143: Ismenius Lacus SW	33.5° N, 342.5° W	EXO: Mamers Valles, fluvial, lake and delta deposits
144: Diacria Southeast	39.5° N, 135.5° W	EXO: lake deposits in Acheron Fossae
145: Memnonia NW	14.5° S, 175° W	EXO: fluvial and lake deposits in crater
146: Phaethontis North-Central	37.5° S, 146° W	EXO: large highland basin, possible hydrothermal system around old volcano, fluvial and lake deposits
147: Sinus Sabaeus NE (White Rock)	8° S, 335° W	EXO: eroded mesa of possible evaporite composition on crater floor, lake deposits
148: Oxia Palus NE	21.3° N, 8° W	EXO: possible evaporites on crater floor, lake deposits
149: Ismenius Lacus South-Central	44.3° N, 333° W	EXO: ground ice, periglacial features in fretted terrain of Deuteronilus Mensae
150: Casius Southeast	36° N, 259° W	EXO: periglacial features, polygonal fractures, ground ice
151: Tempe Terra	40.1° N, 78.2° W	MN: older volcanic materials, faulted terrain
152: Candor Chasma	9.7° S, 73.2° W	MN: landslide deposits, canyon floor plains, layered mesas
153: Daedalia Planum	25° S, 127.5° W	MN: younger basalt flows from Tharsis

Notes: EXO: Exobiology site; **MES**: MESUR mission lander; **MN**: Marsnet site; **PEN**: penetrator; **SS**: small station (Mars 94/96 lander).

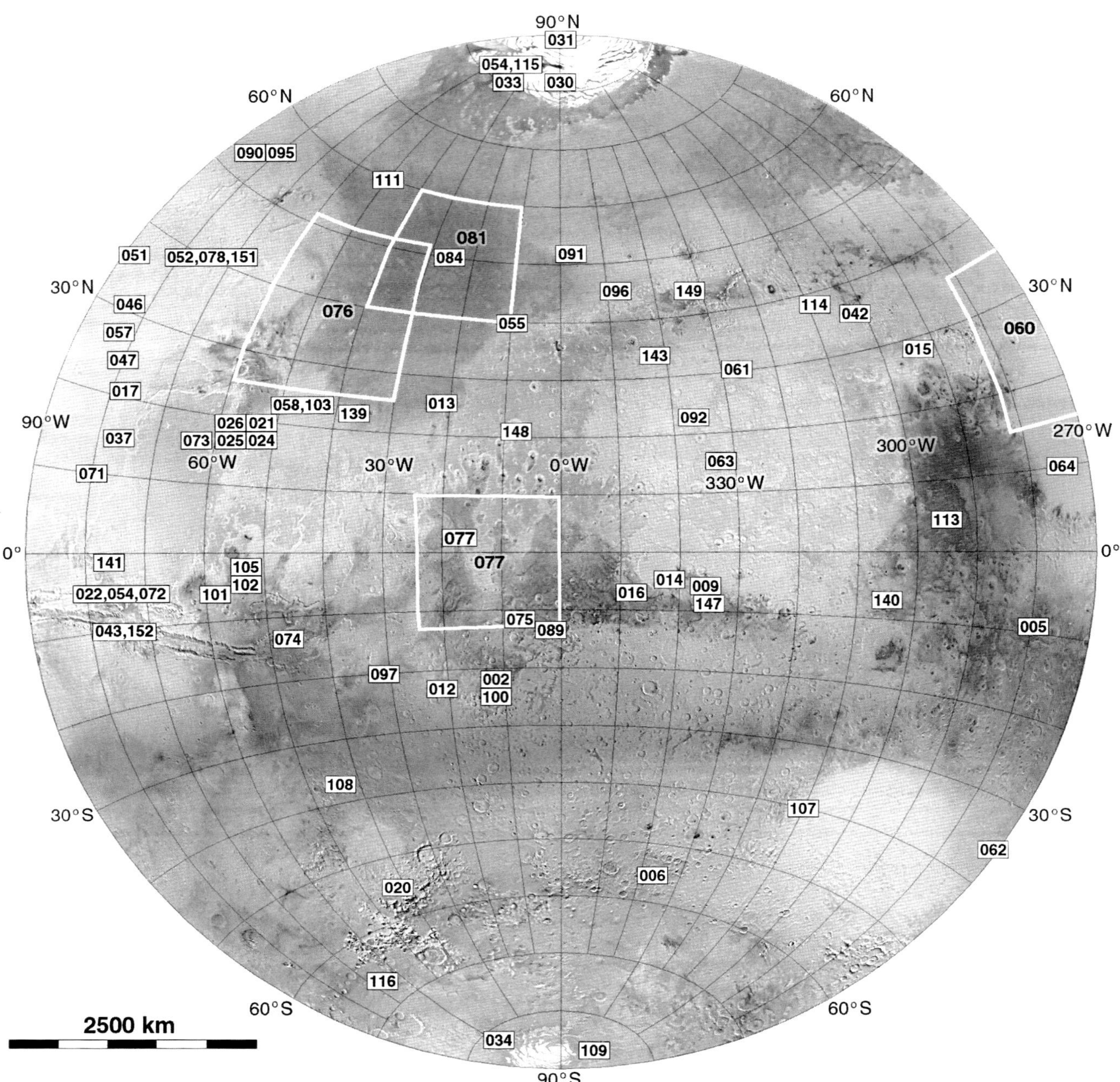

Figure 123 Mars Landing Site Catalog sites, 0° hemisphere (Greeley and Thomas, 1995). The background map is the Viking global photomosaic from Figure 83.

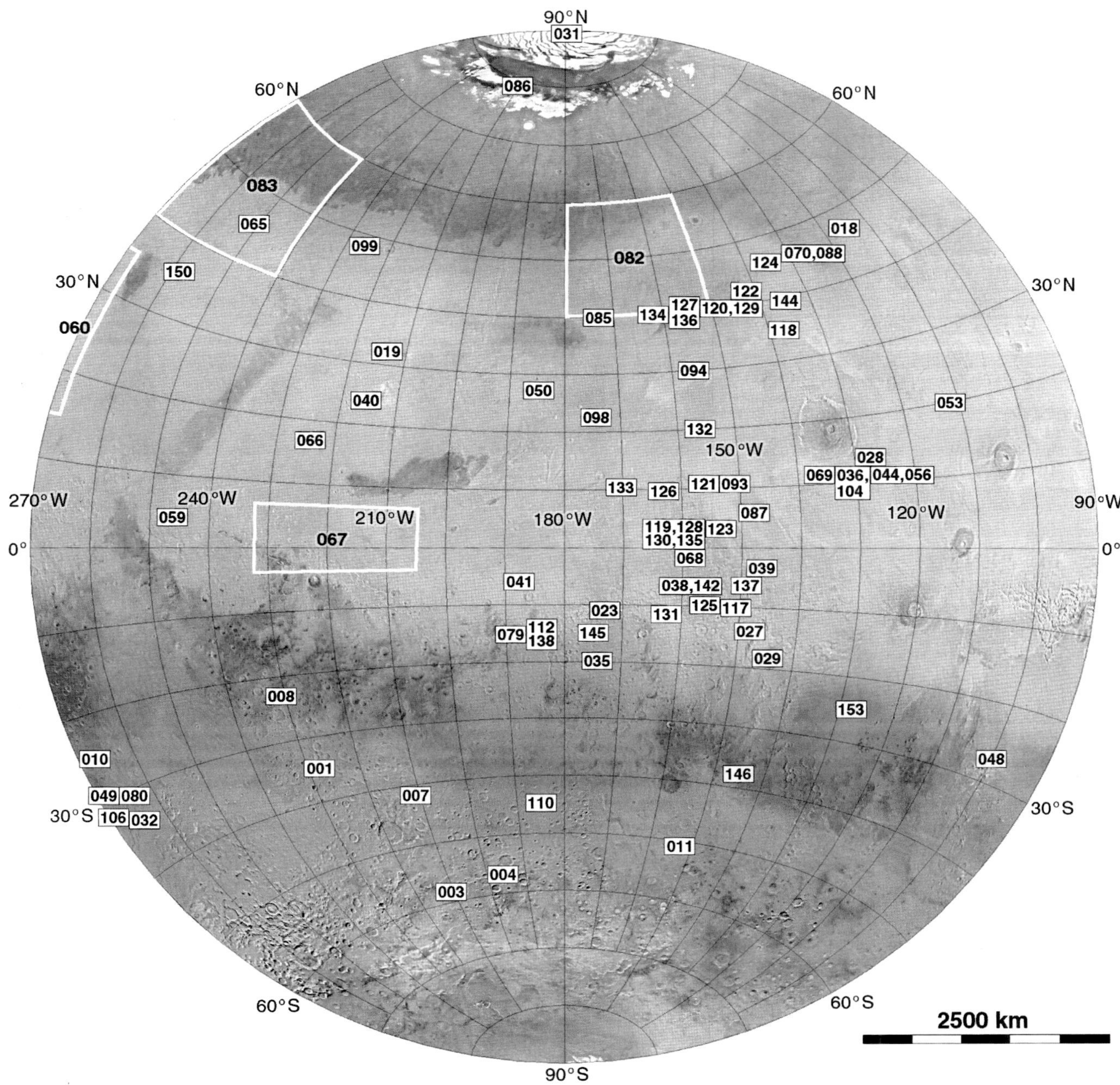

Figure 124 Mars Landing Site Catalog sites, 180° hemisphere (continued from Figure 123).

Table 48. *'Top Ten' List from Landing Site Catalog*

Site	Name	Site	Name
10	Terra Tyrrhena	54 / 72	Candor Chasma
21	Maja Valles	96	West Deuteronilus Mensae
29	Daedalia	112	Gusev Crater
32 / 62	Dao Vallis	115	North Polar Region
36	Olympus Rupes	147	Sinus Sabeus NE
41	Apollinaris Patera		

HST imaged Mars to allow global-scale monitoring of clouds, polar caps and albedo markings at every opposition except that of 1993, when a servicing mission intervened. Mars exploration was aided by observations prior to the arrival of spacecraft at the planet. Figure 141 shows images taken at each opposition during the period 1990–2003, illustrating the varying seasons at each opposition and changes in albedo markings. The north pole is visible in 1995 and 1997, and the south pole in 2003, a difference also reflected in the latitudes of preferred landing sites in those periods. Most of the images show Syrtis Major near or to the right of centre and Sabaeus Sinus extending westwards from it. Features can be identified easily by comparing the June 2001 image with Figure 1. The image labelled Pathfinder was taken on 27 June 1997 to monitor weather conditions a week before the Mars Pathfinder landing. It shows dust clouds in the Valles Marineris which might have threatened the landing but did not spread. The 2001 images show a clear atmosphere in June, apart from a cloudy collar around the north polar cap, and a global dust storm in September. Significant changes in markings around Hellas, the bright circle south of Syrtis Major, occurred between 1990 and 2003.

1990s: **Mars Network Mission Plans**

Several plans for global networks of geophysical or meteorological instruments were formulated in the 1990s and are presented together here. These concepts followed the earlier planning of the Mars Science Working Group. Some of the sites discussed here are included in the Mars Landing Site Catalog (Table 47), which was set up to assist in mission planning. Other site catalogs were also drawn up for the European mission studies, Marsnet and Intermarsnet (Tables 55 and 56).

Early ESA Network

A European Space Agency (ESA) report described a reference study of a network of geophysical and meteorological instruments, one of many network concepts to be discussed in this period (ESA, 1990, Section 9). This concept included a cluster of three small probes and two penetrators forming a triangle on the surface with side lengths of about 3500 km. The probe cluster would land in Lunae Planum, and the two penetrators in Bosporos Planum and just south of Arsia Mons, forming a regional array on the east side of the Tharsis volcanic province. The sites are listed in Table 49 and illustrated in Figure 142. Later, these same sites were identified with Marsnet (ESA, 1991).

Mars Global Network Mission (United States)

Solomon *et al.* (1991) reported on a Mars Global Seismic Network Workshop held in Houston in May 1990 and described a network mission which might deploy up to 20 probes, each with a mass of about 75 kg, distributed globally for seismology, meteorology and geochemical investigations in late 1998. The Global Network Mission would use two orbiters, each equipped with a number of simple landers. Some landers might be deployed during approach to reduce mass before orbit insertion and to reach latitudes not accessible from the chosen orbit. The rest would be deployed from orbit. One configuration for the Global Network Mission involved a 0.2-sol polar orbit with a 275-km periapsis. This would be optimal for lander deployment, allowing easy polar access and global distribution of landing sites with illumination suitable for descent imaging. The polar orbit allows the option of deploying all the landers from orbit. The orbiters would have to wait for 160 days after arrival before deploying landers to optimize illumination.

The network configuration suggested at the Workshop was for triangular arrays, each node of which consisted of its own small triad, or group of three stations. The triads would be about 100 km across. If each station could carry broadband seismometers, sensitive to long- and short-period signals, these three stations would suffice at

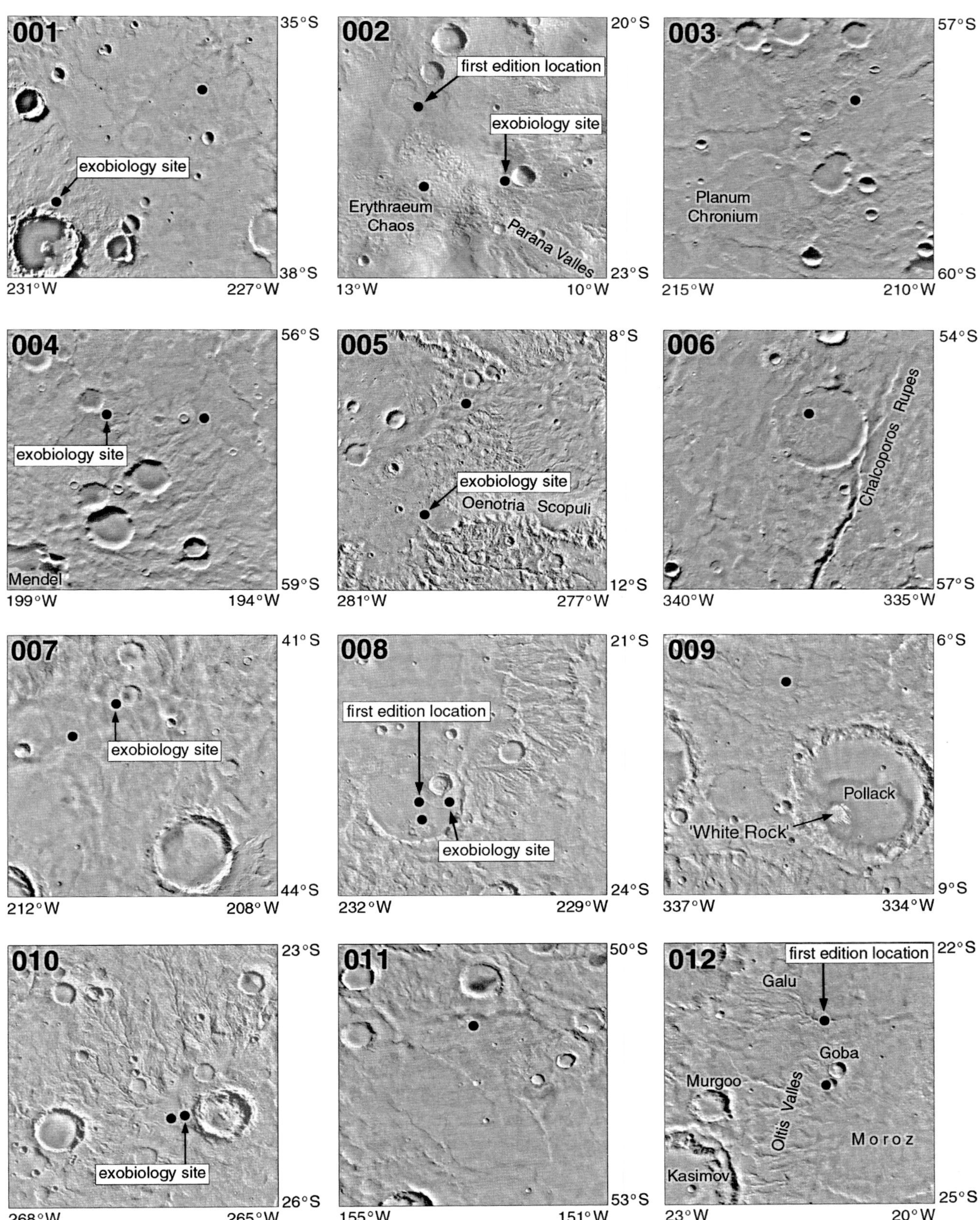

Figure 125 Mars Landing Site Catalog, sites 1 to 12.
Some sites have an alternative exobiology site nearby, and three sites were illustrated differently in the first edition of the catalog. A site near Site 5 is shown in Figure 120A. Site 9 is near a Mars 1984 site (Figure 114B). Site 012 is near the Mars 6 impact site (Figure 35).

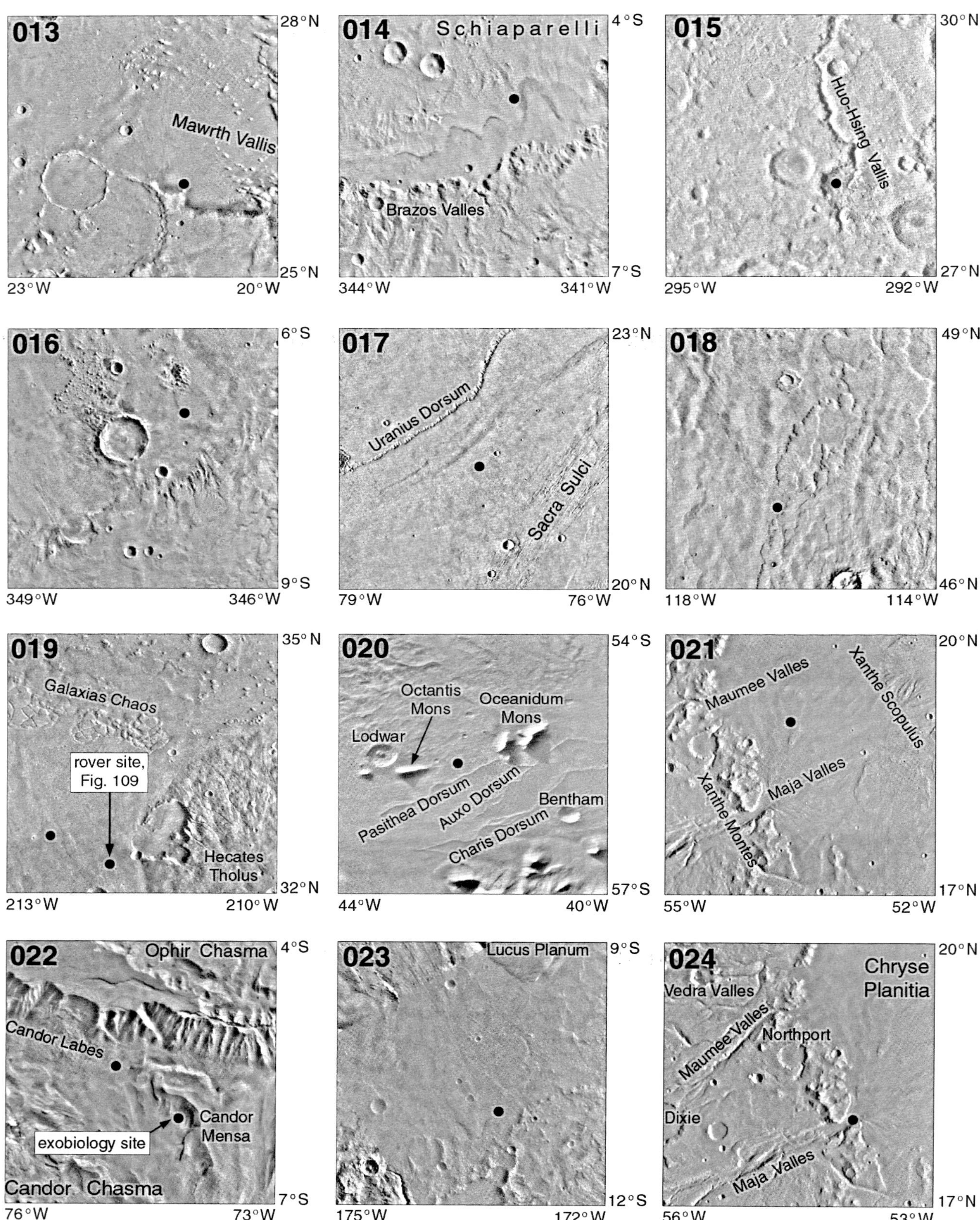

Figure 126 Mars Landing Site Catalog, sites 13 to 24.
Site 22 has an additional exobiology site nearby.

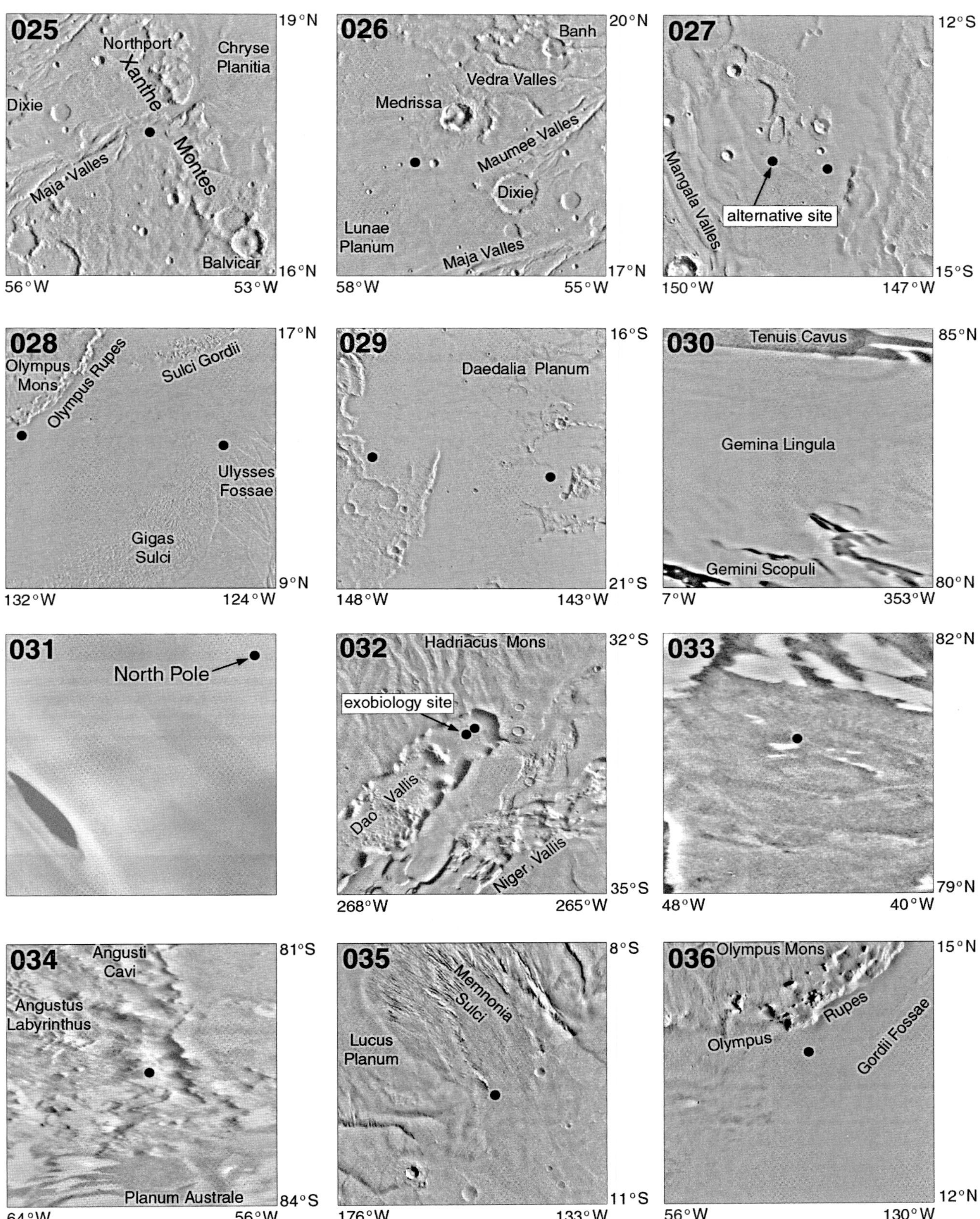

Figure 127 Mars Landing Site Catalog, sites 25 to 36.
Site 32 has an additional exobiology site nearby, and Site 27 has an alternative site. Site 30 could lie anywhere in the latitude band, but a representative location is illustrated.

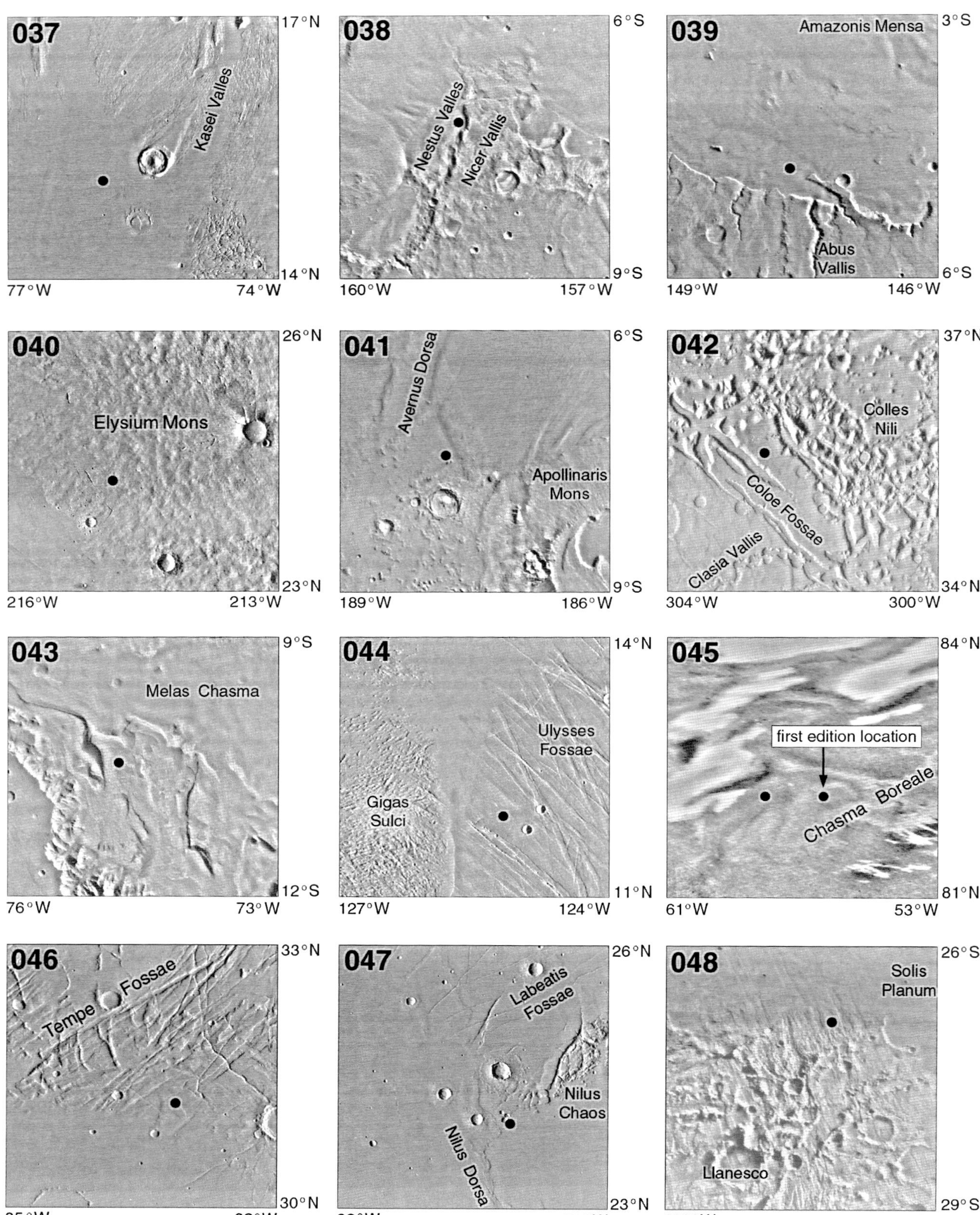

Figure 128 Mars Landing Site Catalog, sites 37 to 48.
Site 41 is near a Mars 1984 site (Figure 114A).

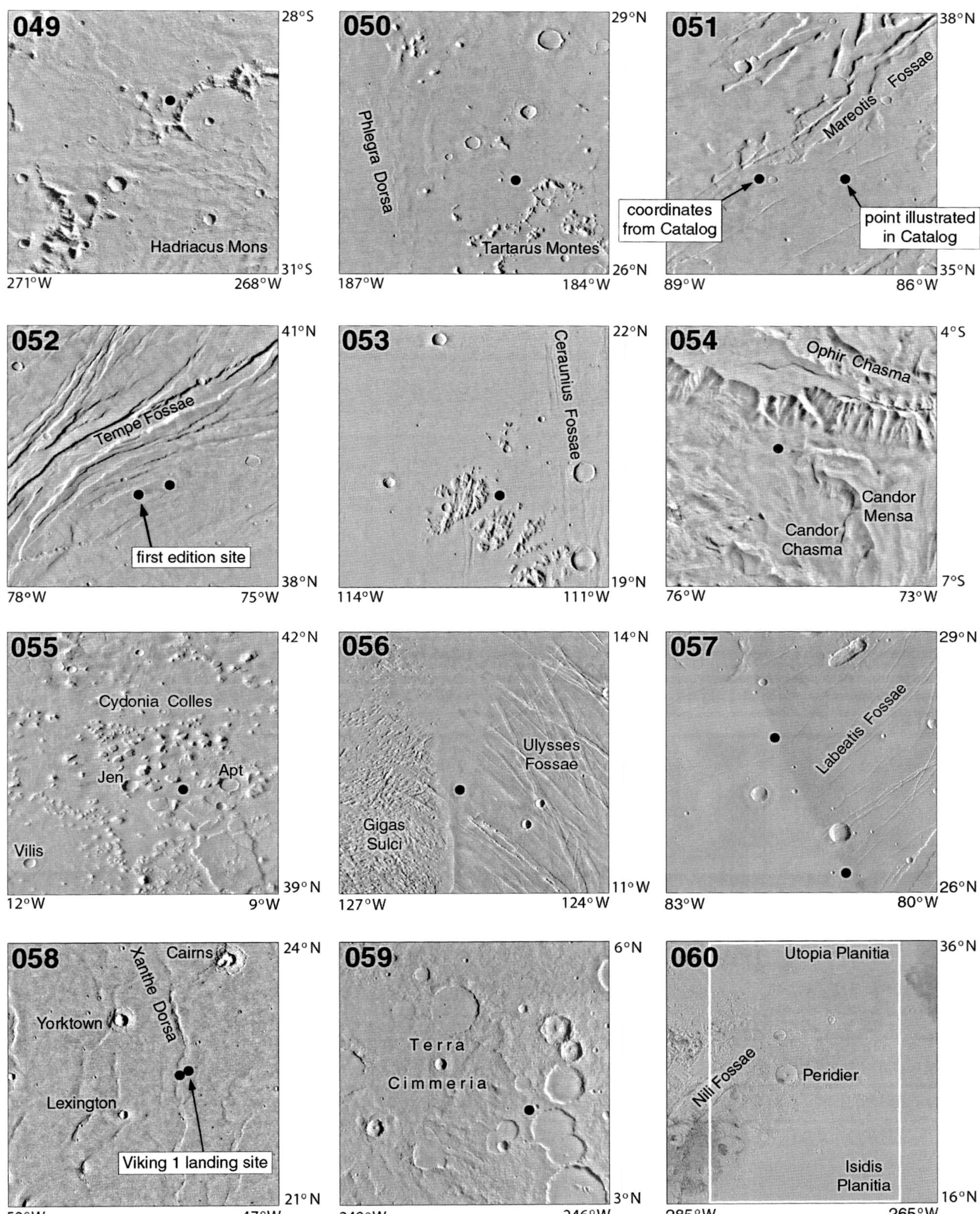

Figure 129 Mars Landing Site Catalog, sites 49 to 60.
Site 58 was at the location then believed to be the Viking 1 landing site (Figure 52).

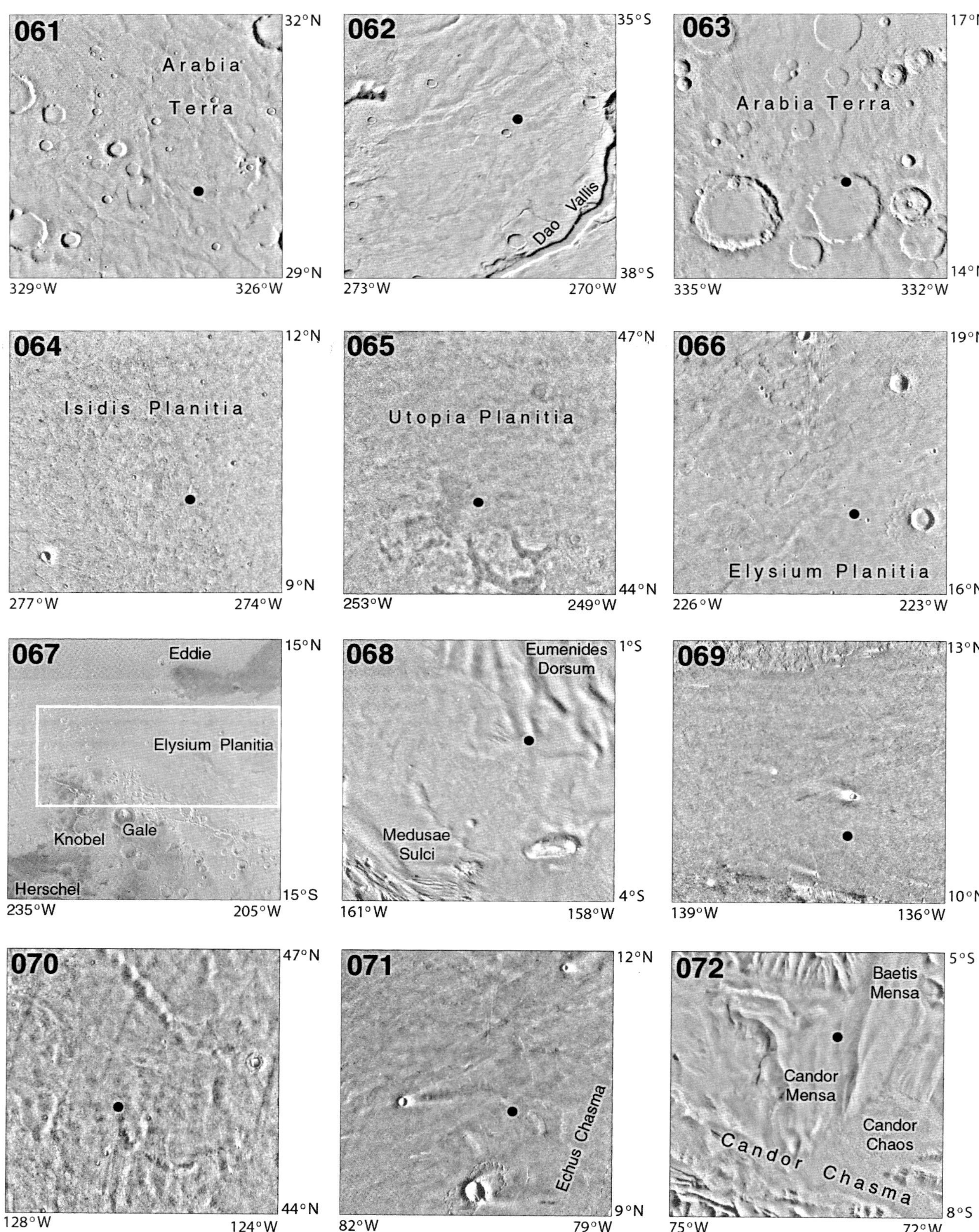

Figure 130 Mars Landing Site Catalog, sites 61 to 72.

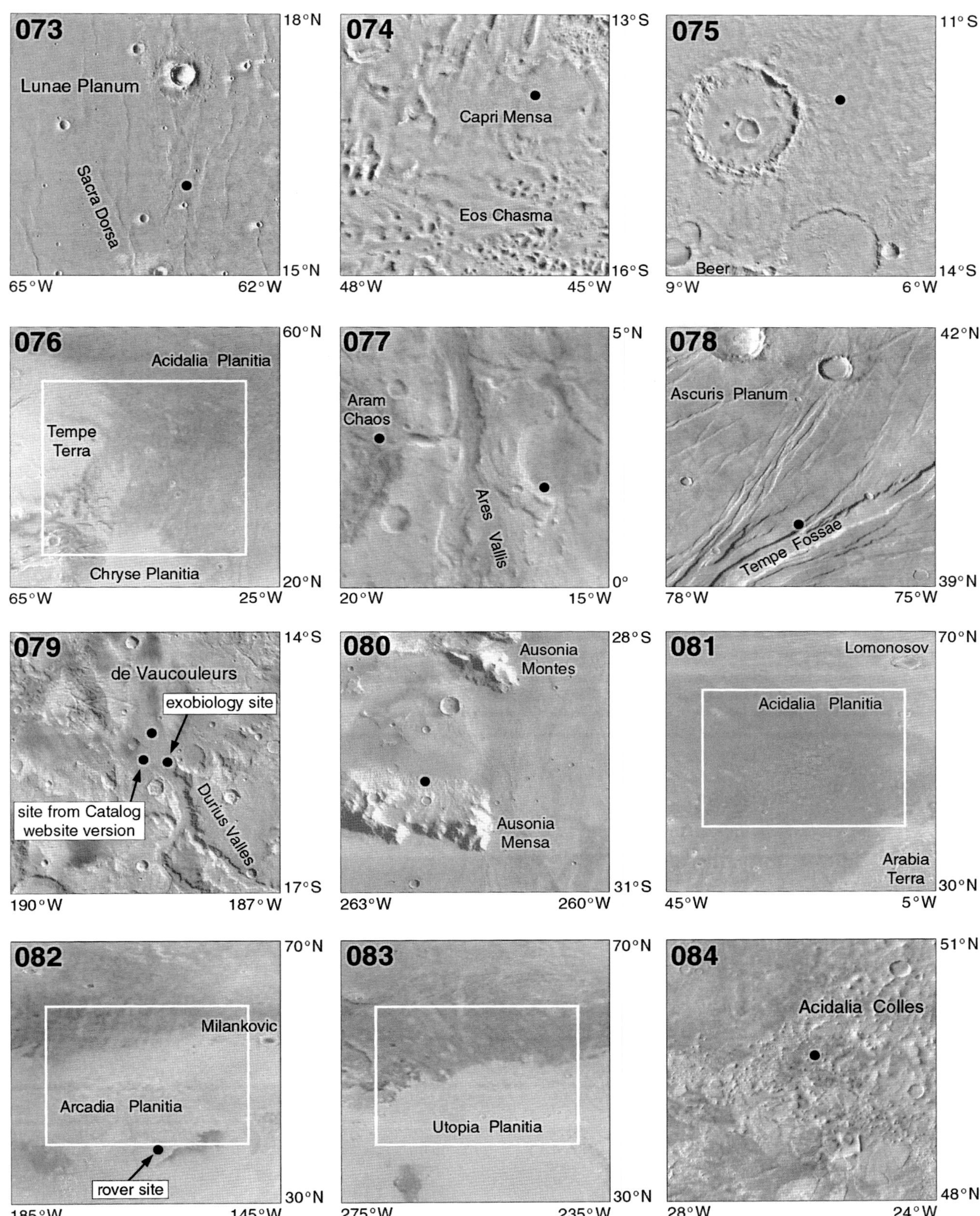

Figure 131 Mars Landing Site Catalog, sites 73 to 84.
Site 77 has two sites, and Site 79 has three, one of which appeared only in a website version of the catalog. Apart from the two specific sites shown for Site 77, the catalog description defined a wider area of interest (Table 47, Figure 123).

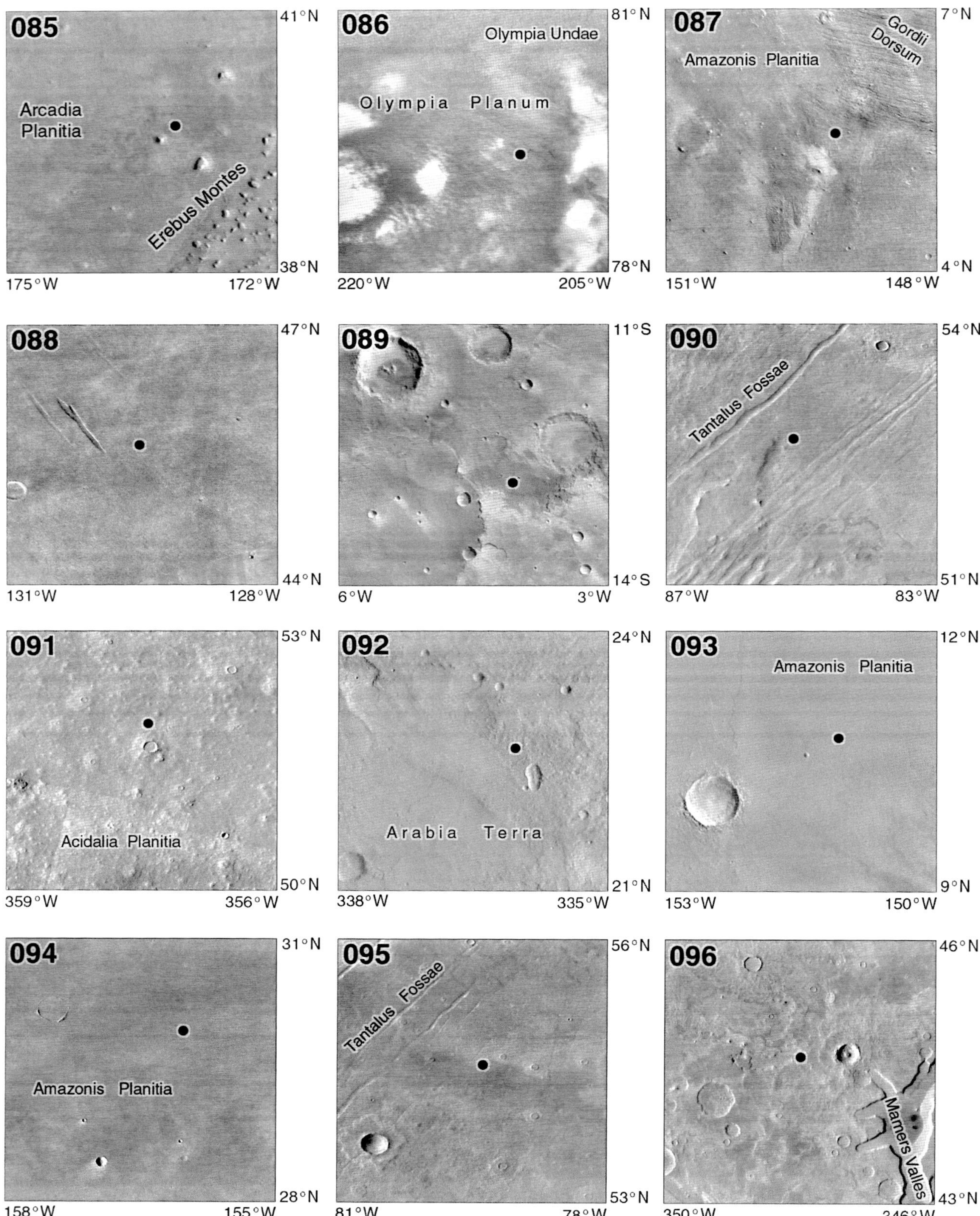

Figure 132 Mars Landing Site Catalog, sites 85 to 96.

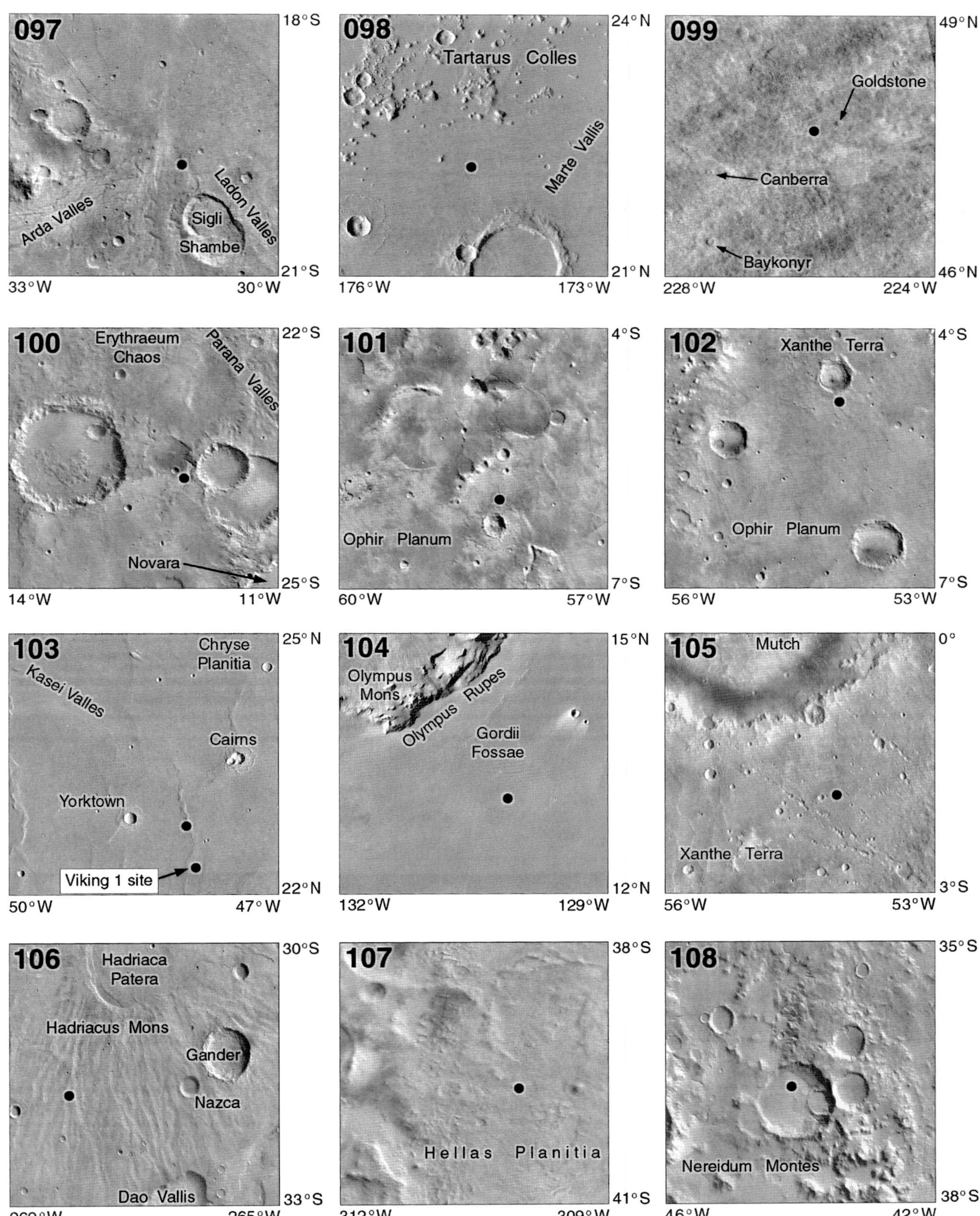

Figure 133 Mars Landing Site Catalog, sites 97 to 108.
Site 099 is the Viking 2 landing site.

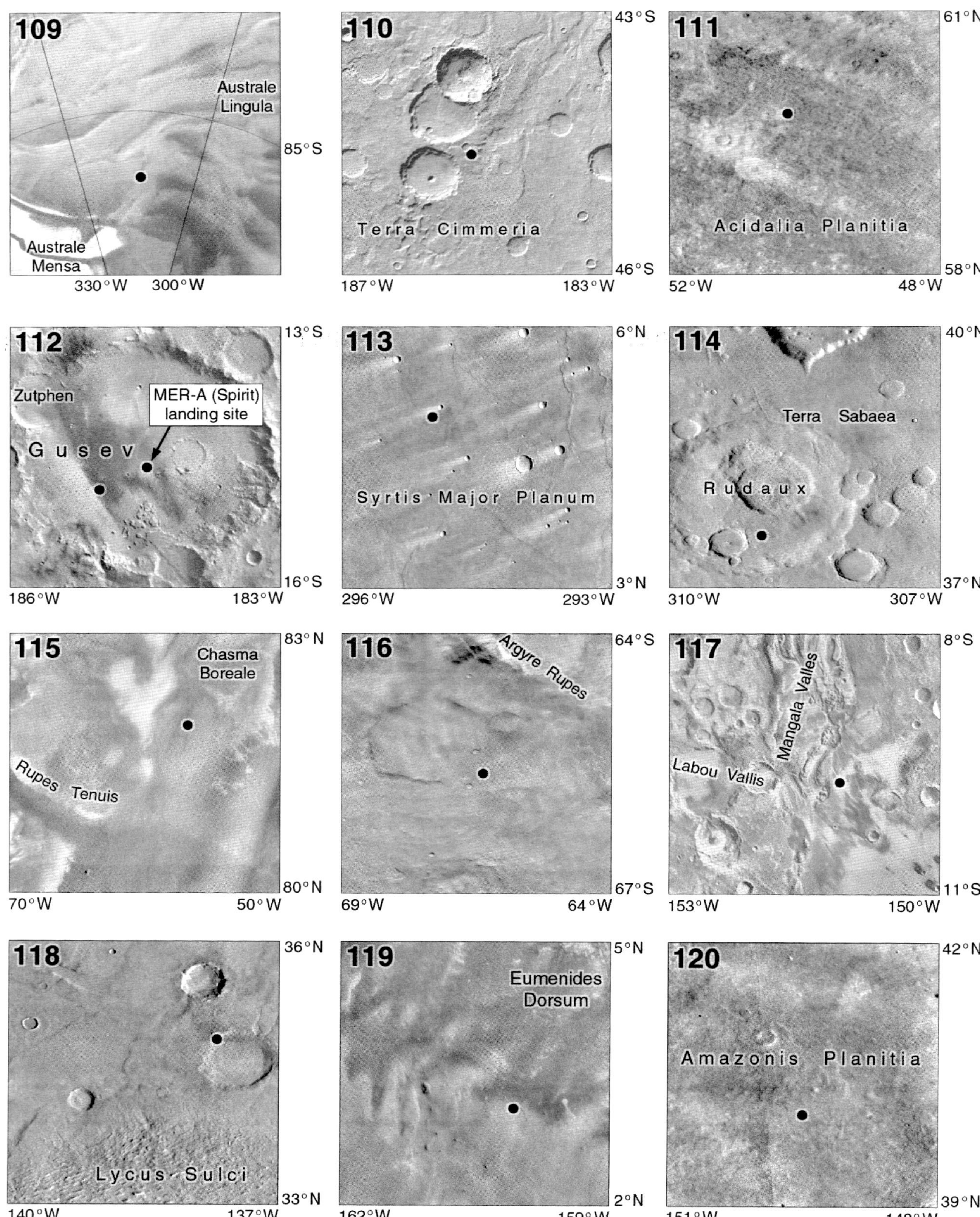

Figure 134 Mars Landing Site Catalog, sites 109 to 120.

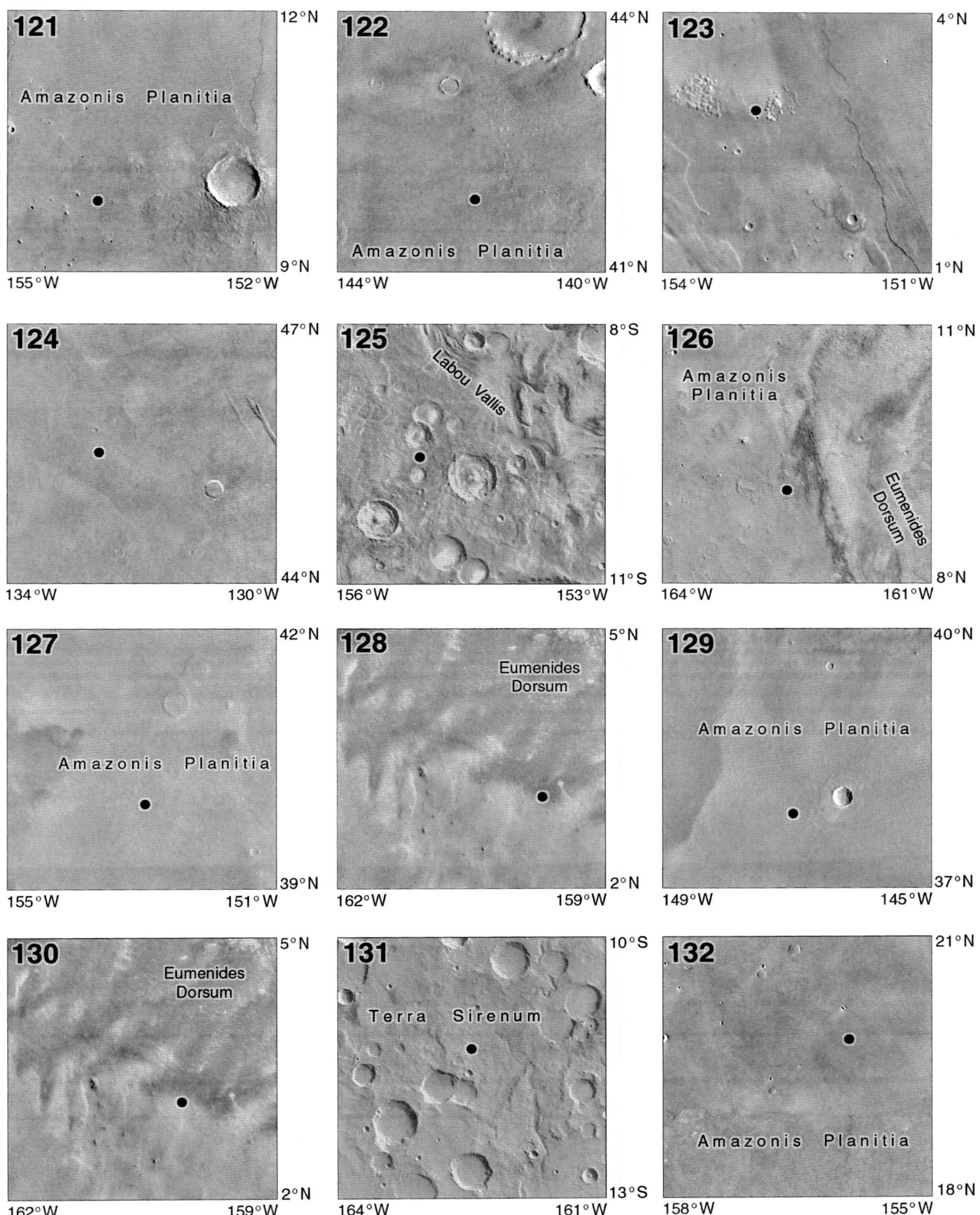

Figure 135 Mars Landing Site Catalog, sites 121 to 132.

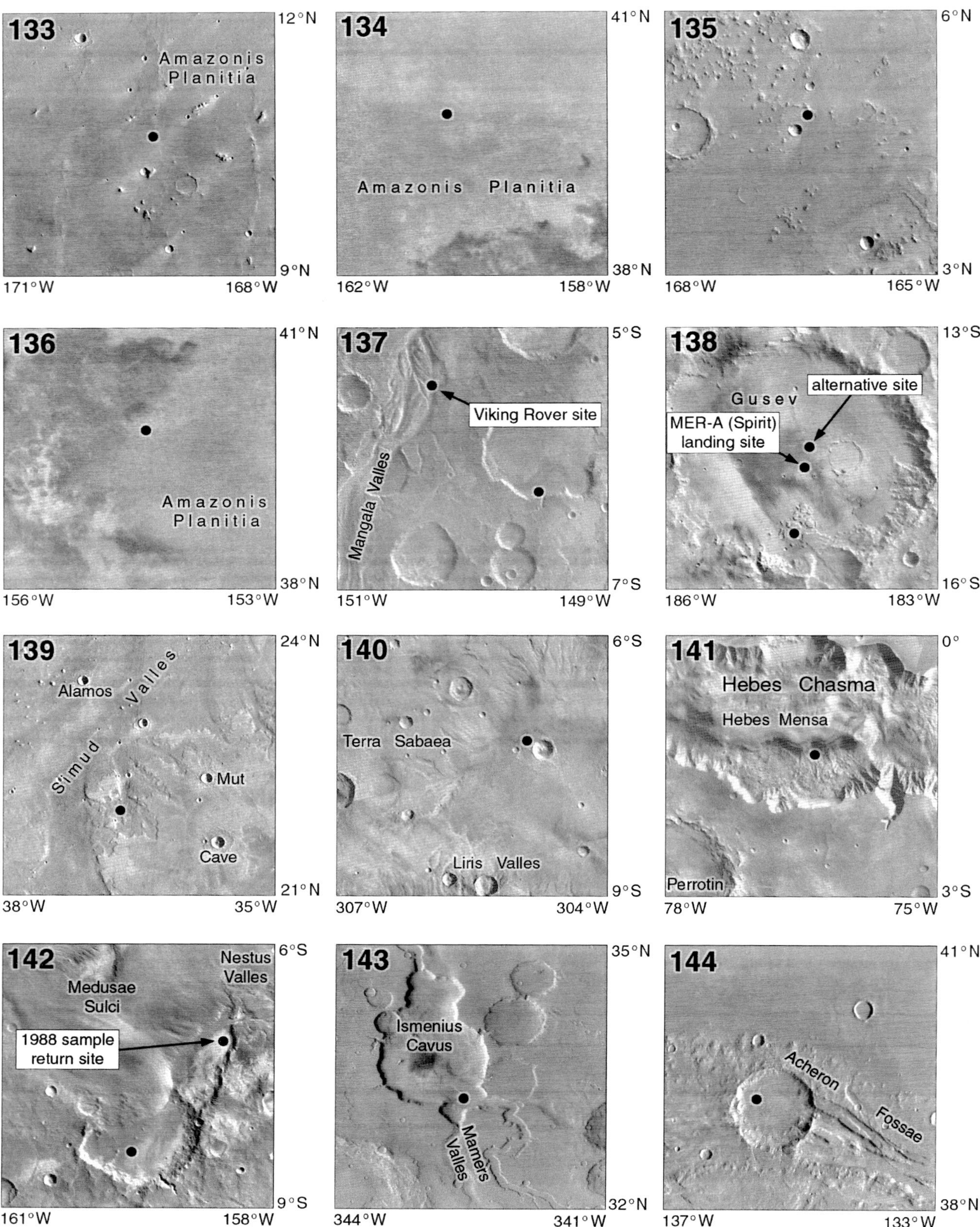

Figure 136 Mars Landing Site Catalog, sites 133 to 144.

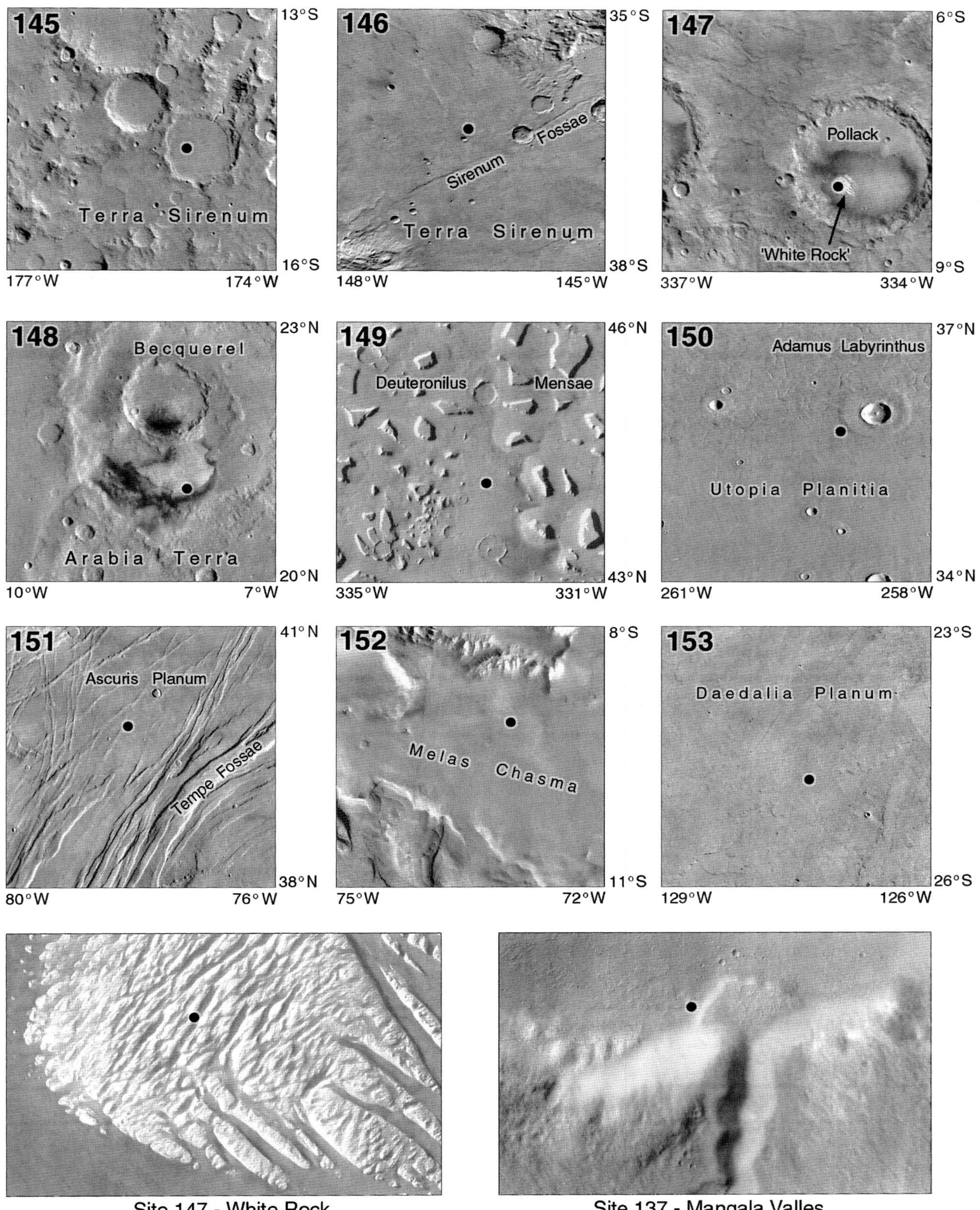

Figure 137 Mars Landing Site Catalog, sites 145 to 153, and higher-resolution images of two sites.

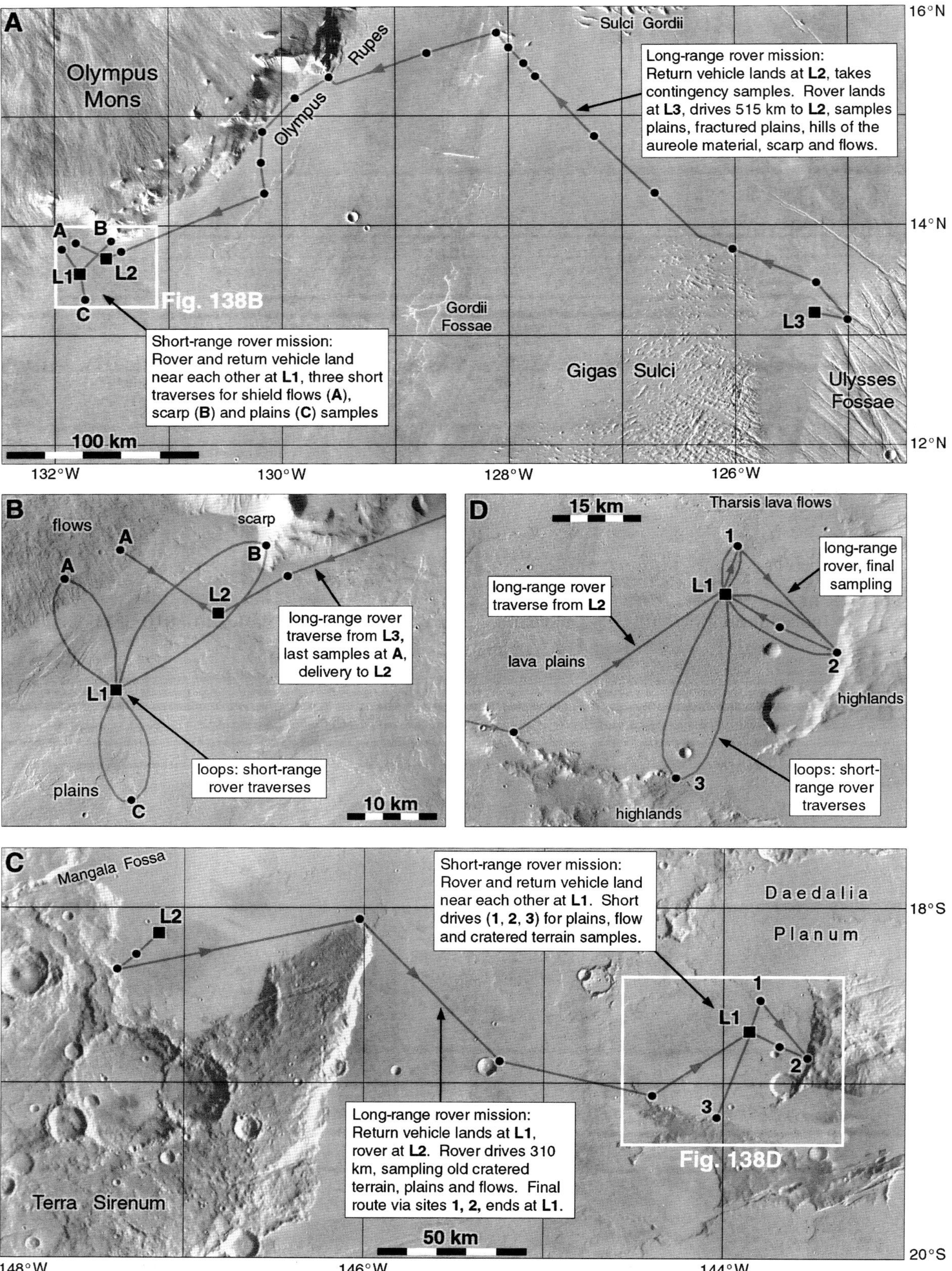

Figure 138 Rover traverses at two Mars Landing Site Catalog locations.
A and B: Site 028, Olympus Mons Southeast. **C and D:** Site 029, Western Daedalia Planum. Two versions of each mission with long- or short-range rovers are illustrated.

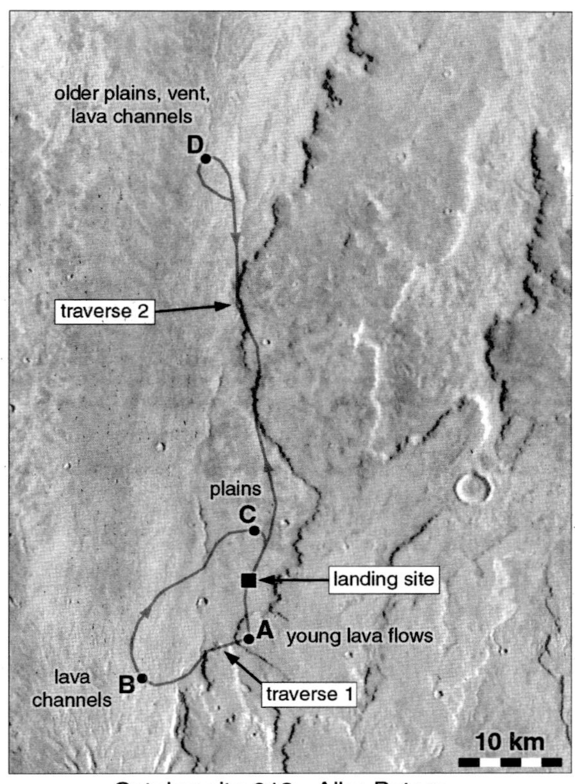

Catalog site 018 - Alba Patera

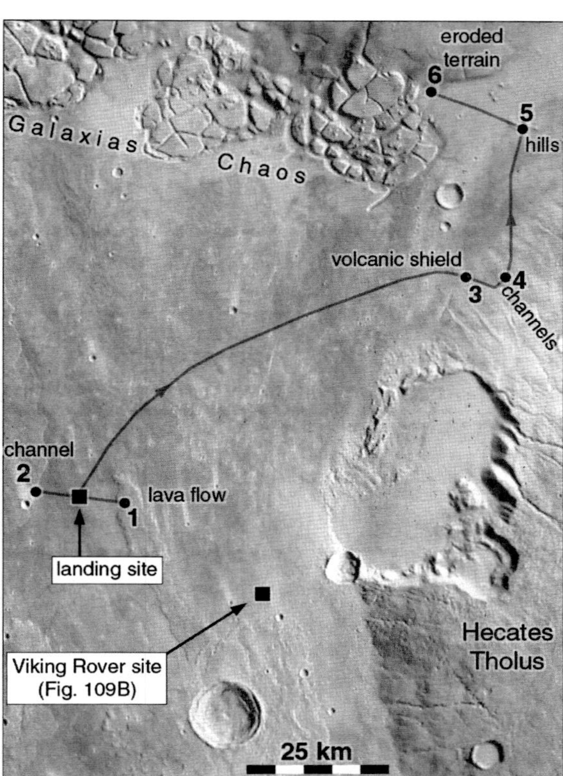

Catalog site 019 - Northern Elysium

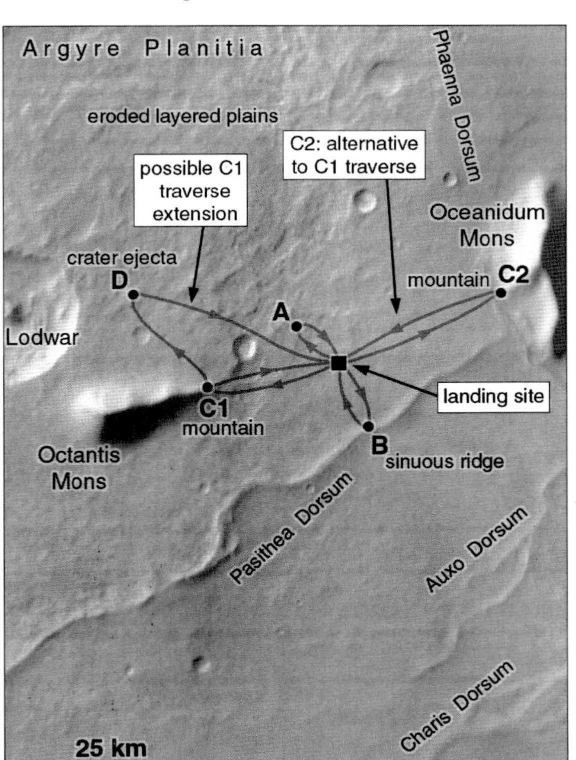

Catalog site 020 - Argyre Planitia

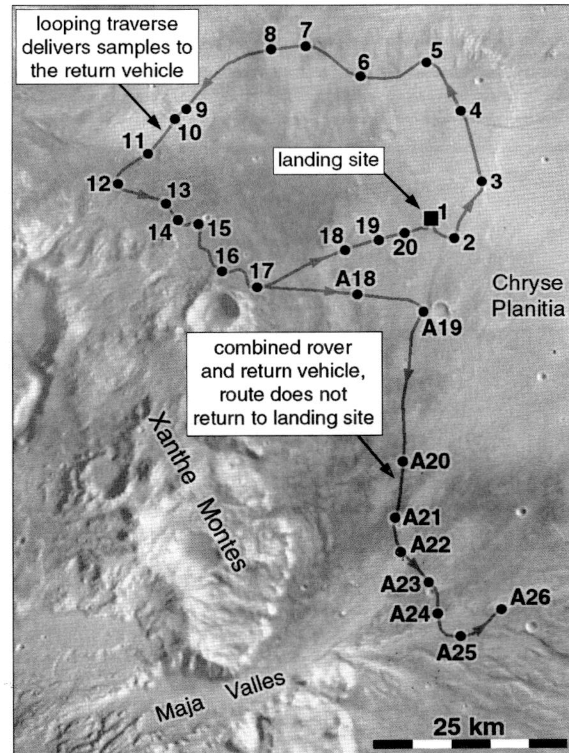

Catalog site 021 - Maja Valles

Figure 139 Rover traverses at four Mars Landing Site Catalog sites.
Context for each map is shown in Figure 125. The images are Mars Odyssey THEMIS infrared mosaics.

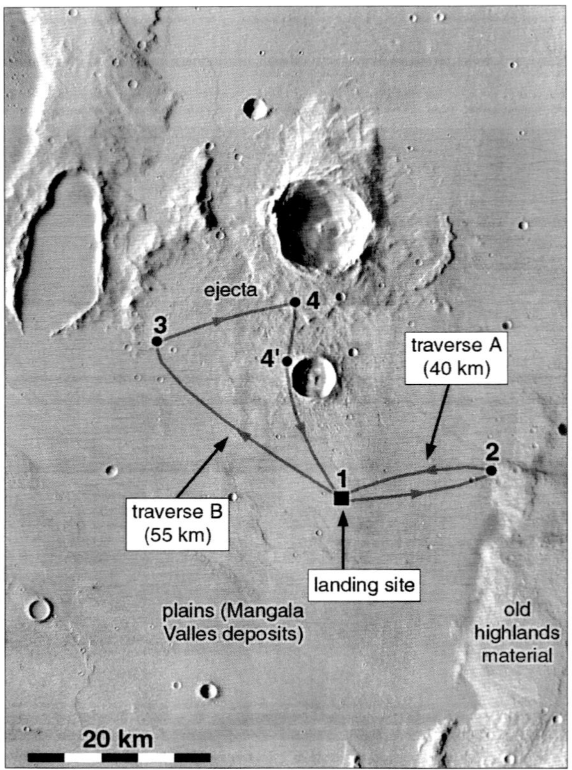

Site 027 - Upper Mangala Valles

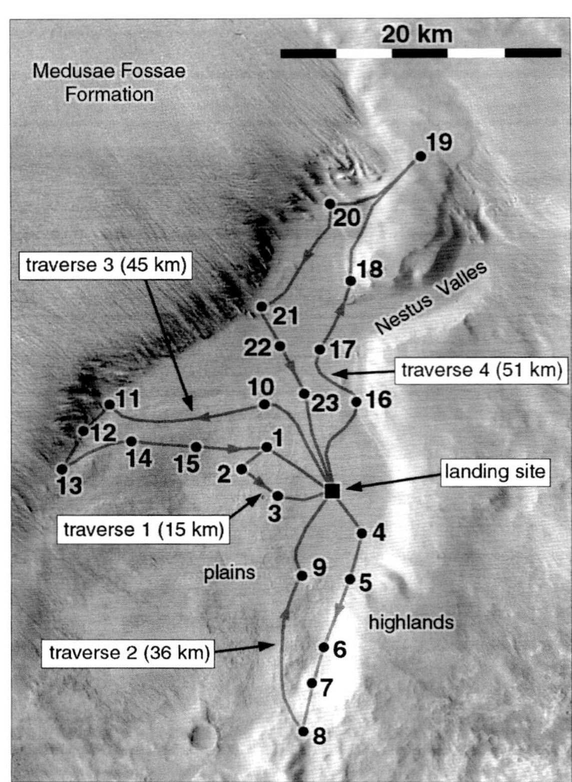

Site 038 - Mangala Valles West

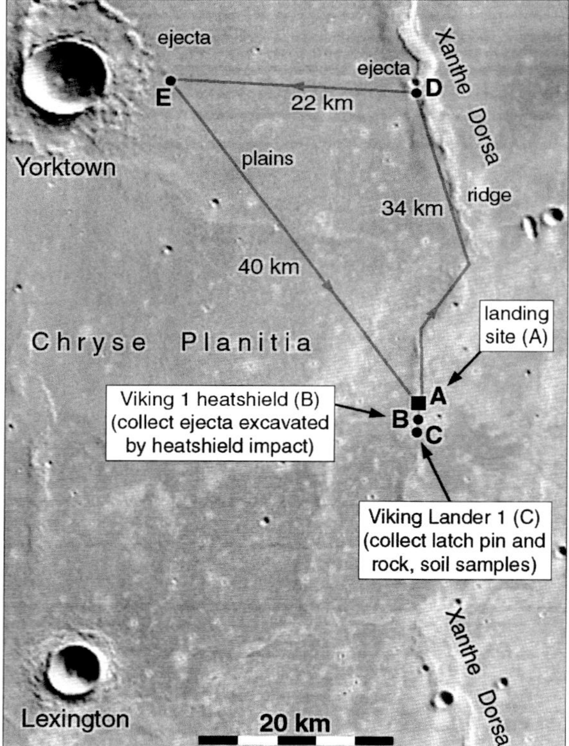

Site 058 - Viking Lander 1

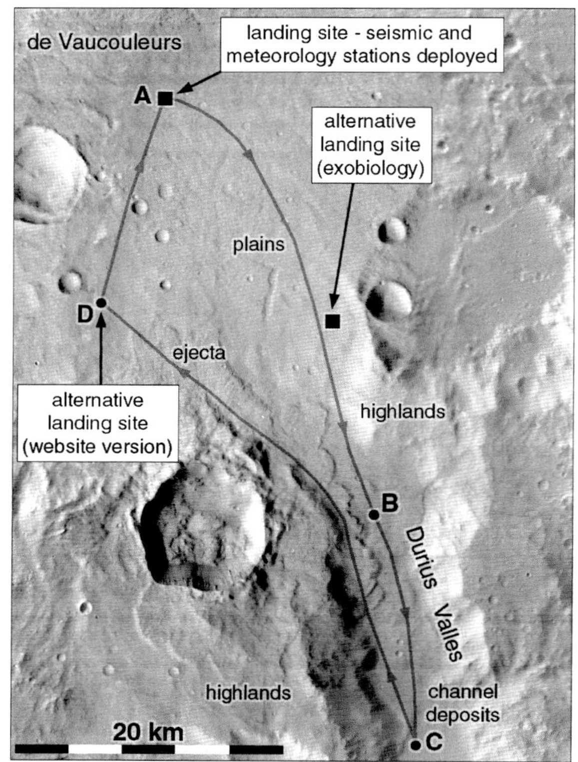

Site 079 - Aeolis Southeast

Figure 140 Traverses at four Mars Landing Site Catalog sites.
The Site 038 traverse is taken from Masursky *et al.* (1988b) and is a variation on Figure 116D. The catalog provides context for each site (Figures 126 to 130). The images are composites of Mars Odyssey THEMIS images and Viking images.

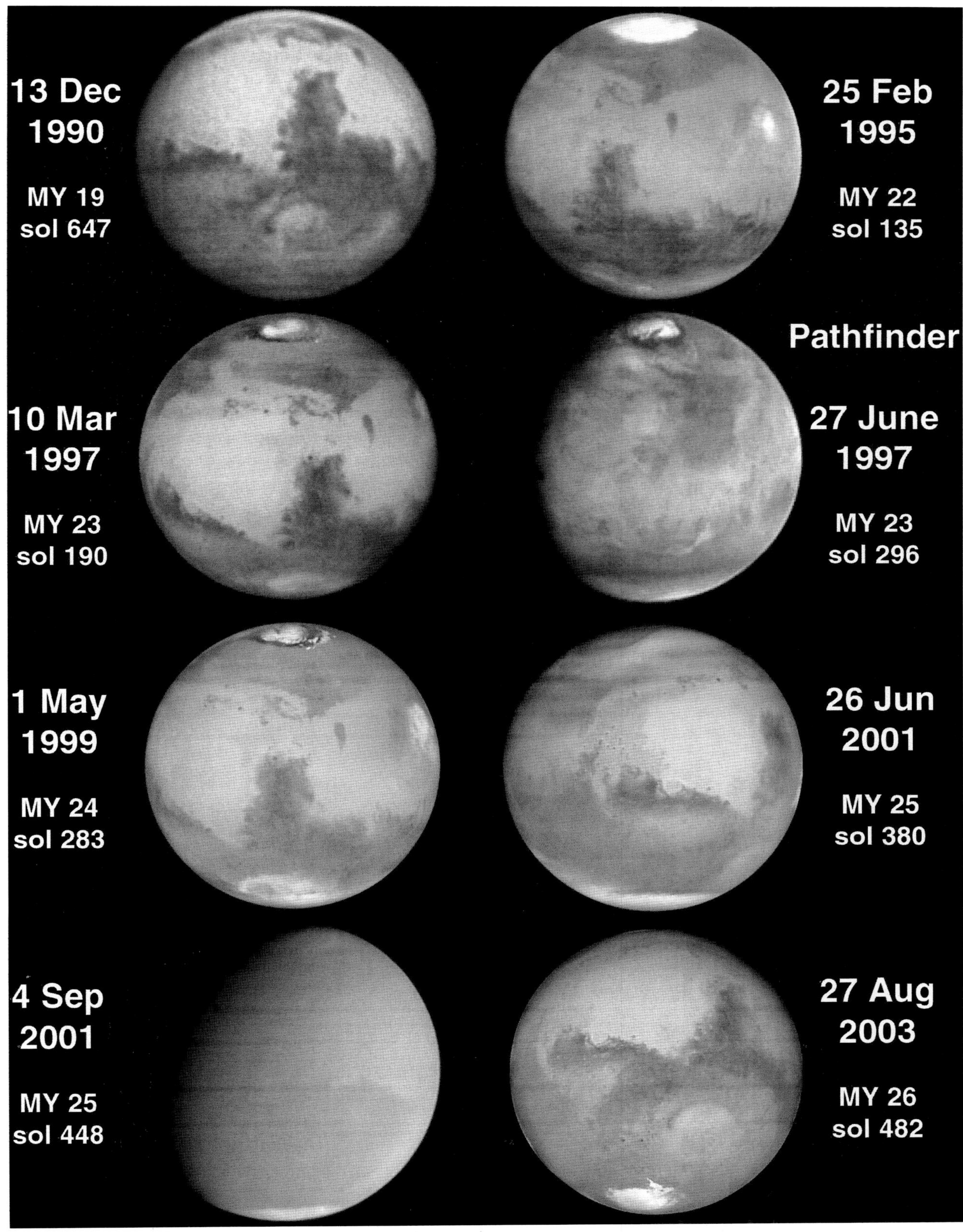

Figure 141 Hubble Space Telescope images of Mars, 1999 to 2003.
HST images by J. Bell, T. Clancy, D. Crisp, P. James, R. Kahn, S. Lee, A. Lubenow, L. Martin, J. Neubert, R. Singer, M. Wolff, R. Zurek, NASA, ESA, the Space Telescope Science Institute and the Hubble Heritage Team. White streaks near the bottom of the Pathfinder image are dust clouds in Valles Marineris.

Table 49. *Network Landing Sites Described in European Space Agency (1990)*

Site	Name	Location	Notes
1	Lunae Planum	20° N, 65° W	Ridged basalt plains near Chryse basin.
2	Bosporos Planum	45° S, 70° W	Cratered plateau near Argyre basin.
3	Arsia Mons	15° S, 120° W	Young lava plains in the Tharsis region.

each node. A less expensive system might include only one broadband system with three short-period instrument stations deployed in a triangle around it. Three nodes like these would form a large network, 3500 km on a side, and two such networks would be emplaced in opposite hemispheres. The total number of stations would thus be 18 or 24. The small triads measure local activity, so several of them should be in active areas such as Tharsis. The sites suggested at the workshop are listed in Table 50 and illustrated in Figure 142.

MESUR (United States)

MESUR, the Mars Environmental Survey, was intended to place a network of small landers on the surface of Mars. The goals were to determine the composition of surface materials, to examine many different geological materials, to study the interior of the planet with a seismic network, and to study local weather and global climate. The network would be built by launching four to eight small landers per launch opportunity, spread over three opportunities in a four-year period beginning in the late 1990s. The intention was to have 16 stations operating simultaneously after the last landing and for them to operate together for a full Martian year. The first landers would have to operate for about seven Earth years. Each lander would fly to Mars separately and would be able to land almost anywhere, carrying a seismometer, a camera, soil and rock composition instruments and meteorology instruments. Because these would be the first NASA Mars landers since Viking, a precursor mission was required to test new technologies for small, inexpensive landers. This MESUR Pathfinder mission became the only part of MESUR to fly, under the name Mars Pathfinder. There might also have been some possibilities for international cooperation, including the incorporation of Marsnet into MESUR, or having other nations drop MESUR landers from their own orbiter or balloon missions. The MESUR project was abandoned before the mid 1990s resurgence in Mars exploration. The description given here is taken from the Mars Science Working Group documents held at the Lunar and Planetary Institute in Houston, Texas.

Table 50. *Mars Global Network Array from Solomon et al. (1991)*

Array	Node location	Notes
Western	15° N, 110° W	Tharsis west of Ascraeus Mons, most recent volcanic activity
	7.5° S, 52.5° W	Valles Marineris, north of east end, volcanism and landslides
	45° S, 105° W	Southern highlands, crustal thickness south of the dichotomy
Eastern	22° N, 211° W	Elysium volcanic province, relatively recent activity
	5° N, 270° W	Isidis basin, mascon basin, lunar analogy suggests seismic activity
	65° N, 275° W	Northern plains, crustal thickness north of the dichotomy

Site selection would be a compromise between many disciplines. Geologists needed landings in varied geological environments. Meteorologists needed a larger number of sites (up to 20 if possible) distributed evenly around the planet, preferably arranged along a limited number of meridians to help separate spatial and temporal variability. Seismologists preferred four widely separated groups of three landers arranged in triangles about 100 km across. The small triangles would reveal local crustal properties; the large pattern would probe as deep as the planet's core. This seismic version of the mission was described by Bruce Banerdt at the Third Mars Science Working Group meeting held on 23 and 24 August 1990 at JPL. A sketch map showed the four 'triads', groups of three landers, with three triads forming a triangle 4000 km across around Tharsis and the fourth between Elysium and Isidis Planitia. The concept is similar to the Mars Global Network plan (Table 50) from the same period. If the network was to be emplaced in a sequence of missions, the large pattern should be established first, and then the triads built

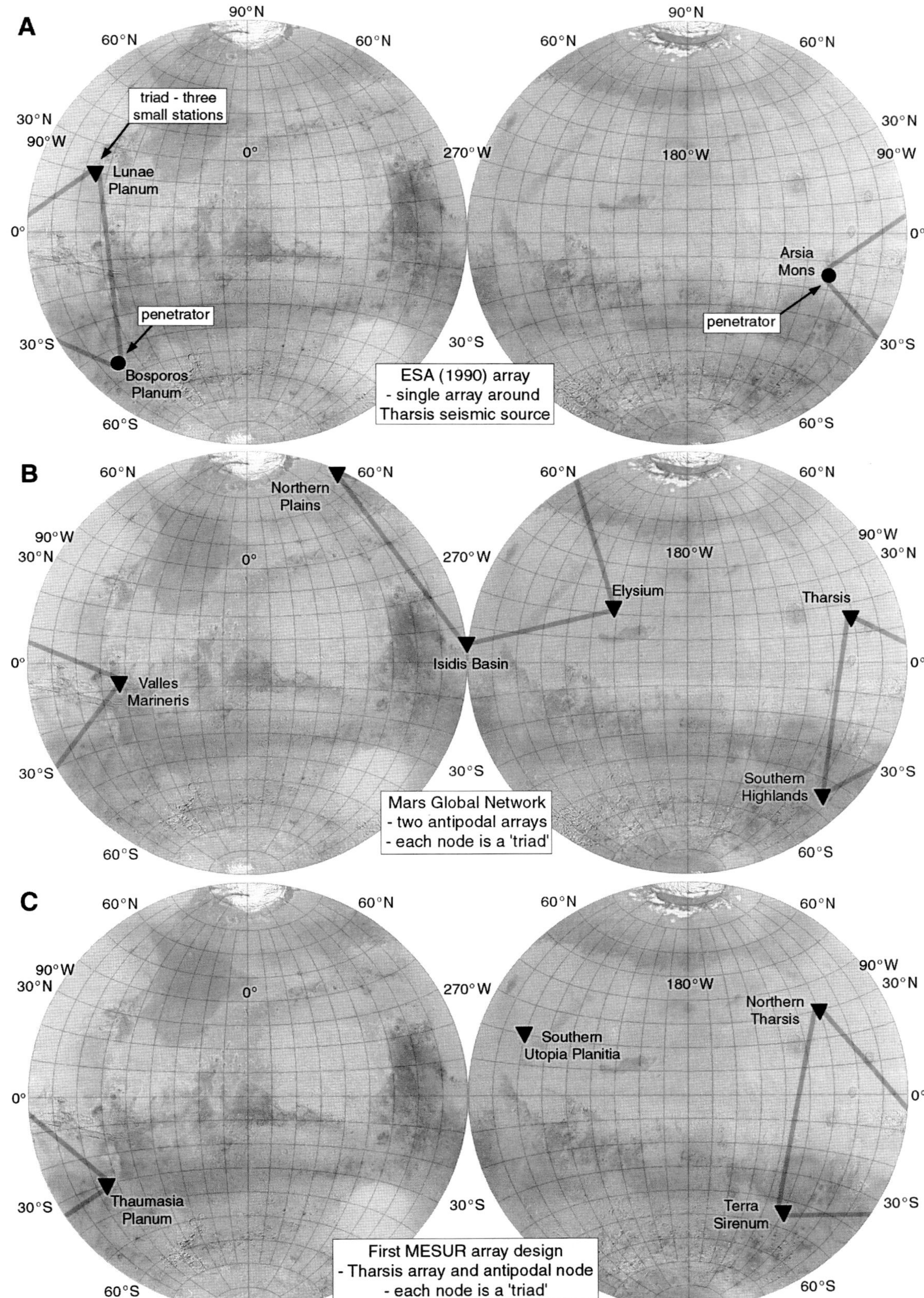

Figure 142 Array designs for network missions.
A: European network from Table 49. **B:** Mars Global Network from Table 50. **C:** MESUR network from Table 51.

Table 51. *First List of MESUR Sites, Third MSWG Meeting*

Region	Location
Northern Tharsis	30° N, 120° W
Terra Sirenum	45° S, 130° W
Thaumasia Planum	30° S, 65° W
Southern Utopia Planitia	20° N, 250° W

up at each site. Table 51 lists the approximate coordinates of these triad locations, based on the third MSWG map. The sites are illustrated in Figure 142C.

The sixth Mars Science Working Group meeting was held in Washington, DC, on 29 and 30 October 1991. A compromise network design satisfying meteorology and geophysics interests was presented by Robert Haberle. The atmospheric scientists needed broad coverage of latitude and longitude, at about 30° and 90° separations, respectively, with stations in areas susceptible to dust storms, at the poles, and at the Viking Lander 1 site for continuity with older data. Seismologists needed groups of stations at various spacings as described earlier, concentrated in areas expected to have above-average seismic activity to increase the chances of obtaining good data. That rationale produced a concentration of sites around Tharsis in the list presented at the meeting (Table 52, Figure 143A).

By the time of the seventh MSWG meeting, held on 17 June 1992 in Washington, DC, planning included consideration of solar elevation at the time of landing and the varying accessibility of sites at different times. A

Table 52. *Second MESUR Site List from Sixth MSWG Meeting*

Location	Deployment opportunity	Location	Deployment opportunity
17° N, 113° W	First	44° S, 117° W	Third
0° N, 60° W		67° S, 52° W	
0° N, 52° W		90° S	
20° N, 48° W		85° N, 5° W	
17° N, 107° W	Second	50° N, 320° W	
30° S, 60° W		40° S, 297° W	
8° S, 56° W		8° N, 290° W	
45° N, 56° W		0° N, 222° W	

'strawman' list of sites was presented, but studies showed that some sites would not be accessible at the suggested flight opportunity. Table 53 lists these sites, which are also shown in Figure 143B.

The final MESUR site list was a modified version of the previous set. The sites found to be unachievable in the second opportunity were transferred to the third, and a few sites were changed or renamed. Table 54 lists the final MESUR landing sites and indicates the order of deployment, and Figure 143C illustrates the sites. This version of the site list was incorporated into the Landing Site Catalog (Table 47).

At the Eighth Mars Science Working Group meeting, held at JPL on 8 and 9 February 1993, Jacques Blamont

Table 53. *Third MESUR Landing Site List from Seventh MSWG Meeting*

Site	Location	Sun elevation*	Deployment opportunity
Valles Marineris	6° S, 58° W	70°	First – 1999
Valles Marineris	5° S, 54° W	70°	
near Viking Lander 1	23° N, 48° W	35°	
Olympus Mons	13° N, 130° W	50°	
Oldest Plateau	2° S, 54° W	65°	Second – 2001
Northern Plains	60° N, 50° W **(NA)**	–	
Hellas	40° S, 310° W **(ND)**	65°	
Polar	90° S **(NA)**	–	
Ancient Argyre Rim	44° S, 55° W	45°	Third – 2003
Noachian Plateau	44° S, 120° W	45°	
Tyrrena Patera	21° S, 254° W	70°	
Lacustrine Deposits	15° S, 185° W	75°	
Syrtis Major	5° N, 295° W	60°	
North Arabia	38° N, 309° W	45°	
Chasma Borealis	82° N, 55° W	–15° (night)	
Polar Plains	66° S, 66° W	25°	

NA: not achievable at that date; **ND:** not desirable.

Table 54. *Final MESUR Landing Sites*

Site	Location	Catalog site	Deployment opportunity
Valles Marineris I	6° S, 58° W	101	First
Valles Marineris II	5° S, 54° W	102	
Chryse Planitia near Viking Lander 1	23° N, 48° W	103	
Olympus Mons	13° N, 130° W	104	
Valles Marineris III	2° S, 54° W	105	Second
Hadriaca Patera	32° S, 268° W	106	
Northwest Hellas	40° S, 310° W	107	
Argyre Planitia	37° S, 44° W	108	
South Pole	86° S, 315° W	109	Third
Sirenum Terra	45° S, 185° W	110	
Northern Plains	60° N, 50° W	111	
Gusev Crater	15° S, 185° W	112	
Syrtis Major	5° N, 295° W	113	
North Arabia	38° N, 309° W	114	
North Polar Region	82° N, 55° W	115	
Aonia Terrae	66° S, 66° W	116	

of CNES suggested that MESUR or similar lander networks could be deployed from balloons. A mission description from that meeting would begin with a single launch in 1998, delivering four long-duration balloons to Mars, each carrying five meteorological stations, one seismological station and a small rover. Global climate models suggested the paths these balloons might follow, dropping probes at intervals. Three maps (Figure 144) illustrated modelled paths these balloons might follow, one across the north pole, and two at northern midlatitude sites across Chryse and Tharsis. A communication relay satellite would be needed in Mars orbit to support this mission. Ultimately the MESUR mission was not funded, and network mission planning moved back to Europe with Marsnet and Intermarsnet.

The MESUR siting strategy becomes apparent when the sites are mapped (Figure 143). In each case the first deployment establishes a regional seismic network around Tharsis. In Figure 143A the second deployment extends that to a larger triangular network, whereas in Figures 143B and C that deployment extends the array in latitude and longitude, including a site roughly antipodal to the first array. The third deployment in each case extends the array to a global scale for meteorological studies, with a very apparent cluster of sites along the 60° W meridian from pole to pole.

Marsnet (ESA)

ESA's Marsnet was a network mission study which would involve three landers, or four if funding permitted. Its goals, described at a meeting held in June 1992 at Mainz, Germany, were to obtain data for meteorology, seismology, geology, exobiology and the study of volatiles (Chicarro *et al.*, 1993). Marsnet studies included hard-landing penetrator designs and semi-hard landers using airbags.

The first sites associated with Marsnet were those listed in Table 49, but new analyses were needed. The landing sites were constrained to latitudes between 45° N to 45° S, elevations below +6 km, in areas where a 100 by 50 km landing ellipse, or alternatively 100 by 20 km (Chicarro, 1993), could be fitted without including any significant obstacles. The ellipses had to be on geologically homogeneous material so that it was clear which geological unit the lander was resting on. Sites where life forms exist or may have existed were to be avoided, to reduce the potential for contamination. The constraints are mapped in Figure 145. The three main sites had to be about 1000 to 3000 km apart to provide an adequate seismic network. The possible fourth site would be roughly antipodal to the first three. At least two sites should be relatively close to a potential seismic source (Tharsis or Elysium). Nilosyrtis Mensae was identified as a desirable location to assess the Martian crustal dichotomy. Meteorological goals would be best met with a wide range of latitudes, longitudes and elevations. Geology goals required sites on a variety of surface materials and ages, including, if possible, ancient crust material from the highlands, volcanic rocks of different ages, fluvial deposits from outflow channels, glacial and polar deposits, dunes and other aeolian deposits. Layered deposits and areas with evidence of water flow or ponding, erosion or periglacial activity were also desired.

Marsnet required a network of stations for global or regional studies of seismology and meteorology. The network design was constrained by the lander delivery system and network geometry requirements, so potential

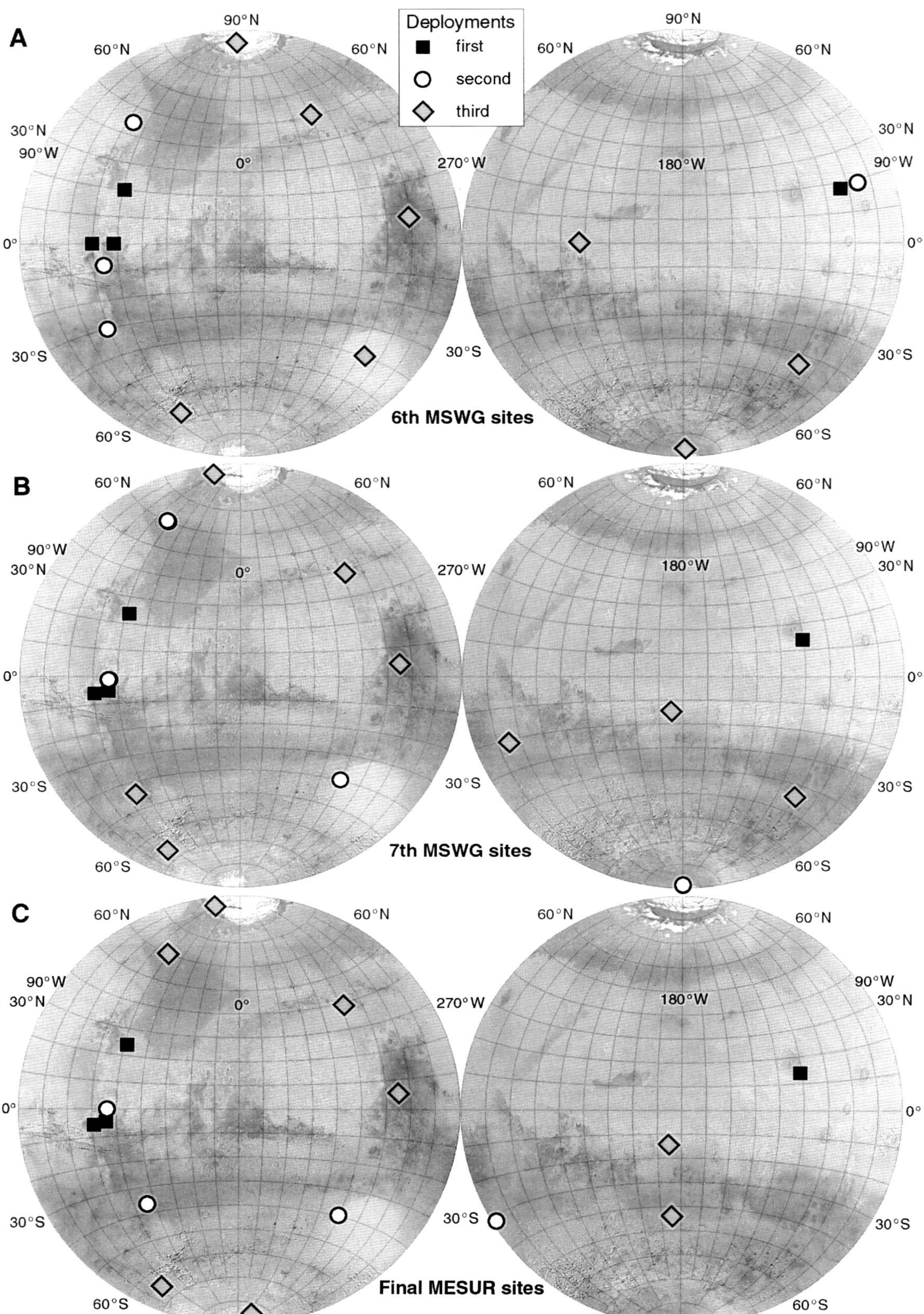

Figure 143 MESUR array designs from the Mars Science Working Group meetings.
A: Sixth meeting. **B:** Seventh meeting. **C:** Final array design.

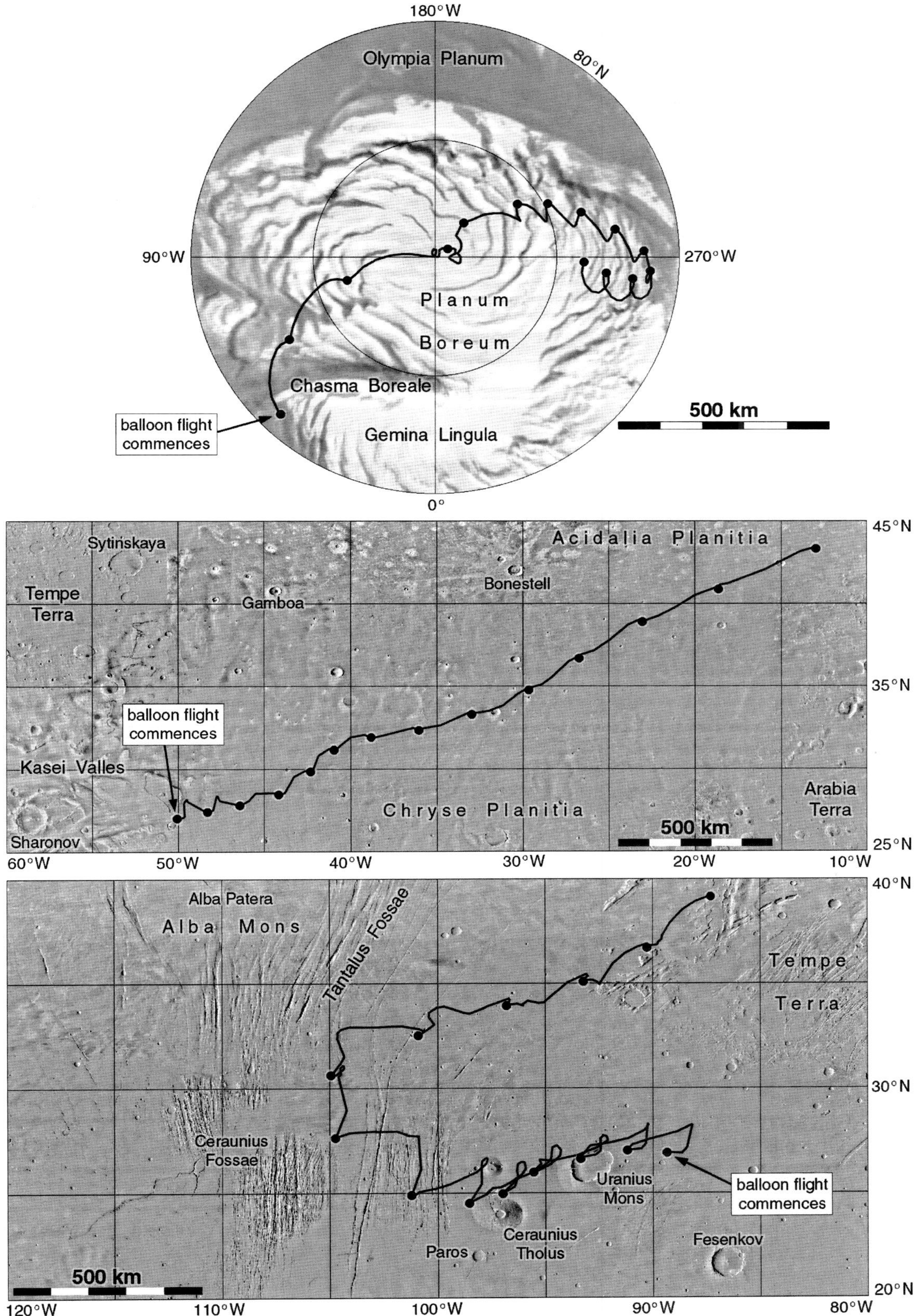

Figure 144 Balloon flight paths modelled by Jacques Blamont, from the eighth Mars Science Working Group meeting. Each black dot is a touchdown point.

landing sites were considered which satisfied these constraints. Individual site characteristics were not considered at this stage, though they would also be studied during the mission. According to De Angelis and Chicarro (1996), 50 sites were selected which satisfied most of the scientific criteria, including roughly 1000-km spacing between stations to study regional internal structure, proximity to probable seismic activity, a variety of surface geology, ages or materials, a wide latitude range for meteorology, and variations in volatiles. The full list is not presented here because ESA will not release the report. The engineering constraints included geological homogeneity within the landing ellipse, latitude limits on solar power, altitude limits for the landing procedure, and compatibility with other proposed network missions.

From this set of 50 sites, three groups of four were selected, each with three regional sites forming a triangle about 1000 km across and one antipodal location. One group was designed to investigate the Tharsis region, one the Elysium region (both expected to be seismically activity) and the third one spanned the Southern Highlands/Northern Plains dichotomy at Nylosyrtis Mensae. In the Marsnet mission Phase-A Final Report (Chicarro, 1993), the landing site set addressing the Tharsis area was selected as the mission baseline or 'strawman' site configuration, a first iteration which might be modified as new data were obtained. Table 55 outlines this network configuration, with references to the Landing Site Catalog (Table 47) where the sites are illustrated. All sites lie between 40° N and 25° S with elevations between +1 and +4 km, on a wide variety of rock types. A fourth lander would be sited at the antipode in the eastern hemisphere. These sites are shown in Figure 145. Marsnet was initially anticipated to launch in about 2001, and the landers were not expected to survive for more than one Mars year, so there was some consideration of delaying launch until 2003 to complement NASA's proposed MESUR mission (Table 54). Eventually ESA abandoned the mission, though other network missions (Intermarsnet, Netlander and Metnet) were considered later.

Table 55. *Marsnet 'Strawman' Landing Sites*

Name	Location	Elevation	Catalog site	Notes
Tempe Terra	40.1° N, 78.2° W	1 km	151	Fractured uplands
Candor Chasma	9.7° S, 73.2° W	3 km	152	Chasma floor
Daedalia Planum	25.0° S, 127.5° W	4 km	153	Lava flows
Antipode	Near 0° N, 270° W		*	Eastern hemisphere

*Catalog sites 5, 59, 64 and 113 are in this general area.

Intermarsnet (ESA)

Intermarsnet was another network mission study, following the work done on Marsnet, and performed from 1993 until 1995 for a possible launch date towards the end of that decade. It involved three landers with a lifetime of at least one Mars year. The objectives were to study geology, seismology and meteorology, placing constraints on the landing sites that were similar to those for Marsnet. Geology required landing on different types and ages of surface materials. Seismology required three sites at least 1000 km apart. Meteorology required a significant range of latitudes. For solar power, the latitude range of the landing sites was between 28° N and 15° S. The entry and landing system required site elevations below +4 km in the Viking topographic model. Landing ellipses should have rock numbers and sizes not exceeding those at the Viking sites.

The sites studied for Marsnet would not necessarily be suitable for Intermarsnet because of the differing mission requirements, instruments, landing dates and delivery methods. A new study was undertaken to assess a large number of sites from which individual network nodes could be selected. Three hundred sites were identified by De Angelis and Chicarro (1995), but the full list of potential sites cannot be identified because ESA will not release the report. From them, a subset of 100, either more scientifically interesting or more easily accessible, was selected by De Angelis and Chicarro (1996). These authors then identified 12 arrays consisting of four sites each for mission planning purposes. These array sites are listed in Table 57a, and the sites are illustrated in Figures 146 and 147. The arrays are concentrated around Tharsis or Elysium, regions expected to be sources of seismic

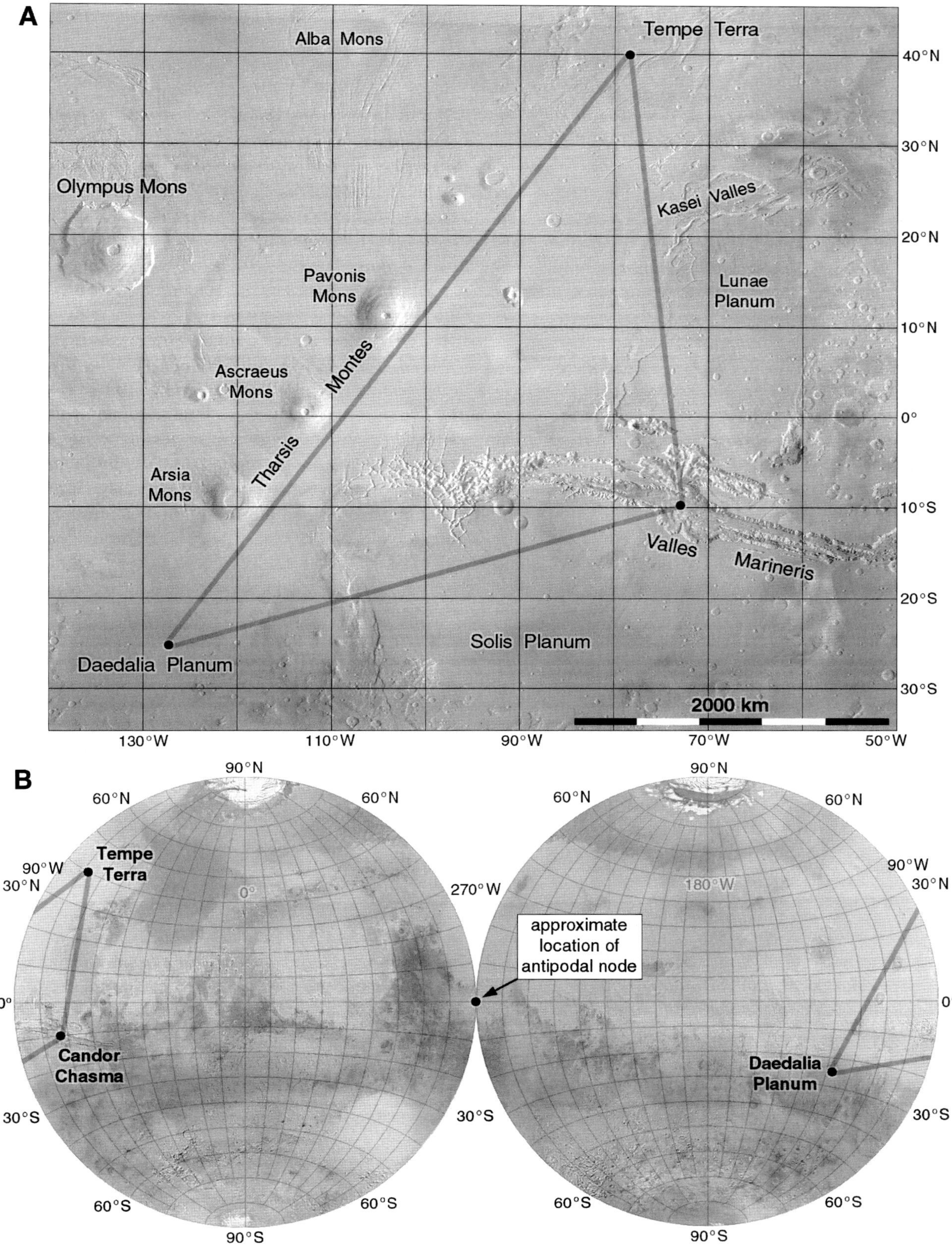

Figure 145 **A:** Marsnet Tharsis array. **B:** Marsnet global array with antipodal point.

events. The targeting uncertainty was assumed to be about 150 by 30 km, but because its orientation could vary considerably, a circle of 150-km diameter was used for the site study. Different delivery systems had different latitude constraints, with a multiprobe carrier having latitude limits of 45° N to 45° S and a single-probe carrier using a different lander having limits of only 22° N to 22° S, mainly restricted by solar power. A mission consisting of four single-probe 'free flyers' could deploy landers independently to any desired sites. The multiprobe carrier would deploy its probes shortly before arrival, with tighter constraints on the network pattern.

The Intermarsnet Phase A study (Banerdt *et al.*, 1996) proposed a three-lander network, dropping the antipodal site. Its sites were a modified version of the single-probe carrier baseline configuration, substituting Uranius Patera for Olympus Mons to give better latitudinal coverage (Table 57b).

These landing sites covered a variety of geological units and surrounded the Tharsis region, which was thought to be the most likely area to still experience significant seismic activity. Site A in Gusev crater was at the end of a long winding valley system (Ma'adim Vallis) which might have been cut by flowing water, so the smooth plains on the crater floor were thought to consist of lacustrine (lake bed) sediments. This site was so enticing that it eventually became the Spirit rover landing site in 2004. Site B was on plains north of Uranius Mons (then referred to as Uranius Patera, a name now restricted to its caldera), with lava flows, low volcanic shields, channels and faults. Site C in the Coprates area was on old cratered plateau materials north of Valles Marineris. The surface materials were expected to be impact breccias containing ancient highlands material, perhaps with younger volcanic and sedimentary material in places. The Phase A study sites are shown in Figure 147C.

Micro-Meteorological Network (United States)

Haberle (1995) suggested that Intermarsnet could be augmented by the addition of 10 to 15 small landers which would measure only atmospheric pressure. Pressure is easy to measure and is of fundamental importance in understanding meteorology. These landers would be more widely distributed than the Intermarsnet landers themselves and were referred to as the Micro-Meteorological Network.

This Discovery-class mission would use a single launcher in June 2003 to deploy 16 Micro-Met (μ-Met) stations on Mars (Haberle and Catling, 1996; Merrihew *et al.*, 1996). They would operate for one year on the surface, providing only surface pressure data to keep costs low. Additional instruments including a descent imager might be added at the expense of fewer stations.

Landing site distribution had to take into account the requirements of different types of study. Zonal mean

Table 56. *Intermarsnet Landing Site Catalog (De Angelis and Chicarro, 1995, 1996).*

Site	Name	Location	Geology	Site	Name	Location	Geology
001	Arcadia Planitia	40.0° N, 176.0° W	Plains	126	E. Elysium	27.0° N, 204.2° W	Lava flows
003	Alba Patera	42.5° N, 122.5° W	Aureole	130	E. Elysium	25.0° N, 195.0° W	Lava flows
006	Arcadia Planitia	34.0° N, 97.5° W	Plains	134	Elysium Planitia	6.0° N, 202.5° W	Alluvium
008	Chryse Planitia	35.0° N, 44.0° W	Lava flows	137	Memnonia N	16.1° S, 153.9° W	Cratered area
018	Tempe Terra	34.0° N, 58.0° W	Ridged plains	139	Daedalia Plan.	13.9° S, 137.2° W	Smooth plain
027	Arabia Terra	32.0° N, 311.5° W	Etched	145	Daedalia Plan.	8.8° S, 144.0° W	Cratered plain
028	Huo Xing Vallis	30.0° N, 295.0° W	Plateau	148	Memnonia	15.0° S, 163.7° W	Cratered area
029	Utopia Planitia	38.0° N, 270.0° W	Plains	153	W. Medusae	5.0° S, 157.5° W	Rolling plain
030	N. Nilosyrtis	39.5° N, 287.0° W	Complex plain	156	E. Mangala	10.0° S, 149.5° W	Alluvium
031	Utopia Planitia	33.0° N, 248.0° W	Plains	160	T. Sirenum	28.5° S, 129.0° W	Lava flows
035	Utopia Planitia	38.5° N, 231.0° W	Complex plain	162	V. Marineris	9.7° S, 73.2° W	Chasma floor

Table 56. *(cont.)*

Site	Name	Location	Geology	Site	Name	Location	Geology
039	Utopia Planitia	34.0° N, 234.0° W	Rolling plains	163	Capri Chasma	14.0° S, 46.0° W	Layered rocks
045	Amazonis Plan.	23.0° N, 157.0° W	Smooth plain	168	N. Coprates	2.0° S, 49.5° W	Cratered area
047	Amazonis Plan.	11.0° N, 138.5° W	Smooth plain	170	Margaritifer	16.7° S, 8.5° W	Smooth plain
050	Amazonis Plan.	2.5° N, 152.5° W	Rolling plains	172	N. Nirgal Vallis	9.2° S, 39.0° W	Chaotic area
052	Olympus Mons	18.5° N, 147.0° W	Aureole	175	N. Noachis	23.5° S, 5.0° W	Smooth plain
053	Olympus Mons	28.0° N, 144.0° W	Aureole	176	Hydraotes	3.5° S, 39.5° W	Cratered area
055	S. Olympus	12.0° N, 132.0° W	Lava plains	182	Schiaparelli	2.5° S, 342.0° W	Aeolian sed.
058	W Ceraunius	23.0° N, 114.0° W	Lava flows	183	Flaugergues*	17.0° S, 341.1° W	Cratered plain
059	Tharsis	26.0° N, 122.5° W	Lava flows	184	Schiaparelli	2.5° S, 345.0° W	Aeolian sed.
060	N. Uranius	29.0° N, 93.3° W	Lava flows	187	Flaugergues S.	21.0° S, 340.4° W	Cratered plains
062	Tharsis	10.0° N, 83.5° W	Lava flows	193	Terra Meridiani	1.0° S, 358.5° W	Rolling plains
065	Tharsis	18.0° N, 87.0° W	Lava flows	196	Serpentis	28.0° S, 317.2° W	Cratered plain
068	N. Kasei Valles	28.0° N, 63.0° W	Alluvium	197	Hellas rim	28.5° S, 296.7° W	Basin rim
069	Tharsis	21.0° N, 82.0° W	Ridged plains	198	Terra Tyrrhena	16.0° S, 302.0° W	Ridged plains
071	Chryse Planitia	25.0° N, 42.5° W	Alluvium	201	Terra Tyrrhena	11.0° S, 292.0° W	Dissected area
072	E. Chryse Plan.	17.3° N, 26.1° W	Layered rocks	202	Terra Tyrrhena	27.2° S, 311.5° W	Cratered area
073	Chryse Planitia	26.7° N, 31.7° W	Volcanics	204	Tyrrhena Patera	20.0° S, 254.0° W	Volcanic shield
079	Chryse Planitia	4.5° N, 24.0° W	Layered rocks	209	Terra Tyrrhena	23.5° S, 266.5° W	Cratered area
083	Chryse Planitia	20.0° N, 37.0° W	Alluvium	210	Terra Tyrrhena	2.5° S, 246.5° W	Cratered area
084	Cassini	24.0° N, 327.5° W	Aeolian sed.	211	Terra Tyrrhena	9.0° S, 264.0° W	Cratered area
085	N. Sabaeus	5.0° N, 347.9° W	Complex plain	214	Hesperia Pl.	12.0° S, 248.5° W	Ridged plains
086	Arabia Terra	28.0° N, 319.5° W	Ridged plains	217	Terra Tyrrhena	28.7° S, 227.0° W	Ridged plains
087	East Cassini	23.5° N, 322.5° W	Cratered plain	221	Gusev	15.0° S, 184.5° W	Lake deposits
093	Syrtis Major	4.0° N, 294.5° W	Ridged plains	226	Aeolis	10.0° S, 215.0° W	Cratered area
094	West Syrtis	12.5° N, 303.0° W	Cratered plain	230	Elysium Planitia	0.1° S, 192.0° W	Alluvium
095	N. Antoniadi	28.5° N, 302.0° W	Cratered plain	231	Elysium Planitia	0.1° S, 200.0° W	Alluvium
098	Syrtis Major	12.5° N, 296.0° W	Ridged plains	237	Terra Sirenum	37.0° S, 132.5° W	Plateau area
100	Isidis Planitia	12.0° N, 274.0° W	Aeolian sed.	238	Sirenum Fossae	34.0° S, 154.0° W	Plateau area
101	Isidis Planitia	8° N, 277.3° W	Aeolian plains	243	Terra Sirenum	33.6° S, 167.5° W	Plateau area
104	Elysium Planitia	14.0° N, 231.1° W	Lava plains	245	Bosphorus Plan.	39.5° S, 70.0° W	Smooth plains
105	Elysium Planitia	16.0° N, 229.0° W	Smooth plains	251	Argyre Quad.	32.5° S, 44.5° W	Cratered area
108	Elysium Planitia	22.0° N, 248.0° W	Smooth plains	257	Argyre Rim	44.0° S, 53.5° W	Basin rim
109	S. Elysium	1.0° N, 253.8° W	Smooth plains	264	Hellas Planitia	34.0° S, 303.5° W	Ridged plains
115	S. Nepenthes	3.0° N, 248.0° W	Cratered area	266	Hellas Planitia	43.0° S, 306.5° W	Hellas floor
116	Amenthes	5.5° N, 263.5° W	Basin rim	268	Hellas Planitia	42.5° S, 297.5° W	Hellas floor
117	E. Nepenthes	3.8° N, 235.0° W	Knobby area	271	Hellas Planitia	41.0° S, 285.0° W	Hellas floor
118	Elysium Planitia	27.0° N, 182.5° W	Alluvium	273	Hellas Planitia	36.0° S, 282.0° W	Hellas floor
119	Elysium Planitia	9.0° N, 222.5° W	Smooth plains	277	Hadriaca Pat.	33.0° S, 269.0° W	Volcanic shield
122	W. Elysium	19.0° N, 217.5° W	Lava flows	285	Terra Cimmeria	43.0° S, 202.5° W	Rolling plains
				289	Terra Cimmeria	31.0° S, 224.0° W	Cratered plain

*The coordinates for site 183 correct an error in the source.

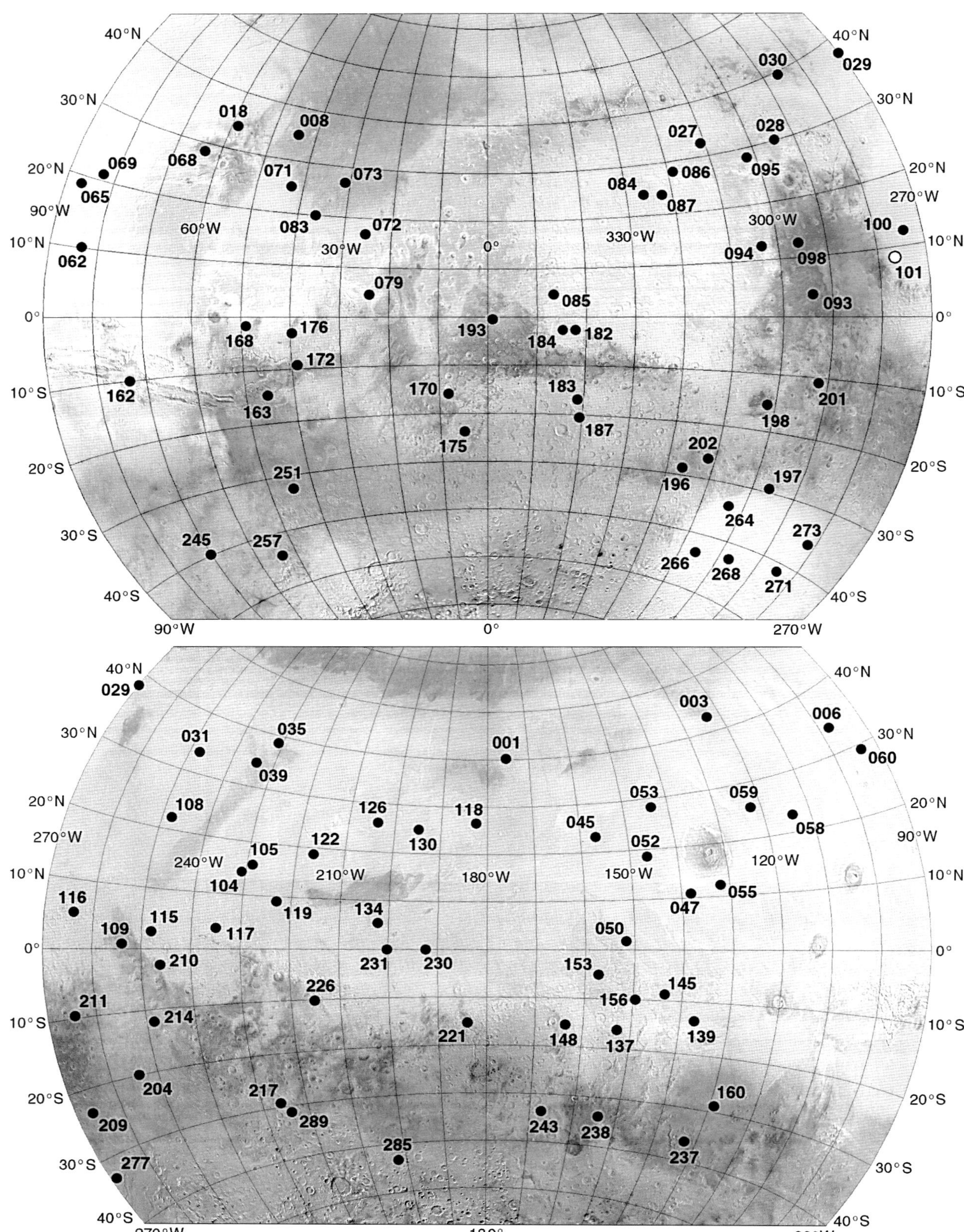

Figure 146 The Intermarsnet candidate sites described by De Angelis and Chicarro (1995, 1996) from Table 56. Black dots are the 100 preferred sites (1996 ref). Site 101, the single white dot in Isidis, is the only additional site from an unreleased larger list to be identified.

Table 57. *Intermarsnet Network Configurations*

57a. Networks from De Angelis and Chicarro (1996)

Single probe carrier			**Multiprobe carrier**		
De Angelis and Chicarro (1996), Table 5			De Angelis and Chicarro (1996), Table 6		
Configuration	Triangle	Antipode	Configuration	Triangle	Antipode
1 (baseline)	055, 168, 221	204	1 (baseline)	003, 062, 237	268
2	052, 148, 172	198	2	045, 059, 238	271
3	062, 156, 162	100	3*	006, 160, 257	210
4	047, 083, 153	201	4	058, 139, 245	209
5	101, 104, 211	162	5	031, 126, 289	251
6	108, 201, 230	163	6	105, 130, 231	083

57b. Intermarsnet Phase A Study landing sites (Banerdt *et al.*, 1996).

Site	Name	Location	Elevation	Notes (site no. from Table 56)
A	Gusev crater	15° S, 184.5° W	0 km	Channel material (Site 221)
B	Uranius Patera	29° N, 93.5° W	+1 km	Lava flows (Site 060)
C	Coprates highlands	2° S, 49.5° W	+2 km	Cratered plateau (Site 168)

Notes: Sites are identified by their numbers in the Intermarsnet Landing Site Catalog (Table 56). Site 101 (single-probe carrier configuration 5) was the only site in this table not included in the 100 preferred sites in De Angelis and Chicarro (1996).

*Multiprobe configuration 3 was referred to as the baseline configuration in De Angelis and Chicarro (1996, Table 7). A variation on this is presented in Table 71 and Figure 178.

circulation would need 15 landers spread between 70° N and 70° S over all longitudes. Midlatitude atmospheric waves would require seven probes at 60° N and five at 50° to 60° S, equally spaced in longitude. Equatorial waves could be studied with three landers at 15° N, 0° and 15° S, widely spaced in longitude. Regional winds would be studied with three probes forming a triangle about 300 km across. The global mean pressure would be determined by 16 widely spaced probes. Dust storms would require eight landers at suitable storm-prone longitudes, two each at 15° N, 0°, 20° S and 35° S. Because many probes would contribute to two or more of these studies, a total of 16 should be sufficient.

A plausible network design, presented by Haberle and Catling (1996), consisted of two sets of eight probes deployed separately by the carrier during approach. Each set of eight would consist of two deployments of four probes, using rotation to provide a centrifugal effect to disperse the probes. After the first eight were released, a short engine burn would speed the spacecraft up slightly. The next two sets of four probes would then be deployed, and as they were travelling faster they would arrive a quarter of a Martian day earlier, providing the longitudinal offset required. These sites are shown in Figure 148A and listed in Table 58. The Set 1 and Set 2 designations in Table 58 are not necessarily indicative of the order of deployment or arrival. Each probe would be directed to a grazing entry near the limb as seen from the approach direction, so the eight probes in each set are distributed around a large circle, distorted by the map projection in Figure 148A.

1990s: **Future Exploration Studies**

In the late 1980s and early 1990s, several groups described ambitious future exploration plans for Mars, including human and robotic components. Nash *et al.*

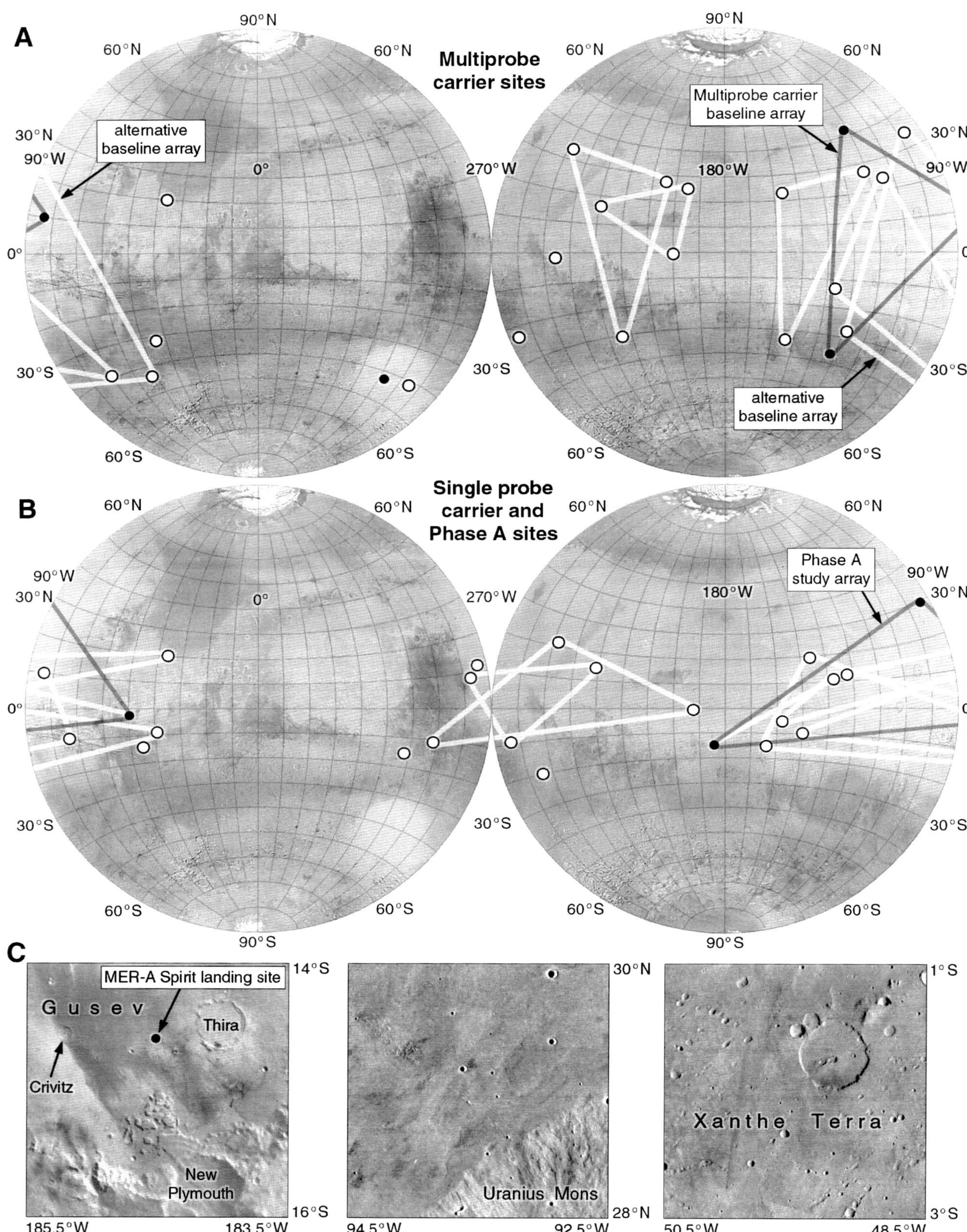

Figure 147 Intermarsnet network configurations listed by De Angelis and Chicarro (1996), from Table 57.
A: Multiprobe carrier arrays, with the baseline array in a dark outline. **B:** Single-probe carrier arrays with the Phase A recommended array in a dark outline. Nodes in A and B are linked to form the triangular arrays, and sites not on an array node are antipodal to another array. **C:** Phase A sites (Table 57b). Each square image is 120 km across and would almost fill the 150-km circular landing area.

Table 58. *Micro-Meteorological Network Sites (Haberle and Catling, 1996)*

Set 1 locations		Set 2 locations	
75° N, 165° W	62° S, 344° W	42° N, 142° W	40° S, 318° W
63° N, 63° W	58° S, 242° W	2° S, 128° W	4° N, 300° W
16° N, 43° W	13° S, 223° W	47° S, 102° W	52° N, 288° W
28° S, 32° W	30° N, 217° W	67° S, 24° W	80° N, 218° W

Table 59. *Human Exploration Sites from Nash et al. (1989)*

Site	Location	Notes
M-1: Terra Tyrrhena	27° S, 265° W	Cratered highlands, ridged plains: composition and ages
M-2: Xanthe Terra	15° N, 35° W	Cratered highlands, large outflow valleys: geology and biology
M-3: Tharsis Montes	1° S, 120° W	Large volcanoes, young lava plains: ages, origin of debris lobes
M-4: W. Daedalia Planum	19° S, 144° W	Cratered highlands, central peak, lava flows, wind streaks
M-5: Planum Boreum	82° N, 58° W	Ice cap, CO_2 seasonal cap, layered sediments: biology, climate
M-6: Candor Chasma	6° S, 76° W	Canyon walls, layered deposits: origin, composition, ages
M-7: Chasma Australe	85° S, 269° W	Layered deposits: origin, composition, biology, erosion

(1989) surveyed potential human activities at the Moon, Mars, Phobos and asteroids, which were all being considered as future targets for human exploration by NASA's Office of Exploration. They described science goals and exploration activities and identified robotic precursor missions needed to support human missions. These robotic missions were assumed to include the Soviet Union's Mars 1994 mission and NASA's Mars Observer orbiter, Mars Rover Sample Return, a penetrator network and an atmospheric science orbiter. Geology, geophysics, meteorology and biological potential would all be examined at Mars, as well as the potential use of local resources to support exploration crews. The study identified seven potential landing sites (Table 59, Figure 148B)

Project Ares was a class project undertaken by Amrine *et al.* (1992) at the Air Force Institute of Technology on the Wright-Patterson Air Force Base near Dayton, Ohio, to design a program of robotic and human exploration of Mars, culminating in a permanent presence on the planet. The first human mission was targeted for the favorable opposition of 2014, following a period of testing of hardware on the Moon and scientific study of Mars. The full program consisted of five phases.

Phase 1, Project Luna, would involve lunar base operations. It would be used to study the Moon and create an observatory, but another role was to test Mars hardware, including the habitat module. Robotic precursors would begin in 1998, and cargo and human missions would begin in 2006, with the base operating at least until 2015.

Phase 2 would see robotic missions to Mars, an orbiter in 2001 and rovers in the 2005–2007 period. This would be followed by Phase 3, a series of cargo missions delivering supplies and equipment at every opposition from 2009 until 2018. The first human flights (Phase 4) would be in 2014 and 2017, and permanent occupation of the base (Phase 5) would commence in 2019 with a series of overlapping missions each lasting about three years with 600 days at a time on the surface of Mars.

The Mars habitat, field tested on the Moon, would be delivered by a cargo lander and powered by a nuclear system with solar power backup. The habitat would use a closed system for life support and environmental control. Many other technologies and human factors would be thoroughly tested on the Moon.

The robotic exploration phase was intended to find an optimum base site, beginning with an orbital mapping

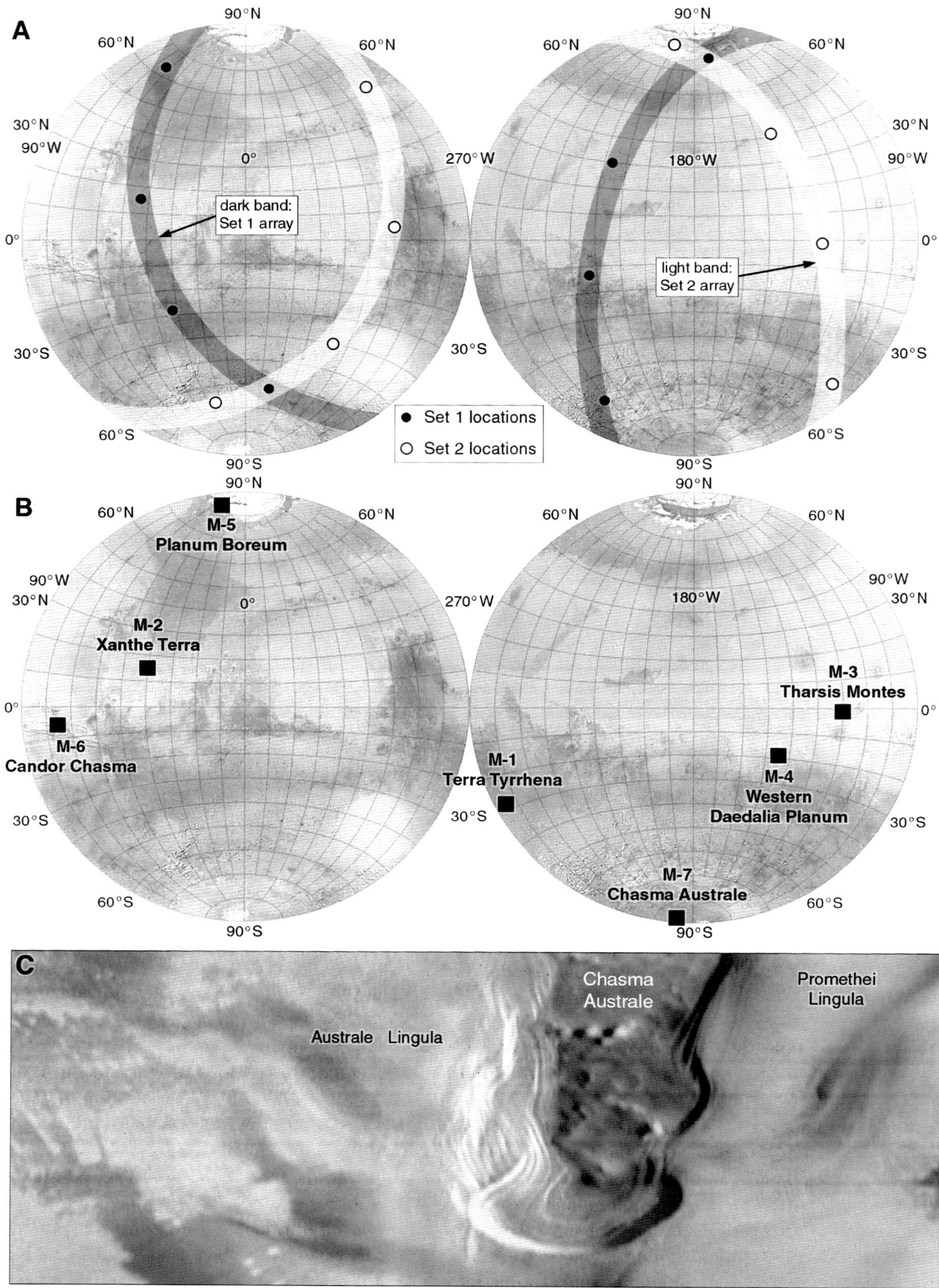

Figure 148 A: The MicroMet meteorological array described by Haberle and Catling (1996), from Table 58. **B:** Human exploration landing sites identified by Nash *et al.* (1989), from Table 59. **C:** The Chasma Australe landing area from Figure 148B (Mars Odyssey THEMIS infrared mosaic).

mission accompanied by two satellites for communication and navigation. The orbiter would launch early in 2001 and arrive late in that year. It might operate in its near-polar Sun-synchronous orbit for a decade, but initial mapping would be completed after three years, allowing 12 candidate sites to be chosen for further investigation. Those sites, 10 km by 10 km in size, would be mapped at 1 m/pixel, and the rest of the planet at medium resolution (200 m/pixel). Other instruments would map the magnetic field, surface elevations and composition, with capabilities very similar to those intended for Mars Observer. The communication satellites, also used for navigation and solar flare detection, would be in equatorial synchronous orbits at 20 425 km orbital radius.

Landers would be sent in 2005 and land in 2006, including 12 small landers and two 'super-Vikings'. The 12 landers would be delivered from a polar orbit by one carrier spacecraft, using parachutes and thrusters to land as Viking had done. They would analyze soil and atmospheric composition, operating for six to 12 months. Only very limited imaging capability was assumed, but the landers would carry navigation beacons. The two most promising sites would receive the large landers about six months later, each carrying a rover with a 12-month lifetime and operating within 30 km of the beacons. The landers would survive for at least ten years. Site selection for the base should follow in 2007 and would emphasize the availability of natural resources as well as proximity to numerous sites of scientific interest, smoothness and lack of large rocks. The old Viking sites might be visited by robots or humans at some point to investigate the long-term effects of exposure to the environment of Mars.

The cargo missions following this phase would deliver the habitat and nuclear power source to the chosen site, where successful operation would commence and be confirmed before people arrived. Additional communication satellites would be included in this phase so the base would always be in contact with Earth. When everything was ready, the first human flight would occur, involving a short stay at Mars, only about 60 days, to complete the base set-up and prepare the base for longer operations later. A later mission would bring an additional habitat to expand the base and allow a stay of as long as 180 days.

Project Ares is described here as an example of academic class studies, often undertaken by groups for experience in major project planning, and an interesting source of ideas for future missions if the report is made public.

Another study like this, Project Aeneas, was conducted in 1993 by Argos Space Endeavours and the University of Texas Department of Aerospace Engineering for the University Space Research Association (USRA) and NASA (Botbyl, 1993). This robotic exploration mission to both Mars and Phobos is only briefly summarized here. The Aeneas mission involved three spacecraft launched on two Soviet Proton rockets. The first launch would deliver one Mars orbiter (Mars-Silva 1) and one Phobos probe delivery spacecraft (Mars-Silva 2). The second Proton launch would deliver the second Mars orbiter, Mars-Silva 3. Each orbiter would carry remote sensing instruments, surface probes and penetrators and an In Situ Resource Utilization (ISRU) device for fuel production. Its objectives were to provide data to help select landing sites for future human missions to Mars, to demonstrate the ability to produce fuel from Martian resources, and to determine the composition of Phobos.

The first objective would be addressed by very high-resolution imaging between latitudes 60° N and 60° S, followed by the delivery of penetrator probes and micro-rovers to the surface of Mars to provide 'ground truth' of the remote sensing data. Additional probes would investigate geology and meteorology. The second objective would be tested by combining carbon dioxide from the atmosphere with hydrogen brought from Earth to produce methane for use as fuel. The third objective would involve landing Mars-Silva 2 inside the crater Stickney on Phobos (Figure 194).

1992: **Mars Rover Reference Mission (France: CNES)**

This mission was designed at a time when France's Centre National des Etudes Spatiale (CNES) was considering a planetary exploration mission, either on its own or as part of an international project. A long-range rover would be landed near Kasei Valles and would drive more than 300 km north into Tempe Terra, making in situ observations and deploying two instrument packages.

The landing was targeted at site 17 of the Mars Landing Site Catalog (Figure 126; Costard *et al.*, 1993). The route is illustrated in Figure 149A. At the landing site an instrument package (fixed station) would be deployed, and samples would reveal the age of the plains and the mode of origin of Kasei Valles. On the traverse, samples would characterize the younger volcanic plains of Tharsis (Uranius Patera flows) and a wrinkle ridge. A second fixed station would be deployed near the ridge. Then a fluidized ejecta crater and finally the older plains of southern Tempe Terra would be examined. A third fixed station would be deployed at the end of the traverse.

25 September 1992: **Mars Observer (United States)**

Mars Observer was intended to build on the Viking-era knowledge of Mars with updated instruments, but this ambitious mission failed just before arriving at Mars. It was initially referred to as the Mars Geoscience Climatology Observer. The spacecraft body had dimensions of 1.1 by 2.2 by 1.6 m, with two 6-m-long booms supporting instruments and another 6-m boom carrying the 1.5-m-diameter communication antenna. Its instrument payload consisted of a gamma ray spectrometer, a thermal emission spectrometer, an infrared radiometer, a laser altimeter for topographic mapping, a magnetometer and an electron reflectometer. It also carried a sophisticated new scanning camera system to provide both wide-angle images of the entire planet each day, for long-term atmospheric and dust storm monitoring, and targeted very high-resolution (1.5 m/pixel) frames. Mars Observer also carried a communication relay supplied by France for the balloons and landers of the Russian Mars 94/Mars 96 missions.

The 2573-kg spacecraft was launched at 17:05 UT and placed on a heliocentric transfer orbit, intended to arrive at Mars on 24 August 1993 (MY 21, sol 268). As it approached Mars and began to prepare its propulsion system for the braking maneuver, all communications suddenly ceased on 21 August 1993 (MY 21, sol 265). It is believed that catastrophic tumbling caused by a propellant leak ended the mission. All the Mars Observer instruments were eventually reflown on future orbiters. Figure 149 includes two Mars images returned by Mars Observer about a month before the failure, taken one hour apart on 27 July 1993 (MY 21, sol 241) at a range of 5 800 000 km. A third colour image from the wide-angle camera, an unresolved point, is not shown here.

1990s: **Mars Discovery Missions**

The Discovery program was established in 1992 to accommodate a desire for small, frequent and relatively low-cost planetary missions. Numerous Mars missions have been proposed to the Discovery program since its inception, beginning with nine which were suggested at the initial Discovery Program Workshop in 1992 (Table 60). Discovery mission proposals are not usually made public as they may contain proprietary information, though some are described in conference abstracts or other literature. Several missions are described here, but no full list is available. After 2000, Mars proposals were removed from the Discovery program and directed instead to Mars Scout, a similar program specifically focussed on Mars.

The first Mars mission chosen under the Discovery program was the MESUR Pathfinder, which was renamed Mars Pathfinder and launched in 1996. It was developed at NASA Ames in 1990 and proposed as a Discovery mission in 1992, before the Mission Concept Workshop.

Proposals in 1994 for a 1998 launch included the Mars Aerial Platform, a balloon mission for atmospheric studies; the Mars Polar Pathfinder (MPP), a polar landing mission; and MACS, the Mars Ancient Climate Surveyor, a lander with a Russian Marsokhod mini-rover. The MPP would land like Mars Pathfinder, using airbags. Its target was a circle 100 km in diameter centred on the north pole, and its goal was to examine the composition and structure of the polar cap to see whether climate history is preserved in the dust and ice layers. Existing images showed no visible features at all in a 100-km diameter circle centred on the north pole, so it could accommodate the 30-km long landing ellipse, as close as possible to the pole. MPP would deploy a small rover like Sojourner, the rover carried by Mars Pathfinder. A radar sounder would examine subsurface layering as much as 5 km deep, and a heated probe on a tether would melt its way down about 100 m, measuring temperature and composition. A drill would penetrate about 50 cm into the surface for microscopic imaging of near-surface layering. MPP would launch in July 2000 on a long 27-month trajectory tailored to the

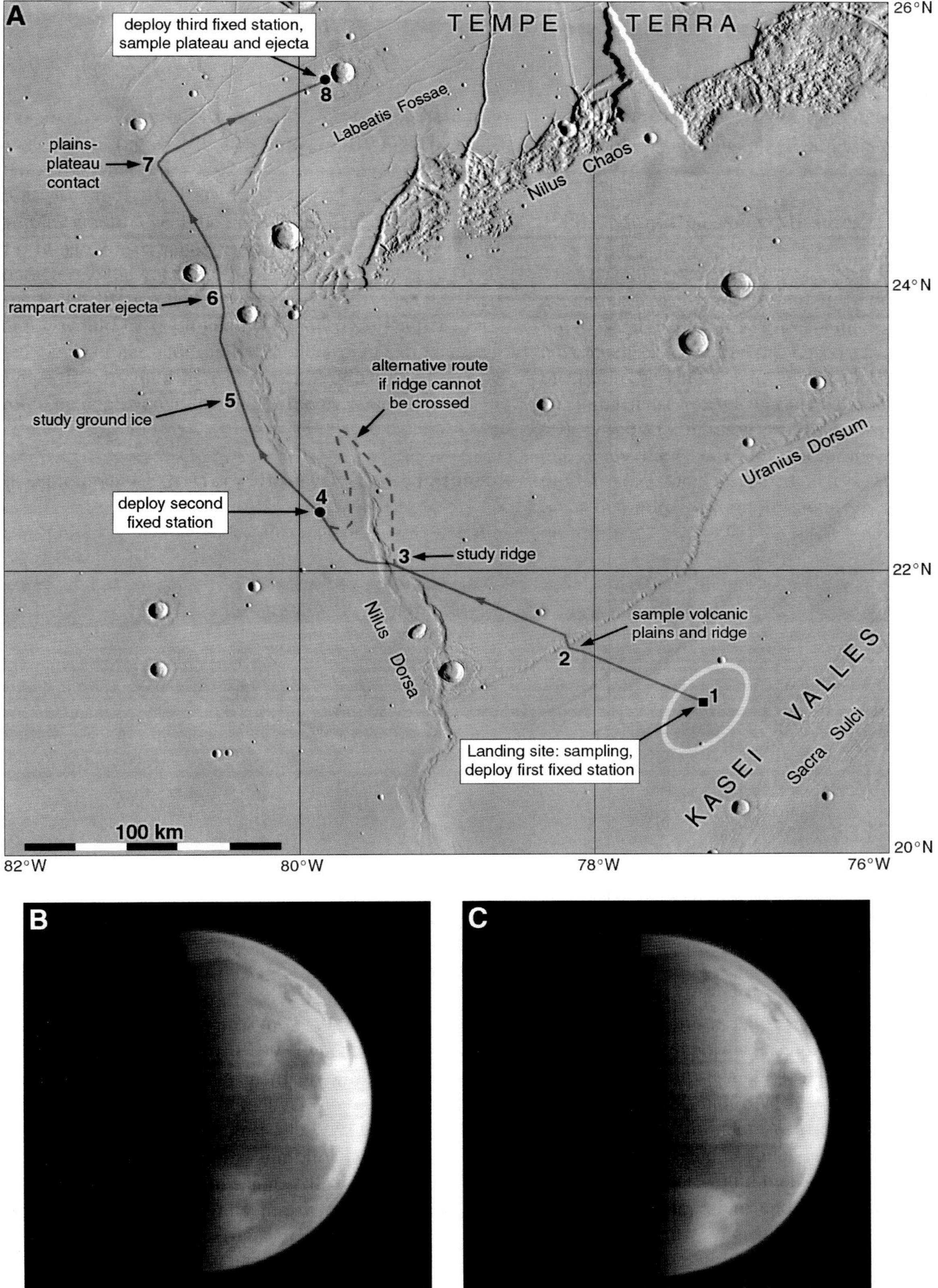

Figure 149 A: Rover route from Costard *et al.* (1993). **B and C:** Two Mars Observer images (NASA/JPL/MSSS). Hellas is the bright area near the bottom, and Syrtis Major is the dark region above and to the right of Hellas.

Table 60. *Mars Missions Proposed at the Discovery Mission Concept Workshop, 1992*

Number	Name	Description
3	Mars Climate Variability	V. Suomi – four microsatellites paced in coplanar Sun-synchronous near-polar orbits to survey the atmosphere by using radio occultations
20	CHEMIN	D. Blake – chemistry and mineralogy of the surface materials using x-ray diffraction and fluorescence from a lander (instrument proposal only)
49	MOES	Mars Operational Environment Satellite: S. Limaye – orbital survey of weather and climate for one Mars year, from a 25° inclination, 2250-km circular orbit
51	MAAP	Mars Atmospheric Aircraft Platform: J. Langford – small aircraft to be carried on MESUR missions, for imaging or other science observations
75	Prospector	B. Edwards – elemental composition and mineralogy data on various targets – this proposal specifically addresses Phobos
79	MUADEE	T. Killeen – Mars Upper Atmosphere Dynamics, Energetics, and Evolution, an orbiter to study the upper atmosphere and ionosphere
80	Little Dipper	D. Lyons – orbiter to study composition and density of the atmosphere, gravity mapping as orbit evolves from highly elliptical to circular
86	MGM/SPAM	Mars Gravity Measurement/Surface Penetrator Assembly Mission: W. Fowler – gravity, composition and subsurface ice measurements. May include one orbiter and three penetrators or two orbiters and two penetrators.
100	Phobos Sample Return	T. Duxbury – joint Russian/US sample return mission, using Russian carrier and lander with NASA instruments and return vehicle

Note: The number in the first column is taken from a list of mission proposals, many of which were to other destinations.

required polar landing. It would land in November 2002 just before the northern summer solstice, to ensure enough solar power for its 100-sol operational lifetime.

Another 1994 proposal was MUADEE (Mars Upper Atmosphere Dynamics, Energetics, and Evolution, sometimes misspelled MAUDEE), one of 11 finalists after the Discovery Mission Concept Workshop. Its goal was to investigate the upper atmosphere of Mars by making low orbital passes through it. The spacecraft carried a combination of remote and in situ sensors mounted in three instrument packages to collect data in the atmosphere at heights between 60 and 200 km during several mission phases. Aerobraking would lower the apoapsis after MUADEE entered its initial elliptical orbit, and the spacecraft used body-mounted solar cells rather than external panels to enable these low passes through the upper atmosphere. The orbit's high inclination would allow these low passes to extend nearly from pole to pole during the initial elliptical orbit phase. Later the low passes would cease and the spacecraft would enter a circular orbit phase, allowing detailed daily remote-sensing measurements. The nominal mission would last one Mars year (Killeen, 1994; Killeen and Brace, 1995).

Several Mars aircraft missions have been proposed, some for the Discovery program (Figure 150A). Proposals in 1996 included a rather complex aircraft called AME (Airplane for Mars Exploration) from NASA Ames (Hall *et al.*, 1997). It would fly over and attempt to land in Gusev crater (Figure 150C), later to be the landing site of the Mars Exploration Rover Spirit, but also considered for Mars Surveyor 2001. The total flight distance was 3400 km, including two flights on two days separated by the time necessary to transmit all its data, which might have been as long as 40 days. AME would be deployed over the plains north of Apollinaris Patera and would fly around the volcano and into Gusev, where it would land. On the second day it would take off and fly around the crater and its vicinity before landing in the highlands. While on the surface the vehicle would operate a robotic arm carrying a camera and analytical instruments. Also in 1996, JPL and AeroVironment Corporation proposed a mission which would deploy six gliders over Vallis Marineris.

Each would have a limited range, perhaps 100 km, but by starting at different points, they might examine much of the giant rift system. A later trajectory analysis (Frumkin and Riley, 2004) identified the Louros Valles area at 8.3° S, 81.2° W as a suitable target for a mission like this.

Proposals in 1998 included the Mars Aircraft for Geophysical Exploration (MAGE), a simpler design than AME. This would deploy from its aeroshell during descent in the atmosphere and fly about 1750 km over Valles Marineris (Figure 150B) on 17 December 2003, the 100th anniversary of the Wright Brothers' first flight (Malin Space Science Systems, 1998). It would launch in May 2003. The aircraft, called Kitty Hawk, would carry a gravity meter, a magnetometer, an electric field sensor, a laser altimeter, an infrared imaging system for surface composition data and six cameras, including a video system. The flight might last only about three hours, ending with a crash which might not be survived.

Other Mars airplane missions have been proposed outside the Discovery program. As early as 1978, Development Sciences Inc. in City of Industry, California, described a remotely piloted vehicle for use at Mars (Development Sciences Inc., 1978). Their concept would have involved 12 aircraft launched in groups of four, each group with its own carrier spacecraft, upper stage rocket and communication satellite. These three missions would launch at intervals of one week on Space Shuttles in the mid 1980s. The planes might be able to land and take off again and would take images and collect magnetic, compositional and other data.

In 1999, NASA and CNES, the French space agency, studied a Mars Airplane Micromission intended to arrive at Mars on 17 December 2003, the Wright Brothers' anniversary. This was one of two inexpensive 'Mars Micromissions' that NASA and CNES considered as auxiliary payloads on the large Ariane V launcher. The second was a communications orbiter. The NASA Langley Research Center would build the Mars Airplane for launch in November 2002. It would carry a 2-kg instrument payload containing cameras, a magnetometer and a miniaturized spectrometer for surface composition data. The spacecraft would first enter a highly elliptical Earth orbit and use subsequent lunar flybys to help propel it towards Mars. An orbiter would serve as a communication relay, so flight time would be limited to about 20 minutes, the time an orbiter would remain in view. The Mars Airplane would fly along the Valles Marineris canyon system. A study of entry trajectories for this mission (Murray and Tartabini, 2001) described three other sites at sufficiently low elevation for successful deployment. These were Parana Valles, a prime scientific site at 25° S, 11° W, Acidalia Planitia at 55° N, 27° W, and Hellas Planitia at 38° S, 297° W.

Discovery proposals for Mars in the year 2000 included Pascal, a network of 24 landers later proposed as a Scout mission; a repeat of the Kitty Hawk (MAGE) proposal; a Mars balloon called Piccard; and MAOS (Mars Atmospheric Oxidant Sensor), a Mission of Opportunity instrument to fly on Beagle 2. Later Mars proposals were directed to the Mars Scout program, not Discovery.

1994: **Mars 94 (Russia)**

This multiple lander and penetrator mission was originally planned for a 1994 launch. A follow-on mission with rovers was scheduled for 1996 and was therefore called Mars 96. The Mars Landing Site Catalog (Table 47) uses these designations, and some additional sites were listed later by Kuzmin *et al.* (1994) at the Mars Pathfinder Landing Site Workshop (Golombek, 1994a). A delay pushed Mars 94 into the 1996 launch window so it became Mars 96, and this necessitated the delay of the 1996 rover mission which became known as Mars 98. This last mission was cancelled as Russia's financial situation deteriorated during the 1990s. A launch delay results in spacecraft arrival at a different season, so landing sites do not automatically transfer unchanged from one opportunity to the next. The original Mars 94 site list (Table 61, Figure 151) was dropped, and a new list was compiled for its reincarnation as Mars 96 (Table 64).

The Mars 94 mission included an orbiter, two small landing stations and two penetrators. Launch would have been in October 1994 and arrival on 2 September 1995. The general landing area was Arcadia and Amazonis, roughly from 15° S to 50° N and 120° W to 170° W, in areas below +2 km elevation.

Mars 94 grew out of a previous mission study called Columbus which would have flown in 1992, the 500th anniversary of the first transatlantic voyage of Christopher Columbus. This would have involved two orbiters, each dropping a balloon and a lander with a rover. When

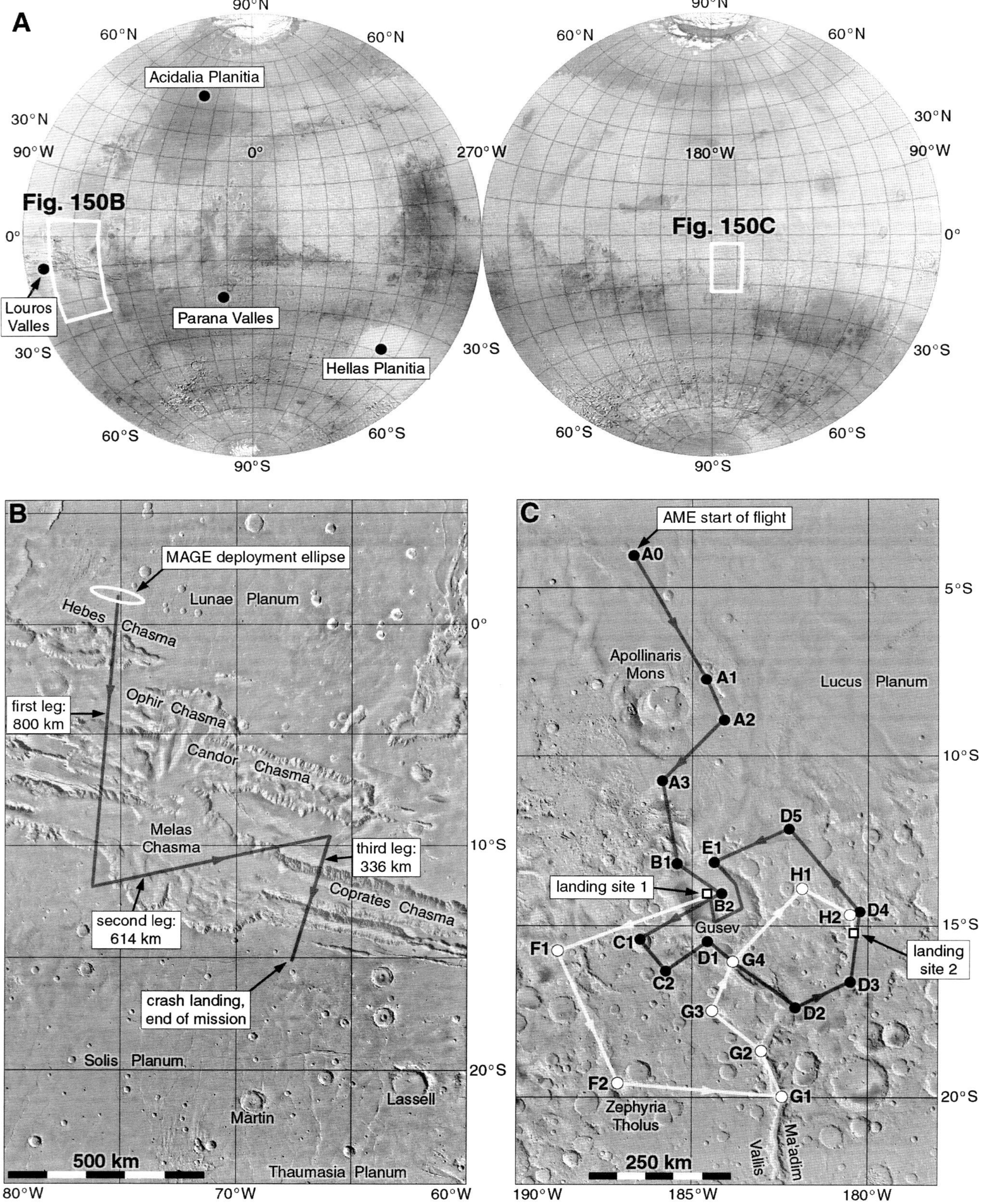

Figure 150 Mars airplane proposals.

A: Sites proposed for Mars airplane missions. **B:** Route of the MAGE airplane over Valles Marineris. **C:** Route of the AME airplane over Apollinaris Mons and Gusev crater. Figure 150C covers the same area as Figure 193E.

Table 61. *Mars 94 Candidate Landing Sites*

Lander	Location	Site	Catalog site	Source
Small stations	3° N, 160° W	1	119, South Amazonia	Mars Landing Site Catalog (Table 47)
	40° N, 149.5° W	2	120, North Amazonia	
	9.5° N, 163° W	3	126, South Amazonia	
	40° N, 153° W	4	127, North Amazonia	
	11° N, 169.5° W	5	133, South Amazonia	
	40° N, 160.5° W	6	134, North Amazonia	
Small station backup sites	10° N, 154° W	7	121, South Amazonia	
	42° N, 142° W	8	122, North Amazonia	
	3° N, 159.5° W	9	128, South Amazonia	
	39° N, 147.5° W	10	129, North Amazonia	
	5° N, 166.5° W	11	135, South Amazonia	
	40° N, 154° W	12	136, North Amazonia	
Penetrators	6° N, 148.5° W	13	87, South Amazonia	
	46° N, 128.5° W	14	88, West Alba Patera	
	3° N, 153° W	15	123, South Amazonia	
	46° N, 133° W	16	124, West Alba Patera	
	3° N, 160° W	17	130, South Amazonia	
	11.5° S, 162.5° W	18	131, Mangala Valles	
	20° N, 156° W	19	132, South Amazonia	
Penetrator backup sites	10° S, 151° W	20	117, Mangala Valles	
	35° N, 137° W	21	118, West Diacria Patera	
	9.5° S, 155° W	22	125, Mangala Valles	
Lander	**Location**	**Site**	**Description**	**Source**
Small stations	40.6° N, 158.5° W	23	Arcadia, young sediments	Kuzmin *et al.* (1994)
	30.5° N, 165° W	24	N. Amazonis, young volcanics	
Penetrators	38° N, 162° W	25	Arcadia, young sediments	
	39° N, 154° W	26	Arcadia, young sediments	

this was delayed to 1994, its association with Columbus was lost so the name became Mars 94. Sites considered for Columbus are not known.

1995: **Exobiology Site Study**

Farmer *et al.* (1995) studied potential exobiology landing sites for future missions. Former river and lake environments or springs where deposition of minerals might preserve fossils were favoured. The two editions of the Mars Landing Site Catalog (Greeley, 1990; Greeley and Thomas, 1995) were reviewed and sites selected for their exobiological interest. The Viking landing sites were added for comparison purposes but ranked low for exobiological interest. Three sites (19, 7 and 17, corresponding to Catalog site numbers 32, 138 and 58) were included in MESUR site plans (Table 54) at that time, and the MESUR site list was modified to incorporate the results of the exobiology study. Table 62a identifies the sites listed by Farmer *et al.* (1995). Greeley and Thomas (1995) contained a separate list of exobiology sites from which this table was derived. It included all the above sites and the additional sites listed in Table 62b. All sites identified as low priority by Farmer *et al.* were classed as moderate by Greeley and Thomas. These locations are illustrated in Figure 152, identified by their site numbers from Table 62.

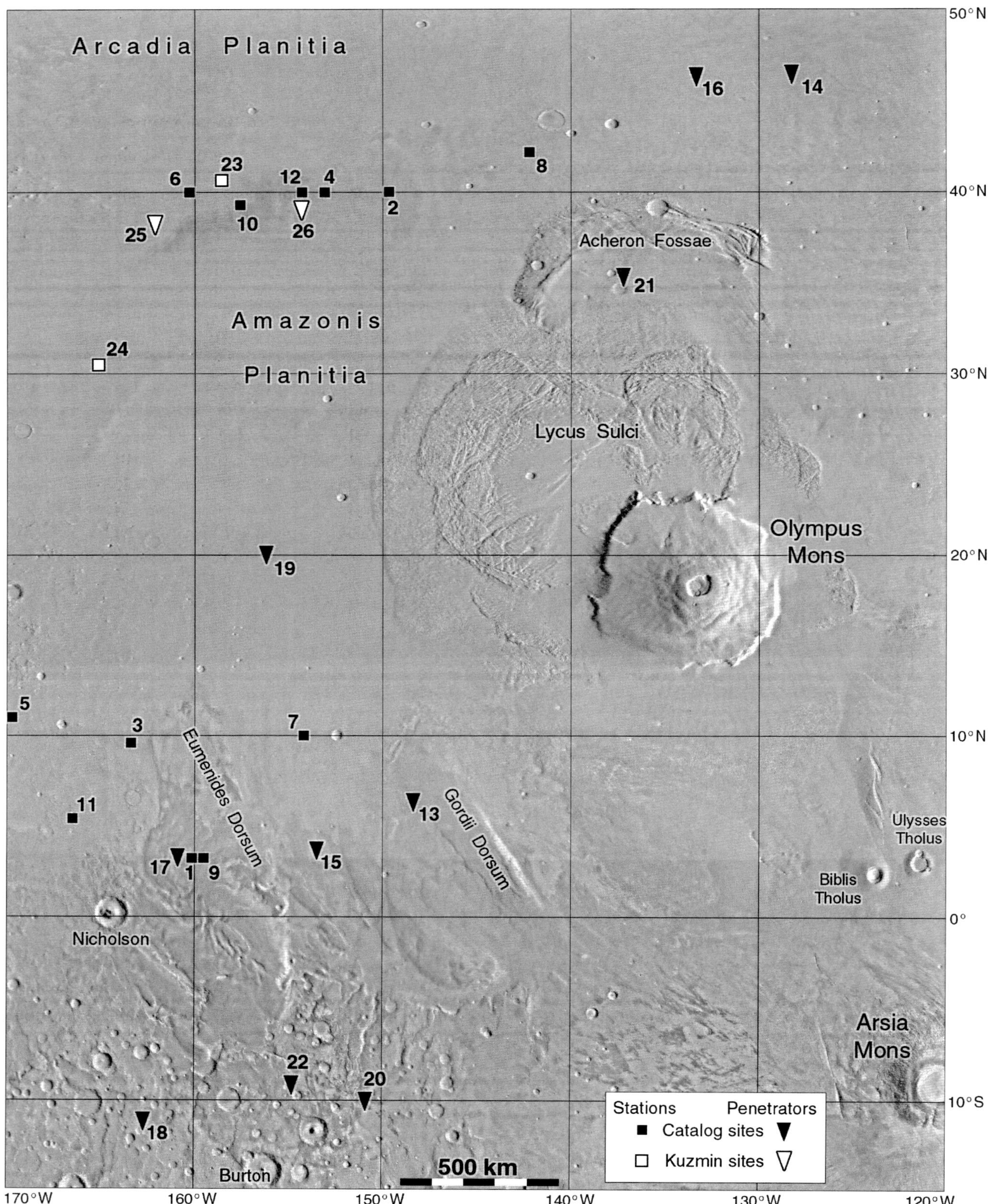

Figure 151 Candidate sites for Mars 94 stations and penetrators, from Table 61.
The background map is a composite of the Viking MDIM-2 mosaic and Mars Global Surveyor MOLA shaded relief.

Table 62. *Exobiology Landing Sites*

62a. Sites from Farmer *et al.* (1995)

Type	Site	Catalog number	Name	Location	Priority
Fluvio-lacustrine	1	1	Eridania NW	37.0° S, 230.0° W	High
	2	2	Margaritifer Sinus SE	22.0° S, 11.0° W	
	3	8	Mare Tyrrhenum SE	22.8° S, 230.6° W	
	4	10	Mare Tyrrhenum SW	24.8° S, 265.8° W	
	5	79	Aeolis SE	15.5° S, 188.5° W	
	6	137	Mangala Valles	6.3° S, 149.5° W	
	7	138	Aeolis NE (Gusev)	15.5° S, 184.5° W	
	8	140	Iapygia NW	7.3° S, 305.0° W	
	9	5	Iapygia NE	11.0° S, 279.5° W	Moderate
	10	7	Eridania NE	43.2° S, 208.1° W	
	11	22	Candor Mensa	6.1° S, 73.8° W	
	12	4	Eridania SE	57.0° S, 197.0° W	Low
	13	21	Chryse Planitia	18.9° N, 53.5° W	
	14	26	Maja Valles	18.1° N, 55.7° W	
	15	77	Ares Vallis	2.0° N, 16.0° W	
	16	141	Hebes Chasma	1.5° S, 76.5° W	
	17	58	Viking 1, Chryse Planitia	22.5° N, 47.9° W	
	18	99	Viking 2, Utopia Planitia	47.9° N, 225.7° W	
Spring	19	32	Dao Vallis	33.2° S, 266.4° W	High
	20	77	Ares Vallis	2.5° N, 19.0° W	Moderate
Periglacial	21	139	Oxia Palus	22.1° N, 37.6° W*	Low

62b. Additional sites from Greeley and Thomas (1995)

Type	Site	Catalog number	Name	Location	Priority
Fluvio-lacustrine	22	142	Mangala Valles	8.5° S, 159.5° W	High
	23	143	Ismenius Lacus SW	33.5° N, 342.5° W	
	24	147	Sinus Sabaeus NE	8.0° S, 335.0° W	
	25	144	Diacria SE	39.5° N, 135.5° W	Moderate/High
	26	145	Memnonia NW	14.5° S, 175.0° W	Moderate
	27	146	Phaethontis NC	37.5° S, 146.0° W	
	28	148	Oxia Palus NE	21.3° N, 8.0° W	
Ground ice (= periglacial)	29	149	Ismenius Lacus SC	44.3° N, 333.0° W	High
	30	150	Casius SE	36.0° N, 259.0° W	Moderate

Notes: (1) *Oxia Palus longitude in the catalog is 36.7° W.

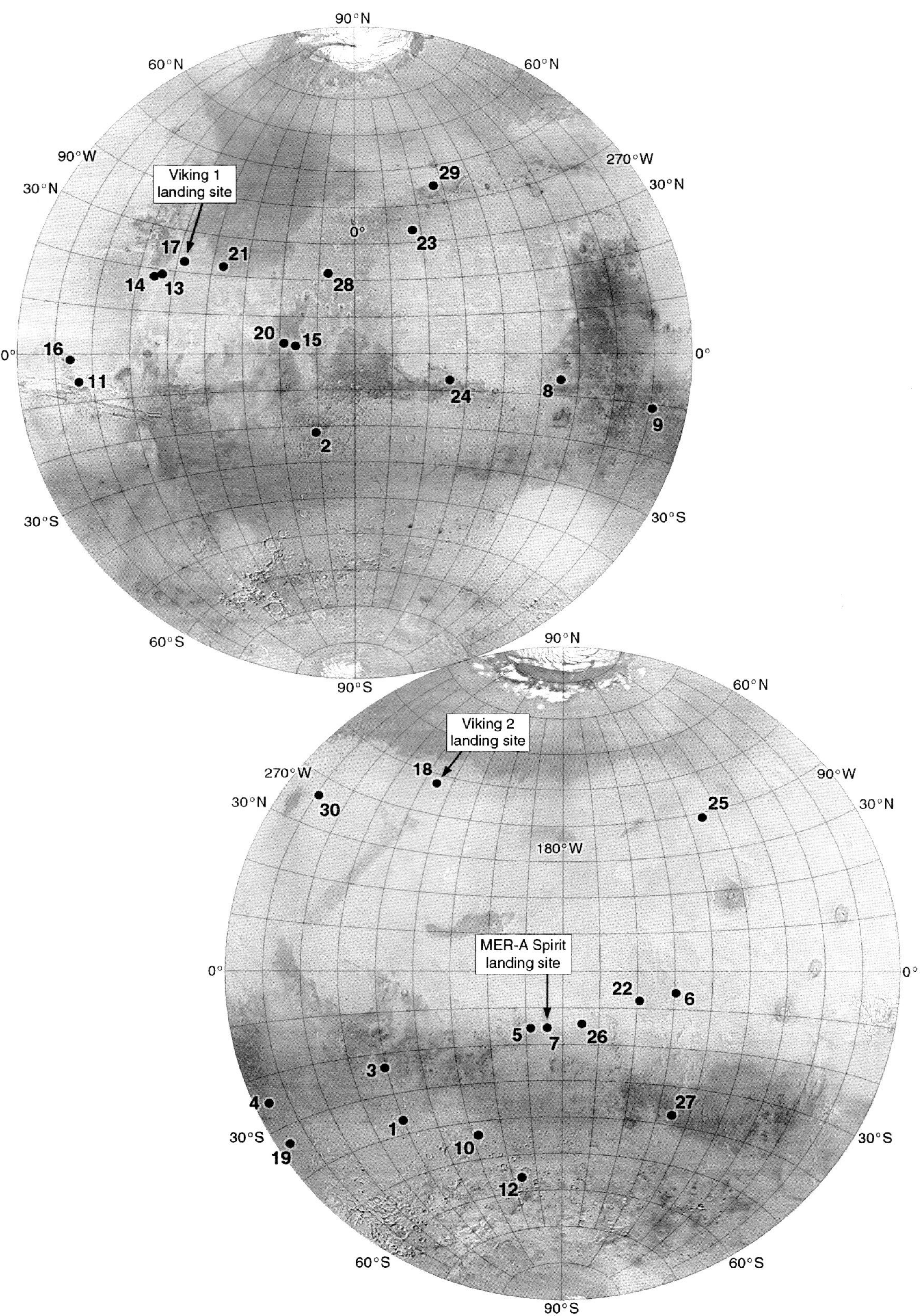

Figure 152 Exobiology sites listed in Table 62.

7 November 1996: **Mars Global Surveyor (United States)**

Mars Global Surveyor (MGS) was intended to recover some of the observations not made by Mars Observer. The science goals were high-resolution imaging, topographic mapping with a laser altimeter, gravity mapping from radio tracking, climate monitoring, the composition of the surface and atmosphere, and studies of the planet's magnetic field. The 1030-kg spacecraft was a 1.2- by 1.2- by 1.7-m box with two long solar arrays, each 3.5 by 1.9 m across. A dish antenna was mounted on a 2-m boom.

MGS was launched at 17:01 UT, and after a 10-month cruise, it began its orbit insertion burn at 01:17 UT on 12 September 1997 (MY 23, sol 371). The initial orbit was 54 021 by 258 km high with a period of 45 hours. For 16 months after orbit insertion, aerobraking gradually modified the original elliptical orbit to make it almost circular at an average height of 378 km with a period of 118 minutes (Albee *et al.*, 2001). Mapping was originally scheduled to begin in March 1998, but a malfunction of one of the solar panel supports slowed the aerobraking process considerably, delaying the start of routine mapping to 9 March 1999. Many observations were still possible during aerobraking, especially magnetic field mapping, which benefitted from the lower altitudes. The mapping orbit was polar and Sun-synchronous, so every image was taken with early-afternoon illumination, except at locations near the poles. After mapping commenced, the spacecraft operated almost flawlessly for nine years, or 3340 days, greatly exceeding its design lifetime of one Mars year. The last communication was on 2 November 2006 (MY 28, sol 277), when a solar panel steering joint apparently failed at the end of a solar conjunction imaging hiatus, probably causing the spacecraft to enter a safe mode from which it did not recover. MGS should stay in orbit until about 2046. Mission events at Mars are listed in Table 63.

MGS carried four instruments to address its science goals. The Mars Orbiter Camera, MOC, produced over 243 000 images (Figure 153), including global coverage at 230 m/pixel (Caplinger and Malin, 2001), daily global weather maps at 7500 m/pixel (Figure 153E), and more than 97 000 very high-resolution (up to 1.5 m/pixel) images of small areas distributed across much of the planet (Malin and Edgett, 2001; Malin *et al.* 2010). Figure 153A shows the Mars Pathfinder landing region two days before the landing, with dust clouds in Valles Marineris, also seen in Figure 141. Total surface coverage at high resolution (better than 12 m/pixel) was about 5.4%. Images of Phobos and Deimos were also obtained (Figure 202). MOLA, the Mars Orbiter Laser Altimeter, produced a high-resolution global topographic map far better than any previous effort (Smith *et al.*, 2001). A shaded relief map derived from MOLA data forms part of Figures 3 and 4. This altimetric map also provided the best positional control then available for Mars, and this became the standard control for future maps. MOLA ceased operations on 30 June 2001 (MY 25, sol 384) after firing 700 million laser pulses, when its precision timing mechanism failed.

The magnetometer system (two magnetometers and an electron reflectometer) provided the first global map (Acuña *et al.*, 2001; Connerney *et al.*, 2005) of magnetic anomalies (Figure 154A). The Thermal Emission Spectrometer (TES) mapped surface composition, establishing the average basaltic composition and identifying interesting anomalies, especially a large deposit of hematite in Meridiani Planum which became the target for the Mars Exploration Rover Opportunity in 2004 (Christensen *et al.*, 2001). In addition, the Radio Science Experiment monitored radio signals to measure orbital velocity, from which gravity and crustal thickness maps could be derived, and occultations probed the atmosphere at the limb.

Paper maps had been produced by USGS from Mariner 9 and Viking data, but after MGS, most cartography of Mars moved into the digital realm. USGS created a global photomosaic of Viking images tied to MOLA control which could be manipulated via the Internet at www.mapaplanet.org, allowing users to create their own maps with customized scale, map projection and extent. The MOLA altimetry and the MOC global photomosaic were also made available on this site. The same data and many other images were also made available through 'Mars in Google Earth', a data layer in another popular online mapping tool from Google.

Mars Global Surveyor data revealed Mars to be more diverse, active and interesting than Viking had suggested. The magnetic anomalies (Figure 154A) suggested intense geological activity in the distant past. The hematite

Table 63. *Mars Global Surveyor Orbital Events*

Date	Orbit	Events
12 September 1997	1	Orbit insertion, 263 by 54 026 km, 44.99-hour period (MY 23, sol 371)
15 September 1997	3	Initial aerobraking period begins
8 October 1997	15	Solar panel flexing caused by drag noted, altitude about 115 km
12 October 1997	18	Orbit raised for assessment, 175 by 45 100 km, 35.4-hour period, data collection begins during hiatus
7 November 1997	36	Aerobraking period 1 begins (MY 23, sol 426)
27 March 1998	201	Science phasing period 1 begins, 175 by 17 800 km, 11.6-hour period, in order to position mapping orbit correctly. Data collection continues
29 April 1998	269	Solar conjunction (MY 23, sol 594)
27 May 1998	328	Science phasing period 2 begins, same orbit, data collection resumes on 2 June
7 August 1998	476	Phobos encounter (MY 24, sol 23)
19 August 1998	501	Phobos encounter
31 August 1998	526	Phobos encounter
12 September 1998	551	Phobos encounter (MY 24, sol 58)
24 September 1998	573	Aerobraking period 2 begins
4 February 1999	1284	Aerobraking period 2 ends, 377 by 454 km, 118-minute orbit (MY 24, sol 199)
19 February 1999	1473	Orbit adjustment, 368 by 438 km, 118-minute period
9 March 1999	1683	Premapping mission ends
9 March 1999	1	Mapping mission begins, orbits renumbered from 1 (MY 24, sol 231)
29 March 1999	247	High-gain antenna deployed, continuous mapping begins
3–30 May 1999		Geodesy campaign, global and colour mapping (north of 70° S)
June 1999		Mars Polar Lander landing site imaging
1999		Mars Surveyor 2001 site imaging
13–20 December 1999		Science campaign C, completion of global mapping south of 70° S
December 1999		Mars Polar Lander search
1 February 2000	8506	End of nominal mission, start of extended mission operations (MY 24, sol 551)
22 June–12 July 2000		Conjunction, imaging hiatus
2001–2004		Mars Exploration Rover site imaging
1–18 August 2002		Conjunction, imaging hiatus
30 June 2001		MOLA timing mechanism fails, end of altimetry data collection
2003–2006		Phoenix site imaging
8–24 September 2004		Conjunction, imaging hiatus
27 August–7 September 2005		MGS in safe mode, no data received
2005–2006		Mars Science Laboratory site imaging
18 October–2 November 2006		Conjunction, imaging hiatus
2 November 2006		End of mission (MY 28, sol 277)

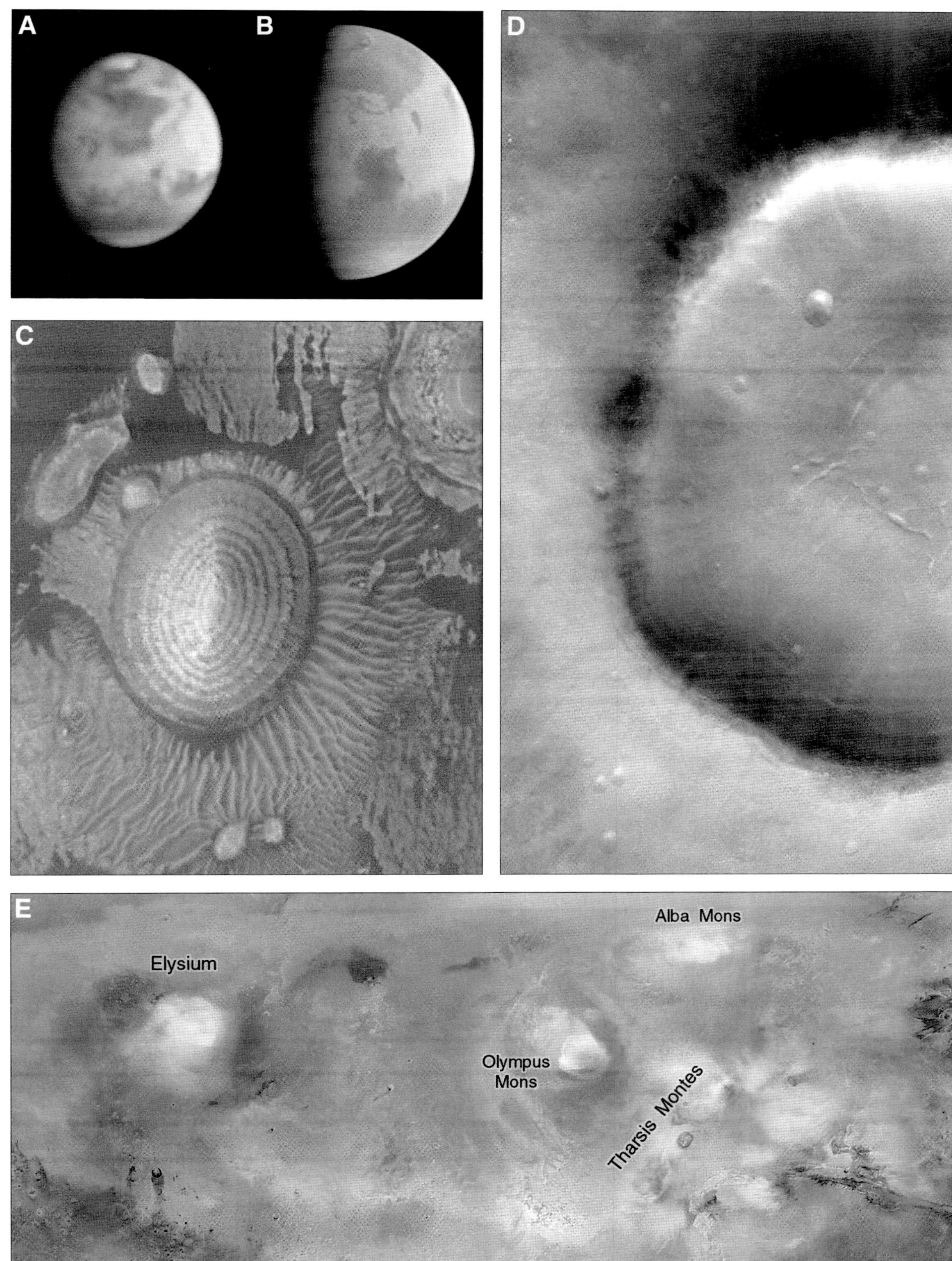

Figure 153 Examples of Mars Global Surveyor data.
A: Approach image taken on 2 July 1997 showing the Mars Pathfinder landing region from 17 million km. **B:** Approach image taken on 20 August 1997 from 5.5 million km. **C:** MOC image of layered sediments and sand dunes on the floor of a crater in Arabia Terra (MSSS release MOC2-1591, 20 September 2006). **D:** MOC image R1102651 of the same area of Tempe Terra shown in shown in Figures 51D, 188C and 190D. **E:** Part of a MOC wide-angle global weather map showing clouds over Elysium and Tharsis in April 1999.

deposits strongly suggested the action of water. Recent gullies on steep slopes hinted at surface water flow, and some were seen to be active during the mission. Fresh impact craters were observed to have formed during the mission, and polar terrains evolved before the gaze of the camera. Landing sites, past and future, were mapped in unprecedented detail, allowing realistic landing site certification and rover route planning for the first time. This was essential for the surface operations of the Mars Exploration Rovers beginning in 2004.

1996: **Marsokhod International Mission**

Lusignan (1996) described a possible international mission which would place a Russian Marsokhod rover on Mars. The 93-kg rover would be able to drive up to 800 m per day and could collect samples for later collection and return to Earth. A carrier spacecraft would deploy the rover in its landing vehicle and then enter orbit to map sites and serve as a data relay. It might also deploy penetrators. Nine landing sites for the Marsokhod rover were suggested and were said to have been 'identified by international scientists'. No details of site selection were given, but the list somewhat resembles the list of US/Russian penetrator sites listed in the Mars Landing Site Catalog (Table 47). The target ellipses for penetrator delivery were 150 by 250 km across, with smaller ellipses for the rover. The sites (Figure 154B) are listed in Table 64 with their approximate Landing Site Catalog equivalents. An ellipse illustrated by Lusignan (1996) on Aurorae Planum adjacent to Capri Chasma is about 40 by 120 km across at 9.7° S, 44.2° W (site G in Table 64).

Table 64. *Marsokhod Landing Sites from Lusignan (1996)*

Site	Name	Location	Catalog site
A	Candor Chasma	6° S, 74° W	72
B	Margaritifer Terra	18° S, 7° W	Near 75
C	Gusev	14° S, 183° W	112 or 138
D	Utopia Planitia	45° N, 250° W	65
E	Maja Vallis	20° N, 54° W	21
F	Promethei Terra	36° S, 248° W	–
G	Capri Chasma	10° S, 44° W	74
H	Meridiani/Aram Chaos	2° N, 17° W	77
I	Arabia Terra	28° N, 328° W	61

16 November 1996: **Mars 96 (Russia)**

Mars 96 was initially planned as Mars 94 but was delayed into 1996, displacing a subsequent rover and balloon mission from 1996 to 1998, which was later cancelled. Mars literature thus refers to two missions as Mars 96, creating a potential for confusion. The original Mars 96 mission with rovers and balloons is described later under the name Mars 98, its new designation before it was cancelled. Landing sites for this mission are listed in Table 65A, as given in the Mars Landing Site Catalog (Table 47). They lie around the Chryse-Acidalia lowlands. Four balloon sites were identified by Kazarian-Le Brun (1996), whose description assumed data relay by the Mars 94 orbiter and therefore clearly drew on studies much earlier than the publication date. All these sites are shown in Figure 155.

Mars 96 as flown consisted of a 3159-kg orbiter, two landers and two penetrators. A third lander, funded by NASA and carrying two US instruments, was considered for a time during mission planning. The objectives were to investigate the evolution of the planet's atmosphere, surface, ionosphere and interior. Some astronomical observations would be made during cruise as on several previous Soviet planetary missions. The spacecraft was launched at 20:49 UT from Baikonur and would have arrived at Mars on 12 September 1997 (MY 23, sol 371). It entered a 165-km parking orbit, but the upper stage burn to place it on a Mars trajectory failed. It re-entered the atmosphere at about 01:00 UT on 17 November and crashed in northern Chile or the Pacific Ocean.

The two small stations would be deployed before arrival, landing on the surface with instruments similar to those carried by the penetrators. Both the penetrators and the small stations were expected to last for a full Mars year. The landing sites were restricted to dayside latitudes between 20° S and 60° N, and specific sites are listed in Table 65B based on the websites of Malin Space Science Systems (MSSS) in San Diego, California, and the Institute for Cosmic Research (IKI) in Moscow. These reflect different stages of site planning, the IKI list

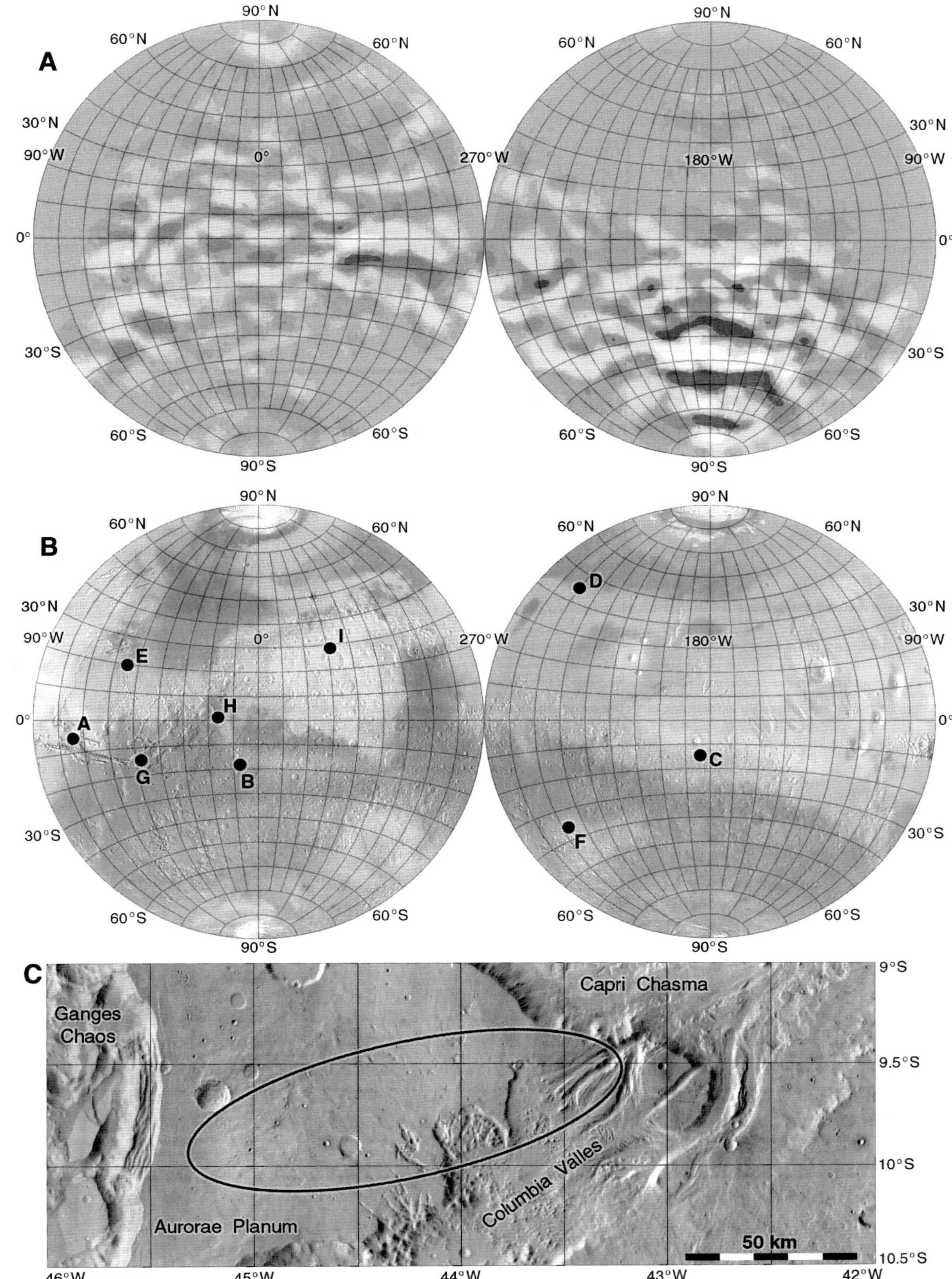

Figure 154 A: Mars Global Surveyor magnetic anomalies adapted from Connerney *et al.* (2005). The light and dark areas represent variations in the radial magnetic field, dark for negative values and bright for positive values. **B:** Landing sites for the Marsokhod International Mission (Table 64; Lusignan, 1996). **C:** Capri Chasma ellipse for the Marsokhod International Mission (Mars Odyssey THEMIS infrared mosaic).

Table 65. *Landing Sites for Mars 96*

65a. Sites Proposed for the Original Mars 96 Rover Mission

Lander type	Location	Notes
Penetrators	13° S, 4° W	Catalog site 89: South Terra Meridiani
	53° N, 85° W	Catalog site 90: North Tempe
	52° N, 357° W	Catalog site 91: Cydonia
	23° N, 336° W	Catalog site 92: West Arabia
Small stations	11° N, 151° W	Catalog site 93: South Amazonia
	30° N, 156° W	Catalog site 94: North Amazonia
	55° N, 80° W	Catalog site 95: North Tempe
	45° N, 348° W	Catalog site 96: Deuteronilus Mensae West
Balloons (Kazarian-Le Brun, 1996)	45° to 56° N, 12° to 40° W	Acidalia Planitia
	47° to 56° N, 248° to 260° W	Utopia Planitia
	38° to 43° N, 254° to 258° W 50° to 56° N, 173° to 183° W	Arcadia Planitia

65b. Sites Proposed for the Mars 96 Mission as Flown

Landers	Location	Notes	Source
Small stations	32.48° N, 169.32° W 41.31° N, 153.77° W	Arcadia Planitia, 600 by 120 km ellipses, azimuths 115° to 145°	MSSS website
Small station backup site	3.65° N, 193° W	Elysium Planitia	
Small stations	37° N, 162° W	Amazonis Planitia	IKI website
	33° N, 169° W	Amazonis Planitia	
Penetrators	36° N, 161° W	Amazonis Planitia	
	36° N, 125° W	Western edge of Alba Mons	
Alternative penetrator site	36° N, 245° W	Utopia Planitia	Surkov and Kremnev (1998)

being the final stage of planning prior to launch. Figure 155 shows those sites, clustered in the Tharsis-Amazonis region. Surkov and Kremnev (1998) identified an alternative penetrator location in Utopia Planitia (Table 65B). The landing ellipses were 240 by 30 km across.

After small station deployment, the orbiter would have been deflected from its Mars collision trajectory to enter a 500 by 52 000 km orbit with a period of 43 hours, an inclination of 106° and periapsis at 26° N. The orbiter carried a high-resolution camera provided by ESA, a near-infrared spectrometer for mineralogy studies, and ground-penetrating radar to observe subsurface layering and buried water or ice. The two penetrators would be deployed after the main spacecraft had entered orbit around Mars. Each would impact at high velocity, burying a forebody to a depth of 1 m or more while the aftbody remained on the surface, the two components attached by a cable. The forebody contained a gamma-ray spectrometer and a seismometer. The aftbody carried meteorology instruments, a camera and a transmitter.

Jacques Blamont of the University of Paris and the French space agency CNES suggested a possible alternative Mars 96 mission at the seventh Mars Science Working Group meeting held in Washington, DC, on 17 June 1992. This was intended to overcome serious budgetary problems in Russia at the time. French space agency

CNES and NASA would purchase some of the hardware to help fund the mission. Blamont suggested two landing sites as part of his proposal, at 10° N, 160° W and at 40° N, 150° W. Both sites are similar to locations chosen for Mars 94 and Mars 96 (Tables 61, 64, 65), and they are shown in Figure 155.

Mars Pathfinder Landing Site Selection

The Pathfinder landing site latitude had to be between 10° and 20° N, where the Sun would be overhead at the time of landing, and its elevation below 0 km based on Viking topography (Figure 157). The landing ellipse size for initial planning was 300 km by 100 km, later reduced to 200 km by 100 km. Matthew Golombek (JPL), the Pathfinder project scientist, organized a meeting on 18 and 19 April 1994 to discuss landing sites (Golombek, 1994a). The open nature of the site selection process became a model for many future missions. Participants suggested the 43 sites listed in Table 66 and plotted in Figure 156, many of them falling outside the acceptable latitude zone because the initial range had been 0° to 30° N.

These sites fell into three categories. At 'grab bag' sites, such as the mouths of large channels, fluvial deposits containing many different types of rock might be found within reach of the small Pathfinder rover. The original sources of the rocks would not be known, but future orbital remote sensing might help locate them. At large uniform sites of unknown origin, such as a smooth plains deposit which might be volcanic or sedimentary, lava or ash and deposited by wind or water, Pathfinder should be able to resolve the uncertainty. At large uniform sites of suspected or known composition, such as a lava flow, Pathfinder would confirm the interpretation and provide more refined composition data. Grab bag sites were most favoured. The Cerberus area was the most attractive uniform site.

All the sites in Table 66 were plotted on the USGS 1:15 000 000 scale geologic maps using the 300 km by 100 km landing ellipse, and those outside the engineering constraints were dropped immediately or moved to nearby locations within the constraints. A few additional sites matching scientific goals expressed at the meeting were added. This produced a shortlist of ten sites (Table 67a, Figure 157) which were ranked and discussed at a meeting of the Mars Pathfinder Project Science Group on 9 and 10 June 1994 (Golombek, 1994b). The prime and alternate sites (1 to 4 in rank order in the table) were chosen by voting. The remaining sites were rejected, and their order in Table 67a is not ranked.

In 1995 additional studies of radar and thermal inertia data provided more information on roughness and rock abundance. Several additional sites were examined (Table 67b, sites A, B, C, D). The final site certification was now undertaken, and Ares/Tiu, Tritonis Lacus and Isidis (B) were the three 'finalists' in this last ranking (Golombek *et al.*, 1997). Tritonis was safe but was rejected as its science potential was unclear. It was almost identical to Viking site A2 (Figure 45). Isidis was the smoothest site but appeared to have little chemically unmodified material to examine. Ares/Tiu, the grab bag site, was favoured as the most interesting location for a small rover and became the Pathfinder target. These sites are illustrated in Figure 158.

4 December 1996: **Mars Pathfinder (United States)**

Mars Pathfinder, the second of NASA's Discovery missions, was a technology demonstrator for the MESUR network mission and was originally called MESUR Pathfinder. Because MESUR needed many surface stations, an inexpensive and reliable landing system would be required. Mars Pathfinder tested a new entry, descent and landing system different from that used by Viking, involving airbags instead of thrusters for the final stage of descent. Despite a very limited budget, it managed to incorporate some science instruments as well, and a very small rover (the Microrover Flight Experiment) initially developed as an independent exercise. The scientific objectives were to perform atmospheric entry science, surface imaging, chemical analyses of rocks and regolith, material properties experiments and meteorology. Mars Pathfinder operated on Mars for 83 sols, greatly exceeding the originally planned 30-sol lander lifetime and 7-sol rover mission. The small rover, the first to operate successfully on Mars, traversed about 100 m in the vicinity of the lander. The rover's route is shown in Figures 165, 166 and 167, and its daily activities are listed in

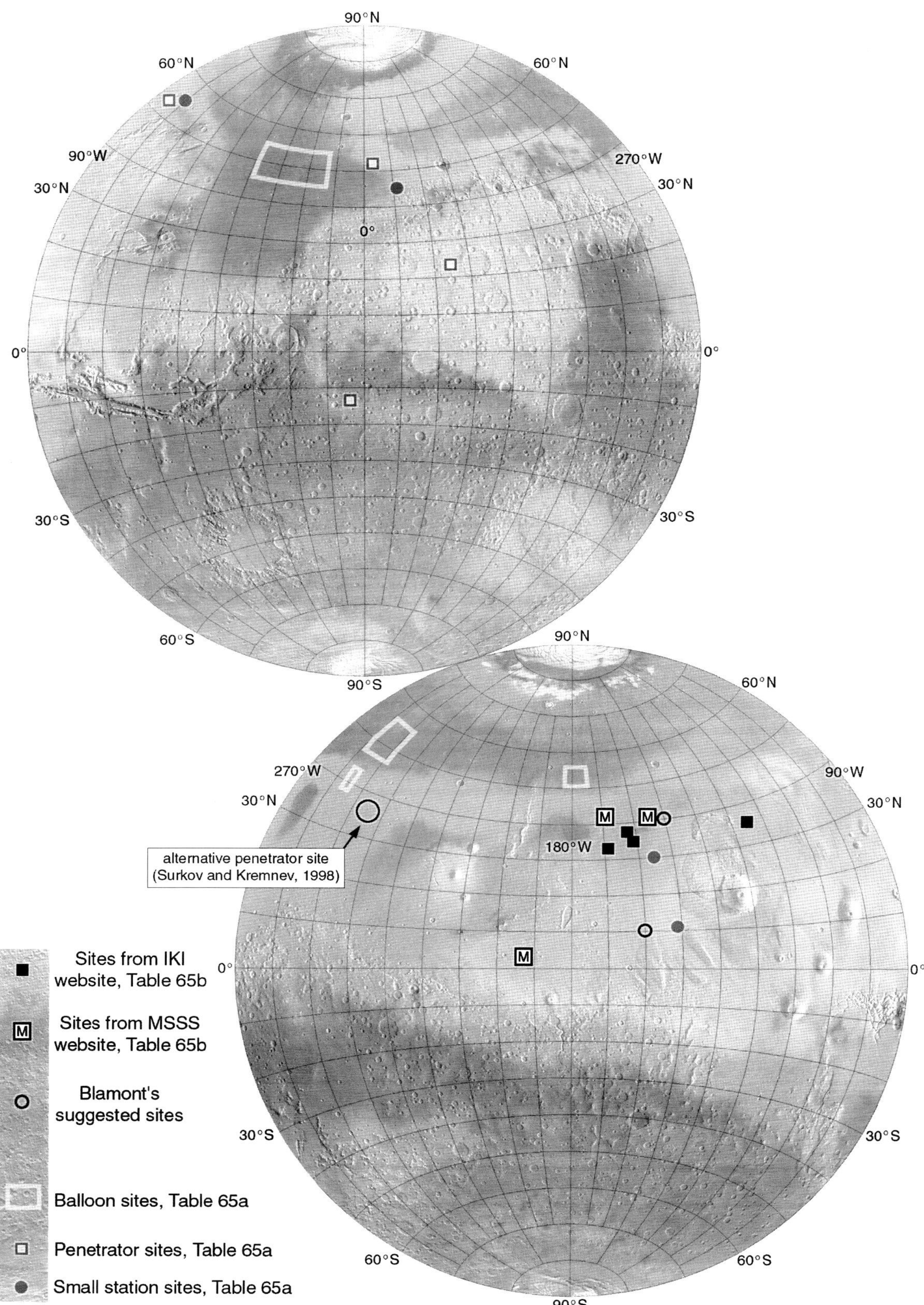

Figure 155 Mars 96 landing sites (Table 65).

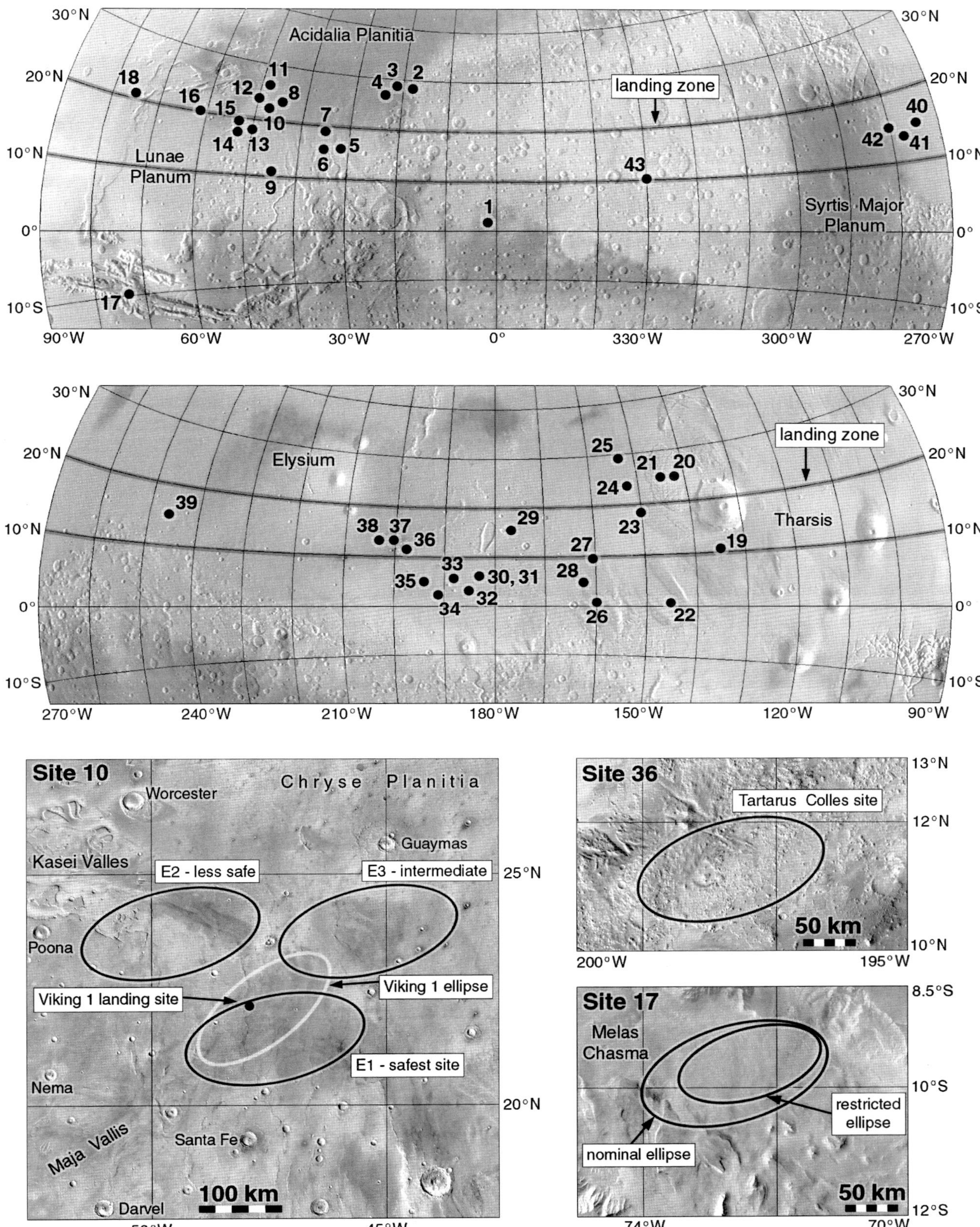

Figure 156 Mars Pathfinder landing sites suggested at the Pathfinder Landing Site Workshop (Table 66). Several specific ellipses illustrated at the workshop are shown below. The Chryse sites were presented and ranked for safety by Larry Crumpler; the other two were from Alan Treiman and Scott Murchie. The Melas site was too far south and could only fit the terrain if a smaller ('restricted') ellipse was approved.

Table 66. *Sites Suggested at the Pathfinder Landing Site Workshop, April 1994*

Site	Location	Notes
1. Mawrth – Ares area	2° N, 2° W	Between two valleys
2. Mawrth Vallis	28° N, 18° W	Plateau near the channel
3. Mawrth Vallis	29° N, 21° W	Valley mouth
4. Mawrth Vallis	27° N, 23° W	Fluvial deposits (Kuzmin rank 10)
5. Ares Vallis	16° N, 32° W	Meteorology site – ranked 3rd of 5
6. Ares – Tiu Valles	15° N, 35° W	outflow channel mouth, grab bag site
7. Ares – Tiu Valles	19.3° N, 35° W	Fluvial deposits (Kuzmin rank 1)
8. Chryse near Viking 1	23.5° N, 45.5° W	Extend Viking results
9. Hypanis Valles	11.5° N, 45.5° W	Fluvial deposits (Kuzmin rank 11)
10. Chryse – Viking 1	22° N, 47.5° W	Extend Viking results, examine VL1 or aeroshell
11. Kasei Valles	26° N, 48.5° W	Fluvial deposits (Kuzmin rank 9)
12. Chryse	23.5° N, 50° W	Near Viking 1, to extend Viking results
13. Maja Valles	18° N, 50° W	Fan at mouth of valley
14. Maja Valles	17.5° N, 53° W	Fluvial deposits (Kuzmin rank 8)
15. Maja Valles mouth	19° N, 53.5° W	Outflow channel mouth, grab bag site
16. Lunae Planum	20° N, 61° W	Ridged plains, widespread unit (Kuzmin rank 5)
17. Melas Chasma	9.75° S, 72.75° W	Too far south, needs smaller ellipse; view canyon walls
18. Kasei Valles	21° N, 75.5° W	Fluvial and eolian deposits (Kuzmin rank 2)
19. Tharsis Lavas	10.8° N, 135.2° W	Ejecta of SNC source crater?
20. Tharsis Lavas	24.8° N, 142.1° W	Ejecta of SNC source crater?
21. Lycus Sulci	25° N, 145° W	View of Olympus Mons scarp
22. Medusae Fossae	1° N, 146° W	Fluvial and volcanic materials (Kuzmin rank 4)
23. Amazonis Planitia	18° N, 150° W	Meteorology site – ranked 1st of 5
24. Amazonis Planitia	23.5° N, 152° W	Ejecta of SNC source crater?
25. Amazonis Planitia	29.5° N, 153° W	Ejecta of SNC source crater?
26. Medusae Fossae	1° N, 160° W	Fluvial and volcanic materials (Kuzmin rank 3)
27. Eumenides Dorsum	9.5° N, 160.5° W	Medusae Fossae
28. Amazonis Planitia	4° N, 162° W	Medusae Fossae Formation
29. Marte Valles	16° N, 177° W	Young outflow channel
30. Cerberus plains	6° N, 183° W	Volcanic (possible SNC source) or fluvial plains
31. Marte Valles	6.5° N, 183.5° W	Fluvial and volcanic materials (Kuzmin rank 6)
32. Elysium SE	3° N, 184.5° W	Climatology site
33. Cerberus plains	6° N, 188° W	'Recently drained ephemeral sea'
34. Marte Valles	2.5° N, 191° W	Fluvial and volcanic materials (Kuzmin rank 7)
35. Cerberus Rupes	5° N, 194° W	190–197, Meteorology site – ranked 4th of 5
36. Tartarus Colles	11.41° N, 197.69° W	Highland materials
37. Cerberus plains	13° N, 200.5° W	Dark mafic sand (box area)
38. Elysium	13° N, 203° W	Lava plains
39. Elysium Planitia	15° N, 246° W	Dark area. Meteorology site – ranked 5th of 5
40. Isidis	15° N, 275° W	Lake deposits?
41. Isidis Planitia	13° N, 278° W	Meteorology site – ranked 2nd of 5
42. Isidis	15° N, 280° W	Lake or volcanic deposits
43. Arago crater	10.2° N, 330° W	Fluvial and eolian deposits (Kuzmin rank 12)

Note: Kuzmin rank: priority assigned by Kuzmin *et al.* (1994). Meteorology sites: priority assigned by Seiff *et al.* (1994).

Table 67. *Pathfinder Site Shortlists, 1994 and 1995*

Site	Name	Location	Elevation	Notes
67a. Ranked List of Sites, 1994				
1	Ares/Tiu Valles	19.5° N, 32.8° W	–2 km	Prime landing site, grab bag site
2	Oxia Palus Dark Highlands south of Trouvelot crater	10° N–17° N, 11° W–20° W	Below 0 km	Alternative site, ancient highlands with dark eolian deposits. Edgett (1995) gives location as 12.4° N, 14.3° W.
3	Maja Valles Fan	18.8° N, 52° W	–0.5 km	Alternative site, grab bag site – fan too small for entire landing ellipse
4	Maja Highlands	13.5° N, 53° W	0 km	Alternative site, highlands with valley networks
5	Dark Hesperian Ridged Plains	14° N, 243° W		Large uniform site, added after workshop
6	Marte Vallis	17° N, 176° W		Large uniform site
7	Hypanis Valles	11.5° N, 45.5° W		Large uniform or grab bag site, poor images available
8	Isidis Planitia	15° N, 275° W		Large uniform site
9	Tartarus Colles	11.5° N, 198° W		Large uniform site
10	Elysium Lavas	13° N, 203° W		Large uniform site
67b. Additional Sites, 1995				
A	Tritonis Lacus	20° N, 252° W		Radar safe site, added after workshop
B	Isidis	18.2° N, 275.3° W		Very young plains
C	Isidis	16° N, 276° W		Possible alternative to B
D	SW Elysium	13.9° N, 232° W		Ridged plains, good images

Notes: If radar data revealed the prime site (1) to be too rough for a safe airbag landing, the alternative sites (2 to 4) would be considered. Sites 5 to 10 were rejected late in the analysis and are not ranked. Sites A, B, C and D are mentioned by Golombek *et al.* (1997) as additional locations studied in 1995 during the site certification process.

Tables 68 and 69. The lander returned 16 500 images and 8.5 million meteorological measurements, and the rover returned 569 images and 16 chemical analyses. Mars Pathfinder was the first planetary mission to take place as the World Wide Web underwent its rapid growth, and the daily rover activities and press conferences drew record web traffic. Previous missions such as Magellan had seen image and status report distribution to limited audiences through the text-based Usenet, but Mars Pathfinder initiated the process of wide public access to data and documents which revolutionized NASA's public relations in the next decade.

Jacques Blamont of the University of Paris and the French space agency CNES suggested a possible addition to Pathfinder at the seventh Mars Science Working Group meeting held in Washington, DC, on 17 June 1992. CNES would purchase from Russia one of the small stations intended to be flown on Mars 94. It had a payload of cameras, spectrometers, a seismometer and a magnetometer, but NASA instruments might be substituted. This would fly on the Pathfinder spacecraft and be deployed during the approach to Mars with no impact on Pathfinder itself. The idea was not adopted.

On the 30th anniversary of the Mariner 4 flyby of Mars (14 July 1965), NASA announced it had selected the name 'Sojourner' for its first Mars rover. The name was chosen after a global competition for school children run by the Planetary Society. They were asked to write an

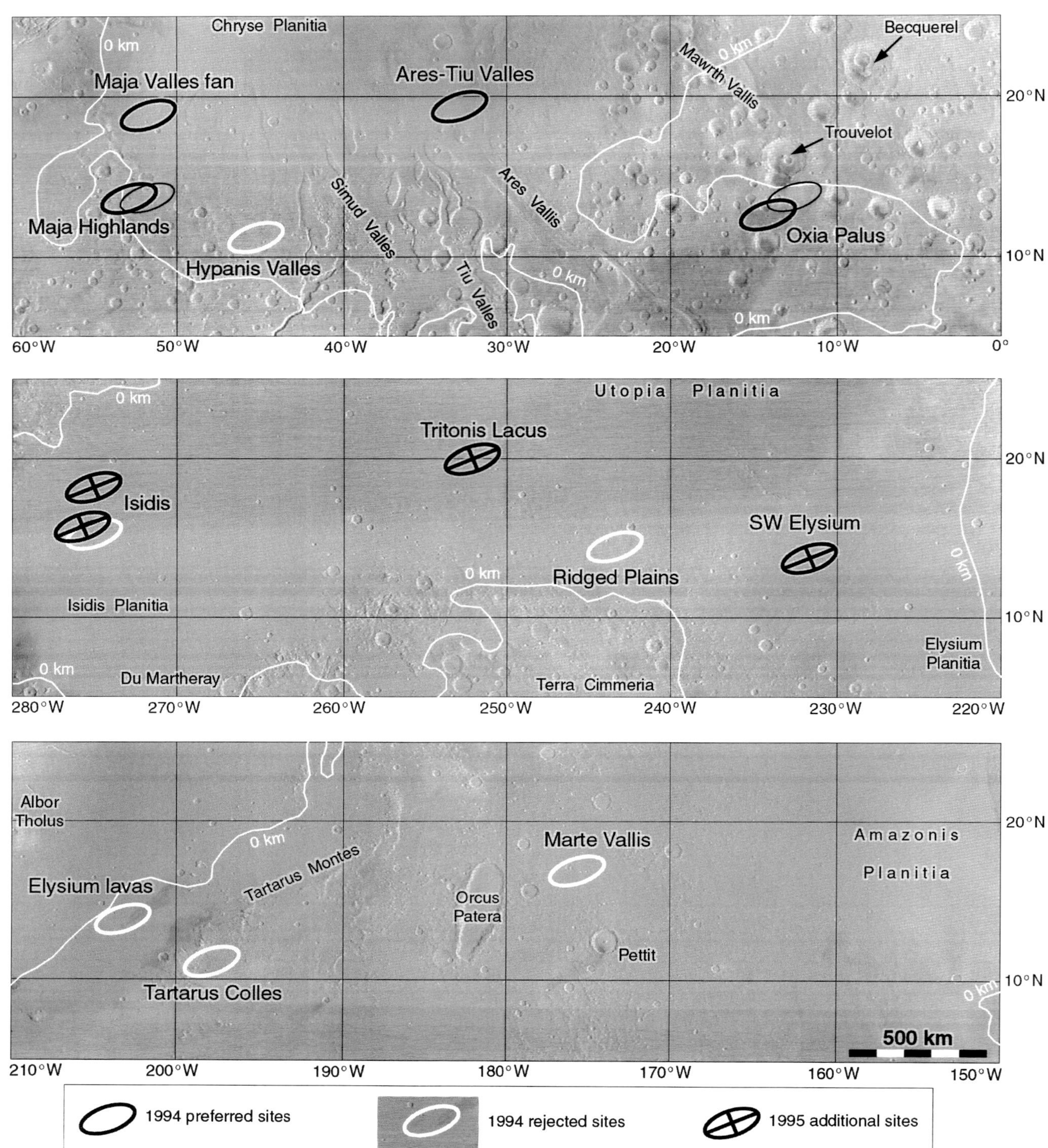

Figure 157 Shortlisted sites for Mars Pathfinder from Table 67.
The white line is the zero elevation limit for Pathfinder sites. Two alternative ellipses (thinner outlines) at Oxia Palus and Maja Highlands are shown at locations illustrated by Golombek *et al.* (1997).

essay about a historically significant woman. Valerie Ambroise, then 12 years old, of Bridgeport, Connecticut, won with an essay about Sojourner Truth (1797–1883), an African American abolitionist and champion of women's rights whose original name was Isabella Baumfree. The name Sojourner also means 'traveler'.

The Pathfinder lander was a tetrahedral structure which unfolded after landing, automatically orienting itself. The three 'petals' which surrounded the base after it opened, spanning 3 m, contained more than 2.5 m^2 of solar panels to generate electricity. The base carried a camera on a 0.8-m-tall mast, a meteorology mast with temperature sensors and wind 'socks', and the rover. The main task of the lander was to document the site in support of rover activities and to serve as the rover's data relay (http://nssdc.gsfc.nasa.gov/planetary/image/marsrover.gif). The rover was a 63- by 48- by 28-cm box, topped with a solar panel and antenna and supported on six wheels. The primary objectives of the rover were to demonstrate rover operations on Mars and to obtain chemical analyses (Foley *et al.*, 2003) of several rocks with an Alpha Proton X-Ray Spectrometer (APXS) initially designed for the Mars 94/96 mission. Earlier plans for rover operations included operating it on a tether and using it to deploy a seismometer or to deliver soil to the lander for analysis.

The following summary of the mission is based on Golombek *et al.* (1999), Edgett (1997) and the status reports released by JPL during the mission, though there are discrepancies between published accounts. Mars Pathfinder was launched from Cape Canaveral Air Station at 6:58 UT on 4 December 1996. A Sun sensor problem was overcome very early in the flight, allowing a successful cruise to Mars. Four trajectory correction burns were made during cruise, on 9 January 1997, 3 February, 6 May and 25 June. The spacecraft entered the atmosphere of Mars on 4 July 1997 (MY 23, sol 304) without going into orbit, unlike the Viking landers. Analyses just before arrival suggested Pathfinder might descend near the southwest end of the landing ellipse, possibly on a streamlined hill, but it landed only 23 km from the centre of the landing ellipse (Figures 158, 159).

Mars was monitored by the Hubble Space Telescope (HST) as Pathfinder approached the planet (Figure 141). Images on 17 May and 27 June were designed to look for dust storm activity, and the June images did observe yellow dust clouds in Valles Marineris. On 2 July, just before arrival, the canyon dust storm persisted and more dust activity was seen around the north polar cap. Mars Global Surveyor was approaching Mars at that time and also took images of Mars on 2 July (Figure 153). The north polar dust was also seen on 9 July by HST, but it did not spread to threaten Pathfinder. The approach trajectory was so accurate that a final course correction was not needed.

The lander was deployed from the cruise stage 30 minutes before entering the atmosphere. The heatshield braked the probe initially, after which a 12.5-m parachute opened to slow it further. Some atmospheric characteristics were measured during entry and descent. The heatshield was released after the parachute opened, a 20-m bridle (cable) was deployed below the spacecraft, and the lander separated from its backshell and descended to the bottom of the bridle. Shortly before landing, as indicated by a radar altimeter, a cluster of air bags inflated to protect the lander. At an altitude of 100 m, three small rockets in the backshell fired to provide the final braking, and 21 m above the ground the bridle was cut, dropping the lander with its protective airbags.

The lander, encased in its airbags, struck the surface about 500 m southeast of the backshell and 2 km north of a crater later informally named Big Crater, the largest in the immediate vicinity. It hit the ground at 16:57 UT on 4 July 1997, bounced at least 15 times over 2.5 minutes and rolled downhill, covering about 10 m in elevation and 1000 m horizontally in a northwesterly direction before coming to rest in a small hollow (Figures 159, 160, 163). Markings in the regolith suggest that in its final seconds of motion, the lander rolled from east to west across its final location, then stopped at the slight rise occupied by the 'Rock Garden', a cluster of larger rocks, and rolled back a few meters before coming to rest (Figure 163). The heatshield appears to have broken into several fragments which were scattered across the landscape and were later seen by the Mars Reconnaissance Orbiter's HiRISE camera (Figure 160). A bright object southeast of the lander and visible in surface images was initially thought to be part of the backshell (Golombek *et al.*, 1999), but Tim Parker has suggested that it is a thermal blanket from the airbags (Figures 159, 160).

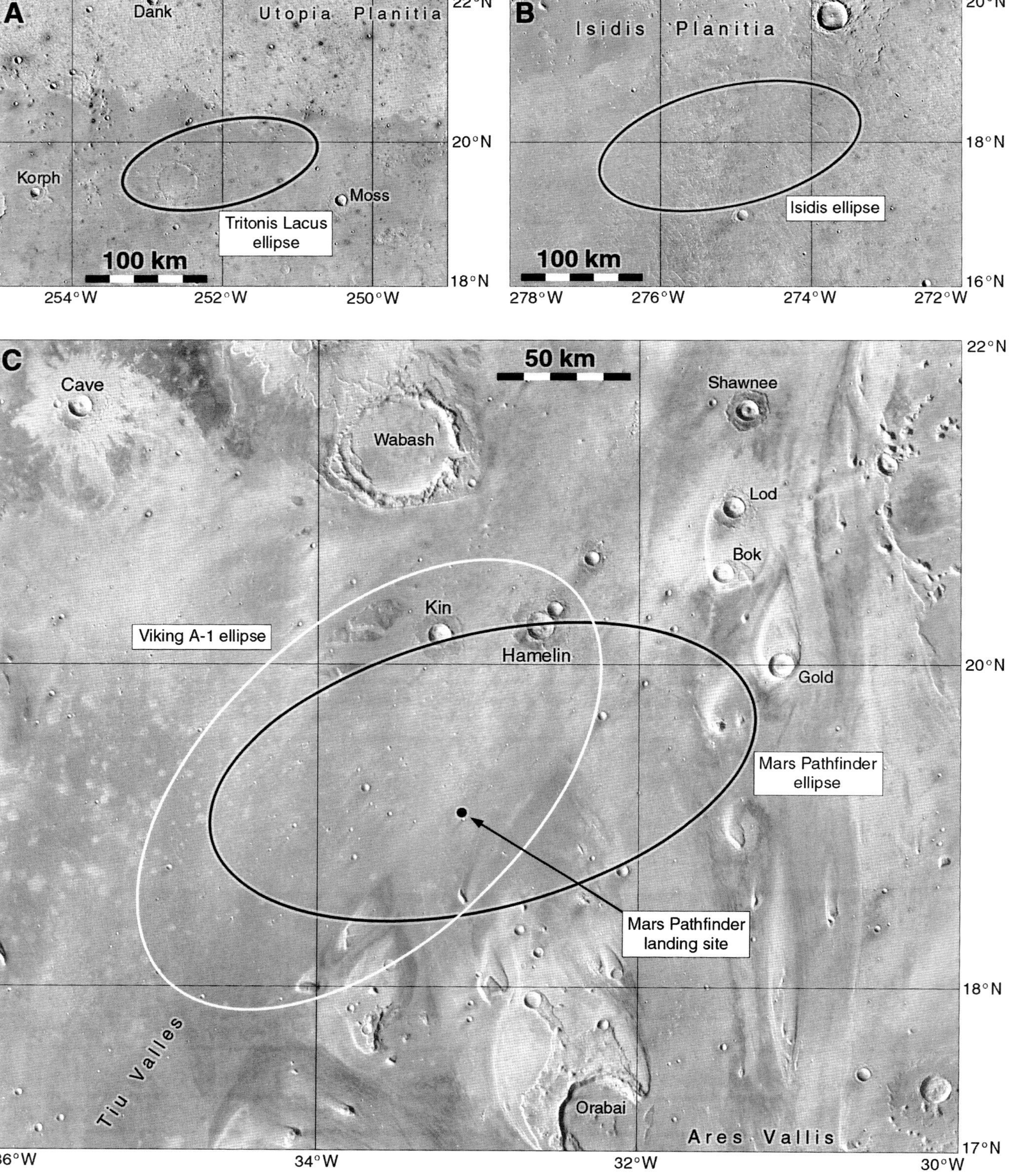

Figure 158 A: The backup Mars Pathfinder ellipse at Tritonis Lacus. **B:** The backup ellipse at Isidis. **C:** the primary Mars Pathfinder landing ellipse at Ares/Tiu Valles. The Viking A-1 ellipse (Figure 43) is also shown. The three Pathfinder ellipse sizes and locations are taken from Golombek *et al.* (1997). The images are all Mars Odyssey THEMIS infrared mosaics.

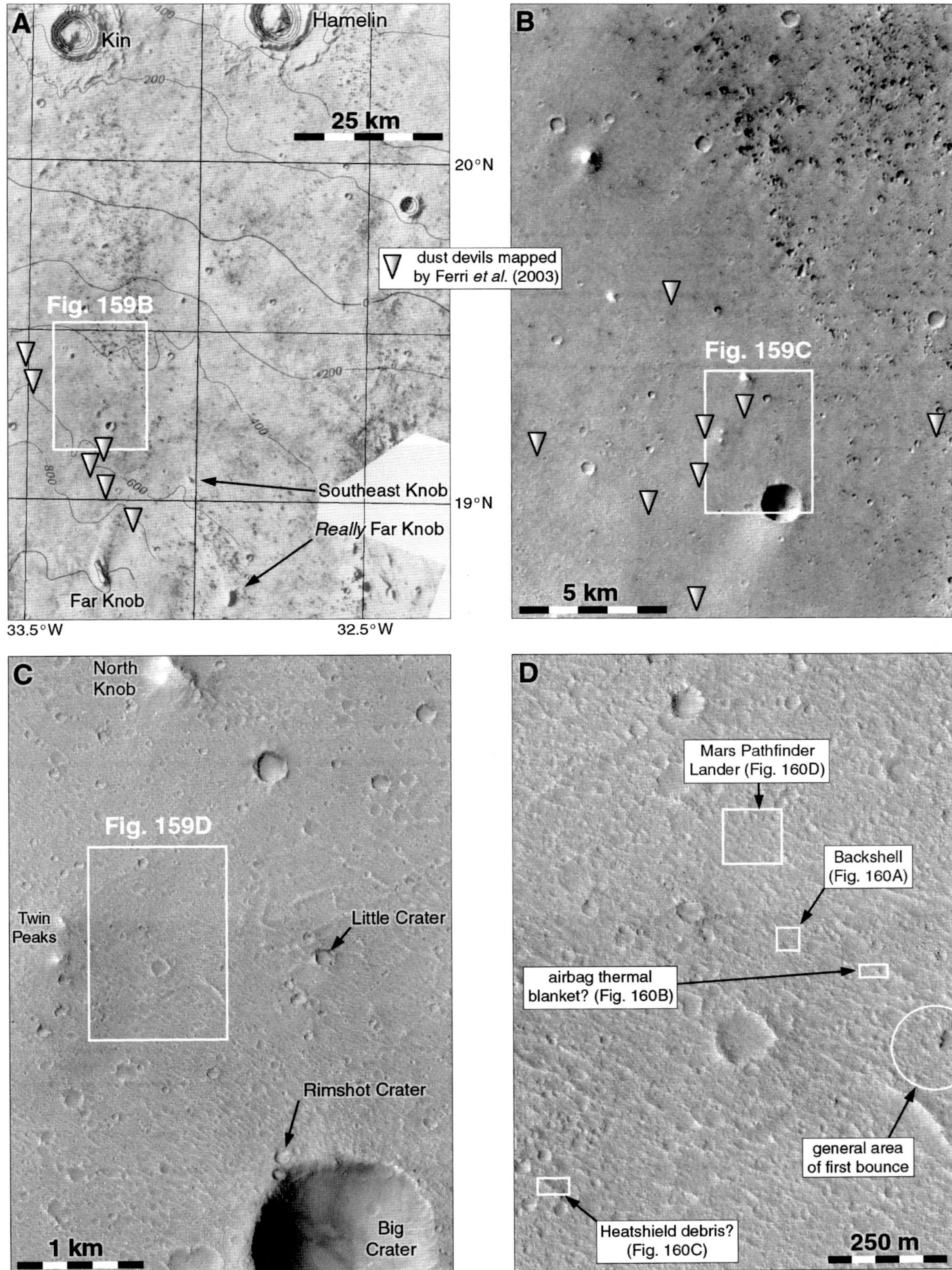

Figure 159 The Mars Pathfinder landing site.
A: detail of USGS map I-1069. **B:** Mars Odyssey THEMIS visible image V02610007. **C and D:** Mars Reconnaissance Orbiter HiRISE image PSP_002391_1995. Hardware elements shown in Figure 160 are indicated in D. In A and B the locations of dust devils mapped by Ferri *et al.* (2003) are indicated by triangles.

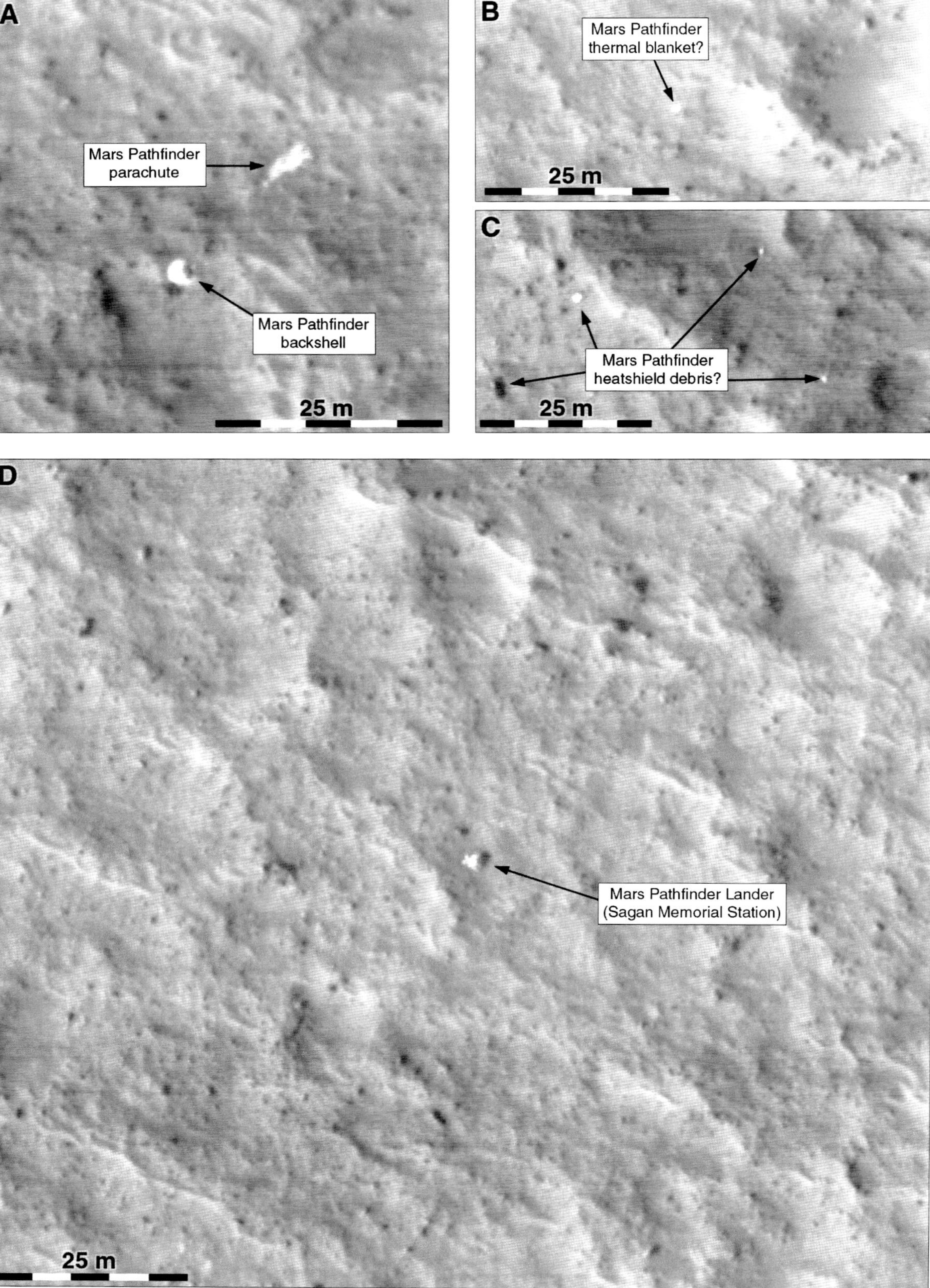

Figure 160 Mars Pathfinder hardware on the surface.
The image is Mars Reconnaissance Orbiter HiRISE image PSP_002391_1995.

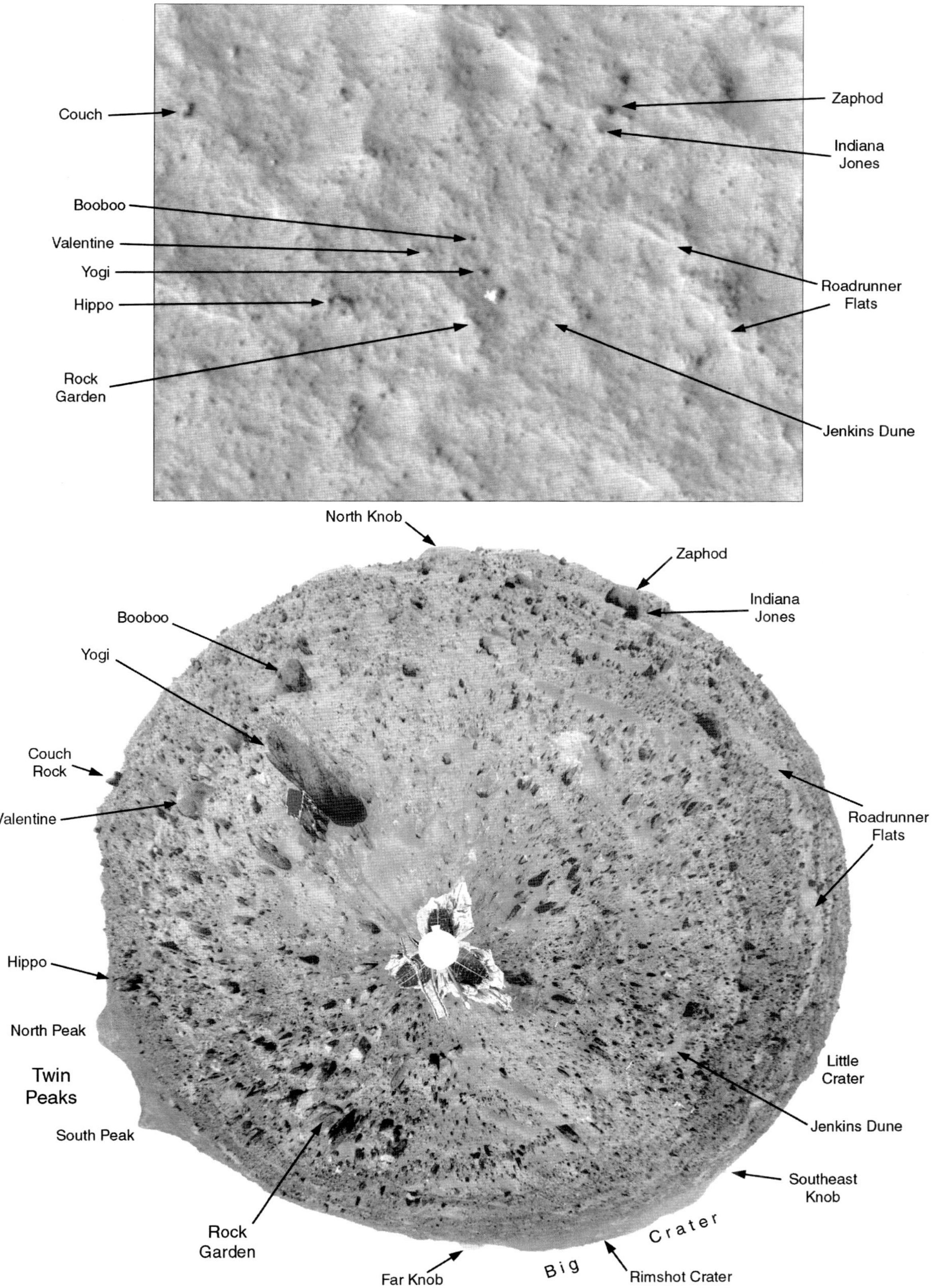

Figure 161 Comparison of HiRISE image PSP_002391_1995 with the Pathfinder panorama projected into a circle with exaggerated horizon relief.
Common features are labelled. More distant features are identified in Figure 159.

Figure 161 shows the area around the Mars Pathfinder lander, with many of the informal names given to rocks and other features. This is a reprojected version of the full panorama acquired later in the mission. The landing site was covered with rocks, providing an ideal area for the small rover to place instruments on a variety of targets. The site was exactly what had been predicted from imaging, radar and thermal data, a major triumph of the site selection process compared with the surprising nature of the Viking sites. Among the rocks were a few small drifts of dust and many well-defined wind tails and moats like those seen at Sponge rock by Viking 1 (Figure 63). The wind tails were aligned with streaks behind obstacles seen from orbit, suggesting prevailing winds blowing from the northeast.

The panorama itself is shown in Figure 162, and Figure 164 is the same panorama projected into map geometry with feature names added. So many informal names were given that not all can be shown here. The region closer to the lander is shown in more detail in Figures 165, 166, 167 and 173. Figures 165 to 167 show the route of the rover, with annotations showing the locations of the various experiments and the rover itself at each stop. Soon after landing, the spacecraft was named the Carl Sagan Memorial Station, commemorating the well-known planetary scientist and writer. Sagan had died on 20 December 1996, and a 100-km-diameter crater near the landing site was later named after him (Figure 49).

After landing, the lander deflated its airbags and retracted them, pulling them up against its tetrahedral body as much as possible. After 81 minutes on the surface, the panels ('petals') of the tetrahedron separated and folded flat against the ground, exposing three triangular solar panels, the small Sojourner rover and the lander's instruments. The landing took place just before dawn, and the lander and its Earth-bound team now waited for sunrise on Mars so the solar panels could start generating power. Engineering and atmospheric science data collected during entry and landing were then transmitted, and the camera was unlatched and rotated to find the Sun, allowing Earth's position to be predicted for high-gain antenna deployment.

The first images were taken soon afterwards, revealing parts of the lander and its surroundings. The initial view of the western horizon included two distinctive small hills, dubbed 'Twin Peaks' after a popular television series of the time. They were identified by Timothy Parker (JPL) on Viking images of the area, allowing the site to be located. Parker predicted that a crater rim would be seen southeast of the lander when the remainder of the panorama was transmitted at the end of the sol. The crater rim appeared in those new images as expected, and the site identification was confirmed. The landing site coordinates were 19.33° N, 33.53° W (Folkner *et al.*, 1997). The lander had touched down about 20 km southwest (downrange) of its target and tilted only 2.5° on the very uneven surface. Apart from Twin Peaks and the rim of Big Crater, three more distant hills were visible. North Knob was a small hill 2 km north of the lander. Far Knob, also called Misty Mountain, was a larger hill 30 km south of the lander, and Southeast Knob was faintly visible about 20 km away (Figures 159, 161).

The lander views showed the airbags had not retracted enough to deploy the rover's ramps, so the lander petal carrying the rover was raised and the airbags were retracted further. Subsequent images showed the ramps could now be deployed. Finally on sol 1, the meteorology mast was deployed, and the camera took and transmitted more compressed images from its pre-deployment position on the lander surface. When combined with the first set of images, these provided a full 'First Look Panorama' or 'Mission Success Panorama' for initial rover planning purposes. Sojourner obtained an APXS atmospheric measurement (referred to as A1) during this first sol while still on the lander.

At noon local Mars time on sol 2, the ramps were unrolled to provide a route off the lander for the small rover. One, facing south, was still prevented by an airbag from descending fully to the ground, but the north-facing ramp was ready to use. The rover was released and driven down the rear (north) ramp onto the surface at 14:40 local time, becoming the first wheeled vehicle on Mars (Figures 165, 168). A brief problem with rover communications was resolved, and Sojourner placed its APXS instrument on the soil to take a composition measurement (A2) over the following night. APXS was mainly used overnight to reduce thermal noise. The camera was deployed on its mast, 80 cm above the lander deck and 100 cm above the surface, late on sol 2 after taking images for a full 'Insurance Panorama', less compressed than the first, in case the deployment failed. These images had a lower priority than the science data taken in the following week and might not all have been transmitted, but in fact all were eventually received on Earth (Figure 173C).

Sojourner performed a soil mechanics experiment (called S1) on sol 3 by locking five wheels and spinning the sixth to see how easily the soil was eroded. Then it approached Barnacle Bill backwards, placing APXS on the rough-surfaced rock. Results of the overnight analysis (A3), announced the next day, showed a rock composition unexpectedly high in silica. On sol 4 Sojourner drove to a patch of soil in a moat-like depression near the large rock Yogi and placed its APXS (A4) on the ground at the end of the sol. Several experiments were performed earlier in the sol. These were a soil mechanics experiment (S2), a technology test of a wheel (T1), and the first wheel abrasion experiments at Barnacle Bill (W1) and near Yogi (W2). These were engineering tests involving the rotation of the right middle wheel of the rover, which held an aluminum plate with thin coatings of nickel, platinum and aluminium on a black substrate. As the wheel scuffed the soil, the amount of coating abraded by friction was estimated from changes in reflectivity as measured by a light sensor. Deimos was observed by the lander on sol 4, among other remote sensing targets. Nightime observations were possible early in the mission before the rechargeable batteries degraded, but after 30 sols they became very limited. During this first week a full stereoscopic 'Monster Panorama' was obtained from the deployed camera for rover planning.

On sol 5 the rover performed another wheel abrasion test (W3) on that soil, imaged Yogi at close range and placed the APXS on the soil again (A5). While attempting to deploy its APXS on the side of Yogi on sol 6, Sojourner approached the rock awkwardly, riding slightly up on it before sensing a problem and halting (Figure 168). Soil disturbed by the rover in the vicinity of Yogi exposed white material resembling a light-toned rock called Scooby Doo, which would be examined later. A communication problem and a lander computer reset on sol 7 delayed an attempt to back off Yogi, approach again and redeploy the APXS, but on sol 8 the rover backed away from Yogi. Pathfinder's receiver had been turned off to save power and was not turned back on to receive its instructions. On sol 9 the lander began the transmission of part of a large three-colour 'Gallery Panorama' of the landing site.

Sol 10 was 13 July 1997 (MY 23, sol 312). The panorama transmission was interrupted again by another computer reset, but despite this Sojourner was able to test APXS noise with the instrument in the air (A6) and then place APXS on Yogi. On the next sol the panorama images were received along with the APXS data (A7) for Yogi, which was a normal basalt with little silica. During the remainder of the mission, the high-quality multispectral 'Super Panorama' was obtained in sections and was 83% complete before the mission ended. Typical temperatures during this early phase of the mission, in the middle of northern hemisphere summer, were as high as 1° C during the day, dropping to –80° C overnight. The atmosphere was becoming dustier as winds carried fine particles from Valles Marineris, where the Hubble Space Telescope had observed a dust storm just before Pathfinder landed. On sol 12 Sojourner drove 3.6 m towards the white rock Scooby Doo and performed a wheel abrasion test (W4) (Figures 165, 168). The lander imaged a sunrise and monitored the set of three conical 'windsocks' hanging from brackets on the meteorology mast, as they did almost every day. Many images were taken in sequence to show their motion as they were moved by the wind, allowing estimates of wind speed and direction. On sol 13 the rover drove another 2.5 m towards Scooby Doo, stopping for soil mechanics experiments in cloddy soil (S3) and among small rocks at Cabbage Patch (S4), and the lander continued panorama imaging. Wheel abrasion test W5 was done at the start of sol 14 before Sojourner moved onto Scooby Doo. APXS was deployed on the rock (A8) at the end of the sol, and just before midnight, the lander imaged Deimos. A sudden darkening of the sky observed at 11:00 local time on sol 14 may have been caused by a nearby dust devil (Ferri *et al.*, 2003). Ground communication problems delayed receipt of sol 14 data on sol 15, but the missing information was received later.

On sol 15 Sojourner used its wheels to try to scuff or scratch Scooby Doo to estimate its hardness (S5) and tried to deploy APXS on the scuffed area, but missed and measured nearby soil instead (A9). Scooby Doo had a composition similar to typical soils but with more calcium and silicon. It might have been a piece of cemented regolith like some of the crusts seen by Viking, but it was too hard to scratch. Sol 16 activities were lost due to communication problems, which were

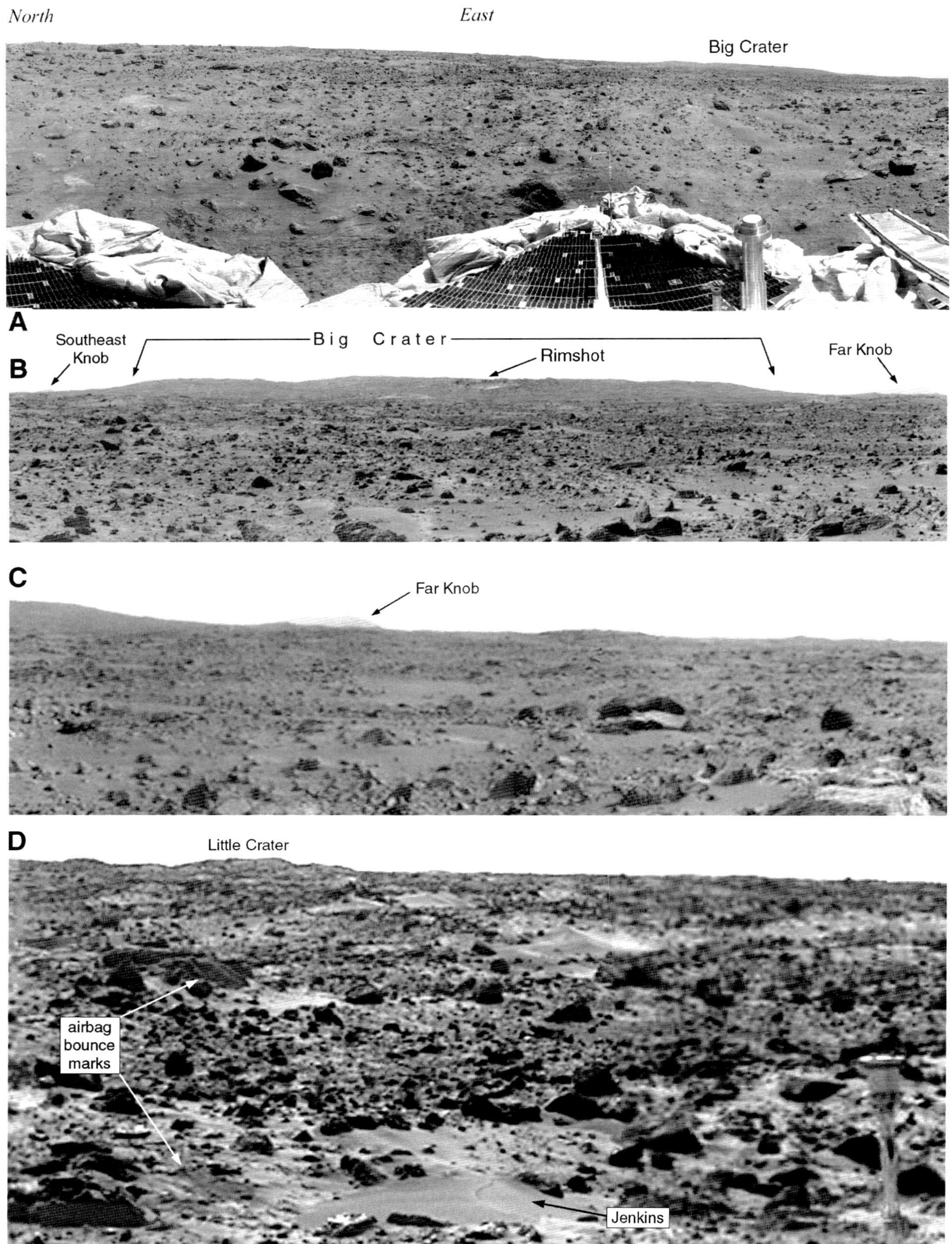

Figure 162 Mars Pathfinder images of the landing site.
A (across both pages): Full panoramic image. **B:** Rim of Big Crater. **C:** Far Knob on the southern horizon. **D:** Jenkins Dune and an airbag bounce mark. **E:** Twin Peaks. **F:** The Rock Garden. Images (B), (E) and (F) (NASA/JPL) were created by Tim Parker (JPL).

E

F

Figure 162 (continued)

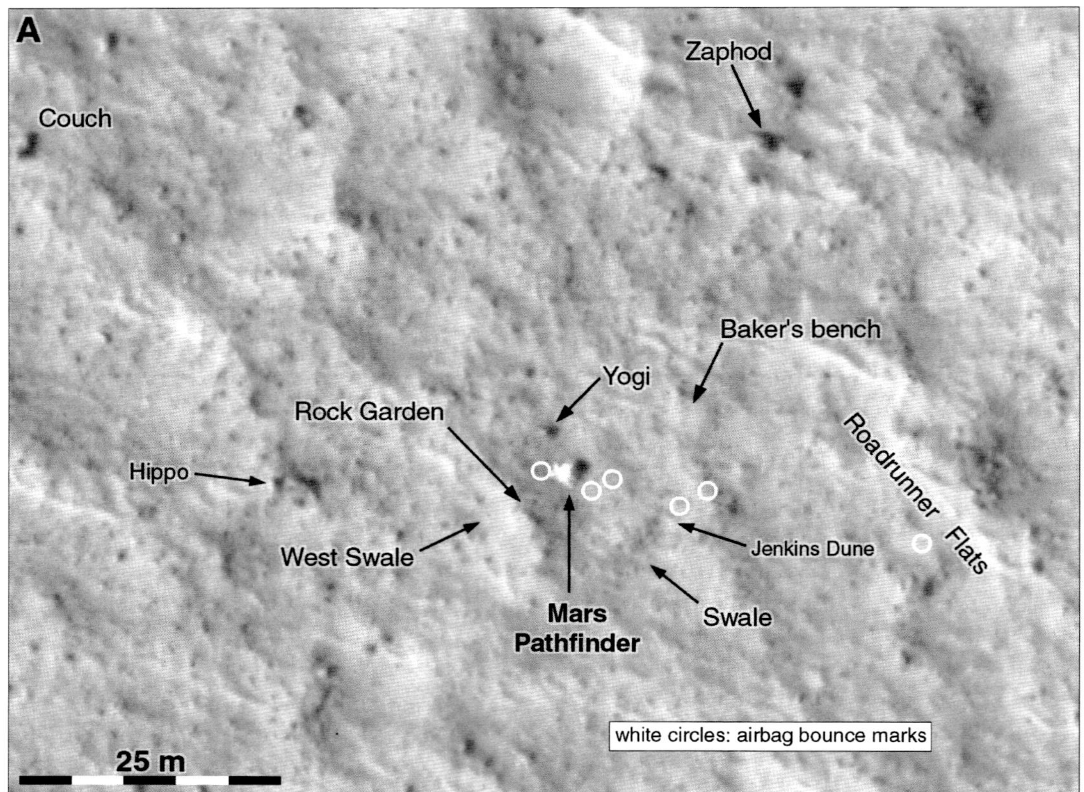

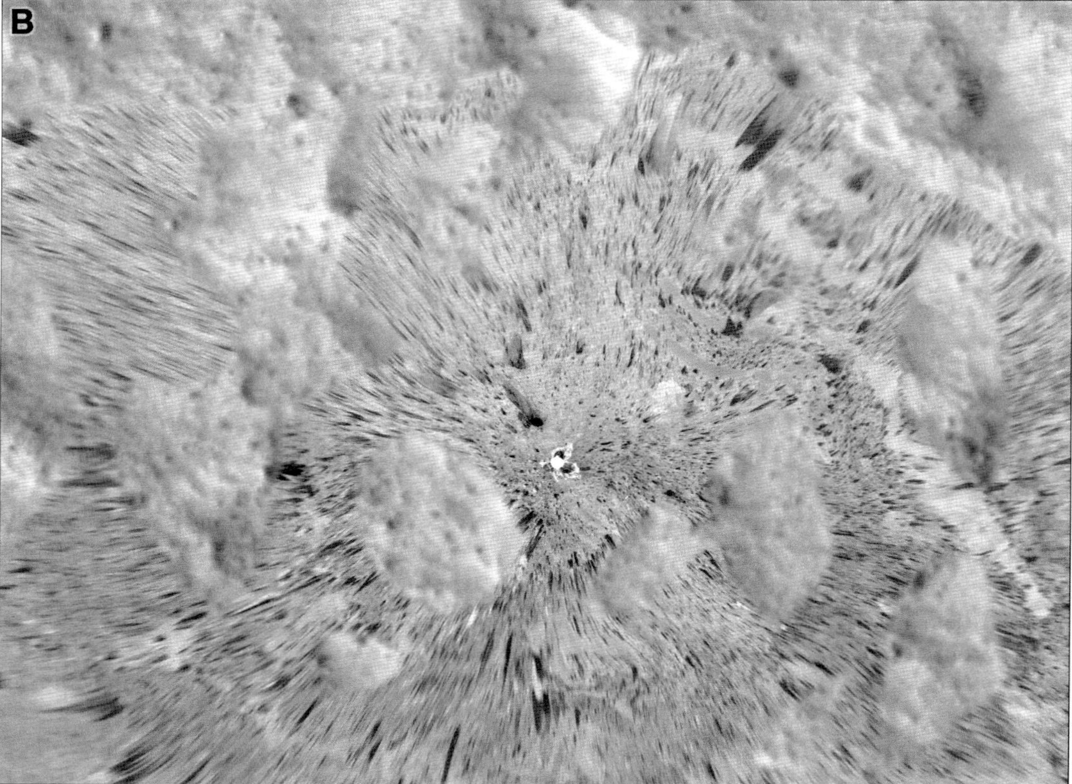

Figure 163 HiRISE image PSP_002391_1995 (top) and the surface panorama reprojected to fit it (bottom). The airbag bounce marks are from the surface panorama (Figures 161, 162) and are not visible in the HiRISE image.

resolved over the next sol. Some images taken before dawn on sol 16 and transmitted later were supposed to show the Earth in the sky, but it was hidden by clouds, so movies were made on several later sols to show the clouds moving across the sky. Atmospheric pressure reached a minimum at about this time in mid July 1997, then rose as the southern polar cap began to sublimate. Sojourner moved back to Cabbage Patch on sol 18 where it conducted soil mechanics (S6, S7) and wheel abrasion (W6) tests (Figure 169). There was no rover activity on sol 19.

On sol 20, or 23 July 1997 on Earth (MY 23, sol 322), the rover performed some more technology tests on the wheels (T2) and then moved to a patch of dark soil near the rock Lamb to APXS on the ground to collect data overnight (A10). Sol 21 included soil mechanics (S8), technology (T3) and wheel abrasion (W7) tests and a 3-m driving experiment in which the rover found its way around a cluster of rocks via two pre-identified way-points, approaching the next rock target, Souffle, using its own hazard-avoidance system. The drive to Souffle was completed on sol 22, but the rover climbed onto the low rock and did not deploy its APXS properly, taking data in the air (A11). On the next sol, Sojourner undertook its longest drive to date, about 6 m, skirting the lander and numerous rocks (Figures 165, 169), and also performed soil mechanics (S9) and wheel abrasion (W8) tests. The soil experiments had revealed that much of the surface material was as fine as talcum powder, notably a very soft fine-grained soil near Casper on sol 23 (S9). The lander observed the sunrise and the small moon Phobos and transmitted parts of two of the panoramas it had been taking during the mission. On sol 24 the lander again imaged sunset, Phobos, the windsocks, and the sky, looking for clouds. Sojourner drove around a group of rocks, including Calvin and Hobbes, and then back towards the lander. It was supposed to approach Mini-Matterhorn, a pyramidal rock near the lander airbags, imaging the rock and the lander nearby, but that sequence and a soil mechanics test were aborted by a computer fault. APXS made three tests of instrument noise (A12, A13 and A14).

The images of Mini-Matterhorn intended to be taken on sol 24 images were postponed until sol 25. Another rock, Squash, was also supposed to be imaged by the rover on sol 25, as its lumpy surface suggested to the geologists that it might be a conglomerate, a sedimentary rock containing fragments rounded during transport in water. The landing site had been regarded as a flood deposit since the time of Viking, and these images might help confirm the hypothesis. Unfortunately this drive and its imaging sequence were again interrupted, and the rover just turned in place. For the lander, sol 25 was devoted largely to meteorology. Water ice clouds were observed on many mornings at this site, dissipating later in the day. The atmospheric pressure fluctuated frequently, sometimes several times a day, as weather systems blew over. Shorter-term fluctuations were attributed to possible dust devils, one of them on sol 25, though at this stage none had been noticed. However, later analysis of images taken for the Gallery Panorama on sols 10 and 11 showed five very faint dust devils, all seen around noon local time (Metzger *et al.* 1999), and more were described by Ferri *et al.* (2003). The approximate locations of dust devils observed around the landing site are shown in Figure 159.

Sojourner resumed driving on sol 26, heading towards the dusty drift Mermaid, which was sometimes called Mermaid Dune, though it was not strictly a sand dune. Its left front wheel jammed late in the drive, probably because a small rock became stuck in the wheel. The rover backed up in a pre-programmed hazard-avoidance sequence and freed the wheel. The drive included a turn to image Mini-Matterhorn. Sojourner turned, moved slightly closer to the rock, took its images, backed up, turned onto its original heading and moved on. The science team released sunrise and sunset images and described ice crystal clouds seen early each morning which evaporated as temperatures increased. On sol 27 the rover performed two soil mechanics tests (S10 and S11) before driving to Mermaid. Sol 28 began with a combined wheel abrasion and technology test (W9 and T4), and later the APXS was placed on Mermaid for analysis (A15). Communications were interrupted on sol 28, but data from that day were stored and transmitted on sol 29. Also on sol 29, Sojourner conducted a soil mechanics test (S12) on the drift, imaged the resulting trench and then left Mermaid (Figures 166, 170), but its drive was cut short by a tilt sensor error. NASA's Planetary Data System assigns the sol 29 activities to sol 30, but the description here is based on Golombek *et al.* (1999).

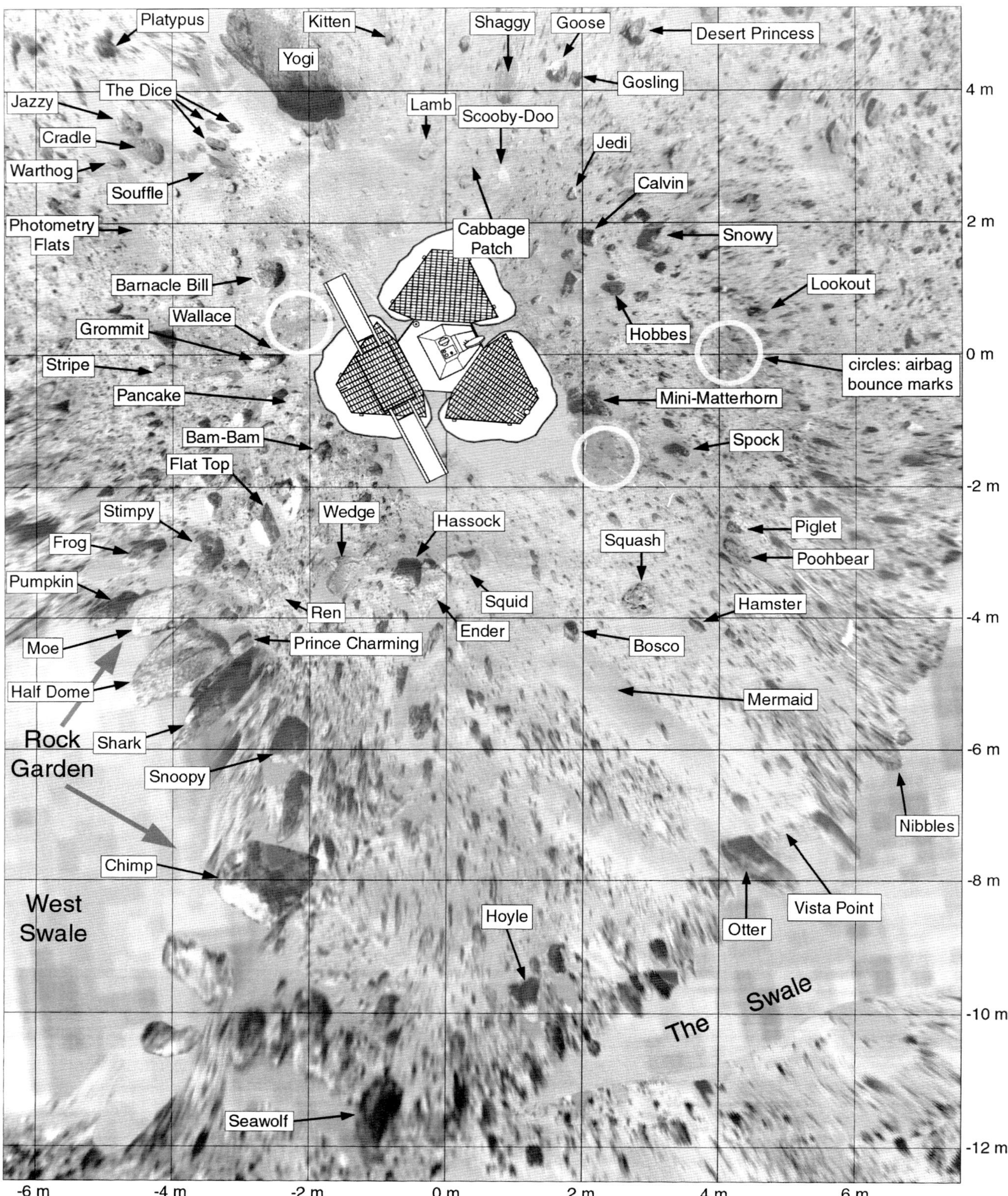

Figure 164 The Mars Pathfinder landing site mapped by reprojecting the surface panorama, using the HiRISE image for geometric control, with informal names for rocks and other features.
The large white circles indicate dark marks in the soil made by the airbags during landing. The lander bounced from east to west, rolled over the rocky area north of the Rock Garden (large rocks between Flat Top and Chimp) and settled in a low point northeast of the Rock Garden.

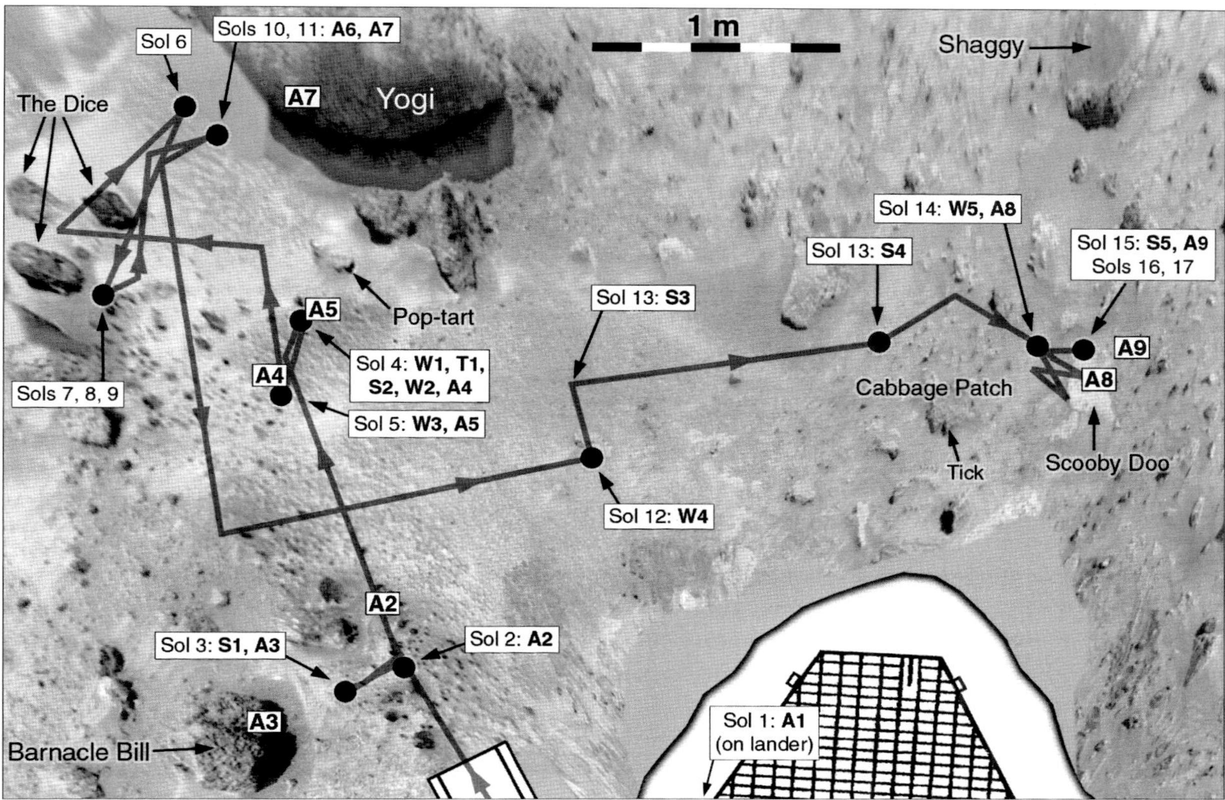

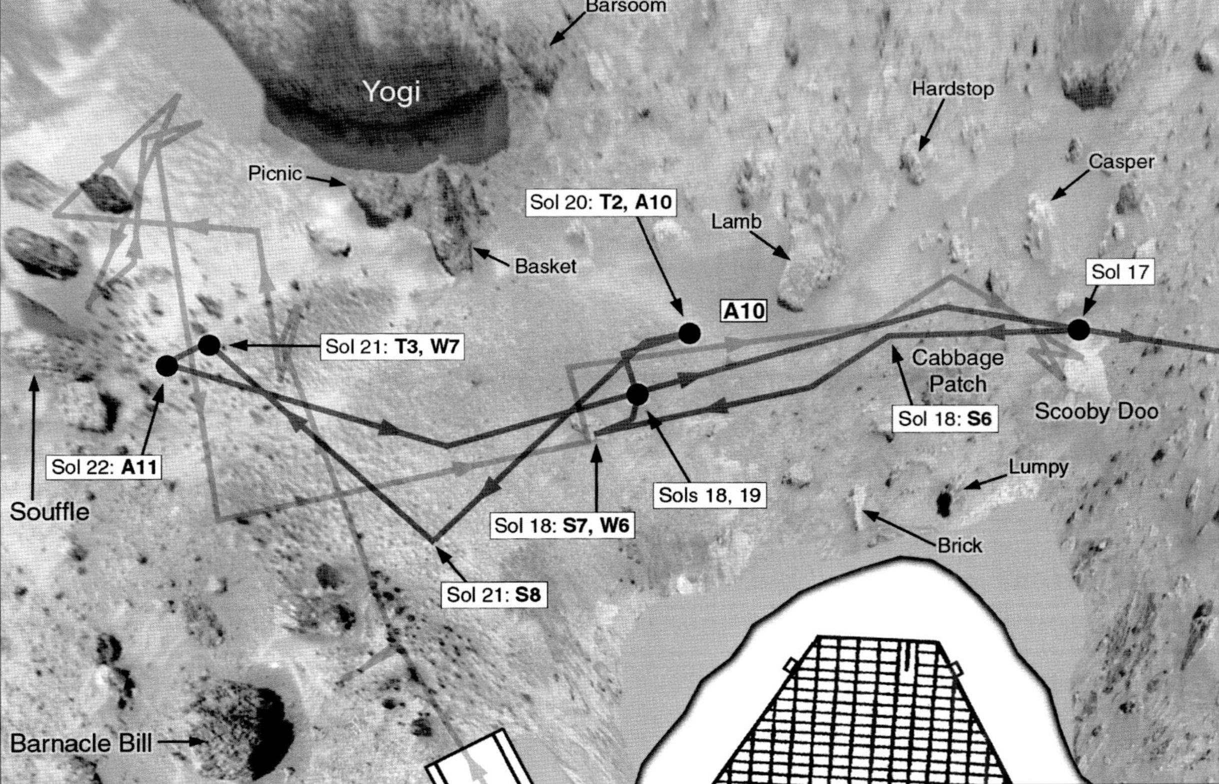

Figure 165 Sojourner rover route and activities up to sol 23.
Black circles show the end of sol locations. Rectangles with A-numbers show the locations of surface APXS measurements. Other experiments, including atmospheric APXS analyses, are indicated in text boxes.

Sol 30 was 3 August 1997 on Earth (MY 23, sol 332). The primary mission of Pathfinder was now finished, and all goals had been fulfilled. The lander and rover were shut down for two days over sols 30 and 31 to allow the lander batteries to recharge to the maximum extent possible. The rocks analyzed by APXS so far had revealed high levels of sulphur, suggestive of dust contamination or chemical weathering, and the degree of mineralogical variability was uncertain. Now the rover would be driven towards the large rocks of the Rock Garden in the hope of finding dust-free surfaces on their vertical faces. On sol 32 operations resumed with a drive northwards off Mermaid Dune and around its western end towards the Rock Garden, with a soil mechanics test (S13) at the end of the drive. The lander was busy collecting data for the Super Panorama throughout this period. On sol 33 the rover attempted to drive to a rock called Wedge, from its shape. It drove south, turned and drove north again to bypass other rocks. The turn angle was too small, probably because of interference by small rocks, and the rover climbed up onto rocks near Hassock. If Sojourner's wheels climbed onto a rock, its sensors usually detected excessive tilt and stopped the drive. There was no drive on sol 34, as the team examined images and decided on a new route.

On sol 35 the rover drove back off the small rocks, turned west and skirted the hazardous area, passing through a narrow gap between Wedge and Hassock. It struck Hassock, which sat loose on the ground rather than being embedded like many of its neighbours, and rotated it (Figure 170), but completed its drive and stopped on the north side of the rock. Sol 36 involved rover imaging of nearby rocks, and a turn which resulted in a wheel climbing onto Hassock. On sol 37 it turned again and approached Wedge, placing its APXS on the rock for a measurement (A16) which continued into sol 38. Then on sol 39 it retracted the APXS and moved back, performing a wheel abrasion experiment (W10) and engineering tests of the wheels (T5 and T6). As the rover backed off Wedge on that sol, it turned and struck the rocks Tigger and Hassock, moving Hassock again until its tilt sensor stopped the drive (Figure 171).

Several problems, including communications, limited Deep Space Network (DSN) time and computer resets, combined to leave Sojourner in this area, with only small moves to the west on sol 40 (13 August 1997, MY 23, sol 342) and sol 42. As it turned south towards the Rock Garden on sol 43, it climbed up onto rocks on the west side of Wedge and stopped moving when the tilt sensor indicated a problem. Images returned on sol 44 showed its precarious position (Figure 171). Careful driving eventually freed the rover from its perch, first with small turns and movements on sols 45 and 47, and then on sol 49 with a drive down off the rocks and over to the Rock Garden at last. Communication problems on sol 45 and lack of DSN coverage on sol 46 had contributed to the delay.

Lander images on sol 50 (24 August 1997, MY 23, sol 342) confirmed Sojourner's new location adjacent to Shark rock. Sol 51 was another day with no planned data transmission. On sol 52 Sojourner began its APXS analysis of Shark (A17), which was found to be silica-rich like Barnacle Bill. On the next sol it completed the overnight APXS measurement, performed a combined wheel abrasion and engineering test (W11 and T7) and drove to the nearby rock Half Dome, climbing partly onto the rock before stopping as it tilted past a preset limit. APXS measurements were made on two different parts of Half Dome to try to estimate the contribution of dust to the composition data. The first (A18) started on sol 54, but on sol 55 the rover position was adjusted slightly, and the A18 data collection continued. Sol 54 included a joint test between the lander and Mars Global Surveyor, which was approaching the planet. Near-simultaneous transmissions could be used for very precise tracking, a process which might increase the accuracy of future landings.

The rover's battery failed on sol 56, limiting activity to the hours of daylight when its small solar panel could generate power. Nevertheless, on that day it retracted APXS from its first deployment site, moved to its second site on Half Dome and deployed the instrument again for analysis A19. Also on sol 56 the lander imaged Phobos and continued its regular remote sensing work, building the large panorama and monitoring the rover, the Sun and the wind socks as it did almost every day. There was no rover command transmission from Earth on sol 57 and no transmission from the rover on sol 59, illustrating the operational difficulties that affected this mission, but sol 58 was successful and allowed the second APXS data collection (A19) on Half Dome to continue.

On sol 60, or 3 September 1997 (MY 23, sol 363), the APXS was retracted and redeployed at the same location to continue data collection, which was now only possible during daylight hours. After another day without any rover activity on sol 61, the A19 analysis continued on sols 62 and 63. The APXS was retracted and redeployed on the next rock, Moe, on sol 64 for two sols of data collection (A20), followed by an atmospheric APXS session (A21) on sol 66. Another atmospheric dataset (A22) was collected by APXS on sol 67 after it failed to deploy properly on the rock Stimpy, as the rover had climbed too high during its approach to the rock. On sol 68 it moved back slightly and successfully deployed the APXS for the last analysis in the Rock Garden (A23). The site was a nearly dust-free corner of Stimpy (Figure 172). Data collection continued on sol 69. The lander imaged Phobos very early in the morning on sol 69, one of its rare overnight observations late in the mission.

Sojourner was now given a new task, to drive out of the Rock Garden and look behind it. It moved away from Stimpy on sol 70 (14 September 1997 or MY 23, sol 373) and turned to conduct a soil mechanics experiment. Because it inadvertently climbed onto Stimpy, its left rear wheel was not touching the ground, so the attempted soil dig became an engineering experiment (T8), driving the wheel without ground contact. On sol 71 it passed Shark and took images of Shark and Half Dome. The lander transmitted atmospheric data, indicating that two dust devils had passed over it, though nothing was seen by the cameras. In mid September, the late northern summer season, daytime temperatures reached as high as –9° C and fell as low as –75° C. Then on sol 72, the rover left the Rock Garden, turned south and took images of Chimp, a rock at the eastern end of this cluster of large rocks. Sol 73 was lost to communication difficulties, but on sol 74 closer images of Chimp were obtained, and a soil mechanics experiment (S14) was performed. Sojourner ended that sol behind Chimp, at the most distant point of its traverse, about 12 m from the lander's exit ramp (Figures 167, 172). Here it could look into two depressions called The Swale and West Swale, the latter possibly a degraded impact or secondary crater behind the Rock Garden.

Images taken on sols 75 and 76 showed these depressions and the Twin Peaks beyond them on the horizon. Large drifts were found in West Swale, hidden from the lander's camera by the ridge which formed the base of the Rock Garden. After taking those images the rover returned to Chimp later on sol 76, but stopped short apparently because a rock was caught in its left rear wheel. An atmospheric APXS dataset (A24) was acquired on the same sol, another on sol 77 (A25) and a third on sol 78 with the APXS just above Chimp (A26). Finally on sol 79 the APXS was deployed on Chimp for the last analysis of the mission (Figure 172).

APXS data (A27) were collected from Chimp on sols 80 (24 September 1997 or MY 23, sol 383), 81 and 82. The last successful transmission of data was on sol 83, or 27 September 1997 on Earth (MY 23, sol 386). The rover was supposed to end its activities at Chimp and drive back towards Mermaid, but Mars Global Surveyor activities early in its aerobraking operation took precedence at the DSN. Brief transmissions were received on sol 87 and sol 90, indicating that the lander was operating on solar power, but its battery had failed. On sol 90 the rover would have started to drive towards the lander and begin circling around it. The last brief transmission was sent on sol 93 (MY 23, sol 396). Despite repeated attempts to communicate, nothing further was heard, and all efforts ended on 10 March 1998. The end of the mission prevented meteorological observations which might have helped plan the Mars Global Surveyor aerobraking campaign. If it had survived beyond this, the hope had been that the lander could last for a year, monitoring the weather and looking for surface changes like those seen by Viking 1.

After analyzing Chimp, the rover was to be driven back to the lander as quickly as possible, counterclockwise and a little farther out from the lander than on its clockwise journey, without any stops along the way. It would then use its APXS on magnetic targets on the ramp near Barnacle Bill to reveal the composition of the magnetic component of the fine airborne dust. The rapid traverse would provide experience with long-range driving of small rovers that might be useful for Mars Surveyor 2001, whose larger rover might drive for more than 10 km during a year-long mission. After that APXS measurement, the rover would take high-resolution images of Barnacle Bill which had been omitted earlier. After that it might have returned to the Mermaid Dune area or headed north in an attempt to reach the nearby horizon, where it would have a better view of North Knob

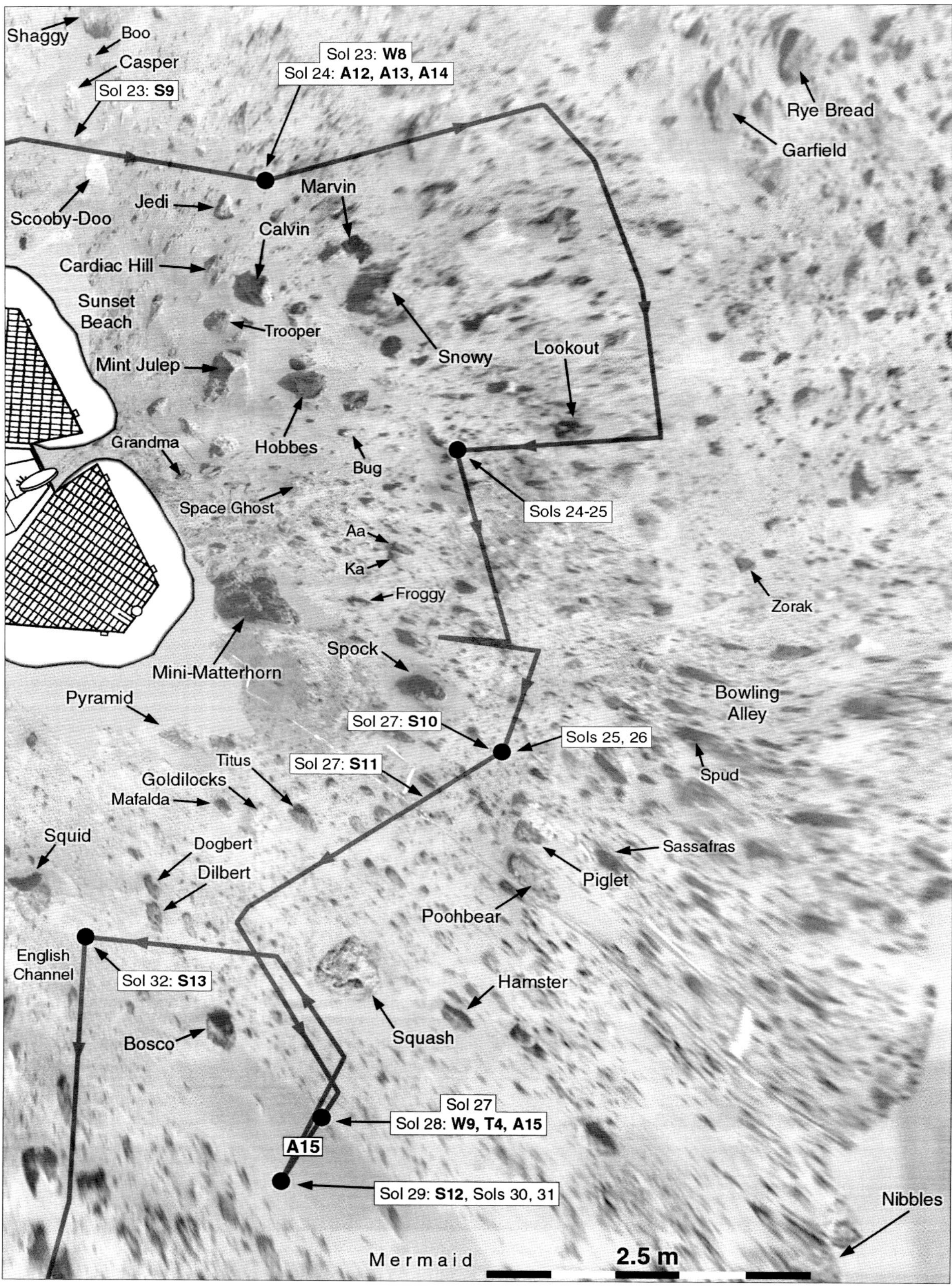

Figure 166 Sojourner rover route and activities from sol 23 to sol 32.

and its surroundings (Edgett, 1997). This might be possible if planners could find a path which was always in view of the lander. These plans were prevented by the sudden end of the mission, but Sojourner may have moved away from Chimp. If it did not hear from Earth it was to circle the lander, but it would probably stop driving as soon as it exceeded a critical tilt angle on the rocky surface. A HiRISE image taken on 29 January 2007 (MY 28, sol 363) showed a small spot which did not correspond with a known rock and might have been the rover, possibly indicating its final resting place (Figures 167, 173). Sojourner was made an honorary member of the Geological Society of America in October 1997.

APXS made atmospheric measurements as well as analyses of rocks and soil. Its first was on sol 1, when the background level of cosmic rays was measured to help calibrate the instrument. Other atmospheric analyses occurred on sols 10, 24 and 25. If the APXS was supposed to contact a rock or other target for analysis and failed to deploy properly, another atmospheric measurement would be made. This happened on sols 22, 66, 67, 76, 77 and 78.

The meteorology instruments made measurements at regular intervals day and night during the first 30 sols. After sol 30, the night operations were reduced, and late in the mission, data were only collected during the mornings. Five exceptions were made with data collection commencing on sols 25, 32, 38, 55 and 68 and continuing for a full sol each time. Meteorological observations by Pathfinder revealed numerous signs, in pressure and in wind data, of passing air vortices similar to 'dust devils'. No obvious sign of raised dust in these vortices was seen during the mission, but a few were reported later by Metzger *et al.* (1999) following careful processing of images from the Gallery Panorama. Ferri *et al.* (2003) described more of these events. The first was found in images taken on sol 2 and estimated to have been about 7 km east of the lander. One was seen on sol 10 in the direction of Couch rock, but about 17 km away, and three more west of the lander beyond Twin Peaks. On sol 11 a dust devil was seen moving southwards on the southern slope of South Twin Peak, another one southwest of the lander, and two more due south of the lander, apparently moving southeast. The closer one was crossing the elevated outer flank of Big Crater and the other one was farther south. One more seen in this direction was estimated to be about 24 km from the lander, the most distant one seen. Approximate locations of all observed dust devils are shown in Figure 159. Dust devils have also been seen from orbit by the Viking (Thomas and Gierasch, 1985), Mars Global Surveyor and Mars Reconnaissance Orbiter spacecraft and from the surface by the Mars Exploration Rovers and Phoenix.

Multispectral images of Deimos were made on sol 4, just after midnight, and sol 14 before midnight, and of Phobos early on sol 56 as it emerged from eclipse and again on sol 69. These provided estimates of albedo as well as spectra to gain a better understanding of satellite composition and origins. The disk of Deimos appeared less than 1 pixel across in these images, and Phobos was only about 3 pixels across, so no surface features were visible. The areas of each satellite contributing to these spectral observations are shown in Figure 201. Images of stars were made during some overnight observations to measure atmospheric clarity at night, and images of the Sun were taken on almost every sol for the same purpose. Over 1000 images of the Sun were taken in the first month alone. The top of Sojourner was covered with solar cells, one of which at the left front corner was equipped to measure obscuration by accumulating dust. It observed some reduction in generated power during its three-month mission, providing information for designers of future rovers.

1997: **Mars Reference Mission (United States)**

During the 1990s a Mars Exploration Study Project was conducted for planning purposes, assuming a six-person mission to Mars launched in 2013 and spending 18 months on the surface (Hoffman and Kaplan, 1997; Hoffman, 1998; NASA, 1998; Hoffman, 2001). The crew would be preceded by the infrastructure they would need with launches beginning in 2007, a bad opposition for Mars because of the planet's elliptical orbit. Because the design suited the worst case, the mission should be able to start in any year.

The process would begin with robotic sample return and In Situ Resource Utilization (ISRU) demonstrations and the launch of a prototype Mars habitat on a Russian Energiya rocket for testing at the Space Station. The first components to arrive at Mars would be an ascent vehicle

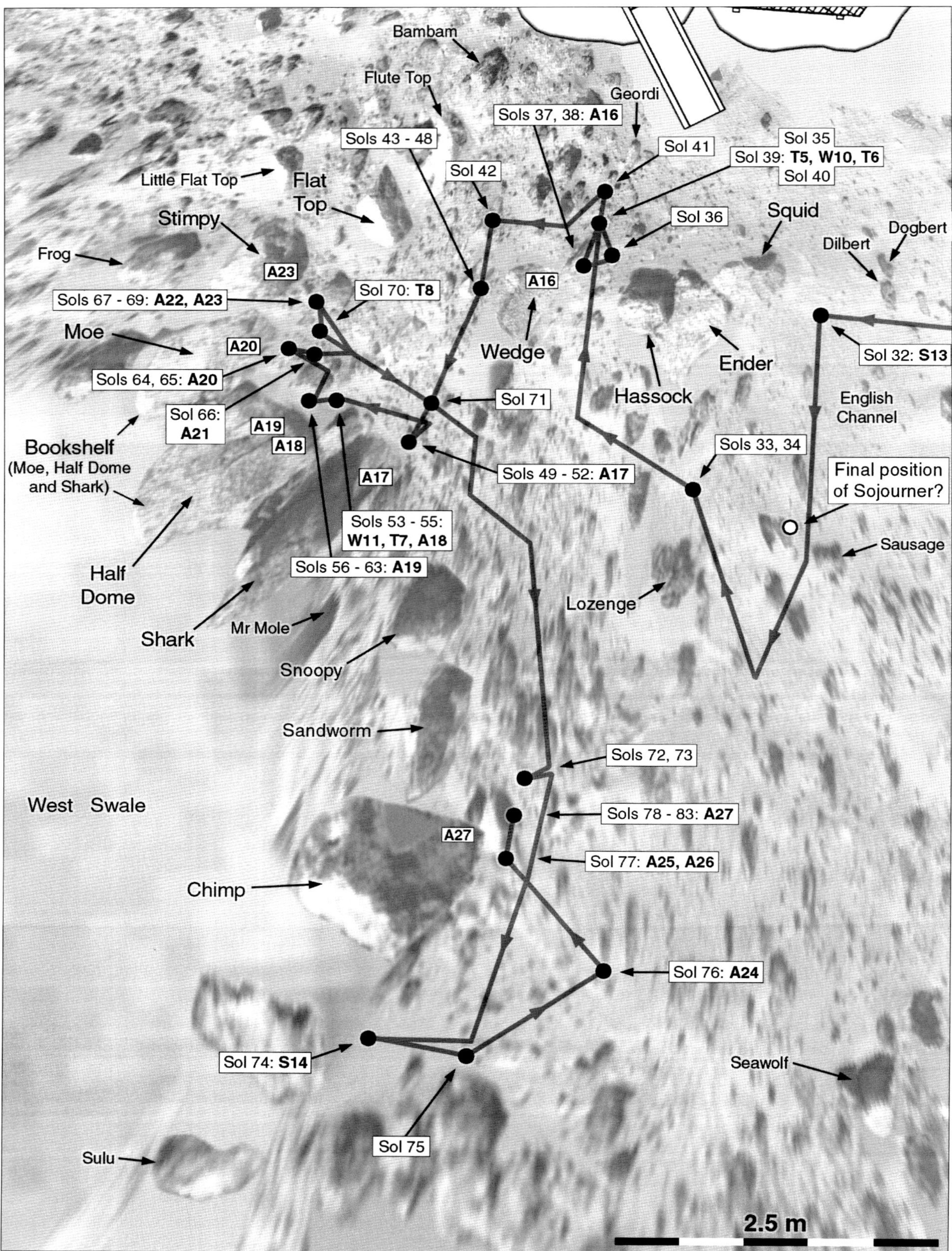

Figure 167 Sojourner rover route and activities from sol 32 to sol 83, the end of the mission.
After sol 83 the Sojourner rover may have driven towards the lander, and its final location may be marked by the white circle as suggested by a feature in HiRISE image PSP_002391_1995.

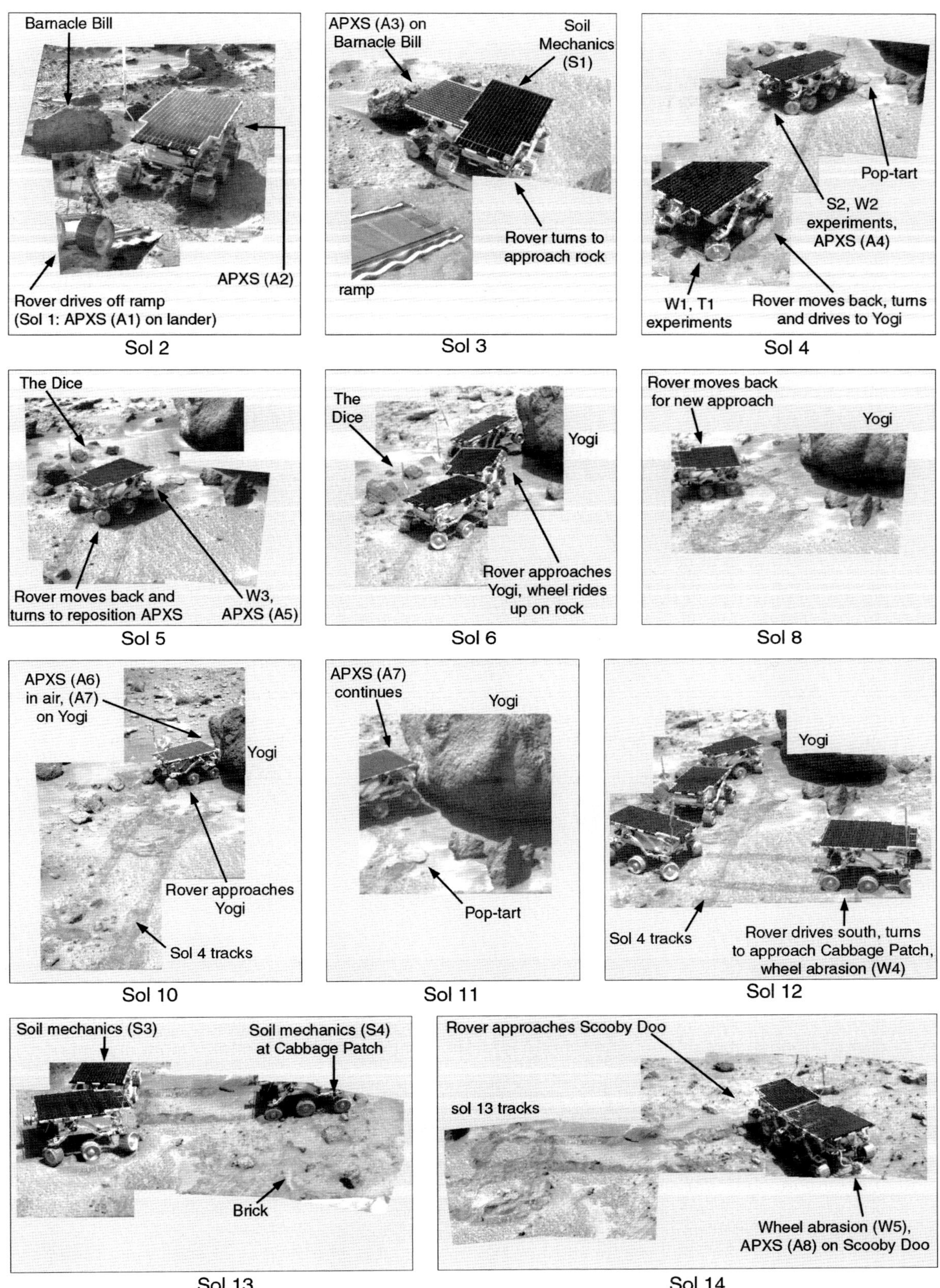

Figure 168 Mars Pathfinder images of the Sojourner rover, documenting its operations.
Figures 168 to 172 illustrate all rover activities. Images of the rover were not taken on sols omitted in this sequence. Some images were taken at the beginning of a sol to document activities on the previous sol.

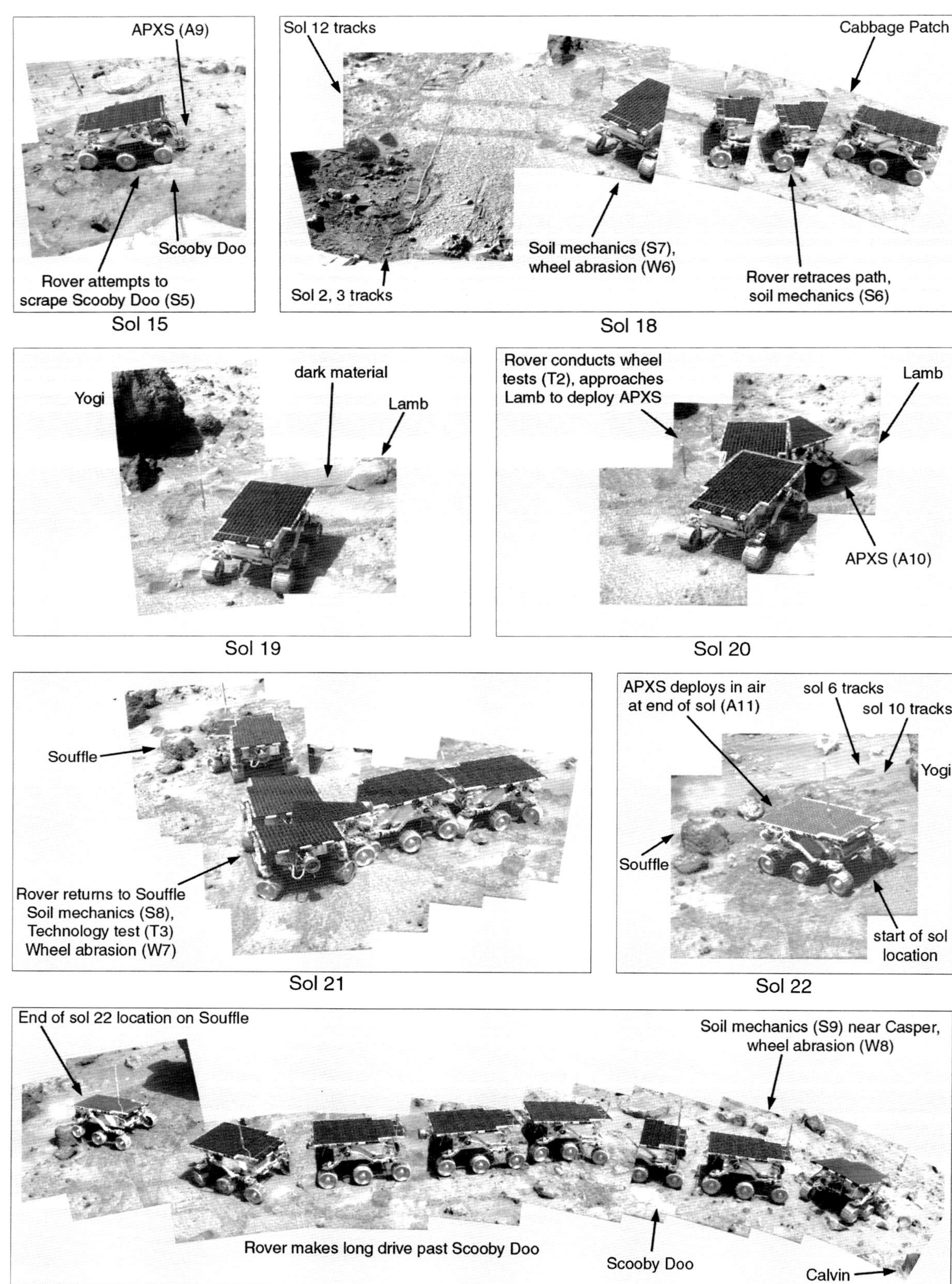

Figure 169 Sojourner activities (continued from Figure 168).
For scale in these figures, Sojourner is 63 cm long, 48 cm wide and its body is 28 cm high.

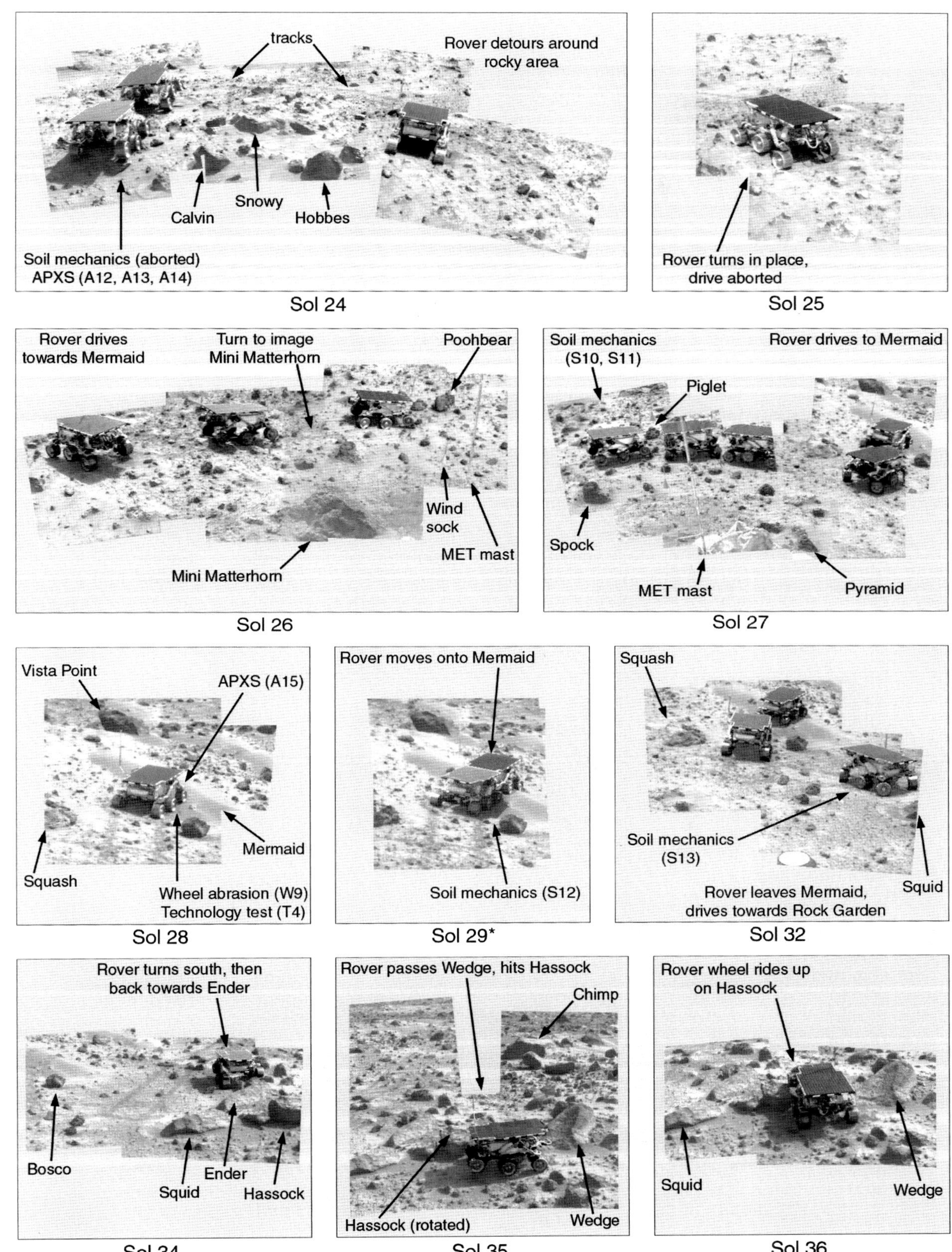

Figure 170 Sojourner activities (continued from Figure 169).
*Sol 29 activities (from Golombek *et al.*, 1999) are assigned to sol 30 by NASA's Planetary Data System.

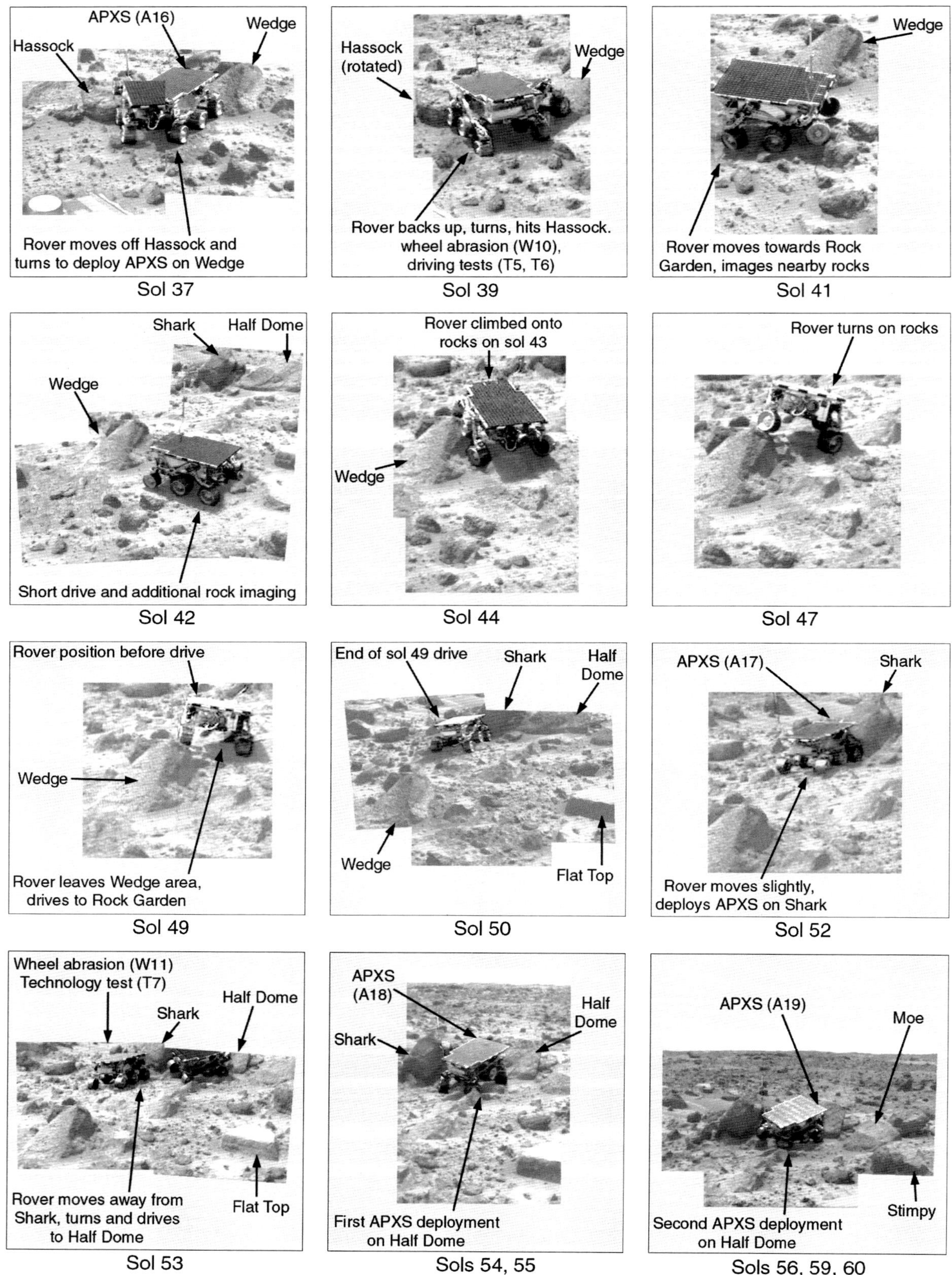

Figure 171 Sojourner activities (continued from Figure 170).

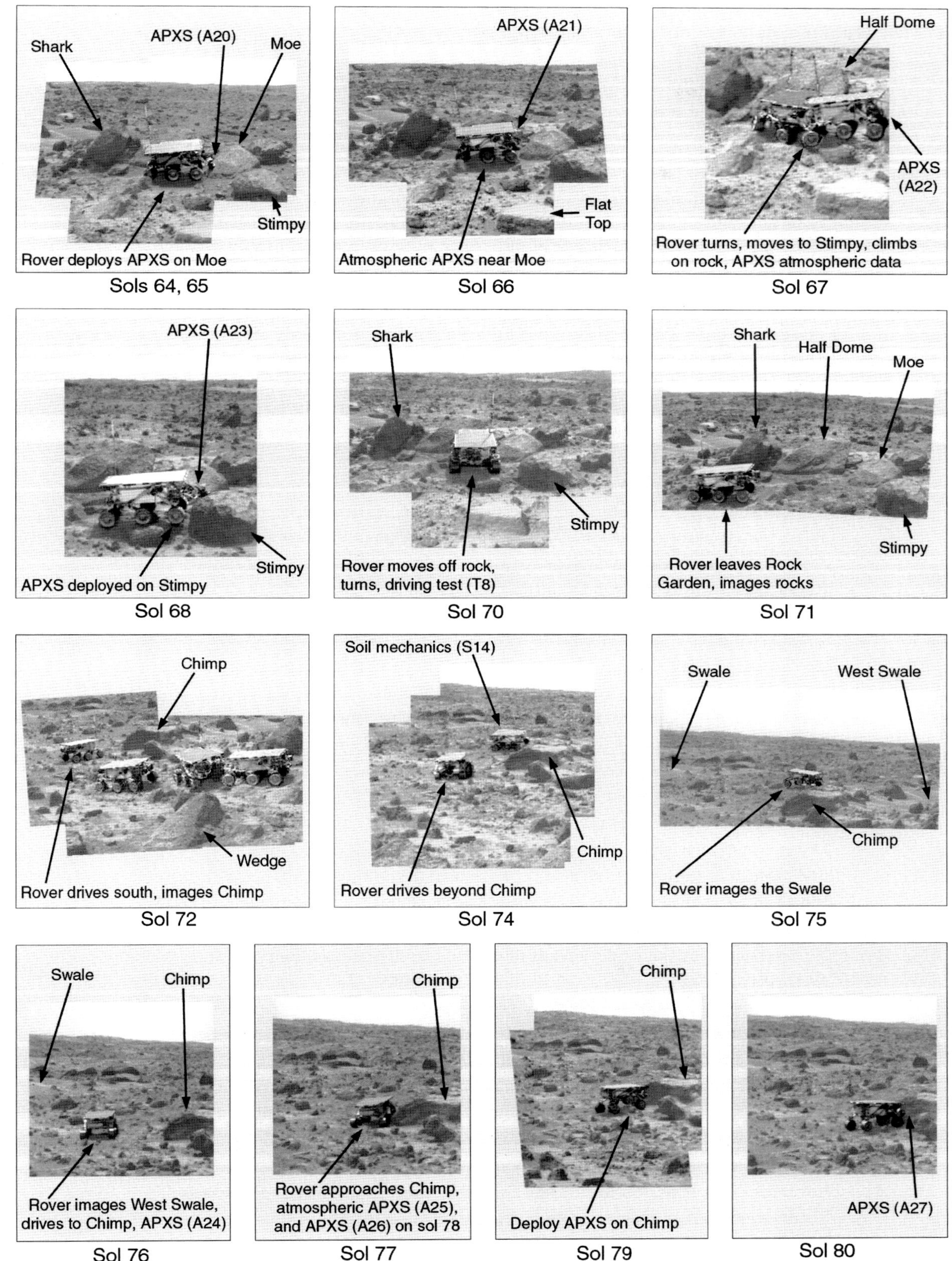

Figure 172 Sojourner activities (continued from Figure 171).

Table 68. *Mars Pathfinder Operations, Sols 1 to 45*

Sol	Activities
1	Landing, APXS background data (A1), landing site imaging (MY 23, sol 304)
2	Ramp deployment, Sojourner rover egress, APXS placed just above soil near ramp for analysis (A2)
3	Rover soil mechanics experiment (S1), APXS on Barnacle Bill rock (A3)
4	Wheel abrasion (W1, W2), driving test (T1), soil mechanics (S2), deploy APXS on soil near Yogi (A4)
5	Wheel abrasion experiments (W3), image wheel trench, deploy APXS on soil closer to Yogi (A5)
6	Position APXS on Yogi, but not well placed, image rocks and rover tracks
7	Rover image of APXS, then reverse to get to a better APXS position on Yogi
8	Rover images of nearby rocks, try to redeploy APXS
9	No rover activities
10	APXS noise test (A6), then reposition APXS on Yogi rock (A7), begin APXS analysis
11	End APXS analysis of Yogi
12	Rover wheel abrasion experiment (W4) and remote sensing
13	Soil mechanics experiments S3 (cloddy soil) and S4 (Cabbage Patch), image S4 results
14	Wheel abrasion experiment (W5), APXS on Scooby Doo (A8), image rocks, tracks and Deimos
15	Soil mechanics experiment on Scooby Doo (S5), then deploy APXS on disturbed soil (A9)
16	Attempted Earthrise imaging, but no data transmission
17	No rover activities
18	Soil mechanics in Cabbage Patch (S6) and south of Yogi (S7), wheel abrasion south of Yogi (W6)
19	No rover activities
20	Driving tests (T2), deploy APXS on dark soil near Lamb rock (A10)
21	Soil mechanics experiment (S8), driving tests (T3), wheel abrasion experiment (W7)
22	Deploy APXS near Souffle (intend to touch it but miss) and begin taking atmospheric data (A11)
23	Soil mechanics near Casper (S9), image resulting trench, wheel abrasion experiment (W8)
24	Aborted soil mechanics test, rover imaging of lander, APXS noise tests (A12, A13, A14), long drive
25	Rover turns in place, drive and imaging aborted
26	Drive, turn to image Mini Matterhorn, continue driving to Poohbear area
27	Soil mechanics tests in layered cloddy/drift material at Mermaid (S10 and S11), rover images lander
28	Wheel abrasion and driving tests (W9 and T4), deploy APXS on Mermaid drift (A15)
29	Soil mechanics experiments on Mermaid (S12), image resulting disturbance, image lander and rocks
30	No rover activities, lander documents rover activities at Mermaid. End of Primary Mission.
31	No rover activities
32	Leave Mermaid, images of Squid, Hassock, Bookshelf, and soil mechanics experiment (S13)
33	Rover imaging of Rock Garden, Ender, Hassock and lander, drive ends on rocks near Hassock
34	No rover activities, but lander documents sol 33 drive
35	Rover drives around Hassock, pushing and rotating it, imaging of Wedge and Flat Top rocks
36	Rover imaging of Ender, Chimp and Snoopy rocks, wheel climbs up onto Hassock
37	Deploy APXS on Wedge rock (A16), rover imaging of lander
38	No rover movement
39	Rover backs up, turns, hits and rotates Hasssock, wheel abrasion and driving tests (T5, W10 and T6)
40	Rover turns in place, imaging of Rock Garden and Shark rock
41	Rover imaging of Flat Top, Shark and lander
42	Rover moves towards Rock Garden, imaging of Flat Top, Stimpy, Wedge, Hassock
43	Rover tries to pass Wedge, climbs onto rocks
44	No rover motion, lander imaging documents sol 43 drive
45	Small rover motion not documented in images

Table 69. *Mars Pathfinder Operations, sol 46 to End of Mission*

Sol	Activities
46	No rover activities
47	Small rover motion leaves rover angled on Wedge rock
48	No rover activities
49	Rover drives off Wedge rock and into Rock Garden, close imaging of Shark rock
50	No rover activities, lander documents sol 49 drive
51	No data transmission or rover activities
52	Deploy APXS on Shark rock (A17), imaging of Shark
53	Wheel abrasion and driving experiments (W11 and T7), rover imaging of Half Dome rock
54	Turn to face Half Dome, deploy APXS on Half Dome (A18), begin data collection
55	Start of a full-day meteorology observation, continuing overnight
56	Rover battery dies, operations continue under solar power. APXS repositioned on Half Dome (A19).
57	APXS remains deployed, but no rover data transmission
58	A19 APXS analysis continues
59	APXS remains deployed, but no rover data transmission
60	APXS retracted and redeployed at same location to continue analysis (A19)
61	No rover activity
62	A19 APXS analysis continues
63	A19 APXS analysis continues
64	Rover moves to Moe rock, APXS deployed on Moe (A20)
65	A20 APXS analysis continues
66	APXS used for atmospheric measurements (A21)
67	Rover climbs onto Stimpy rock inadvertently, APXS deployed in air (A22)
68	Rover moves to place APXS on Stimpy rock (A23)
69	A23 APXS analysis continues
70	'No load' driving test (T8) with wheel suspended in air, rover imaging of Moe and Stimpy rocks
71	Rover imaging of Half Dome and Shark rocks
72	Rover imaging of surroundings and Chimp rock
73	No rover activity or data transmission
74	Close imaging of Chimp rock, soil mechanics experiment (S14)
75	Rover images Swale, West Swale and North Peak
76	Rover images West Swale, Twin Peaks, APXS deployed in air for atmospheric measurement (A24)
77	A24 continues, then APXS deployed for another atmospheric measurement (A25)
78	A25 continues, then APXS deployed above Chimp rock for new atmospheric measurement (A26)
79	A26 continues, then APXS deployed on Chimp (A27), data collection begins
80	A27 data collection continues
81	A27 data collection continues, rover images Chimp
82	A27 data collection continues
83	Partial downlink, last useful results of the mission. Intermittent contact with lander continues
93	Last contact with lander, effective end of mission (MY 23, sol 396)

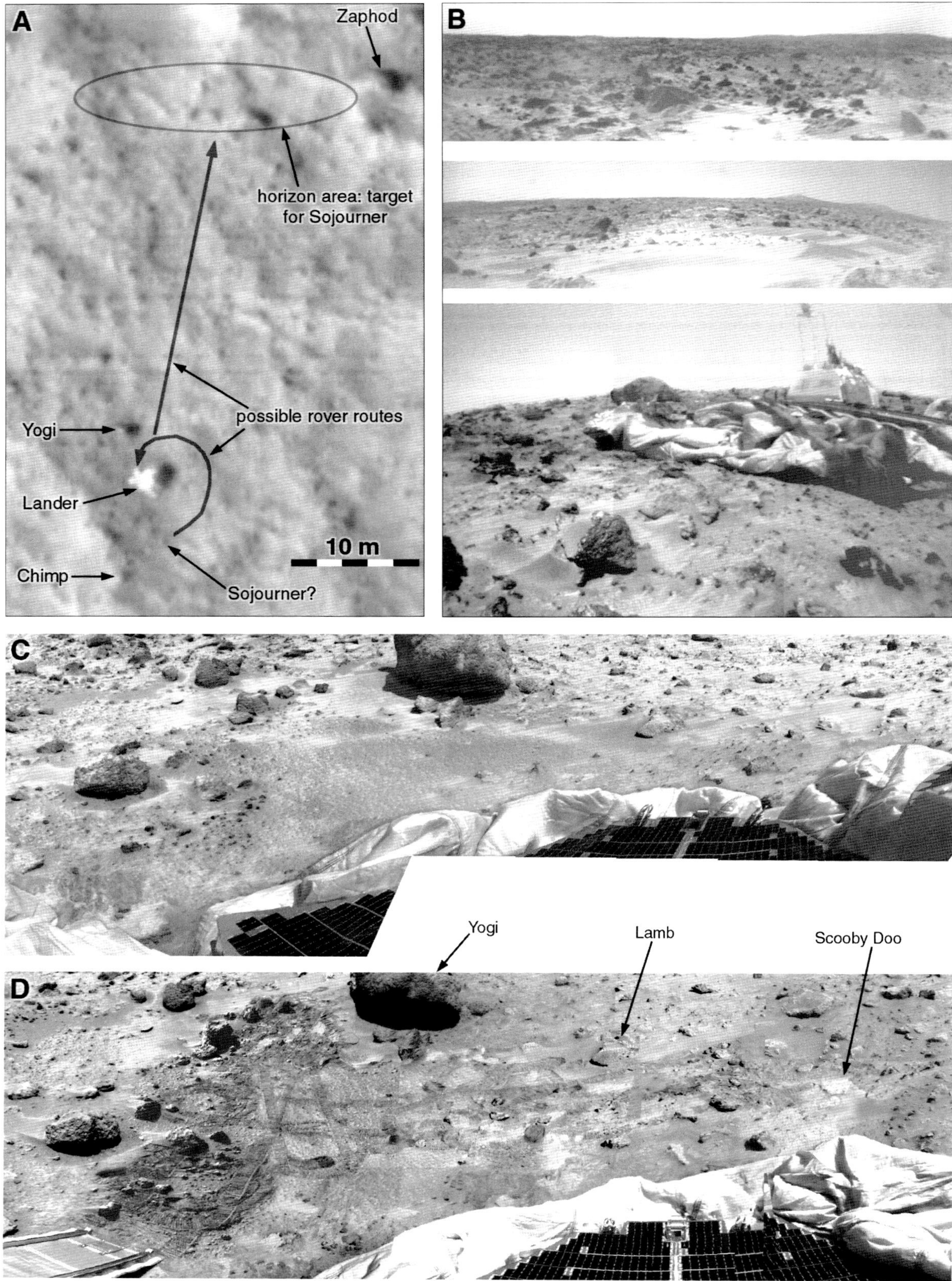

Figure 173 **A:** Possible routes for Sojourner after sol 83. **B:** Sojourner images. From the top: View to the southeast with Big Crater on the horizon; view to the southwest with drifts in West Swale; the lander seen from the Rock Garden with Yogi behind it. **C and D:** The area north of the lander seen before and after rover operations in the area.

with a power system, a propellant production (ISRU) system and other cargo. The ISRU unit would manufacture propellant from the atmosphere, using hydrogen brought from Earth, to fuel the ascent vehicle. Two habitat modules would follow, as well as an Earth return spacecraft in Mars orbit and a backup ascent vehicle. The crew would not leave Earth until everything was prepared and confirmed to be operational. After 500 days on the surface, the crew would return to orbit, board the return vehicle in orbit and begin the 180-day trip back to Earth. Every element would be duplicated for full redundancy. Research would include geology and meteorology, sample collection and analysis on site, and experiments to support future Mars missions. Samples might be brought to the outpost from distant locations by robotic rovers delivered in advance.

Landing site selection would involve the usual safety and scientific concerns, but also the ability of the site to support ISRU, especially the availability of water. Hoffman and Kaplan (1997) suggested landing in Candor Chasma in the Valles Marineris, from 2.5° to 7.5° S and 70° to 75° W (Figure 174A). In a separate study, James *et al.* (1998) considered the availability of subsurface ice and geothermal heat sources, as estimated at the time, as well as thermal inertia, elevation and other factors which might influence local resource use. They chose three areas for potential human landings, at 45° N, 120° W in Arcadia, 45° N, 220° W near Viking 2 in Utopia, and 45° S, 275° W in eastern Hellas (Figure 178E).

Several university groups conducted design studies based on the Reference Mission and presented them at a HEDS-UP (Human Exploration and Development of Space – University Programs) meeting at LPI in 1998 (Budden and Duke, 1998). A group from the University of Maryland led by Dr. David Akin proposed large-scale exploration of the Tharsis region (Figure 174B) from a base camp near Ascraeus Mons, using a pressurized rover called MERLIN (Martian Rover for Long-range Investigations). Major targets for scientific investigation were Olympus Mons, the other large volcanoes of the Tharsis Montes, Valles Marineris, Fortuna Fossae and Lunae Planum. Specific targets in Valles Marineris were Tithonium, Hebes and Melas Chasmata. Four candidate base camp locations were considered, 4° N, 136° W near Gigas Sulci, 8° N, 85° W near Fortuna Fossae, on the equator at 90° W near Tithonium Chasma and 6° N, 101° W near Ascraeus Mons. All were flat and near interesting targets, but the latter was preferred for its central location.

After arrival MERLIN would be tested with two or three 'break-in sorties'. The first would consist of several small loops around the base camp. The second, to Ascraeus Mons at a distance of 500 km, would include the first scientific studies. Assuming 200 km per sol of driving and 14 days at the site, this would take 19 sols, with another 11 sols added for contingencies. If a third break-in sortie was required, it would target Fortuna Fossae, 650 km from the base camp and lasting 35 sols with contingency time. Otherwise Fortuna Fossae would be visited during the last traverse. The first full science sortie would be a 70-sol, 5660-km round trip to Tithonium and Melas Chasmata, driving through the Valles Marineris, with 14 sols at each science site. The canyon system would be entered at the 'Gate of Valles Marineris' near 9° S, 97° W in Noctis Labyrinthus. The second sortie would be a 55-sol, 5500-km round trip to Olympus Mons and the third a 60-sol, 4090-km visit to Arsia and Pavonis Montes, with 14 sols of work at each volcano. The last of these long trips would be a 4620-km, 65-sol trip to Hebes Chasma and Lunae Planum, passing through Fortuna Fossae in each direction. These long trips would be supported by three communication satellites and the prior deployment of a network of seven automated ISRU fuel-producing plants called CRYSTAL BALL (Cryogenic Storage And Local Ballistic Lander). They would produce methane and oxygen from the atmosphere and stored hydrogen.

HEDS-UP was the first in a series of four annual meetings at LPI, after which the name changed to RASCAL (Revolutionary Aerospace Systems Concepts Academic Linkage). They included numerous innovative concepts and lunar and Martian plans, not described here, in addition to the MERLIN mission.

1998: **Mars 98 (Russia)**

This spacecraft was initially planned for launch in 1996 and referred to as Mars 96, but when Mars 94 was delayed to 1996, this mission was postponed to 1998. It would have consisted of an orbiter to study Mars and to relay data from the surface and atmospheric components,

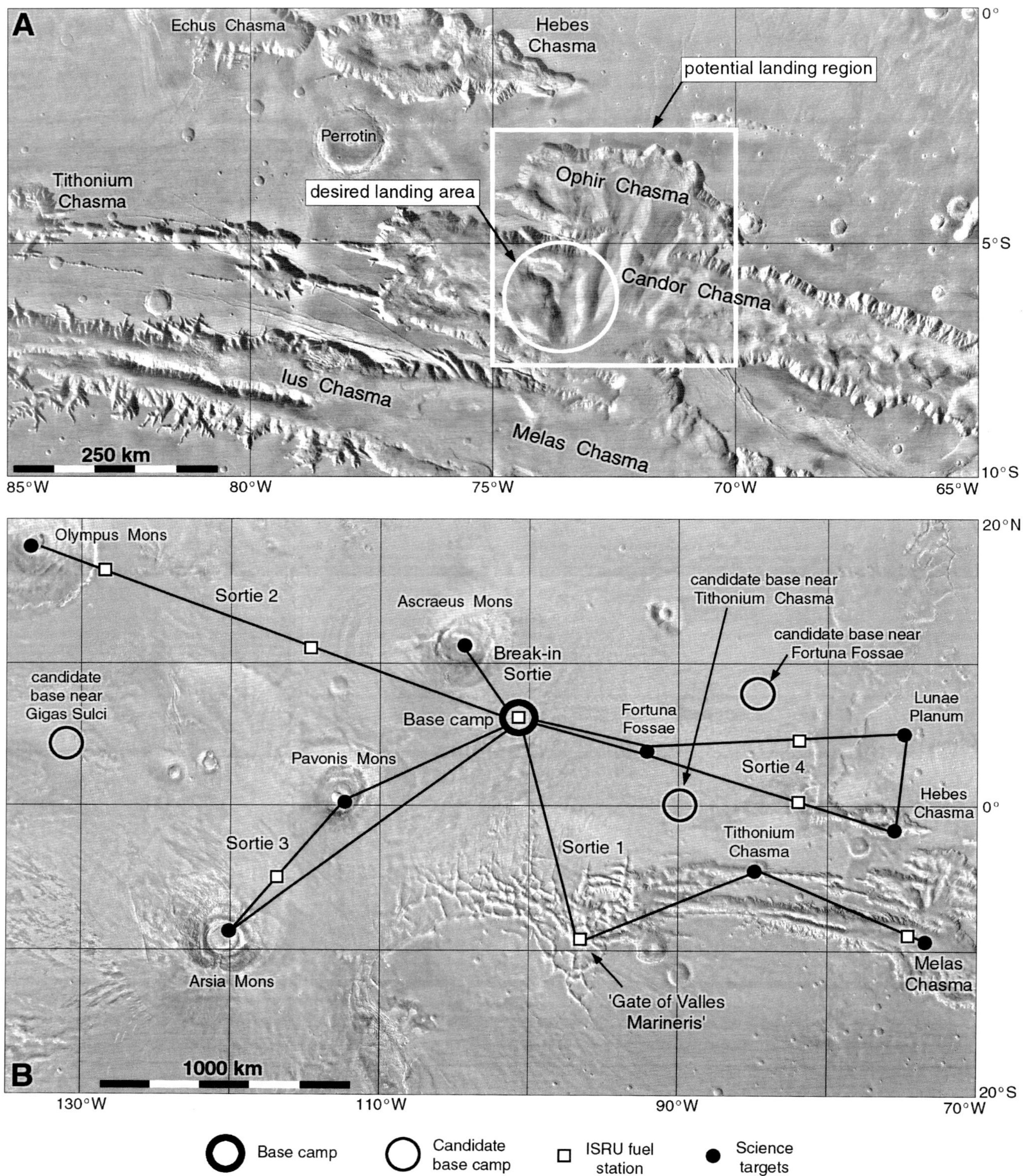

Figure 174 Human Reference Mission studies.
A: Landing area in Candor Chasma from Hoffman and Kaplan (1997). **B:** University of Maryland plans detailed in Budden and Duke (1998). The traverse routes are schematic and do not take topographic obstacles into account.

which were a 'Marsokhod' rover and a balloon system. The orbiter might also deploy several penetrators to the surface. Early plans suggested the mission would include two identical spacecraft.

The balloon and the rover would have been deployed together in a single descent module. The preferred target for this module was at high latitudes, between 50° and 60° N primarily because of the balloon's requirements. The balloon was to be 30 m high and would settle onto the surface as it cooled each night and lift off to float during the day when the Sun warmed it. It was expected to survive for at least ten sols and might travel as far as 1000 km. The balloon would carry a camera system for multiresolution imaging and instruments to study the soil, seek water, monitor the weather and measure the magnetic field. Each night it would drape an instrumented cable on the ground to make in situ measurements. Sterilization of the balloon might have been a serious problem for this mission.

The Mars 98 rover was to weigh about 100 kg and to be about 1 m high. It would carry cameras and would analyze soil chemistry, minerals and perhaps organic materials; measure the regolith's water content; and trace gases in the atmosphere. A drill would collect samples from as much as 2 m below the surface. The rover would operate for one Martian year and travel as much as several hundred kilometres. Table 64 and Figure 155 identify the intended Mars 96 lander and penetrator sites as described in the Mars Landing Site Catalog (Table 47) and balloon sites as identified by Kazarian-Le Brun (1996). Sites planned for one mission date do not necessarily work for a later date because of changing approach and illumination conditions, so these would probably have been changed for Mars 98 if work on the mission had continued.

3 July 1998: **Nozomi (Planet-B) (Japan: ISAS)**

Nozomi (Japanese for 'hope', referred to before launch as Planet-B) was a 258-kg Mars orbiter (540 kg with fuel) intended to study the upper atmosphere and its interaction with the solar wind and to test technology for future planetary missions. Instruments would study the ionosphere, the magnetic field and magnetosphere; the escape of atmospheric constituents, including water or hydrogen; and atmospheric and orbital dust. The mission would also return images of the surface of Mars and its satellites.

Nozomi was a 0.6 by 1.6 by 1.6 m box with two solar panels, an antenna on top and a rocket engine underneath. A 5-m mast and a 1-m boom extended from the sides, and two pairs of long-wire antennae spanned 50 m from tip to tip. It was launched at 18:12 UT from Uchinoura Space Centre into a 400 000 by 340 km parking orbit, and lunar gravity assists on 24 September and 18 December 1998 increased its apogee and provided images of the Moon. It made an Earth flyby on 20 December 1998, 1000 km above the surface, and that combined with an engine burn placed Nozomi in a heliocentric orbit. It was to arrive at Mars on 11 October 1999 at 7:45 UT, but a problem during the burn gave the spacecraft insufficient velocity to reach Mars. Two trajectory corrections on 21 December salvaged the mission, making a Mars encounter possible, but with limited fuel. A distant flyby occurred in August 1999, and one point-like image of Mars was obtained from 5]560]000 km range at 11:00 UT on 28 August (MY 24, sol 399).

The new trajectory would take four more years to reach Mars, with Earth gravity assists in December 2002 and June 2003, and would finally arrive at Mars in December 2003. However, powerful solar flares on 21 April 2002 damaged the spacecraft's communication system electronics. The second Earth flyby on 19 June 2003 approached within 11 000 km, but the health of the spacecraft was declining. On 9 December 2003, efforts to prepare Nozomi for orbit insertion on 14 December failed, and the mission was abandoned. Thrusters were used on December 9 to ensure that Nozomi would miss Mars by 1000 km to avoid contamination. The spacecraft passed Mars on 14 December 2003 (MY 26, sol 588) and entered a roughly 2-year heliocentric orbit. No data were transmitted.

The planned Mars operations were to start with Nozomi in a retrograde (170° inclination) 300 by 50 000 km orbit. Soon after entering this orbit, the mast and antennas would be deployed and the periapsis dropped to 150 km. The orbital period would be 38.5 hours. Nozomi was spin stabilized with its dish antenna facing Earth. Near periapsis the spacecraft would study the upper atmosphere and image the surface. In the outer parts of the orbit, it would study atmospheric escape of gas and ions and the solar wind. It would also have made close passes by Phobos and Deimos for imaging. The nominal mission would last one Mars year.

24 October 1998: **Deep Space 1 (United States)**

The New Millennium program was a series of technology development and verification missions designed to provide some science data while flight testing new spacecraft components and sensors. Deep Space 1 was an asteroid and comet flyby mission, Deep Space 2 flew penetrator probes to Mars, and other New Millennium missions were Earth orbiters. Deep Space 1 also observed Mars from great distances, providing whole-disk spectra which were superior to any other similar data at the time. The original plan for Deep Space 1 would have involved a launch in July 1998 and a Mars gravity assist on 28 April 2000 (MY 24, sol 636) which could have included a close flyby of Phobos or Deimos (Rayman *et al.*, 1999). A launch delay prevented that, requiring the selection of new targets and the loss of the Mars gravity-assist option.

The spacecraft was launched from the Cape Canaveral Air Station at 12:08 UT. It tested an ion drive, new solar panels, control software and experimental sensors and obtained data from asteroid 9969 Braille on 29 July 1999 and from Comet Borrelly on 22 September 2001. Mars data from the MICAS (Miniature Integrated Camera and Spectrometer) instrument were obtained on 7, 9 and 23 May 1999 from 115 million km and on 9 to 11 November 1999 (MY 24, sols 470 to 472) from 55 million km. The November data were collected at intervals over two full rotations of Mars and suggested the presence of clay minerals and sulfates (Soderblom and Yelle, 2001). A MICAS image processed by Ted Stryk is included in Figure 177.

11 December 1998: **Mars Climate Orbiter (United States)**

NASA's intention after Mars Pathfinder was to launch a relatively inexpensive orbiter and lander pair to Mars at every opportunity. This program was called Mars Surveyor, and its first missions were referred to as Mars Surveyor 1998 since the launch period commenced in 1998. They were given more distinctive names prior to launch, Mars Climate Orbiter (MCO) and Mars Polar Lander. The orbiter would monitor dust storms, weather systems, clouds and hazes; record surface changes; measure atmospheric temperature profiles; and seek evidence of past climate change. Its instruments were a Mars Color Imager (MARCI) for daily global weather images and medium-resolution (40 m/pixel) surface images and a Pressure Modulated Infrared Radiometer (PMIRR) to measure atmospheric temperature, water vapor and dust. The orbiter would also serve as a communications relay for the Mars Polar Lander and future landers. MCO had a body about 2.1 by 1.6 by 2 m in size with a 5.5- by 2.0-m solar panel on one side and a 1.3-m-diameter high-gain antenna on a mast at the top of the body. The instruments and batteries were mounted at the other end of the body.

MCO was launched at 18:46 UT from Cape Canaveral Air Station. After a brief period in parking orbit, the upper stage put the spacecraft on its Mars trajectory. Fifteen days after launch a trajectory correction was performed, with three smaller ones on 4 March, 25 July, and 15 September 1999. The 338-kg spacecraft (629 kg including fuel) reached Mars on 23 September 1999 (MY 24, sol 424) and began a 16.4-minute orbit insertion burn at 09:01 UT, during which it passed behind Mars. It was to emerge from behind the planet and contact Earth at 09:27 UT, ten minutes after the burn was completed, but no contact was ever made. An investigation revealed that a navigation error stemming from confusion over units caused it to miss its target altitude at closest approach. It entered the atmosphere during the orbit insertion maneuver, reaching a height of only 57 km, much lower than the intended altitude of 140 to 150 km, and according to the pre-mission plan it probably failed or broke up somewhere near 34° N, 170° W (M. Caplinger, personal communication, 14 November 2005) while moving from north to south (Figure 4).

One very small MARCI image of Mars (Figure 177F), barely resolving the planet's disk in a half-illuminated phase resembling a first quarter Moon, was obtained on 7 September 1999 from a range of 4.5 million km (Malin Space Science Systems, 1999).

Heavy components of the spacecraft, or most of its body if it stayed largely intact, may have escaped Mars or made one full elliptical orbit before final entry on the next periapsis. Lighter or more fragile components such as insulation or the solar panel might have separated and fallen to the surface.

If everything had worked properly, the spacecraft would have entered a 150- by 21 000-km capture orbit with a period of 14 hours. Aerobraking would shrink the orbit to 90 by 405 km with the periapsis at 89° N by about 22 November 1999. Then the thrusters would raise the periapsis, giving a circular 421-km near-polar orbit with a 2-hour period by 1 December 1999. The orbit would be nearly Sun-synchronous, crossing the equator in late afternoon local time. The initial phase of mission operations was support for Mars Polar Lander from its landing on 3 December to the end of the lander's primary mission on 29 February 2000. MCO science operations with MARCI and PMIRR would begin on 3 March 2000 and continue until 15 January 2002 (one Mars year). After the mapping mission ended, MCO would be moved into a stable orbit to serve as a relay for Mars Surveyor 2001.

Mars Polar Lander Landing Site Selection

The purpose of the Mars Polar Lander (MPL, Mars Surveyor 1998) mission was to study volatile materials (water ice, carbon dioxide frost, etc.) near the south pole of Mars. The lander might have been able to contact these volatiles directly and would also examine the polar layered deposits which may contain a long history of climate variations. The landing site was to be somewhere in the broad Planum Australe plateau which extends northwards from the south polar ice cap to almost 70° S between longitudes 135° W and 225° W. A formal landing region extended between 72° S and 78° S, and between 170° W and 230° W (Figure 175). The northern limit was moved to 73° S (Figure 176) by the time of the landing press kit to better match the extent of polar layered deposits. Early mission documents showed a potential landing site at 77° S, 215° W, but this was subject to change as more topographic and image data became available from Mars Global Surveyor.

Table 70. *Mars Polar Lander Candidate Ellipses*

Ellipse	Location of ellipse centre
Site 1	75° S, 180° W
Site 2	76° S, 195° W
Site 3	75° S, 198° W
Site 4	76° S, 206° W
Site W	76° S, 218° W

JPL illustrated five candidate landing ellipses across the landing area, each about 250 km long and 30 km wide and oriented roughly 15° west of north (Table 70, Figure 176). Vasavada (2000) illustrated four of these, omitting Site 3. On 25 August 1999, NASA announced the final selection. The primary landing ellipse (Site 2 in Table 70) was centered at 76° S, 195° W (Figure 176). The secondary site, which would be used if subsequent imaging revealed problems in the primary ellipse, was Site 1 in Table 70, centered at 75° S, 180° W. Figures 176B and 176C are infrared images with shading reversed to resemble the albedo shown in Figure 176A. Because of that reversal, topography appears to be illuminated from the south in those images.

3 January 1999: **Mars Polar Lander (United States)**

Mars Polar Lander was launched at 20:21 UT from Cape Canaveral Air Force Station. It entered a 191-km circular parking orbit, and 30 minutes later its upper stage burned to place it on a Mars trajectory. Five trajectory corrections were made before the spacecraft arrived at Mars on 3 Dec 1999 (MY 24, sol 494). Six minutes before entering the atmosphere, the spacecraft oriented itself for entry, separated from the cruise stage, dropped its Deep Space 2 microprobes and entered the atmosphere at 20:10 UT. It would have been slowed by its heat shield and then by a parachute before descending to the surface using thrusters. A camera would have taken images during the descent. It was intended to land at 20:15 UT. The landing target was at 76.0° S, 194.9° W, about 800 km from the pole and 4 km above the elevation datum (6.1-millibar pressure level). The landing ellipse (99% probability) was 200 km by 20 km, with the lander approaching from slightly east of north.

No signal was received after the expected landing time. Tracking initially suggested a landing or impact at 76.3° S, 195.0° W, later refined to 76.57° S, 194.8° W. Mars Global Surveyor MOC images were taken to search for evidence of the lander or its parachute. One possible candidate was found by Malin Space Science Systems (2005a) in Mars Global Surveyor images near 76.4° S, 195.3° W (Figure 177B), assuming a successful descent with failure at touchdown. Subsequent higher-quality imaging showed that the feature was not Mars Polar Lander (Malin Space Science Systems, 2005b), and its true location is not known. A suggestion that signals were detected in January and February 2000 is now known to be incorrect (Squyres, 2005).

The landed mission was scheduled to last for 87 Earth days, ending about 29 February 2000. The lander was to touch down during late southern spring when the Sun would be above the horizon throughout the day. As the southern summer progressed, the Sun would first climb higher, but then drop towards the horizon, eventually reducing power output by the solar panels. At the same time, the air would become colder as autumn approached, requiring more energy to keep the lander warm. The lander would spend more time each sol in an energy-conserving mode in which it could not collect or transmit data. Eventually, it would completely lose the ability to collect data or communicate, bringing an end to the mission.

The suite of instruments carried by MPL was called MVACS, the Mars Volatiles and Climate Surveyor. The MVACS instruments were designed to search for water, as a gas or ice, and to study its behaviour. MVACS consisted of five separate instruments: a meteorology package to measure atmospheric temperature, pressure, humidity, winds and composition, as well as subsurface soil temperatures; a Surface Stereo Imager to obtain multispectral images of the site; a Thermal and Evolved Gas Analyzer (TEGA) to study the composition of the soil; a Robotic Arm to dig trenches and collect samples to place in the TEGA; and a camera on the arm to obtain very high-resolution images of the holes and soil samples in the arm's scoop. The MVACS payload was built by an international team, including members from the United States, Finland, Germany and Denmark. Instruments very similar to these eventually flew on the Phoenix mission, landing on Mars in 2008 at high northern latitudes.

3 January 1999: **Deep Space 2 (DS2) (United States)**

Like Deep Space 1, this mission was part of the New Millennium program of small technology development missions. Deep Space 2 was intended to flight test two small penetrators, a class of mission discussed for many years but not yet flown successfully as of 2011. Each probe consisted of an aeroshell 27.5 cm high and 35 cm across and a penetrator (forebody) 10.6 cm long and 3.9 cm in diameter. An aftbody, a component intended to remain on the surface as the penetrator buried itself in the surface material, was 10.5 cm high, topped by a 12.7-cm antenna, and 13.6 cm in diameter. The whole probe weighed 3.5 kg. The penetrator contained a device to collect a sample of surface material and test it for water, a thermal sensor, and accelerometers to characterize the atmospheric descent and impact.

The Deep Space 2 microprobes were launched on 3 January 1999 and were carried to Mars by Mars Polar Lander. A public contest was held to name the probes, which received the names Amundsen and Scott. These were suggested by Paul Withers, then a graduate student at the University of Arizona. The probes arrived at Mars with Mars Polar Lander on 3 December 1999 (MY 24, sol 494) and were supposed to be deployed from the MPL cruise stage before MPL itself entered the atmosphere. No braking other than friction with the aeroshell was employed. The probes would strike the surface at a speed of about 644 km/h, experiencing a deceleration of 60 000 *g*. The aeroshell would shatter, the penetrator would bury itself up to 1 m deep, and the aftbody would remain on the surface to transmit data to Earth via Mars Global Surveyor. As with Mars Polar Lander, nothing was ever heard from either probe.

Figure 176C shows the area in which the probes were expected to have fallen. A mission summary at the National Space Science Data Center in 2009 suggested that the target site for DS2 was at 73° S, 210° W (Figure 176A), but this had not been updated to reflect the final MPL site selection. The impact site was near 75.2° S, 196.0° W.

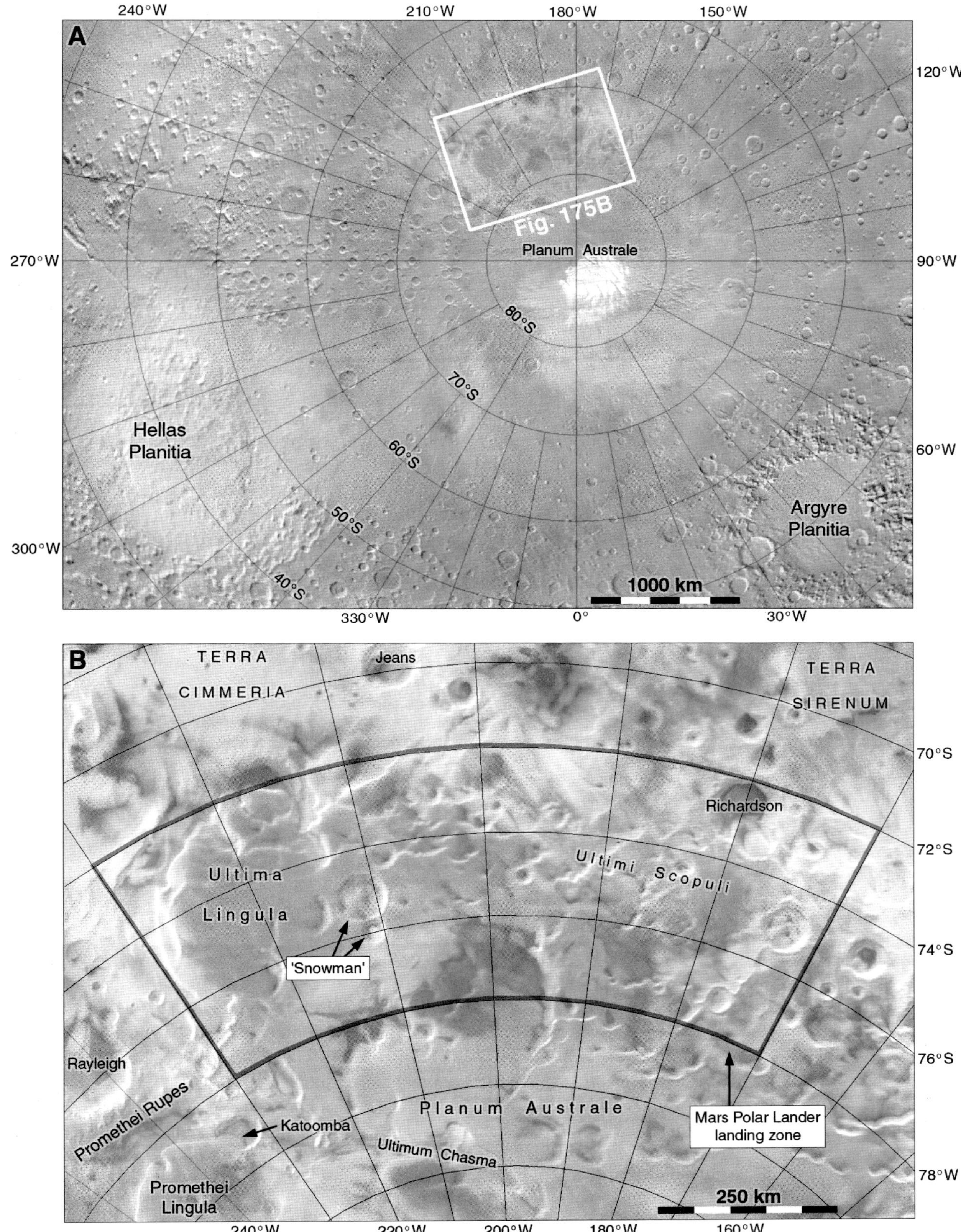

Figure 175 **A:** The Mars Polar Lander landing region near the south pole, on a Mars Global Surveyor map. **B:** The landing zone (dark outline, extending to 72° S) for Mars Polar Lander, plotted over a Mars Global Surveyor MOC wide-angle mosaic. A prominent pair of craters nearby was informally named Snowman. Mission maps of the time were usually oriented south up, enhancing the appearance of a snowman.

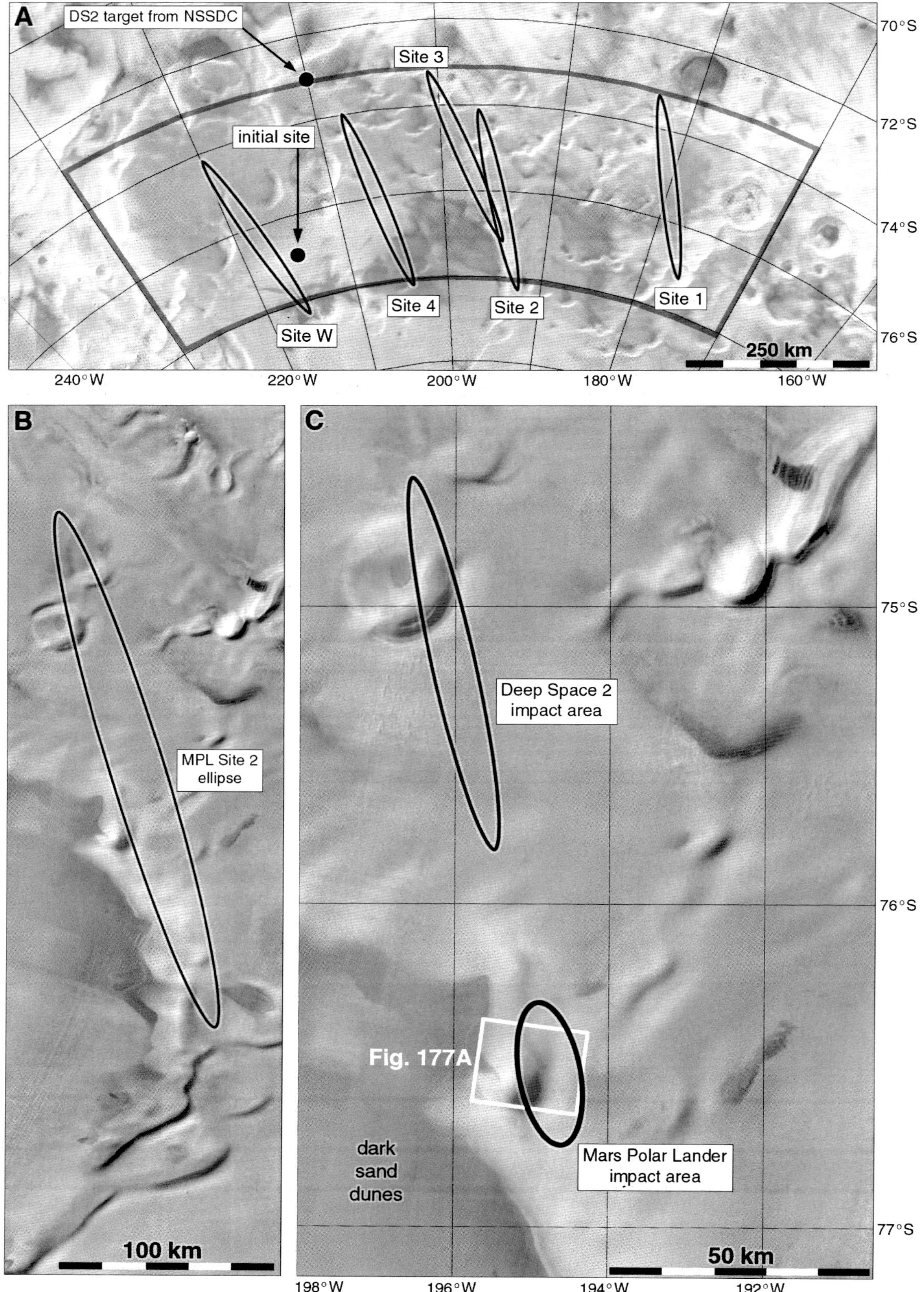

Figure 176 A: Candidate landing ellipses for Mars Polar Lander. The 'initial site' was illustrated in early mission documents. The landing zone now extends only to 73° S. **B:** The Site 2 ellipse selected for the landing. **C:** Estimated impact locations for the Deep Space 2 penetrators and Mars Polar Lander. The backgrounds for B and C are Mars Odyssey THEMIS infrared mosaics with shading inverted to approximate visible albedo.

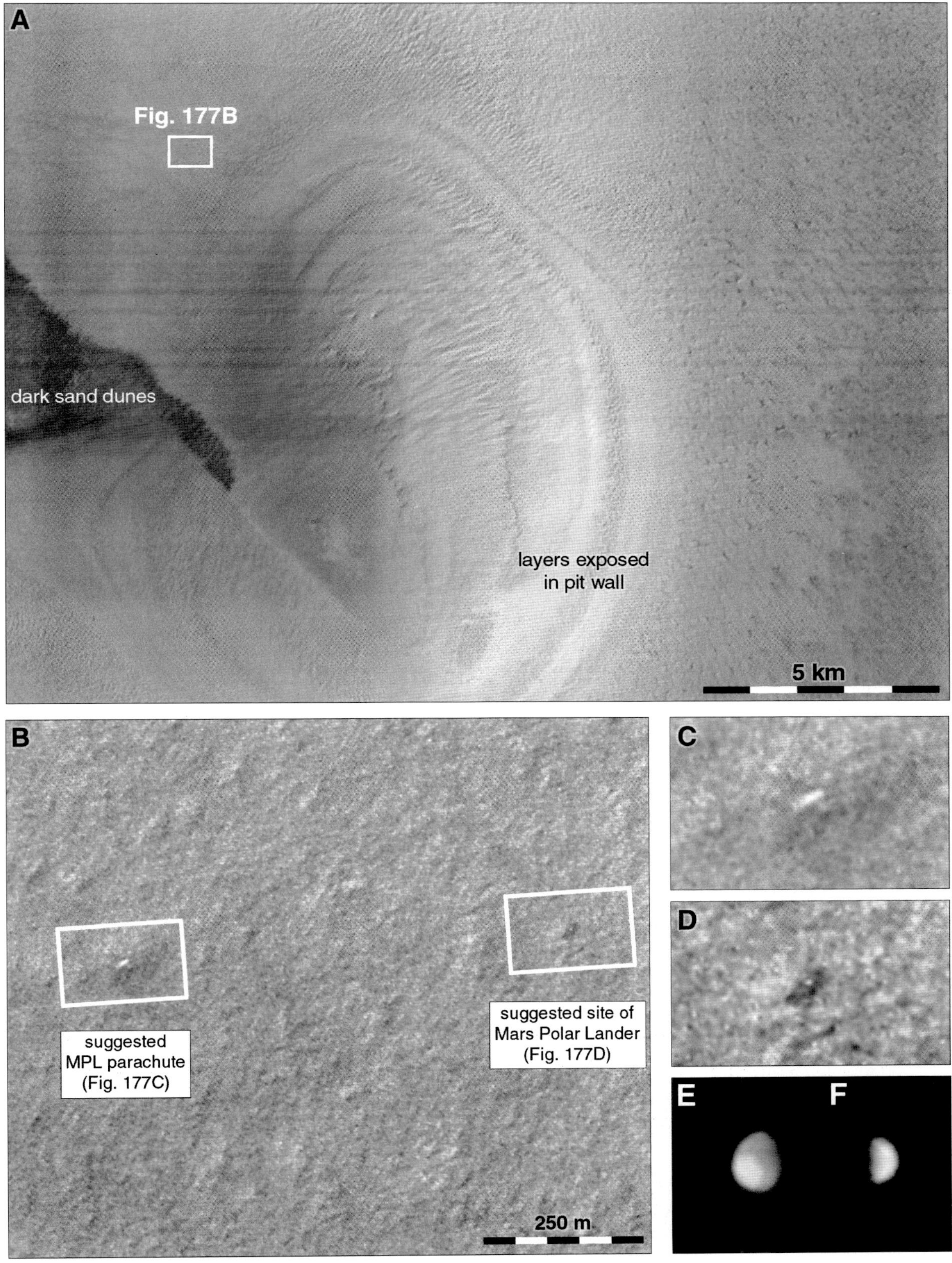

Figure 177 A: Mars Odyssey THEMIS visible image V17718013 showing the layered pit near the Mars Polar Lander impact site. **B, C, D:** MGS MOC images showing objects once suggested to be parts of Mars Polar Lander (Malin Space Science Systems, 2005a). **E:** Deep Space 1 image of Mars, courtesy of Ted Stryk. **F:** Mars Climate Orbiter MARCI image of Mars (Malin Space Science Systems, JPL, NASA).

1999: **Mars Stratigraphy Mission**

Budney *et al.* (2000) described a mission studied in 1999 by JPL's Advanced Projects Design Team. This Mars Stratigraphy Mission would land a spacecraft similar to Mars 2001 and the later Phoenix mission near Valles Marineris. The proposed landing site was at about 14° S, 68° W, roughly 10 km from the southern edge of the canyon (Figure 178). The target site was a circle 20 km across which just touched the canyon rim. Launch would be in April 2007, with landing taking place in October 2009. The lander would release a small rover which would spend up to 50 days driving to the very rim of the canyon, where it would anchor a long tether to the ground. The rover would descend the steep canyon wall, lowering itself on the tether like a climber rappelling down a cliff. Its wheels would be large balloon-like structures quite unlike the small rigid metal wheels of Sojourner or the MER rovers. It would be powered by solar panels and would communicate through an orbiter in an equatorial orbit. Over a primary mission of 200 sols it could descend about 2 km, imaging stratigraphy, measuring composition with Raman and x-ray instruments, and, at intervals of about 100 m, taking short cores for in situ age dating. The rover might descend the full 6-km depth of the canyon and reach the floor in an extended mission. The specific descent area is shown in Figure 178D, on the only Mars Global Surveyor MOC image of that part of the canyon rim.

1999: **Mars Network Design**

Mocquet (1999) analyzed network design for a future geophysical network mission to Mars to determine the minimum number of stations needed for an effective seismic network. The analysis included estimates of the expected seismic activity and properties of the planet's core and mantle. Several network designs were considered, but only one was described in the paper. It included four highly sensitive sensors located at the nodes identified in Table 57 as the 'alternative baseline array' network for Intermarsnet (De Angelis and Chicarro, 1996) (Figure 147). Three less sensitive sensors would be deployed around each sensitive instrument node (Table 71, Figure 178E). This arrangement resembles the 'triads' proposed for the Mars Global Network Mission (Table 50, Figure 142) and MESUR (Table 51, Figure 143), but in this case the triads are much larger. The large network is about 5000 km across, and the four smaller local networks are each roughly 2000 km across.

2000: **Hydrothermal Sites**

Hydrothermal activity occurs when subsurface heat and water interact, as seen in geothermal regions on Earth such as Yellowstone in the United States, Beppu in Japan or Deildartunguhver in Iceland. The heat may be provided

Table 71. *Seismic Network Design from Mocquet (1999)*

Array	Region	High-sensitivity location	Low-sensitivity location
1	Arcadia Planitia	34.0° N, 97.5° W	23.0° N, 75.0° W
			20.0° N, 110.0° W
			42.5° N, 122.5° W
2	Terra Sirenum	28.5° S, 129.0° W	40.0° S, 110.5° W
			14.0° S, 137.0° W
			34.0° S, 154.0° W
3	Argyre basin rim	44.0° S, 53.5° W	44.0° S, 35.0° W
			30.0° S, 60.0° W
			44.0° S, 80.0° W
4(Antipodal)	Terra Tyrrhena	2.5° S, 246.5° W	16.0° N, 229.0° W
			23.5° S, 229.0° W
			0.0° N, 270.0° W

by volcanic intrusions, masses of molten rock which approach the surface but do not reach it, or more briefly by ash deposits or lava flows at the surface. In a planetary context, heat can also be generated by a large impact, so large craters in ice-rich areas on Mars might also cause hydrothermal activity for a limited time. These areas may be conducive to biological activity and so have been considered important targets for landers.

Dohm *et al.* (1998) described several sites in the Thaumasia region in which volcanism, intrusions and impacts might have heated subsurface ice to create liquid water. In turn this could produce minerals characteristic of these environments which might be detected from orbit, pointing the way to landing sites with biological and fossil-preservation potential. Thirteen sites were described in that region (Table 72). Taking the idea further, Dohm *et al.* (2000) listed 23 more widely distributed sites of potential hydrothermal activity which could be targets for future missions, both robotic and human (Table 73). These sites included some from Table 72 as well as new sites from the Tharsis region. All these hydrothermal sites are shown in Figure 179.

Table 72. *Hydrothermal Sites in Thaumasia from Dohm et al. (1998)*

Site	Location	Notes
1	42° S, 110° W	Hill in Icaria Planum with volcano, valley networks and fractures
2	37° S, 108° W	Thaumasia Highlands boundary, volcano, flows, valley networks, fractures
3	40° S, 100° W	Claritas Fossae, volcano, rift system, isolated valleys
4	41° S, 94° W	Warrego Valles, complex vallley networks, fractured uplands
5	38° S, 88° W	Claritas Fossae, volcano, rift system, isolated valleys
6	34° S, 88° W	Thaumasia Highlands, volcano, rift system, isolated valleys
7	36° S, 79° W	Lampland impact crater, ejecta and impact melt, valley networks
8	31° S, 78° W	Coracis Fossae, rifts, ridges, valleys and valley networks
9	32.5° S, 76° W	Voeykov impact crater, ejecta and impact melt, valley networks
10	49° S, 82° W	Lowell impact basin, ejecta and impact melt, valley networks
11	28° S, 62° W	Nectaris Fossae, rifts, ridges, valley networks, possible pyroclastics
12	21° S, 61° W	Coprates Rise, volcano, summit depression, fractures, valley networks
13	18° S, 60° W	Coprates Rise, volcano, summit depression, fractures, valley networks

Table 73. *Hydrothermal Sites from Dohm et al. (2000)*

Site	Location		Site	Location	
14	Claritas	27° S, 106° W	26	South Tempe	30° N, 82° W
14	Syria Planum	14° S, 106° W	27		30° N, 80° W
16	Tempe	35° N, 81° W	28	West Valles Marineris	5° S, 84° W
17	Central Valles	16° S, 80° W	29	East Valles Marineris	15° S, 49° W
18	Marineris	12° S, 78° W	30	Northeast Valles	3° S, 61.5° W
19	West Thaumasia	35° S, 111° W	31	South Kasei	3° N, 78° W
20	Warrego Valles	39° S, 95° W	32	Alba Patera	37° N, 107° W
21	Southern Coprates	28° S, 62° W	33		42° N, 104° W
22	Central Thaumasia	38° S, 88° W	34	West Kasei	25° N, 75° W
23	Southwest Thaumasia	41° S, 100° W	35	Southwest Arsia	7° N, 106° W
24	Northwest Syria	4° S, 107° W	36	North Olympus Mons	28° N, 135° W
25	Planum	3° S, 108° W			

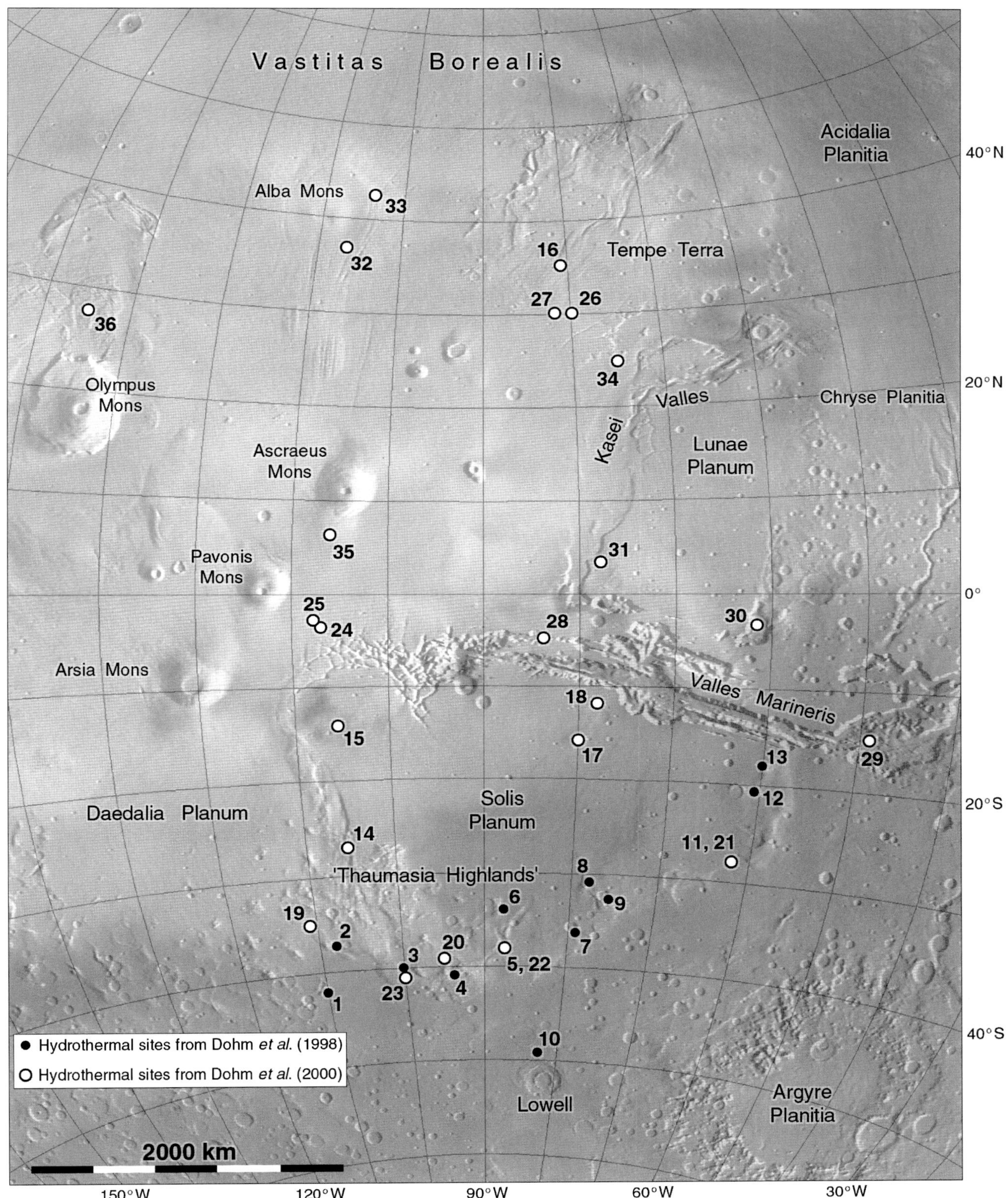

Figure 179 Hydrothermal sites in the Tharsis region identified by Dohm *et al.* (1998; 2000), from Tables 72 and 73. Their biological potential makes these sites good candidates for future landings. Site 33 is near the Mars 1984 site at Alba (Figure 112).

Mars Surveyor 2001 Landing Site Selection

The goal of Mars Surveyor 2001 was to examine materials which might preserve a record of ancient water, climate and possibly biology on Mars with a large rover. Engineering factors (temperature and available solar power) constrained the landing site latitude, initially from 30° N to 15° S, later from 15° N to 15° S and finally to the zone between 3° N and 12° S. These constraints are mapped in Figure 180. Site elevation was also constrained by engineering considerations. Above 2.5 km elevation, the parachute would not be effective. Below –3 km, the higher atmospheric pressure could hinder solar panel deployment. Within those areas, sites had to be neither too dusty nor too rocky and should have acceptable slopes and local relief.

Two site selection workshops were held to enable broad science community consultation. The 1998 workshop (Gulick, 1998) included suggestions for 68 sites (Table 74, Figure 180A). Some sites illustrated in workshop abstracts are illustrated in Figure 181. Suggested rover traverses of several tens of kilometres were illustrated for some sites (Figure 182), assuming that the mission would deploy a rover substantially larger and more capable than Mars Pathfinder's small Sojourner. The landing ellipse was 30 by 10 km at 30° N, shrinking to a 10-km circle at 15° S.

An initial site for planning purposes was on an old valley network in Terra Meridiani at 6.5° S, 358° W. One additional site suggested by Daniel Berman (Planetary Science Institute) during a contemporary landing site study was not presented at this workshop, but was mentioned in the second workshop in 1999. The study incorporated Phobos 2 Termoscan infrared data. This was Hydraotes Chaos at 0° N, 34° W, a relatively young surface in an area shaped by water flows. Both of these sites are included in Figure 181.

Cost and technical problems soon reduced the scale of the mission, and the large rover was replaced by a smaller vehicle called Marie Curie, similar to Sojourner and not expected to traverse more than about 1000 m. The sites suggested at the second workshop in June 1999 reflected this change. A list of sites of astrobiological interest (Table 75, Figure 184A) was compiled by Farmer *et al.* (1999). Farmer and his colleagues had studied astrobiology (or exobiology) sites earlier (Table 62), but this new list was restricted to sites fitting the Mars Surveyor 2001 engineering constraints and incorporated new data from Mars Global Surveyor. These and many other sites were proposed at a second landing site workshop (Gulick, 1999). They are listed in Table 76 and plotted in Figure 180B. This listing is compiled from several sources, with some duplicate sites removed and inconsistencies corrected or noted. The largest list is compiled directly from workshop abstracts (Column A in Table 76). A condensed list (Column B) was included with the published abstract volume, and another version of the list (Column C) was provided on the contemporary mission site selection website. Some individual sites are illustrated in Figures 184 through 186, including site 58, the only one for which an example of a possible rover traverse was illustrated. Locations shown in the illustrations are not always consistent with the published coordinates.

Another workshop was held at LPI in Houston on 2 through 4 October 1999. Only two specific sites were described (Figure 185), new locations in Melas Chasma (Costard *et al.*, 1999) and bright deposits interpreted as evaporites in the Brazos Valles area south of Schiaparelli (Cabrol and Grin, 1999). Those bright, fractured deposits had been seen in MOC image AB102306 in a depression on the plateau above the nearby valley floor. Soon after that meeting, the choice of landing sites was narrowed to eight 'top candidate' sites later in October 1999 (Table 77), listed on the contemporary mission website. The first five sites in Table 77, in two distinct areas, were preferred. These sites are illustrated in Figure 187 with approximate ellipse positions added at the closest reasonably smooth locations near the given coordinates, as they were not illustrated at the time. Discrepancies between ellipse positions and coordinates in Table 77 are caused by the updated coordinate system used in Figure 187. At this point the mission was cancelled and site selection work ended.

2001: **Mars Surveyor 2001 (United States)**

The Mars Surveyor program was a series of missions with the goal of launching an orbiter and a lander in every Mars launch window. In 1999 the program suffered a severe setback when its first missions, Mars Climate Orbiter and Mars Polar Lander, were both lost on arrival

Table 74. *Mars Surveyor 2001 Sites from the First Landing Site Workshop (1998)*

Site	Name	Location	Notes
1	Central Sinus Meridiani	3.2° S, 3.0° W	Aqueous sediments
2	Western Arabia Terra	5° N, 5° W	Cratered plateau
3	Becquerel crater	22° N, 7° W	Evaporite deposit, crater lake
4	Western Arabia Terra	1° S, 8.5° W	Alternative to above, cratered plateau, ridged plains
5	Margaritifer Sinus	1.5° S, 8.5° W	Plains, channels
6	Crommelin crater	5° N, 10° W	Concentric crater fill
7	Double crater – Ares	12° N, 26° W	Crater lake
8	Galilaei crater	6° N, 27° W	Sediments in crater
9	Tiu Vallis	12.5° N, 32.5° W	Masursky – dissected crater on Tiu Valles
10	Chasma confluence	5–12° S, 31–41° W	Junction of Ganges, Capri, Eos, flood sediments
11	Da Vinci crater	2° N, 39° W	Crater floor and lacustrine materials
12	Argyre Planitia	55–56° S, 41–43° W	Large lake basin, too far south, use for future mission
13	Xanthe Terra	10–15° N, 45–54° W	Highlands, valley network
14	Ganges Chasma	7–9° S, 47–52° W	Layered materials, canyon walls and floor, dunes
15	Maja Valles fan	18–22° N, 53–56° W	Highland boundary, fluvial sediments
16	Lunae Planum	23.5° N, 59° W	Ridged plains
17	Melas Labes	9° S, 72° W	Landslide deposits, canyon floor, wall materials
18	Melas Chasma	10° S, 74° W	Layered deposits, canyon floor materials
19	Candor Mensa	5.5° S, 74.6° W	Wall rock, layered deposits, landslide
20	Kasei Valles	21° N, 77.5° W	Traverse like Figure 149, ridge, plains, channel
21	Tempe-Mareotis area	c. 30° N, 80° W	Unspecified site, variety of rocks, small volcanoes
22	Labeatis Fossae	24–27° N, 80–81° W	Fractured plains, lava flows
23	Olympica Fossae	23° N, 123° W	Lava flows, fractures, channels, hydrothermal area?
24	Sirenum boundary	4–5° S, 145–149° W	Dichotomy boundary, fluvial deposits
25	Mangala Valles	14.1° S, 149° W	In channel
26	Olympus Mons	c. 20° N, 150° W	Unspecified location on outer edge of aureole
27	Mangala Valles	8° S, 151° W	Channel deposits
28	Mangala Valles	2.0° N, 152.5° W	Channel mouth
29	Labou Vallis	7.5° S, 156° W	Old crust, channel deposit
30	Mangala Valles	9.0° S, 156.5° W	Highland outside channel, old channels
31	Medusae Fossae	9° S, 159.5° W	Depositional basin
32	Elysium	5° S–10° N, 170–220° W	Channel, lake deposits
33	Memnonia channel	11° S, 173° W	Medusae Fossae Formation, outflow channel
34	Orcus Patera	c. 15° N, 181° W	Unspecified site, hydrothermal, acustrine?
35	Apollinaris Patera C1	8° S, 184° W	Channel source
36	Apollinaris Patera C2	7° S, 184.5° W	Water ponding in channel
37	Ma'adim delta-Gusev	15° S, 184.6° W	Lacustrine deposits
38	SE Elysium A	2.5° N, 184.8° W	Origin of smooth plains (water or lava)
39	SE Elysium D	6.3° N, 185.5° W	Origin of smooth plains (water or lava)
40	Apollinaris Patera B	12° S, 185.5° W	Debris fan, channels, possible thermal springs
41	Gusev – Thira craters	14.5° S, 186° W	Lacustrine deposits

Table 74. *(cont.)*

Site	Name	Location	Notes
42	Kayne crater	16° S, 187° W	Ejecta blanket
43	Apollinaris Patera A	8.6° S, 187.5° W	Ejecta and chaotic terrain
44	SE Elysium B	2.2° N, 188.7° W	Origin of smooth plains (water or lava)
45	SE Elysium C	5.1° N, 188.9° W	Origin of smooth plains (water or lava)
46	Ma'adim delta in Elysium basin	4° S, 189° W	Major estuary and delta west of Apollinaris Mons
47	Southern Elysium Planitia	1.5–3.5° S, 195–198° W	Shoreline?
48	Aeolis basin	14° S, 217.5° W	Sedimentary deposits
49	Gale crater	5° S, 222° W	Central mound, terraces, sediments, hydrothermal
50	Gale crater	5° S, 223° W	Channel mouth, sediments, hydrothermal deposits
51	Mare Tyrrhenum	4.0° S, 242.0° W	Plains, channels
52	Mare Tyrrhenum	10.0° S, 245.0° W	Plains, channels
53	Mare Tyrrhenum	6.4° S, 253.3° W	Plains, channels
54	Mare Tyrrhenum	2.2° S, 255.0° W	Plains, channels
55	Amenthes Fossae	2° N, 258° W	Aqueous sediments, traverse west into channel
56	Libya Montes	0–4° N, 270–280° W	Highlands, plains, valley networks
57	Southern Utopia	28.9° N, 270.5° W	Thumbprint terrain, plains, shoreline? Drive 30 km NW to ejecta, or to 'butterfly crater' (30° N, 273° W), possible Mars meteorite (SNC) source
58	SW Isidis B	1.8° N, 276.2° W	Plains, highland boundary
59	SW Isidis Traverse B	8.4° N, 277.6° W	Plains, sediment, hills, ejecta or landslide
60	SW Isidis A	3.4° N, 277.8° W	Channel, highland (3.9° N, 272.5° W on map)
61	SW Isidis Traverse A	9.6° N, 277.8° W	Plains, sediment, hills, ridge (possible mud volcano)
62	Arabia Terra	5° S–25° N, 310–342° W	Ridged plains, highlands, channels
63	White Rock	8° S, 335° W	Pollack mound, crater lake, evaporite or carbonate?
64	Schiaparelli	3.9° S, 340.0° W	Crater rim, gully, drive to gully or crater 10 km north
65	Schiaparelli	3° S, 343° W	Alternative to White Rock (Pollack) site
66	E. Sinus Meridiani	7.6° S, 346.9° W	Lacustrine deposit
67	E. Sinus Meridiani	0.8° S, 349° W	Varied units in close proximity
68	Northern Meridiani	0–3° N, 350–2° W	Land on highlands adjacent to sediments

at Mars. The immediate result was the cancellation of the 2001 landing mission, just known as Mars Surveyor 2001, because it used the same landing system as Mars Polar Lander. The spacecraft was put into storage and was later modified for use in the Phoenix mission in 2007. The orbiter flew successfully under the name 2001 Mars Odyssey.

Mars Surveyor 2001, which probably would have been given a more distinctive name closer to its launch in April 2001, was to carry a rover designed to survive for at least a year and to drive at least 10 km. It would explore a site in ancient terrain, looking for materials that preserved a record of water, climate and possibly biology early in the planet's history. The rover would collect rock and soil samples with a coring drill and scoop and store them for possible collection by a 2005 sample return vehicle. The rover instrument payload, called Athena, was developed at Cornell University and eventually flew on the Mars Exploration Rovers in 2003. It consisted of high-resolution cameras, an infrared spectrometer and two other instruments for in situ measurements of surface composition. After deploying the

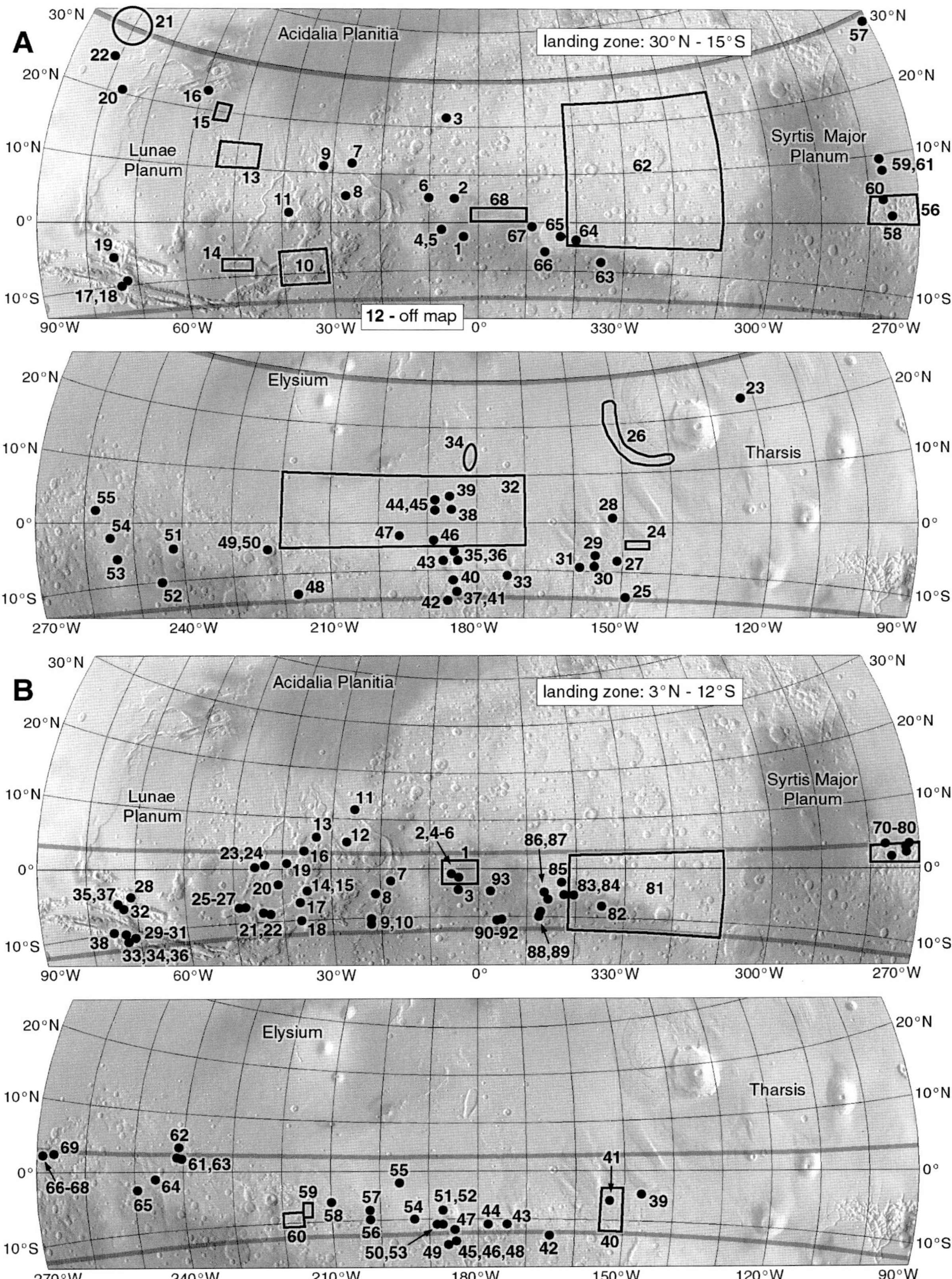

Figure 180 Mars Surveyor 2001 landing sites suggested at (A) the first landing site workshop, (B) the second landing site workshop.

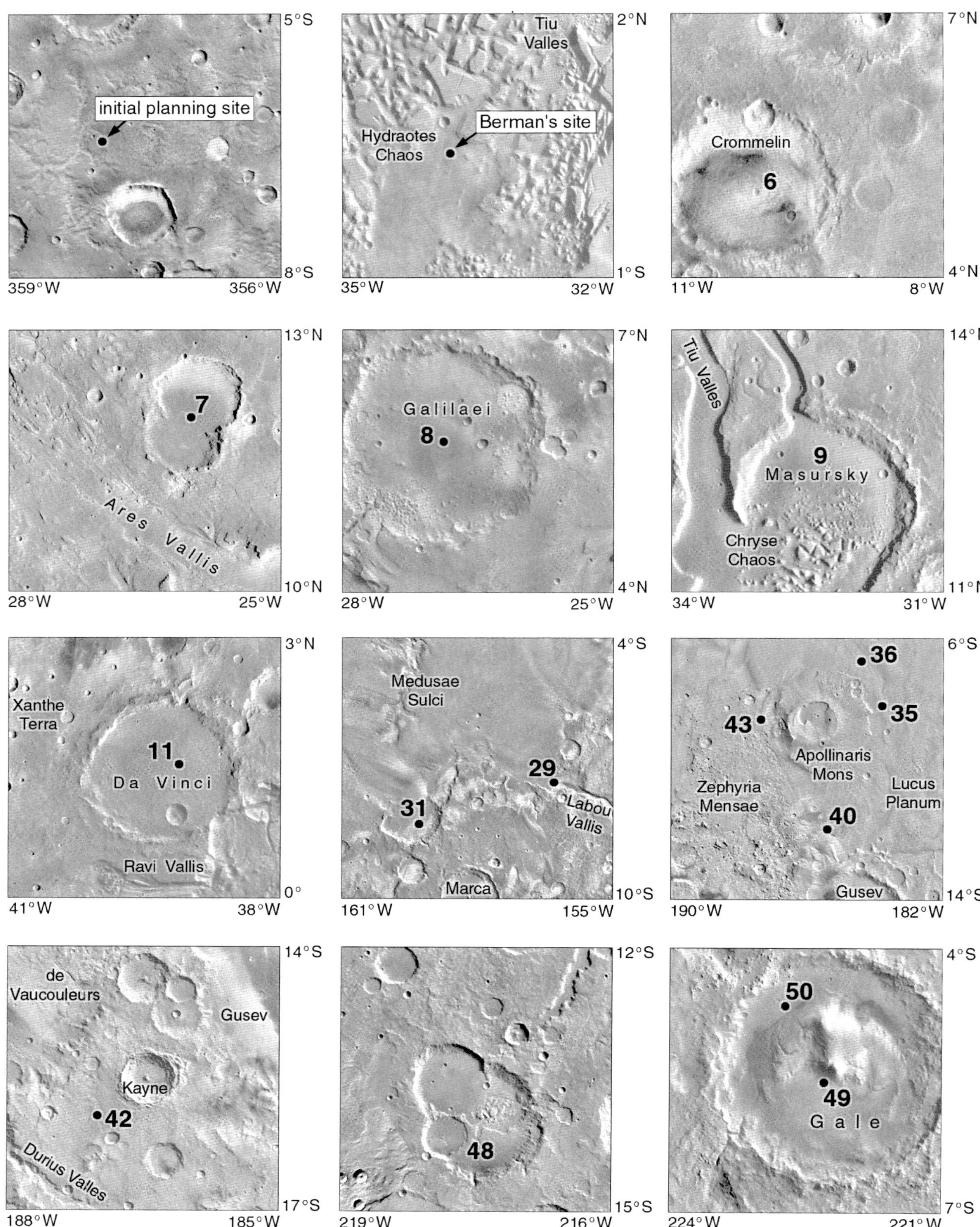

Figure 181 Examples of Mars Surveyor 2001 sites from the first Landing Site Workshop or contemporary sources referred to in the text.

Site numbers refer to Table 4. Black dots are exact locations illustrated in the Workshop abstracts. Sites without black dots were only specified by coordinates. Site 50 in Gale crater was chosen for the Mars Science Laboratory rover, launched in 2011. All maps are 3° (180 km) across, for scale.

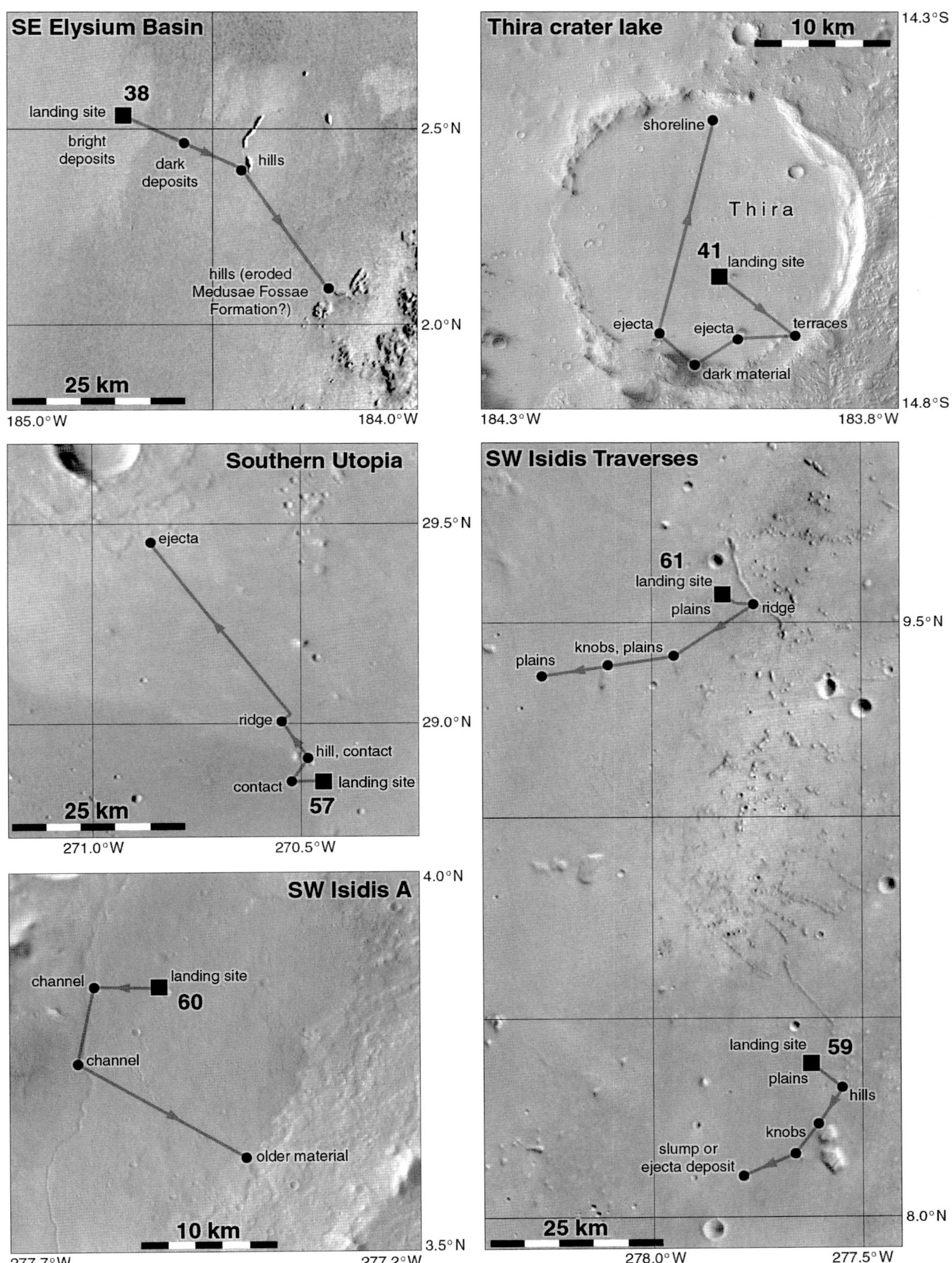

Figure 182 Traverses illustrated or described at the first Mars Surveyor 2001 Landing Site Workshop. Site numbers refer to Table 74.

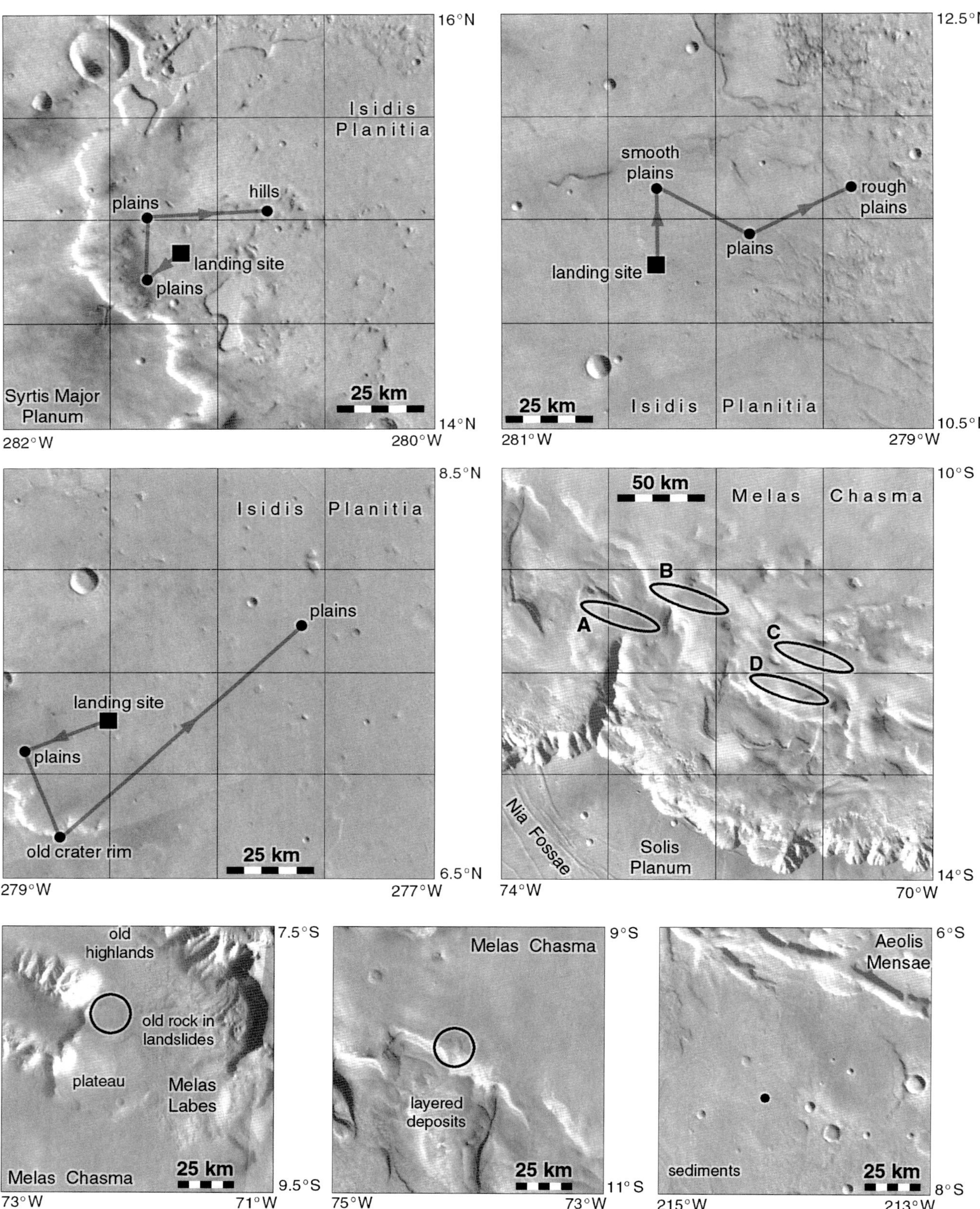

Figure 183 Additional Mars Surveyor 2001 sites and rover traverses suggested at the 30th Lunar and Planetary Science Conference in March 1999.

Table 75. *Mars Surveyor 2001 Astrobiology Sites (Farmer et al., 1999)*

75a. High- to Moderately High-Priority Sites Meeting All Criteria

Site		Location	Type of site*	Priority
1	Terra Cimmeria	8–11° S, 216–220° W	1, 2	Highest
2	Mangala Valles	3–12° S, 150–155° W	1, 2	Highest
3	Xanthe – Da Vinci	0–3° N, 40–44° W	2	Moderate to high
4	S. Ares Vallis	0–3° N, 17–19° W	2	Moderate to high
5	Libya Montes	1–3° N, 272–274° W	1,2	Moderate to high
6	SE Xanthe – Iani Chaos	9–12° S, 27–29° W	3	Moderate to high
7	SE Xanthe – Iani Chaos	9–12° S, 29–30° W	3	Moderate to high
8	SE Xanthe – Iani Chaos	8–12° S, 30–31° W	3	Moderate to high
9	Iani Chaos	0–3° N, 13–15° W	3	Moderate to high
10	Apollinaris Chaos	12–4° S, 188–190° W	3	Moderate to high
11	NE of Gusev crater	14–9° S, 180–181° W	1, 2	Moderate to high
12	Al-Qahira Vallis	15–14° S, 194–196° W	1, 2	Moderate to high

75b. High-Priority Sites With Marginal Rock and Image Data

Site		Location	Type of site*	Priority
13	Amenthes Rupes	2.9° S, 249.5° W	1, 2	Highest
14	Apollinaris Chaos	11.1° S, 188.5° W	2, 3	Highest
15	Da Vinci crater	1.2° N, 39.1° W	2, 3	Highest
16	Ganges Chasma (1)	8.5° S, 43.9° W	1, 2	Highest
17	Ganges Chasma (2)	8.8° S, 42.5° W	1, 2	Highest
18	Libya Montes region	1–3° N, 270–280° W	1, 2	Highest
19	N. Memnonia Terra (1)	11.3° S, 174.2° W	1, 2	Highest
20	N. Memnonia Terra (2)	11.2° S, 178.2° W	1, 2	Highest
21	Nicholson crater	0° N, 164° W	2, 3	Highest
22	Reuyl crater	9.9° S, 192.8° W	2, 3	Highest
23	Shalbatana source (1)	0.2° N, 46.3° W	1, 2, 3	Highest
24	Shalbatana source (2)	0.7° N, 44.5° W	1, 2, 3	Highest

*Site types are (1) grab bag site, (2) fluvial-lacustrine site and (3) potential hydrothermal site.

rover, the lander would function for at least 100 days with a soil and dust analysis instrument, a radiation sensor for future human Mars planning and a Mars In-Situ Propellant Production Precursor experiment. Low power at higher northern latitudes at the time of landing would have delayed any lander science for 30 sols and rover deployment for 50 sols, so the initial latitude range (30° N to 15° S) was soon reduced at its northern boundary to 15° N and then to a still narrower limit (3° N to 12° S) for final site planning.

The mission was reduced in scope in 1999 for technical and budgetary reasons. The Athena instruments and more capable rover were replaced with APEX, the Athena Precursor Experiment, and a small rover called Marie Curie, very similar to Mars Pathfinder's Sojourner. This could only be expected to travel about 1000 m at most. APEX consisted of a panoramic camera and infrared spectrometer mounted on a mast on the lander, an Alpha Proton X-ray Spectrometer on the rover, and a Mossbauer Spectrometer on a robotic arm on the lander.

Table 76. *Mars Surveyor 2001 Sites from the Second Landing Site Workshop (1999)*

Site	A	B	C	Name	Location	Notes
1	x	x	x	Hematite 2, Meridiani	3° S–2° N, 0–7° W	Hematite, eroded layered deposit, paleolake, highlands (also said to be 353° W–5° W)
2	x			Hematite	c. 2° S, 4° W	South edge of hematite, layering
3	x	x	x	S. Terra Meridiani	4.75° S, 4.75° W	Channels in highlands (also 4.5° S, 5° W)
4	x			Hematite	1.2° S, 5.5° W	Edge of hematite exposure
5	x	x	x	Hematite 1, N. Meridiani	1.5° S, 5.5° W	Hematite, smooth deposit (also 6° W)
6	x			Hematite	2.8° S, 5.8° W	Layering and hematite
7	x	x	x	Ares Vallis headlands	2° S, 18° W	Outflow channel (also 2.5° S, 18.5° W)
8	x	x	x	Iani Chaos	5° S, 21° W	Old, smooth lava flows
9	x	x	x	Valley confluence	10.85° S, 21.62° W	Diverse sediments from several sources
10	x	x	x	Margaritifer basin	10.0° S, 21.79° W	Fluvial sediments in depositional basin
11			x	Crater near Ares Vallis	12° N, 26° W	(Too far north?)
12			x	Galilaei crater	6° N, 27° W	Crater floor
13			x	Tiu Vallis crater	7° N, 33° W	(Too far north?)
14			x	Valles Marineris	4.1° S, 35.2° W	Outflow channel environment
15	x	x		Ganges-Eos Chasmata C	4.1° S, 35.2° W	Canyon floor, fluvial deposits, wall debris
16	x			N. Hydraotes Chaos	3° N, 36° W	Source of Tiu and Simud Valles
17	x	x	x	Ganges-Eos Chasmata B	6.5° S, 37° W	Canyon floor, fluvial deposits, wall debris
18	x	x	x	Ganges-Eos Chasmata A	10.5° S, 37.1° W	Canyon floor and fluvial deposits
19			x	Da Vinci crater	1.2° N, 39.1° W	Fluvio-lacustrine, possibly hydrothermal
20	x	x	x	Ravi Vallis	2.8° S, 40.8° W	Lava flows, ejecta
21	x	x	x	Ganges Chasma 2	8.8° S, 42.5° W	Fluvial sediments
22	x	x	x	Ganges Chasma 1	8.5° S, 43.9° W	Outflow sediments, landslides
23	x	x	x	Shalbatana Vallis 2	0.7° N, 44.5° W	Channel floor and wall debris
24	x	x	x	Shalbatana Vallis 1	0.2° N, 46.3° W	Channel source and deposits, ancient crust
25	x	x	x	Ganges Chasma 3B	8.1° S, 48.0° W	Sand sheet, images of layered mesa
26	x	x	x	Ganges Chasma 3A	8.0° S, 49.3° W	Sand sheet, view layered mesa (also 7.9° S)
27	x	x	x	Ganges Chasma 3C	7.4° S, 49.5° W	Layered material, remote sensing of slopes
28	x			Ophir Chasma	4.2° S, 70.8° W	Walls, landslides, layered mesa
29	x	x	x	Melas Chasma 1A	11.5° S, 71.0° W	Layered deposits
30	x	x	x	Melas Chasma 1B	11.5° S, 71.4° W	Foot of canyon wall material
31	x	x	x	Melas Chasma 1C	11.6° S, 72.1° W	Floor material with layers nearby
32	x			Candor Chasma	7.7° S, 72.5° W	Layered deposits, remote sensing of walls
33	x	x	x	Melas Chasma 2	10° S, 73° W	Canyon floor materials
34	x	x	x	Melas Chasma 1D	10.4° S, 73.2° W	Floor material with layers nearby
35	x	x	x	Candor Mensa	6.5° S, 73.5° W	Top of layered plateau

Table 76. *(cont.)*

Site	A	B	C	Name	Location	Notes
36	x	x	x	Candor Mensa	11.0° S, 73.5° W	Lower plateau layers, canyon walls images
37			x	Candor Mensa 2	5.5° S, 74.6° W	Variety of rocks, possible lake bed
38	x	x	x	Candor Mensa	9.7° S, 75.3° W	Lower levels of layered plateau
39	x			Terra Sirenum	4.9–5.2° S, 147.0–147.5° W	Dichotomy, sediments, ejecta, lava flows
40			x	Mangala Valles 2	3–12° S, 150–155° W	Fluvial, lacustrine
41	x	x	x	Mangala paleolake	6.0° S, 153.2° W	Paleolake sediments in northern crater floor
42	x	x	x	Magnetic anomaly B (2)	13.5° S, 166° W	Channels, ejecta (also 13.2° S, 165.2° W)
43	x	x	x	N. Memnonia 1	11.3° S, 174.2° W	Recent fluvial sediments, ejecta
44	x	x	x	Memnonia valley	11.2° S, 178.2° W	Depositional basin, channel mouth
45	x	x	x	Gusev crater	14° S, 184° W	Impact crater, paleolake basin
46	x			Gusev-Ma'adim delta	15° S, 184.6° W	Fluvio-lacustrine, highlands
47	x	x	x	Apollinaris Patera 2	12.0° S, 185.5° W	Eroded fan, hot springs (also 185.75° W)
48	x			Gusev-Thira	14.5° S, 184° W	Lava flows, lake beds, frost mounds
49			x	Kayne crater	15.6° S, 186.4° W	Crater
50		x		N. Memnonia 2	11.2° S, 187.2° W	Paleofluvial – lacustrine
51	x	x	x	Apollinaris Patera 1	8.6° S, 187.5° W	Ejecta, chaotic terrain (also 8.5° S)
52		x	x	Apollinaris Patera 3	8.5° S, 188° W	Dichotomy, lake or ocean, hydrothermal
53	x	x	x	Apollinaris Chaos	11.1° S, 188.5° W	Smooth depression in chaos area
54	x	x	x	Reuyl crater	9.9° S, 192.8° W	Lake sediments, evaporites
55	x	x	x	Ibishead peninsula (S. Elysium)	1.5–3.5° S, 195–198° W	Eroded highlands, Elysium basin deposits (also 2° S, 196° W or 2.5° S, 196.5° W)
56	x			Magnetic anomaly A	9° S, 202° W	Highland-lowland boundary, channels
57		x	x	Magnetic anomaly 1	8.29° S, 202.1° W	Dichotomy positive anomaly, fluvial
58	x	x	x	Dichotomy boundary	6° S, 210° W	Plains, highlands (also 6.5° S, 210.5° W, and illustrated at 7.1° S, 208.8° W)
59	x			Aeolis Mensa	6–9° S, 214–216° W	Layered material
60			x	Terra Cimmeria	8–11° S, 216–220° W	Fluvial, lacustrine
61	x	x	x	Escalante basin	2° N, 240.5° W	Channel deposits
62	x			Terra Cimmeria	4° N, 241° W	Ancient terrain eroded by water
63	x	x	x	Amenthes boundary	2.5° N, 241.5° W	Lava flows, ejecta, channel deposits
64	x	x	x	Amenthes trough	1.7° S, 246.4° W	Ancient sediments
65	x	x	x	Amenthes Rupes (Palos)	2.9° S, 249.5° W	Fluvial, lacustrine (also 2.5° S or 4.0° S)
66	x			Terra Tyrrhena A	2.0° N, 268.5° W	Cratered highlands
67	x			Terra Tyrrhena B	2.4 N, 268.6 W	Cratered highlands

Table 76. *(cont.)*

Site	A	B	C	Name	Location	Notes
68	x			Terra Tyrrhena C	2.4 N, 268.2 W	Cratered highlands
69	x			Terra Tyrrhena D	2.6 N, 267.3 W	Cratered highlands
70			x	Libya Montes 3	1–3° N, 270–280° W	Fluvial-lacustrine
71	x	x	x	E. Libya Montes – 1D	3.1° N, 272.0° W	Fluvial deposits in highland basin
72		x	x	Libya Montes – 1C	3.0° N, 272.0° W	Fluvial deposits in highland basin
73	x			Libya Montes middle	2.4° N, 273.0° W	Fluvial deposits in highland basin
74	x	x	x	Libya Montes NW – 1B	3.0° N, 272.9° W	Fluvial deposits in highland basin
75	x	x	x	Libya Montes SW – 1A	2.8° N, 273.0° W	Fluvial deposits in highland basin
76	x	x	x	Libya Montes 2	2° N, 273° W	Fluvial deposits in highland basin
77	x			Libya Montes B	1.76° N, 276.29° W	Lacustrine sediments
78		x	x	Libya Montes – Isidis 2	2° N, 276.5° W	Highland-lowland boundary, channels
79	x			Libya Montes A	3.73° N, 277.54° W	Fluvial deposits (also 3.4° N, 277.8° W)
80		x	x	Libya Montes – Isidis 1	3.77° N, 277.6° W	Highland-lowland boundary, channels
81		x	x	Arabia Terra	12° S–3° N, 310–342° W	Ancient cratered highlands
82	x	x	x	White Rock basin	7.8° S, 334.85° W	Lacustrine sediments, fans, evaporites
83	x		x	Meridiani – Brazos lakes	5.5° S, 340.7° W	Paleolake bed, evaporites (also 341° W)
84	x			Schiaparelli floor	5° S, 342° W	Fan and floor deposits
85	x	x	x	Schiaparelli S, SE	2.5° S, 343° W	Brazos channel deposits, evaporites?
86	x	x	x	Terra Meridiani	7° S, 346° W	Old terrain, channels, layers (also 347° W)
87	x		x	E. Terra Meridiani	5° S, 347° W	Evaporites, sulphates? (or 5.5° S, 347.7° W)
88	x	x	x	Sinus Sabaeus	9.12° S, 347.81° W	Channel deposits
89	x	x	x	Evros paleolake	10° S, 348.5° W	Lacustrine, evaporites, fans (also 348° W)
90	x			Madler crater	10.6° S, 356.5° W	Preferred site – floor deposits, wall imaging
91	x			Madler crater	10.3° S, 357.4° W	Floor deposits, wall imaging
92	x			Madler crater	10.9° S, 357.6° W	Floor deposits, wall imaging
93	x			Meridiani Sinus	5° S, 358° W	Crater floor materials

Note: Site numbers in this table and Figure 180B are ordered by longitude. A: sites from individual Workshop abstracts. B: sites from a list in Gulick (1999). C: sites from the contemporary landing site website.

7 April 2001: **2001 Mars Odyssey (United States)**

2001 Mars Odyssey, often simply called Mars Odyssey, was the orbiter part of the Mars Surveyor 2001 mission. Its goals were to conduct a mineralogical survey of the surface of Mars from orbit using a thermal emission spectrometer (TES) and a gamma-ray spectrometer (GRS) to characterize the climate and geology of Mars using infrared and visible multispectral imaging (Thermal Emission Imaging System, THEMIS) and to assess the radiation hazard to future astronauts with MARIE (Mars Radiation Environment Experiment). In later years it acted as a communications relay for the Mars Exploration Rovers, a function added to all subsequent orbiter missions.

The spacecraft was a 2.2 by 1.7 by 2.6 m box with a large solar array on one side and a dish antenna on a

Table 77. *Mars Surveyor 2001 Top Candidate Sites*

Name	Location	Notes
Western Hematite	0–3° S, 2–7° W	**Preferred site:** 1.5° S, 4.5° W
Isidis rim	3° N–1° S, 270–280° W	**Preferred sites:** Isidis Rim 1: 1.6° N, 271.9° W Isidis Rim 2: 2.4° N, 274.5° W Isidis Rim 3: 2° N, 276.5° W Isidis Rim 4: 3.77° N, 277.6° W
Amenthes highlands	2.4° N, 241.5° W	
Hesperia paleolake	2.5° S, 249.5° W	(Palos crater)
NE Meridiani	2.0° S, 350.2° W	

mast. The instruments and equipment, mounted on a separate science deck, consisted of star cameras for attitude control, the THEMIS instrument, the MARIE radiation sensor and the GRS, which had two neutron detectors on the science deck as well as a gamma-ray detector on a 6-m-long boom. MARIE seemed to fail during cruise but was revived in March 2002. THEMIS could operate over the day and night hemispheres, investigating both composition and thermal inertia (Christensen *et al.*, 2004). The latter, estimated from the rate of cooling of the surface at night, provided information on the rock distribution at the surface and helped to identify relatively rock-free areas for future landing site studies. There was some hope that night temperature data might reveal areas of enhanced subsurface heat flow related to volcanic activity or heated groundwater, but nothing like this was found.

The 376-kg spacecraft (725 kg when fueled) was launched at 15:02 UT from Cape Canaveral and reached Mars on 24 October 2001 (MY 25, sol 494), entering orbit at 02:26 UT. The initial orbit had an 18.6-hour period, but by aerobraking until 30 January 2002, the spacecraft reached its 201 by 500 km, 2-hour period mapping orbit (Saunders *et al.*, 2004). The nominal mission ended in July 2004 but was extended several times, and the spacecraft was still operating in 2011. On 15 December 2010 (MY 30, sol 404), Mars Odyssey surpassed the 3340-day lifetime of Mars Global Surveyor to become the longest operating Mars orbiter.

When Mars Odyssey arrived at Mars, its orbit crossed the dayside equator at 16:00 local time, but the orbit plane was allowed to drift until it crossed the equator at 17:00 to give the GRS instrument better observing conditions, though this was less than optimal for THEMIS. GRS would have overheated if used closer to local noon. On 30 September 2008 (MY 29, sol 288), a small maneuver changed the orbit period slightly so that the equator crossing would happen a little earlier each day. By 22 June 2009, the spacecraft was over the equator at 15:45 local time, and another maneuver fixed the orbit plane again. This provided better conditions for the THEMIS spectral observations, because the day-night temperature variations were stronger. The instrument was also pointed off nadir at times to fill gaps in mapping or to provide stereoscopic images. Global mosaics of THEMIS infrared images were made available online at the Arizona State University using dynamic mapping software at http://jmars.mars.asu.edu/maps, providing easy access to the data. THEMIS images are used throughout this atlas, including Figures 52C, 85B, 139, 140 and 187.

Figure 188 shows images from THEMIS. Figure 188A is a daytime infrared mosaic of the area illustrated as a Viking image mosaic in Figure 51D. Figure 188B is the nighttime infrared mosaic of the same area. Night temperatures show how much the surface has cooled since sunset, revealing information on surface properties, including rock concentrations. A higher resolution view (Figure 188C) is from the visible imager. Figure 188D shows areas of enhanced hydrogen around the poles, assumed to be in the form of ice in the regolith. Near-global coverage was obtained by the infrared imager in both day and night viewing, and a significant fraction of the surface was mapped at up to 18 m/pixel by the visible imager. THEMIS identified deposits of clays, carbonates,

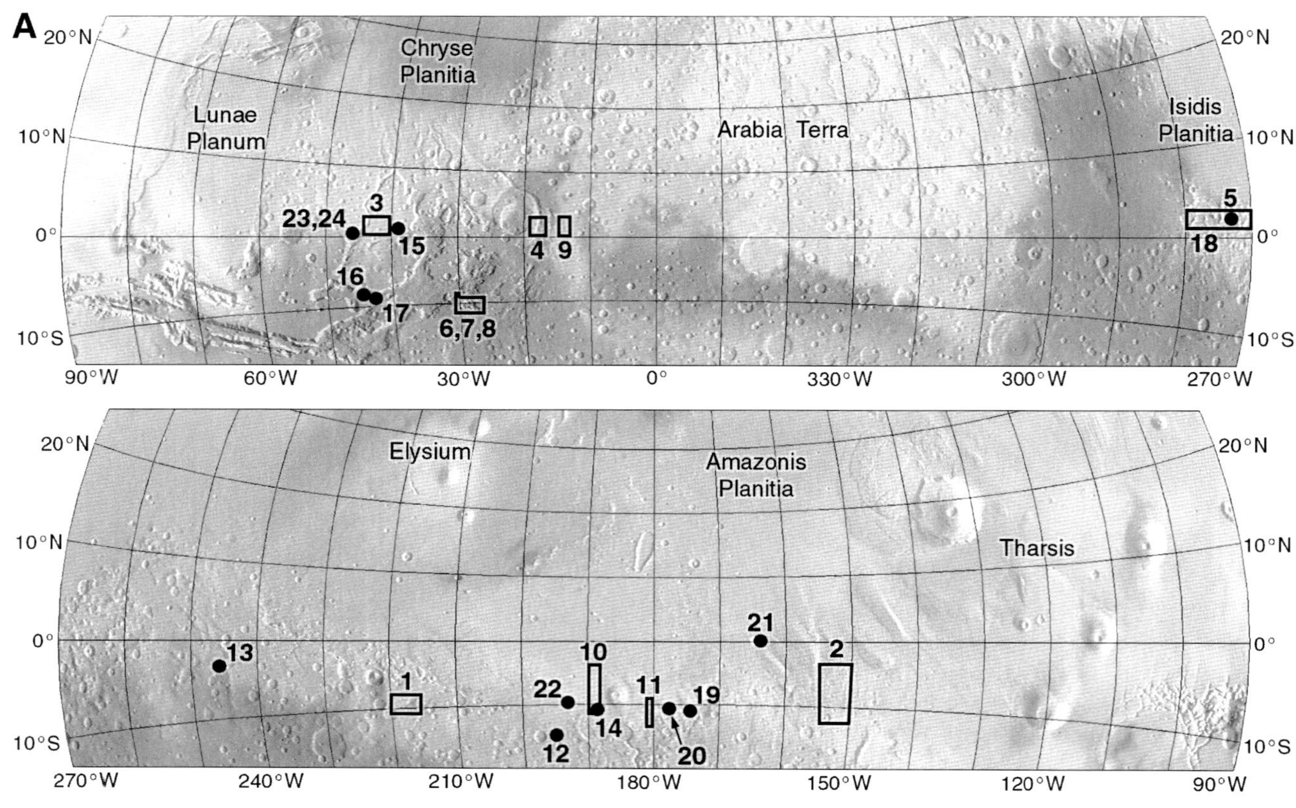

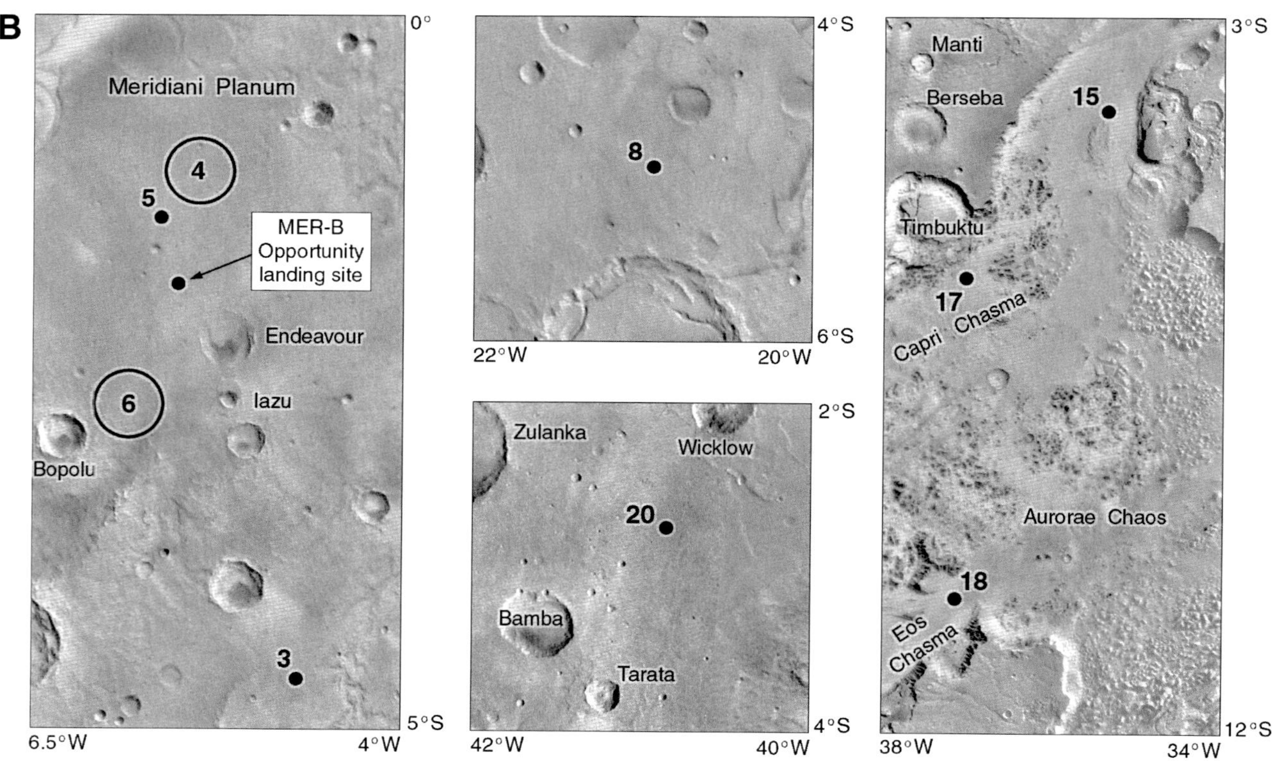

Figure 184 Mars Surveyor candidate landing sites from the Second Landing Site Workshop (Gulick, 1999). **A:** Astrobiology sites from Farmer *et al.* (1999). **B:** Specific sites illustrated in the meeting abstracts, continued in Figure 185. Site numbers are from Table 76. For scale, 1° spans 60 km.

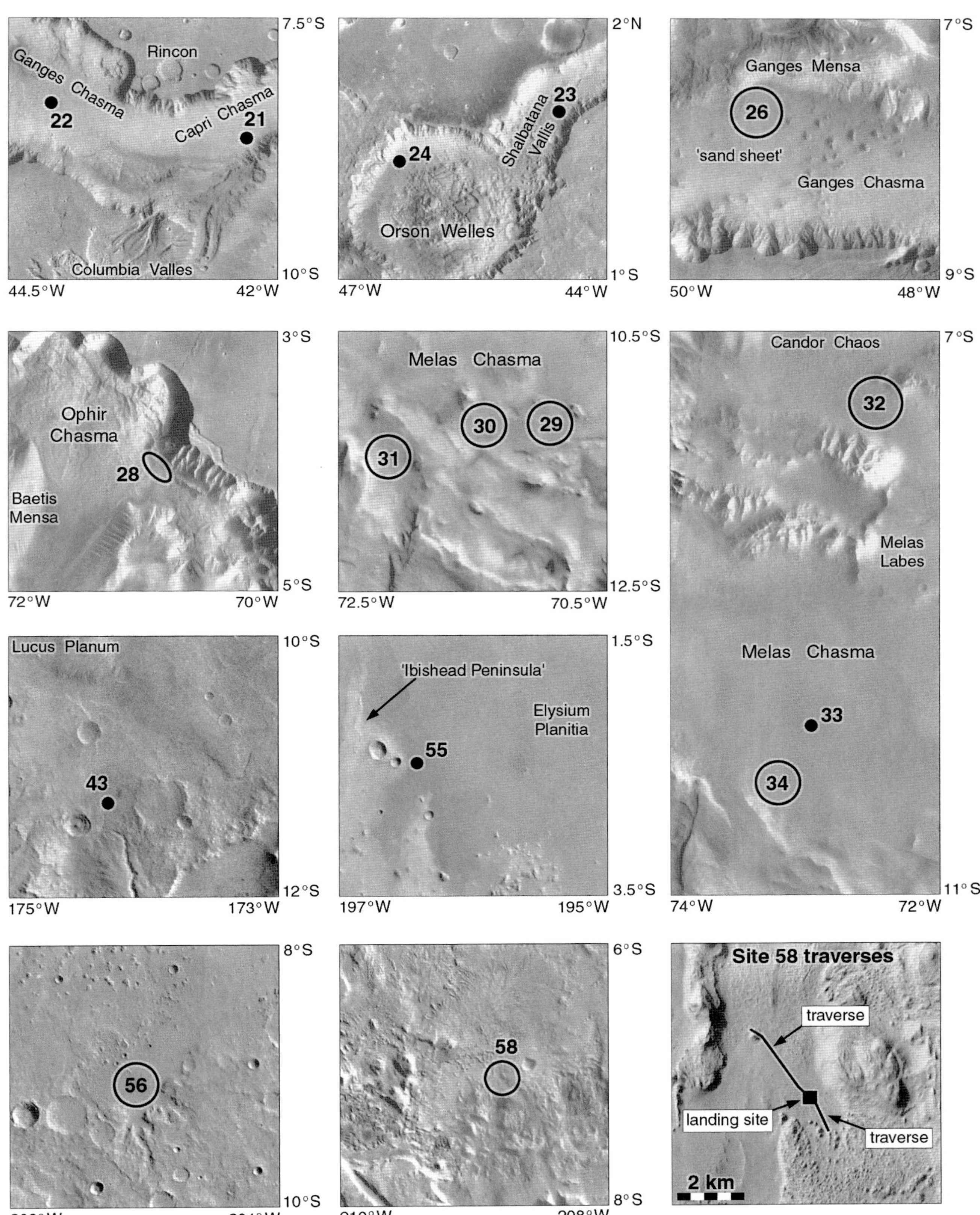

Figure 185 Mars Surveyor candidate landing sites from the Second Landing Site Workshop (Gulick, 1999). For scale, 1° spans 60 km, and most images are 2° or 120 km across. The Site 58 traverses are drawn on Viking image 762A48. They are only representative of possible traverses as the exact landing point could not be predicted in advance.

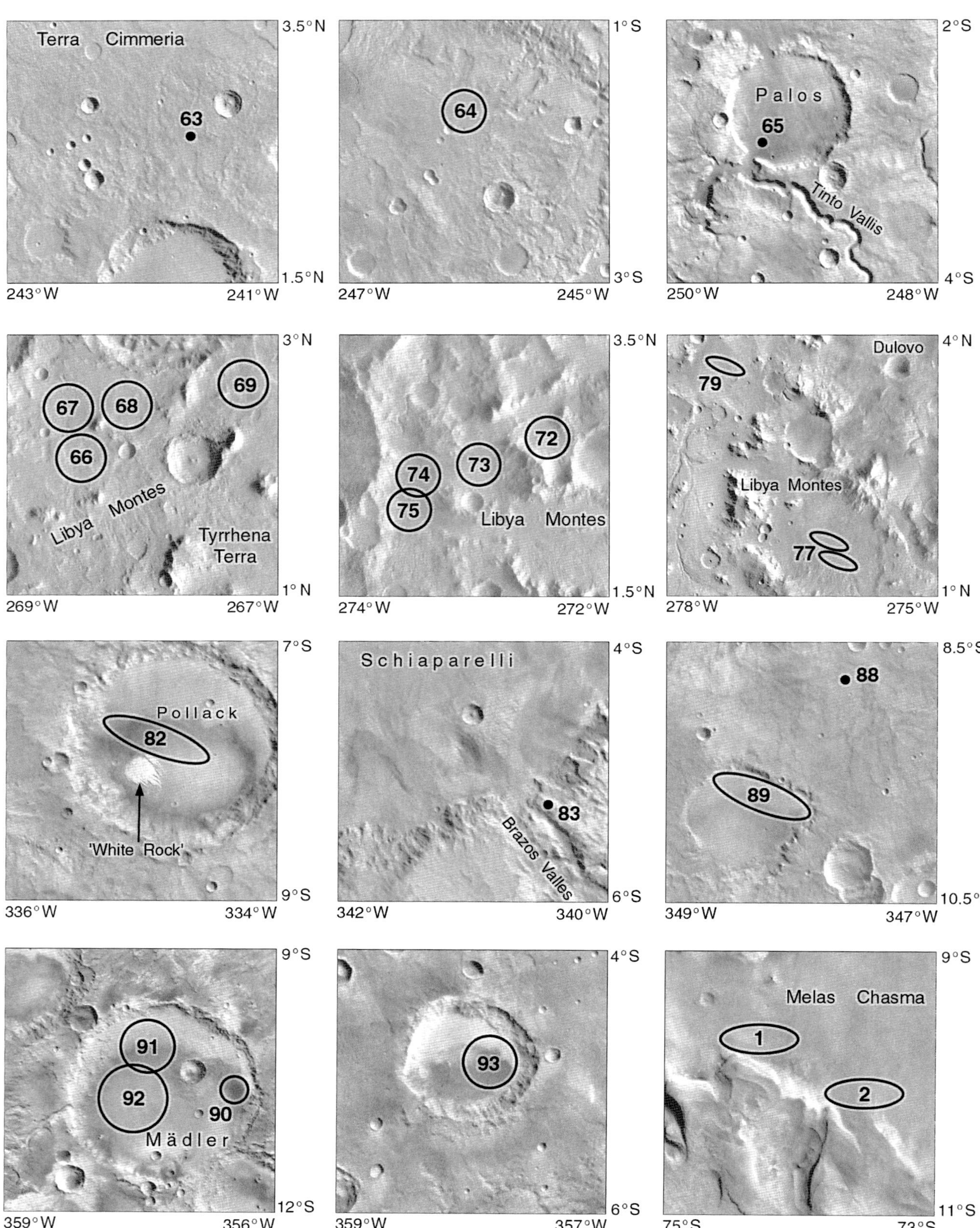

Figure 186 Mars Surveyor candidate landing sites from the Second Landing Site Workshop (Gulick, 1999). Most images are 2° or 120 km across. The last image (lower right) with two sites in Melas Chasma is from Costard *et al.* (1999). The Brazos Valles site (site 83) was also described by Cabrol and Grin (1999). Site numbers for all locations except the last image are from Table 76.

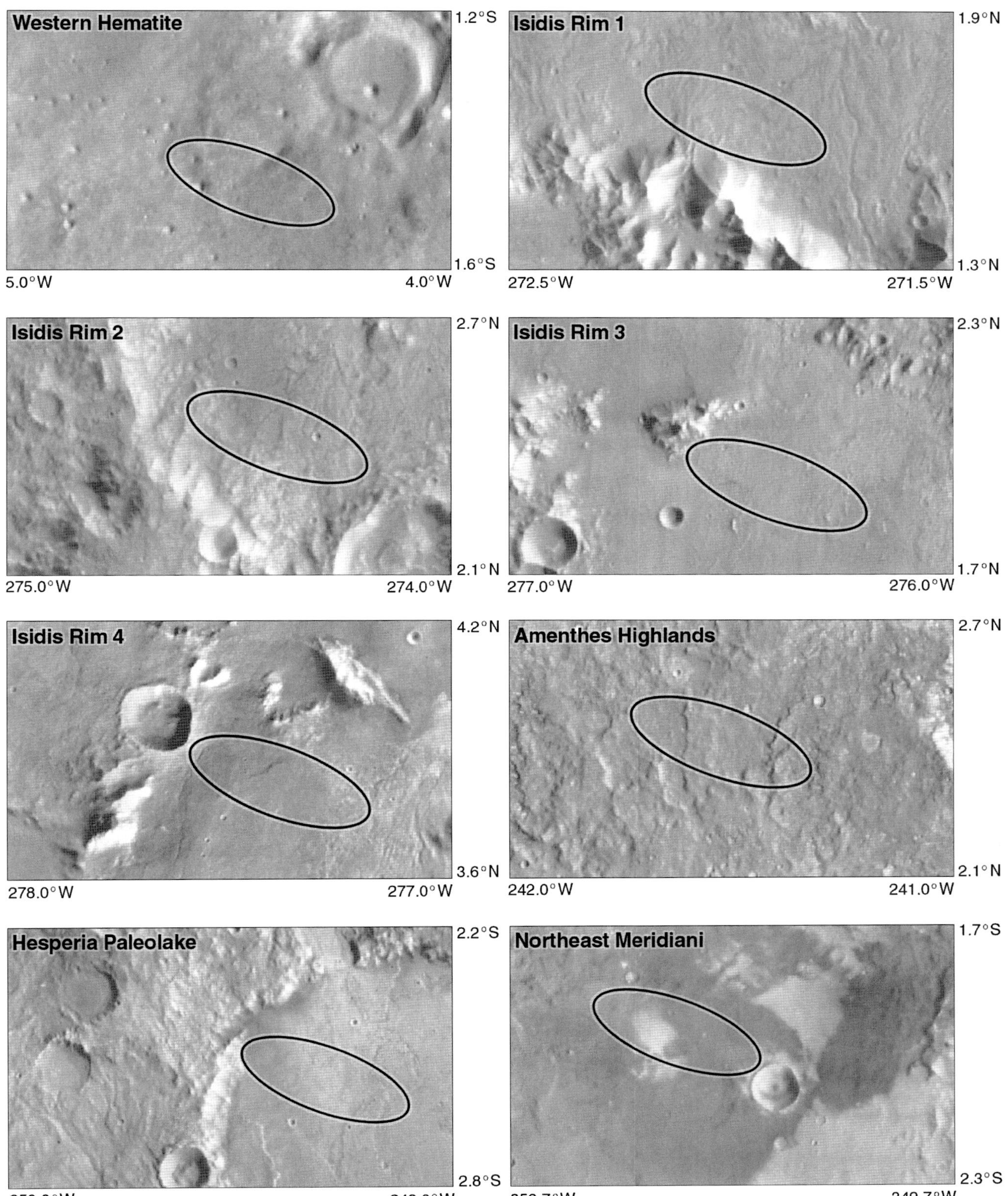

Figure 187 Shortlisted sites for Mars Surveyor 2001 from Table 77.
Each image is a Mars Odyssey THEMIS daytime infrared mosaic from the *Jmars* website at Arizona State University (see text) with shading inverted to resemble visual albedo (especially significant in Northeast Meridiani), and each spans 1° (60 km) from left to right. Isidis Rim 4 is almost identical to SW Isidis A in Figure 182.

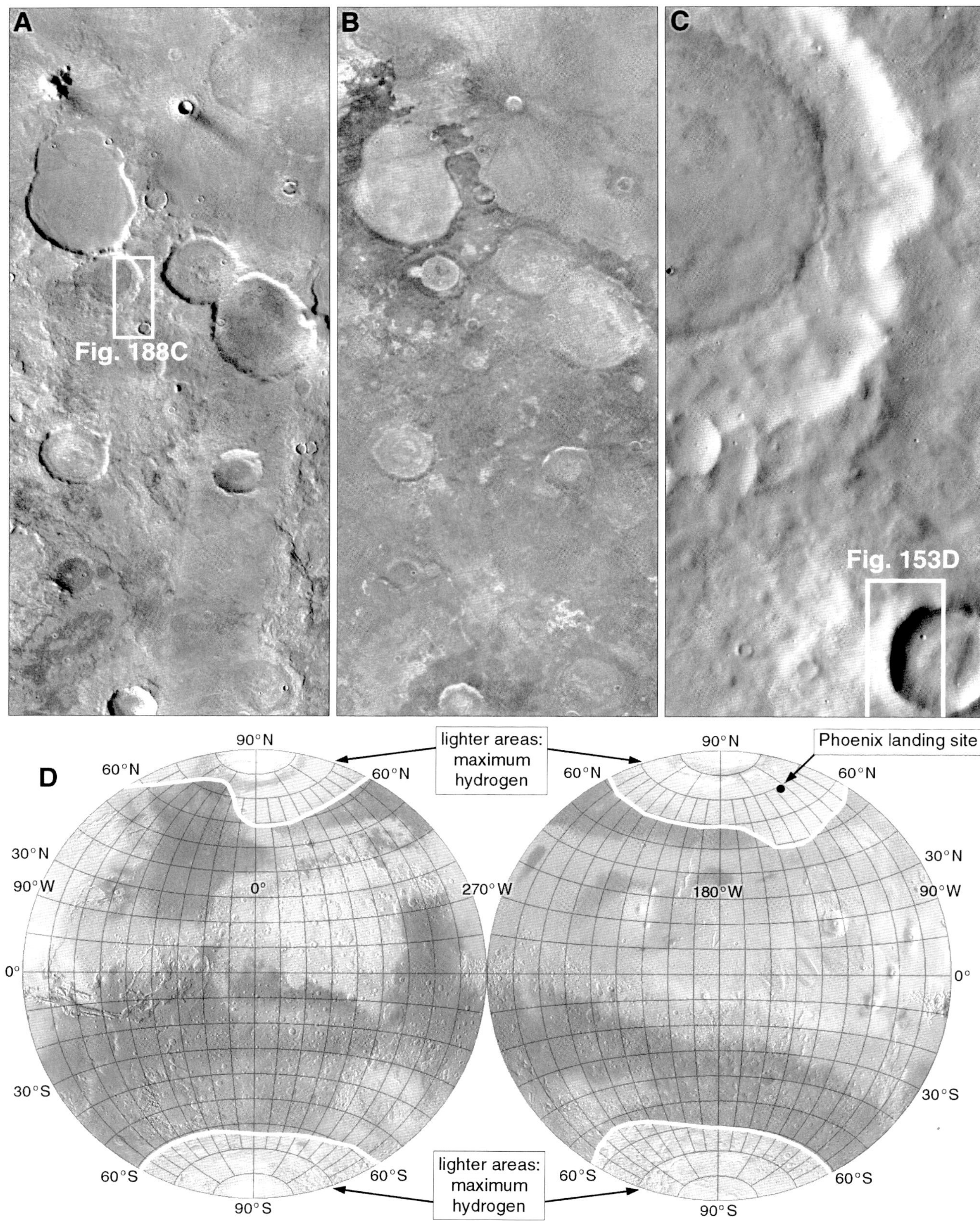

Figure 188 Mars Odyssey images.
A: THEMIS daytime IR mosaic. **B:** THEMIS nighttime IR mosaic. **C:** THEMIS visible image V30005004, approximately 49.5° N, 52.5° W, in the same area as Figures 51D, 153D and 190D. **D:** Hydrogen concentrations around the poles of Mars, detected by the Gamma Ray Spectrometer on 2001 Mars Odyssey. The outlined areas contain at least 25% water by weight in the soil, inferred from the effects of hydrogen on neutrons emitted from the surface.

chlorides, sulphates and other materials which had a profound effect on future landing site selection work. More limited compositional data had been obtained by TES on Mars Global Surveyor, and the future Mars Reconnaissance Orbiter would add higher resolution data from its CRISM instrument. Since the time of Mariner 9, and to a lesser extent even Mariner 4, geomorphology had been the main source of information for site selection, but in future landing site studies, the surface composition played an equal or greater role. Water ice concentration and rock distribution data from Mars Odyssey were used to select the Phoenix landing site in 2008.

2000s: **Seeking Water**

WOBBLE (Water Observations from a Balloon Borne Light Explorer) was a conceptual study for a small balloon probe to explore a region of Mars showing evidence for the presence of water (Udrea *et al.*, 2002). The type of feature chosen for investigation was a gully in a crater wall, interpreted by Malin and Edgett (2000) as the result of fluid flow, some of which appeared to be very recent or currently active. The concept might have been developed for a Discovery, Mars Surveyor or Mars Scout mission.

A lander equipped with airbags like Mars Pathfinder would deploy a 24-m-diameter hydrogen balloon which could reach otherwise inaccessible gullies. Its instruments would be high-resolution cameras, infrared and Raman spectrometers, a laser-induced fluorescence instrument, electric and magnetic field probes, ground-penetrating radar and meteorology sensors. The spectrometers would seek evidence of salts which might lower the freezing point of water to allow flow under current conditions. The radar would seek underground water or ice. Two low-resolution cameras on the lander would observe pilot balloons launched before the main balloon, monitor the inflation of the large balloon and observe its departure.

The lander would set down in a crater known to have gullies and release its balloon to fly over the crater wall, making its observations over a short period. The flight would take place in several hops between touchdown sites, only flying when winds would carry it in the right direction. After data were transmitted from the gully overflight, an extended mission might continue as long as the balloon remained functional. The landing ellipse was reduced to only 30 by 10 km by making a very late trajectory correction during the approach to Mars. The crater had to contain this ellipse, to lie within appropriate elevation and latitude limits, and to have a smooth flat floor and numerous gullies. The chosen site was at 65° S, 15° W in a crater imaged by MGS during its aerobraking (orbit circularization) phase, which provided one of the first views of gullies. Figure 189A shows the crater, then informally referred to as the Aerobraking Crater but now named Sarh, with a possible ellipse at an arbitrary orientation. The final ellipse orientation would depend on the arrival date. This crater floor is probably too rugged for a safe landing.

Figure 189B shows an area in Solis Planum interpreted by Koroshetz and Barlow (1998) and Barlow (2000) as having a near-surface reservoir of ice. Studies of crater ejecta morphology suggested that ice lay near the surface at high latitudes, but might be buried under hundreds of metres of dry material near the equator. This area, however, showed signs of ice at shallow depth which might supply water resources for future human operations. Morphological studies like these were supplemented in later years by Mars Odyssey GRS data which could detect ice directly if it was close to the surface.

2 June 2003: **Mars Express (Europe: ESA)**

Mars Express was the first European Space Agency (ESA) mission to Mars. It consisted of an orbiter and a lander called Beagle 2 and weighed 1123 kg complete with fuel and Beagle 2 attached or 666 kg for the empty orbiter alone. Mars Express was conceived initially to recover European science lost with Russia's Mars 96 mission, which had carried a camera provided by ESA. The goals of Mars Express were to obtain near-global high-resolution images at 10-m resolution, to map mineralogy at 100-m resolution, to study subsurface structures with a sounding radar, and to measure the atmospheric composition and circulation.

Mars Express had a 1.5 by 1.8 by 1.4 m body with two solar panels mounted on opposite sides, spanning a total of 12 m. A 1.8-m-diameter antenna was mounted on one

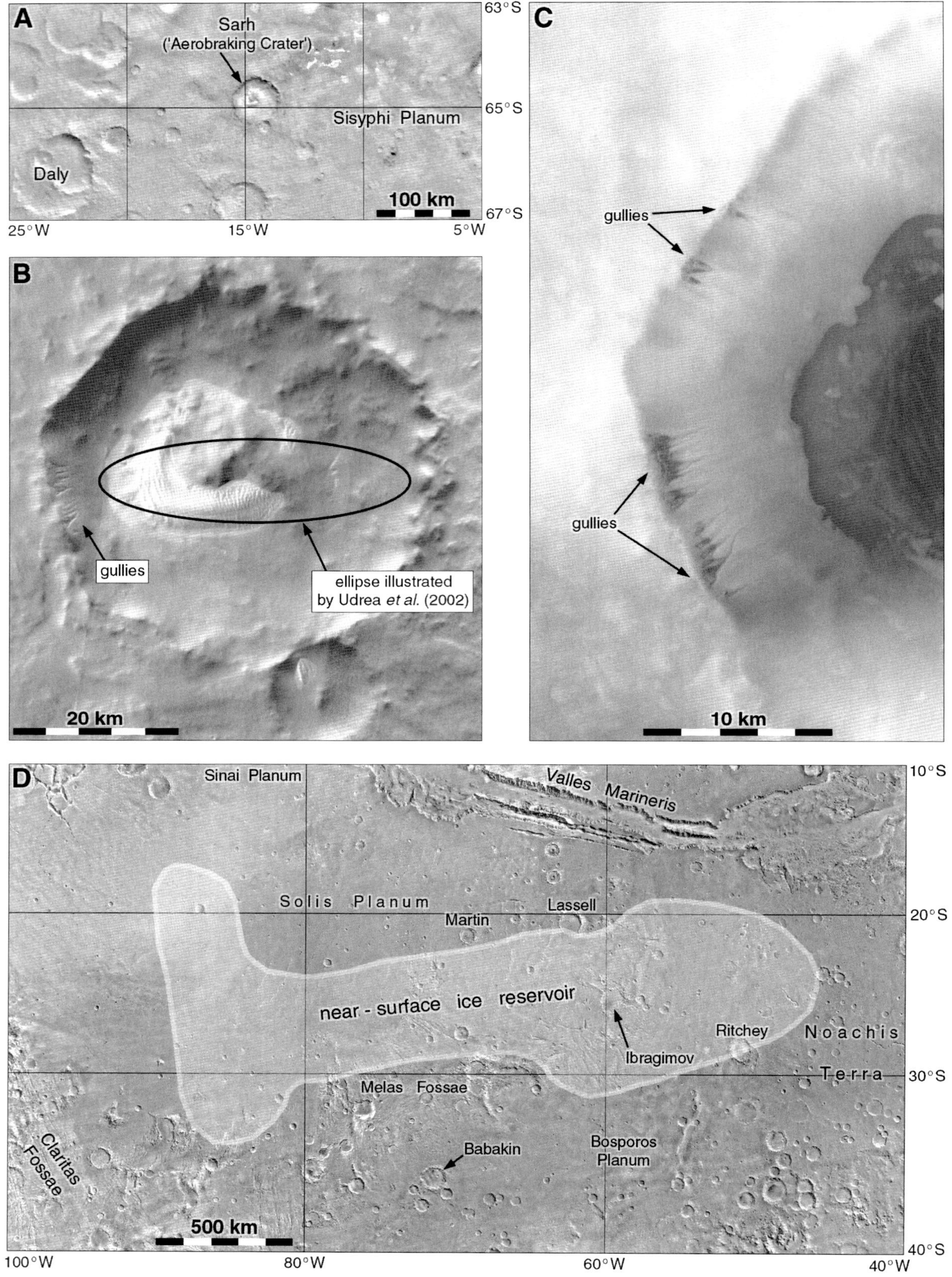

Figure 189 A: The WOBBLE landing site region. **B:** THEMIS infrared mosaic of Sarh ('Aerobraking Crater'), the landing site for WOBBLE. In infrared the dark areas seen in Figure 188C appear brighter. **C:** Gullies in Sarh's western wall, the WOBBLE science target (MOC image AB107707). **D:** Possible subsurface ice reservoir identified by Koroshetz and Barlow (1998) and Barlow (2000).

end of the body, pointing in the same direction as the solar panels. Two 20-m-long antennae for the radar sounder extended from opposite sides of the body perpendicular to the solar panels, and a 4-m-long low-gain antenna extended from the top of the body. Its instruments were a high-resolution stereoscopic camera (HRSC, the instrument lost on Mars 96), a visible and near-infrared spectrometer (OMEGA), an infrared spectrometer (PFS) and ultraviolet spectrometer (SPICAM), a neutral and charged particle sensor package (ASPERA) and the subsurface radar system (MARSIS). A radio science experiment made use of the main communications system.

Mars Express was launched from Baikonur at 17:45 UT and entered a 200-km parking orbit. The upper stage fired at 19:14 UT to place the spacecraft on a Mars transfer orbit, and the upper stage separated at 19:17 UT. When spacecraft are sent to Mars, the initial trajectory is designed to miss the planet, so the unsterilized upper stage will not impact and contaminate the planet. A later trajectory correction retargets the spacecraft for Mars. This retargeting took place on 4 June, and another correction was made on 10 November. The spacecraft survived a potentially damaging solar flare on 28 October. A distant image of Mars was taken on 1 December from 5.5 million km. Beagle 2 was released on 19 December, a last trajectory correction was made the next day to prevent the orbiter from entering the atmosphere with Beagle 2, and on 25 December 2003 (MY 26, sol 599) at 2:47 UT, the main engine burned for 37 minutes to place it in a 250 by 150 000 km capture orbit inclined 25° to the equator. Four more burns beginning on 30 December brought Mars Express to its 259- by 11 560-km mapping orbit with an inclination of 86° and a period of 7.5 hours. Near periapsis on each orbit the instruments faced Mars, and near apoapsis the high-gain antenna was pointed to Earth for communications. The first science data from orbit were taken over Valles Marineris on 14 January 2004. After 440 days the orbit was modified to 298 by 10 107 km to give an orbital period of 6.7 hours. The spacecraft, its instruments and the early mission operations are described by Wilson (2004).

The nominal mission was to last for a full Mars year, but Mars Express far surpassed its goals and was still operating successfully in 2011; however, operations were interrupted several times. Conjunction periods occurred in September 2004, October 2006, December 2008 and February 2011. Eclipse seasons occurred twice each Mars year, limiting power generation and curtailing operations. The second eclipse season in early 2005 brought power levels down to critical levels but the spacecraft recovered successfully, and experience during this event helped plan for survival during future eclipse seasons. On 27 April 2005 science operations were interrupted while the orbit was adjusted to prevent a collision with Mars Odyssey. Another orbit adjustment in late November 2005 returned the spacecraft to its nominal mapping orbit. A combination of eclipses, increased distance to Mars and then conjunction in October precluded science operations between 22 August and 5 November 2006. In December 2006 Mars Express tried, but failed, to image Mars Global Surveyor which had failed on 2 November.

Another challenging eclipse season at aphelion was survived late in 2008. The orbit was adjusted several times after this, first in December 2008 to improve data quality, which was threatened as the orbit gradually evolved, and again in September and October 2009. The periapsis latitude varied considerably over time, from 16° N at the beginning of September 2009 to 87° N in December, moving south again after that, so the latitude of maximum image resolution and other data characteristics varied and had to be taken into account. Also in October 2009, bistatic radar experiments were performed as well as communications tests with the Mars Exploration Rover Spirit. In February and March 2010 a series of Phobos encounters occurred, and after they were completed, a final orbit change was made to set the spacecraft up for several years of future observations. Problems began to appear in the data-handling systems on the spacecraft in August 2011, precipitating several safe modes in which operations were suspended while the problems were solved. Safe modes are designed to protect spacecraft from harm, but also can result in greater fuel use than during normal operations. These events occurred on 13 August, 23 September and 16 October 2011, and after the last shutdown, science operations were suspended until the problems could be explained and further safe mode events avoided. Operations were restored step by step during November 2011.

MARSIS deployment was originally scheduled for 20 April 2004 (MY 27, sol 44) but was delayed until 2005 after simulations suggested it might harm the spacecraft. On 4 May 2005 (MY 27, sol 413), one long boom was deployed, but did not fully lock in place. Orienting the spacecraft so the Sun would heat a hinge corrected the problem on 11 May. The second boom deployed successfully on 14 June 2005 (MY 27, sol 453). MARSIS revealed subsurface structures, including layering in the polar caps and buried impact structures in the plains. A special north polar mapping campaign was conducted in the summer of 2011. The instrument was also designed to study the ionosphere and unexpectedly proved to be able to investigate the planet's complex magnetic field through its interaction with electrons near the spacecraft.

HRSC was a scanning camera with 12 m/pixel resolution at periapsis and multispectral and stereoscopic capabilities (Jaumann *et al.*, 2007). It also included a framing camera system called the Super-Resolution Channel (SRC) with 3-m resolution, slightly compromised by thermal deformation of its mirror. HRSC image and topographic data were incorporated into Google Earth's online Mars map and in a new series of large-scale topographic maps of regions of interest made at the Technische Universität Berlin (Albertz *et al.*, 2005). HRSC images covered 25% of the surface of Mars during the nominal mission and approached full surface coverage in 2011. Its stereoscopic images permitted topographic mapping at much higher resolution than MOLA over broad areas. Another small camera called the Visual Monitoring Camera was intended to observe the separation of Beagle 2 and did so on 25 December 2003 (MY 26, sol 599). Then it was turned off until February 2007, when tests were performed to evaluate its use as a public outreach tool. In October 2007 regular imaging began, with rapid online release of the images to the educational and amateur astronomy communities. Figure 190 shows several images and other data from Mars Express.

Mars Express contributed substantially to knowledge of the planet and mapping of potential landing sites, especially through its topographic mapping and subsurface imaging. Its results also produced a new way of thinking about the planet's geological history. Since Mariner 9, geological mapping based on surface morphology had used the historical framework developed by USGS, in which old cratered terrain formed in the Noachian era, ridged plains in the Hesperian era and smooth plains in the Amazonian era. Compositional data from OMEGA inspired a different approach based on chemistry (Bibring *et al.*, 2006). An early Phyllosian era with abundant water produced clay minerals, taking its name from phyllosilicate clays. A later Theiikian era had a dry climate in which salts, especially sulphates, were precipitated as groundwater evaporated. Finally, a cold dry climate in the Siderikian era produced the ubiquitous iron oxide dust seen today. The relationship between these and the geological eras was unclear, though the Phyllosian and Noachian would roughly coincide.

The orbit of Mars Express crosses the orbit of Phobos at intervals of 5.5 months, allowing repeated very close flybys of Phobos. Images from these encounters were used to extend Viking mapping and to help locate potential Phobos-Grunt landing sites (Figures 204, 205). Mars Express could only obtain distant images of Deimos.

Beagle 2 Landing Site Selection

The initial landing site constraints for Beagle 2 were a latitude range of 30° N to 30° S and elevations below 0 km in Viking elevation maps, or –1.6 km in the Mars Global Surveyor MOLA data, for effective use of the parachute (Figure 190). The earliest set of candidate sites (Pullan *et al.*, 2004) were craters in highland terrain associated with channels, such as Gusev (15° S, 185° W) or Becquerel (22° N, 8° W), areas near the highland/lowland boundary such as the Elysium basin (0° to 20° N, 170° to 210° W) and the margin of Amazonis (4° N, 150° to 151° W); and the previous landing sites of Viking-1 in Chryse Planitia (22.4° N, 48° W) and Pathfinder in Ares Vallis (19.5° N, 32.8° W). Bridges *et al.* (2003) also mention Sinus Meridiani, which was safe and interesting but too high, and the Cerberus plains north of the equator at 210° to 230° W and plains in southern Elysium Planitia (0° to 10° N, 190° to 200° W), which were both too rough.

In early 2000, two sites in Chryse and Tritonis Lacus were favoured (ESA, 2000), both near 20° N (Table 78). The season when Beagle 2 landed would be late spring, with adequate sunlight for solar power and thermal control at that latitude, according to early studies. The landing ellipses used for planning were 240 km long by up to 50 km wide. The Chryse site was at the mouth of Maja

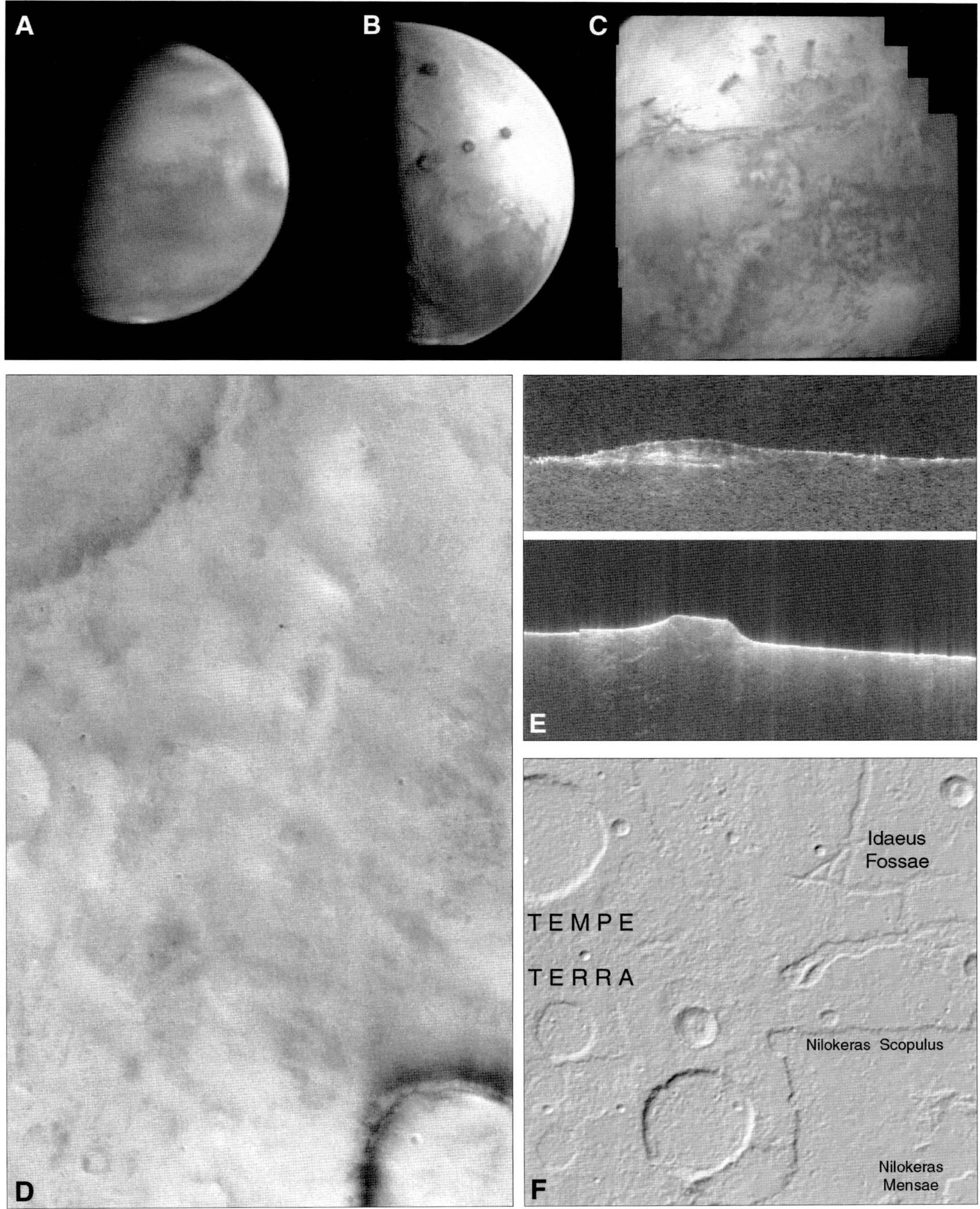

Figure 190 Mars Express Mars data.
A: HRSC approach image from 1 December 2003 showing Margaritifer Sinus at a distance of 5.5 million km. **B:** VMC image showing the Tharsis volcanoes on 25 May 2010. **C:** VMC image mosaic from 23 December 2009 of Valles Marineris and Argyre by Mike Malaska. **D:** Part of HRSC image h5408_0000_nd2 showing the same area in Tempe Terra seen in Figures 51D, 153D and 188C. **E:** MARSIS radargrams 4338 (top) and 3732 showing subsurface structure. **F:** Digital shaded relief from HRSC stereoscopic data, orbit 1568, southeastern Tempe Terra.

Table 78. *Beagle 2 Landing Ellipses*

Ellipse (Figures 190, 191)	Location	Dimensions (km)	Azimuth
Chryse – Maja Vallis	19° N, 51.5° W	240 by 50	
Tritonis Lacus	20° N, 246° W	240 by 50	
Simud – Tiu Valles	6° N, 37° W		
Isidis 1	10.6° N, 270° W	495 by 94	77.6°
Isidis 2		387 by 94	70.4°
Isidis 3		328 by 94	63.4°
Isidis 4 (final)	11.6° N, 269.6° W	174 by 106	74.9°

Valles and included layered sediments, very similar to the 'Maja Valles Fan' site for Mars Pathfinder (Table 67). ESA used the older term 'Vallis' for this site (Table 78), not the plural form 'Valles', which is now preferred. The Tritonis Lacus site (close to the Viking A2 site, Figure 45) was on the edge of the Elysium plains and had a more eroded landscape. Isidis Planitia was added to the list later. Sites in Ganges Chasma and Candor Chasma, especially an area of layered materials imaged by Mars Global Surveyor, were also possible (Figure 191).

Later in 2000, the 20° N sites were dropped when new analyses showed they would be too cold, and the layered region was thought to be too rough. The latitude range was narrowed to 12° N to 12° S. By December 2000, the preferred sites were Simud–Tiu Valles at the southern edge of Chryse Planita, not far from the Mars Pathfinder site (Figure 157), and Isidis Planitia (Table 78). The Simud area was too small to contain a satisfactory ellipse. The first Isidis location studied was centred at 10.6° N, 270° W, with three ellipses differing in size and azimuth, depending on the approach direction, which varied with the arrival date. The final landing ellipse (Figure 192A) was 174 by 106 km across, centred at 11.6° N, 269.5° W (Bridges *et al.*, 2003). After landing, timing of solar eclipses by Phobos in February 2004 would contribute to knowledge of the exact location of the lander.

2 June 2003: **Beagle 2 (Europe: ESA)**

Beagle 2, named after Charles Darwin's ship, was launched with Mars Express from Baikonur at 17:45 UT. It was mounted on the top surface of the Mars Express Orbiter and released on 19 December 2003 at 8:31 UT (MY 26, sol 593). The 60-kg lander's objectives were to characterize the geology of the landing site and the atmosphere and search for possible signatures of life. It entered the Martian atmosphere and was expected to land at 02:54 UT on 25 December 2003 (MY 26, sol 599). No signals were ever received, and the lander was assumed to have crashed.

Beagle 2 was a 33-kg bowl-shaped structure, 65 cm across and 25 cm deep, with a top cover which could unfold to form an array of solar panels. It carried instruments for soil and gas composition analyses and cameras for panoramic and detailed imaging. A robotic arm with a stereoscopic camera system, a small sampling drill, analytical instruments and a microscopic imager could study the surface and collect samples for analysis. The lander could also deploy a tethered 'mole', a mobile device which could move on the surface or dig into loose soil for subsurface samples. The samples would be returned to the lander for analysis. Beagle 2 was intended to operate for about six months.

After the failed landing, a search for evidence of the spacecraft on the ground was undertaken by MGS and later MRO. The final target coordinates just before probe release were 11.6° N, 269.26° W. After separation, the predicted landing site ellipse was 57 by 7.6 km across, centred at 11.53° N, 269.47° W (Sims, 2004). Beagle 2 scientists interpreted a dark patch in a MOC image as the landing site, possibly including a pixel pattern which suggested that the spacecraft might have landed successfully and opened its solar panels, but not been able to transmit data. These features were later seen more clearly in a MRO HiRISE image and revealed to be ordinary

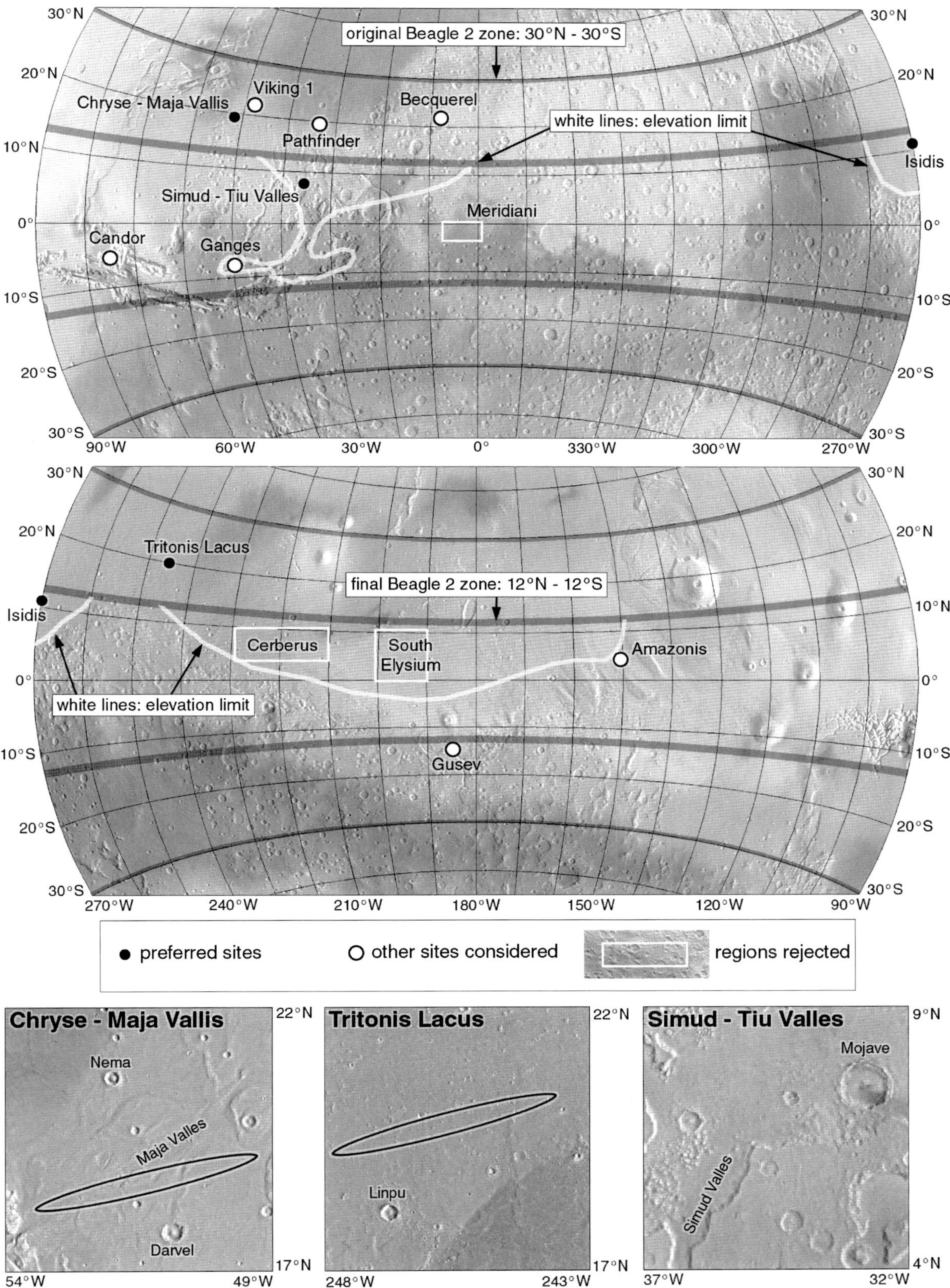

Figure 191 Beagle 2 landing site constraints and candidate sites (top), and three shortlisted sites.
The three small maps are 5° or 300 km across. The Simud – Tiu ellipse would have been on the plateau above the valley floors, but no obstacle-free site could be found.

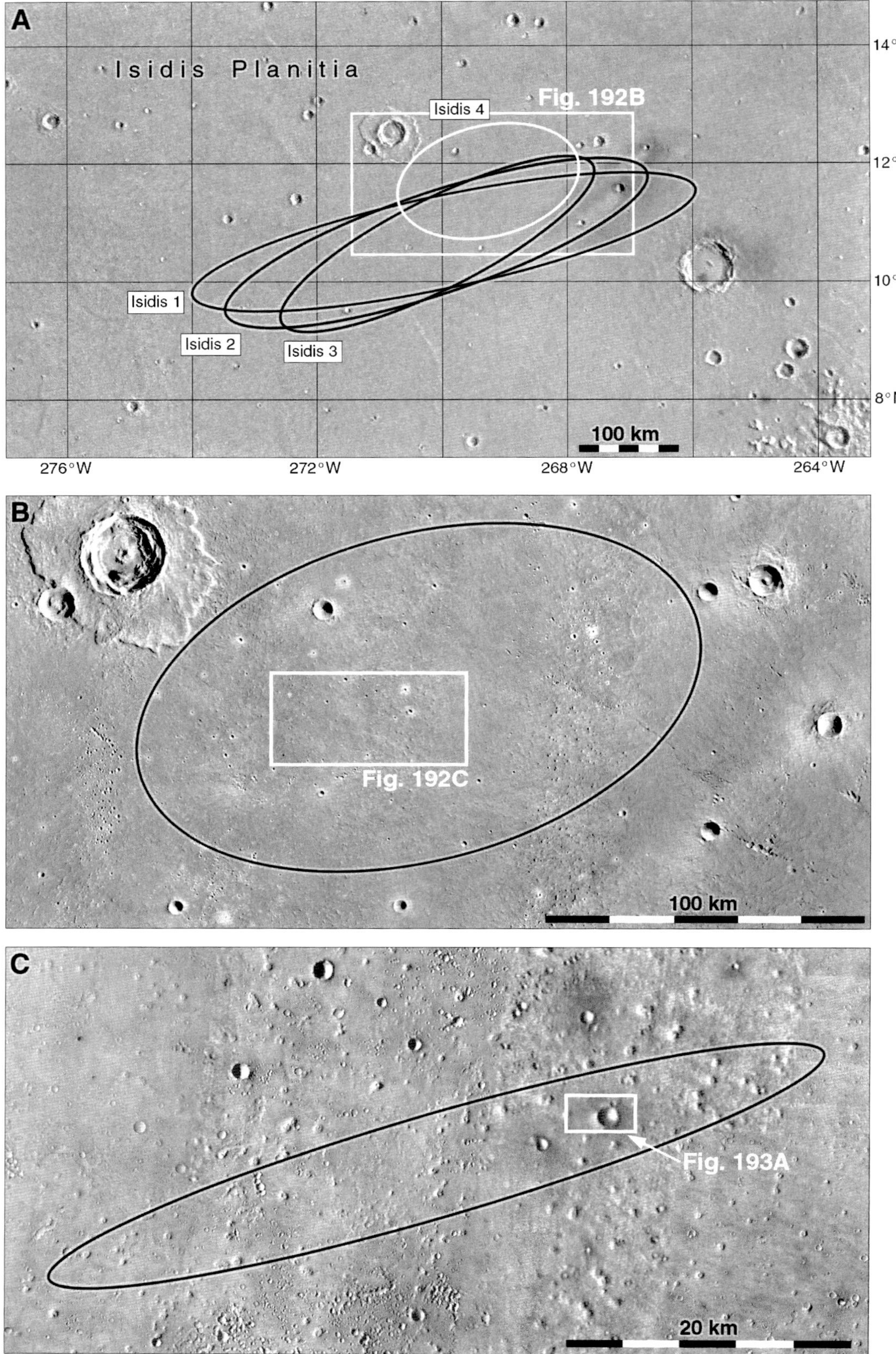

Figure 192 A: Beagle 2 ellipses in Isidis Planitia. **B:** The final target ellipse (Isidis 4, Table 78). The background is a Mars Odyssey THEMIS infrared mosaic with shading inverted to resemble the albedo pattern in (A). **C:** The tracking ellipse in which Beagle 2 struck the surface, on a mosaic of THEMIS visible and infrared images. Dark ejecta in the infrared portions of the mosaic appear bright in (B).

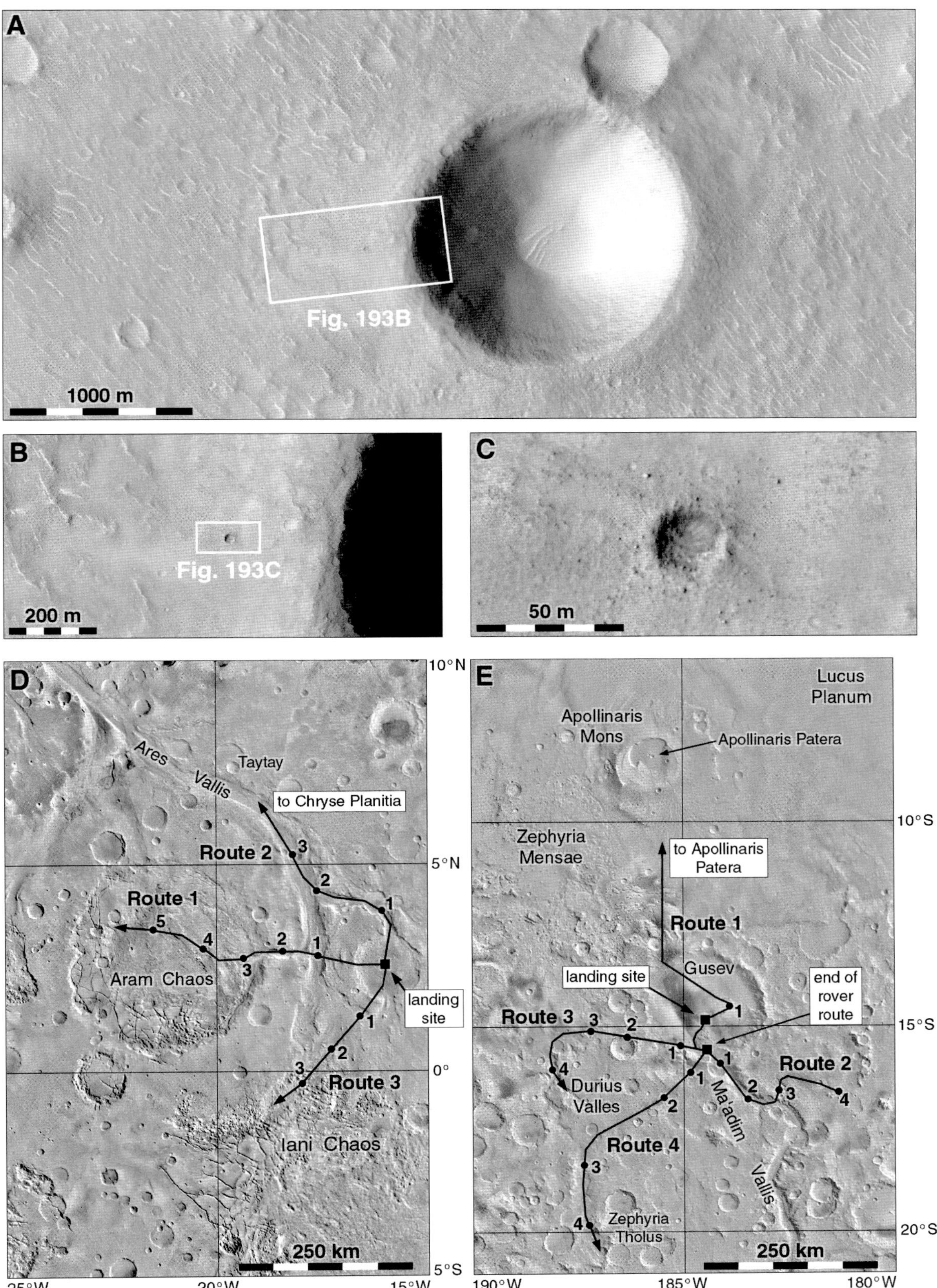

Figure 193 **A:** Candidate Beagle 2 impact site, suggested by MOC images but rejected by examination of HiRISE data. HiRISE image PSP_002347_1915. **B, C:** The small crater suggested as a possible impact site. **D, E:** Bioinspired Engineering of Exploration Systems (BEES) sites at Aram and Gusev. Another aircraft route is shown in Figure 150C.

rocks (Figure 193A). The Beagle 2 crash site has not yet been identified.

2003: **Mars Bioinspired Aircraft**

Thakoor *et al.* (2003) described a concept for Mars exploration using small flying devices called BEES (Bio-inspired Engineering of Exploration Systems). The bio-inspired concept borrowed ideas for the design or control of a mechanism from various plants or animals. Flight systems, for instance, might incorporate ideas based on winged plant seeds, butterflies and eagles. This study applied the concept to small aerial platforms which could explore areas hostile to rovers. Valles Marineris was one example, mentioned but not illustrated. Small gullies high on crater walls might be another suitable target. This study looked in more detail at two examples, one in Aram Chaos, an area with exposed hematite like that detected at Meridiani Planum, and the other at Gusev crater (Figures 193D and 193E).

At Aram, the mission lands in a crater on the path of Ares Vallis. A rover may explore the crater itself, but the small flyers travel to rougher and more distant destinations. Route 1 extends through a connecting channel into Aram Chaos, examining the channels, a hematite-rich delta area (Site 3) and layered sediments (Site 5). Data from the flyer might help direct the rover to targets near site 1 and characterize a future instrument drop site at site 3. Route 2 would explore the giant channel Ares Vallis, and an extension of it could fly over the Pathfinder and Viking 1 landing sites. Route 3 would fly over the rough terrain of Iani Chaos, the source of Ares Vallis, possibly dropping instrument packages along the route.

At Gusev the landing site is on smooth plains just north of the eroded remnants of a possible delta at the mouth of Ma'adim Vallis. This is about 10 km southwest of the Mars Exploration Rover site where the rover Spirit landed in January 2004. The lander deploys a flyer and a rover. The flyer follows route 1, crossing the floor of Gusev and arcing around the inner rim of Thira, a crater within Gusev which was thought to show evidence for a lake shoreline or layered sediments. It continues across possibly ice-related features in Gusev and might extend as far as the large volcano Apollinaris Mons, 300 km north of Gusev. Meanwhile the rover would drive down to the delta area and examine it. When the terrain became too rough, the rover could deploy three more flyers to explore more distant areas. Route 2 would allow a flyer to explore the wall terraces of Ma'adim Vallis (sites 1 and 2), a tributary valley (site 3) and possible small volcanoes (site 4). Route 3 would examine the nearby Durius Valles, also seen in Figure 140. Route 4 would explore the area between Durius and Ma'adim on the way to the unusual highland volcano Zephyria Tholus.

2 Phobos and Deimos

Introduction

These two small satellites of Mars were discovered by Asaph Hall at the US Naval Observatory in Washington, DC (Hall, 1877), Deimos on 11 August 1877 and Phobos six nights later. Phobos measures 27 by 21 by 18 km and Deimos is about half as large, 15 by 12 by 10 km, with their longest axes on the prime meridian (0° longitude) and the shortest axes between their poles. Both satellites have nearly circular equatorial orbits, Phobos with a period of 7.6 hours at an altitude of 5980 km (9400 km orbital radius), and Deimos with a 30-hour period and an altitude of 20070 km (23460 km orbital radius). They are in synchronous rotation with the 0° meridian always facing Mars, 90° W facing forwards along the orbit (leading side), 180° W facing away from Mars and 270° W facing backwards along the orbit (trailing side). Phobos and Deimos were the first small irregularly shaped bodies in the solar system to be observed in detail, and as such they have been important test cases for the development of new mapping techniques for nonspherical worlds. These small moons are likely to play an important role in any future human exploration of Mars.

Figures 194 and 195 are reference maps of each satellite with feature names and labelled grids. The Phobos map (Figure 194) is a USGS shaded relief drawing based on Viking images. The original map (Greeley and Batson, 1997) used a preliminary form of positional control, but this version has been modified by P. Stooke to fit the improved control derived by Simonelli *et al.* (1993). It is presented in the same map projection used for Mars throughout this atlas, Azimuthal Equidistant, but in a form modified to represent the shape of Phobos (Stooke, 1998). The shape in this case is the convex hull of the satellite's topography. The Deimos map is a mosaic of Viking images with contributions from Mars Reconnaissance Orbiter and, near the north pole, Mariner 9. These two satellite maps have different scales, and the map projection reveals the difference in their shapes. Blunck (1977, 1982) describes the naming of the satellites and their surface features, and Stooke (1989), Stooke and Keller (1990) and Blunck *et al.* (1993) illustrate evolving ideas about the cartography of irregular objects with maps of Phobos and Deimos by several authors. The first names given to features on Phobos commemorated people associated with the moons, including Hall and his wife Angelina Stickney. Later names on Phobos were taken from Jonathan Swift's book *Gulliver's Travels*, which included a pre-discovery description of two Martian satellites (Swift, 1726).

1969: **Mariners 6 and 7**

The first spacecraft images of Phobos were obtained by Mariner 6 and 7 in 1969 (Smith, 1970). Mariner 6 may have observed Phobos and its shadow, but the very small objects are hard to identify in the far encounter images. Mariner 7 obtained several images of Phobos. The highest resolution Phobos image (frame 7F91) was taken by Mariner 7 on 4 August 1969 (MY 9, sol 451), revealing its size and shape but no surface features (Figure 196). The little moon was roughly 18 by 22 km across, spanning about 6 pixels at a range of 130900 km, and had a very low albedo. The even smaller satellite Deimos was not found in the images from these missions.

1971–1972: **Mariner 9**

Surface features on Phobos and Deimos were first observed by Mariner 9 (Pollack *et al.*, 1972). Neither satellite's surface was completely covered by the images, but their sizes and shapes were measured, and the first maps were constructed. All resolved images are included in Figures 196 and 197, with photomaps showing the extent of the image coverage. The satellites were irregularly shaped, as expected for such small objects, and were covered with craters, though Deimos appeared smoother than Phobos.

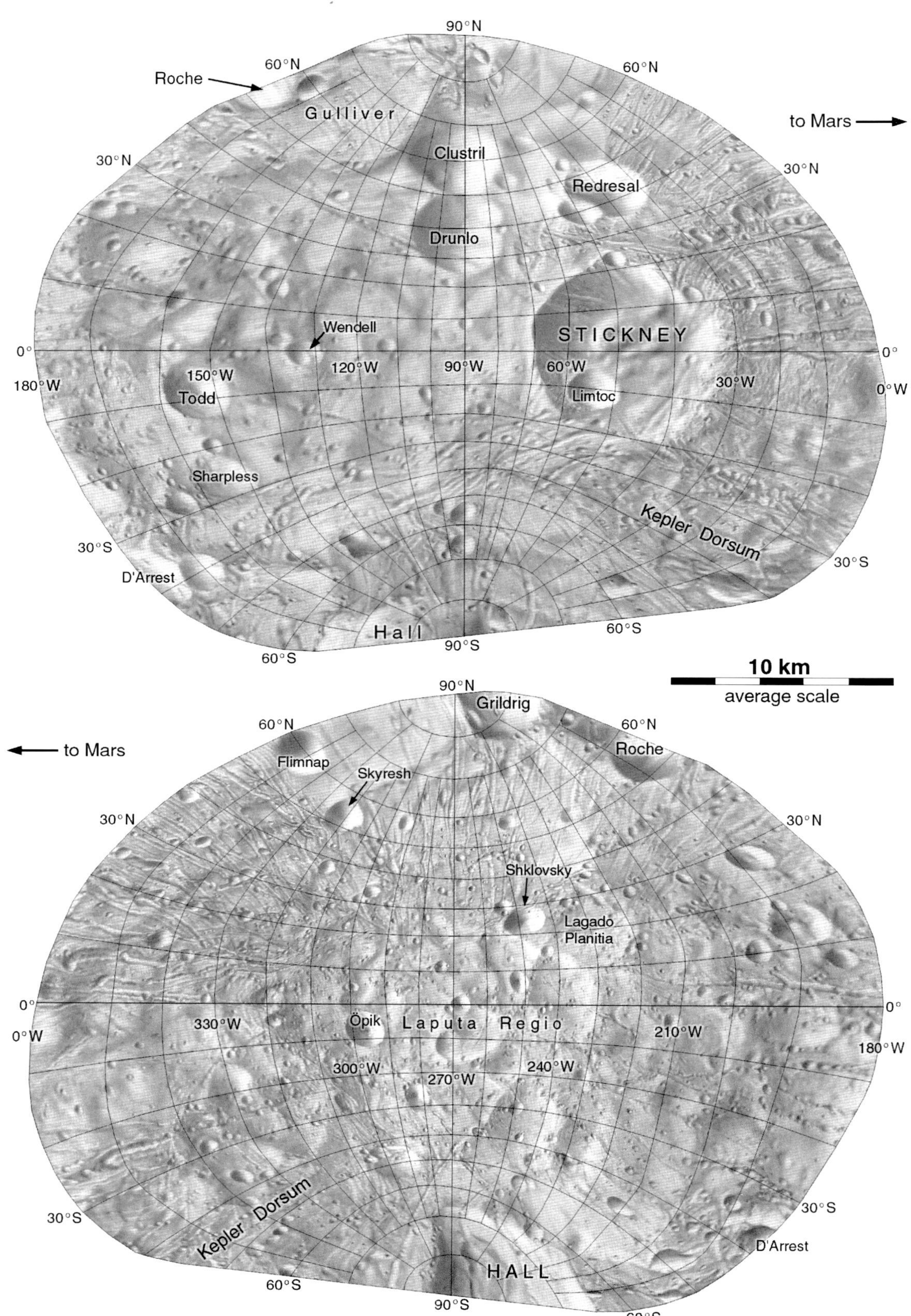

Figure 194 Reference map of Phobos.
The shaded relief drawing made by USGS has been adjusted to match positional control from Simonelli *et al.* (1993). The map projection is Morphographic Equidistant (Stooke, 1998).

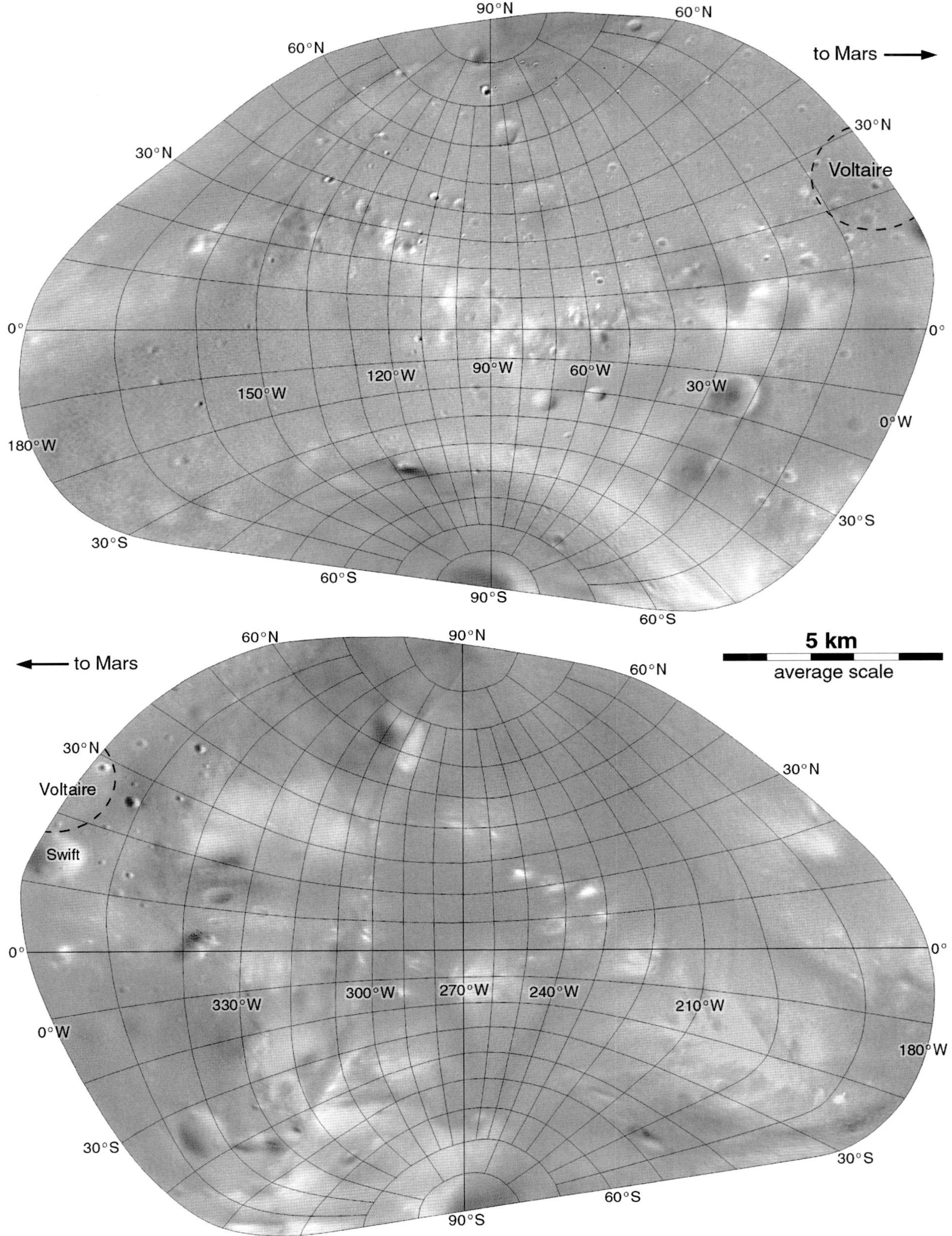

Figure 195 Reference map of Deimos.
The image is a photomosaic of Mariner 9, Viking and Mars Reconnaissance Orbiter images using a mosaic by Thomas (1993) for positional control. The map projection is Morphographic Equidistant (Stooke, 1998).

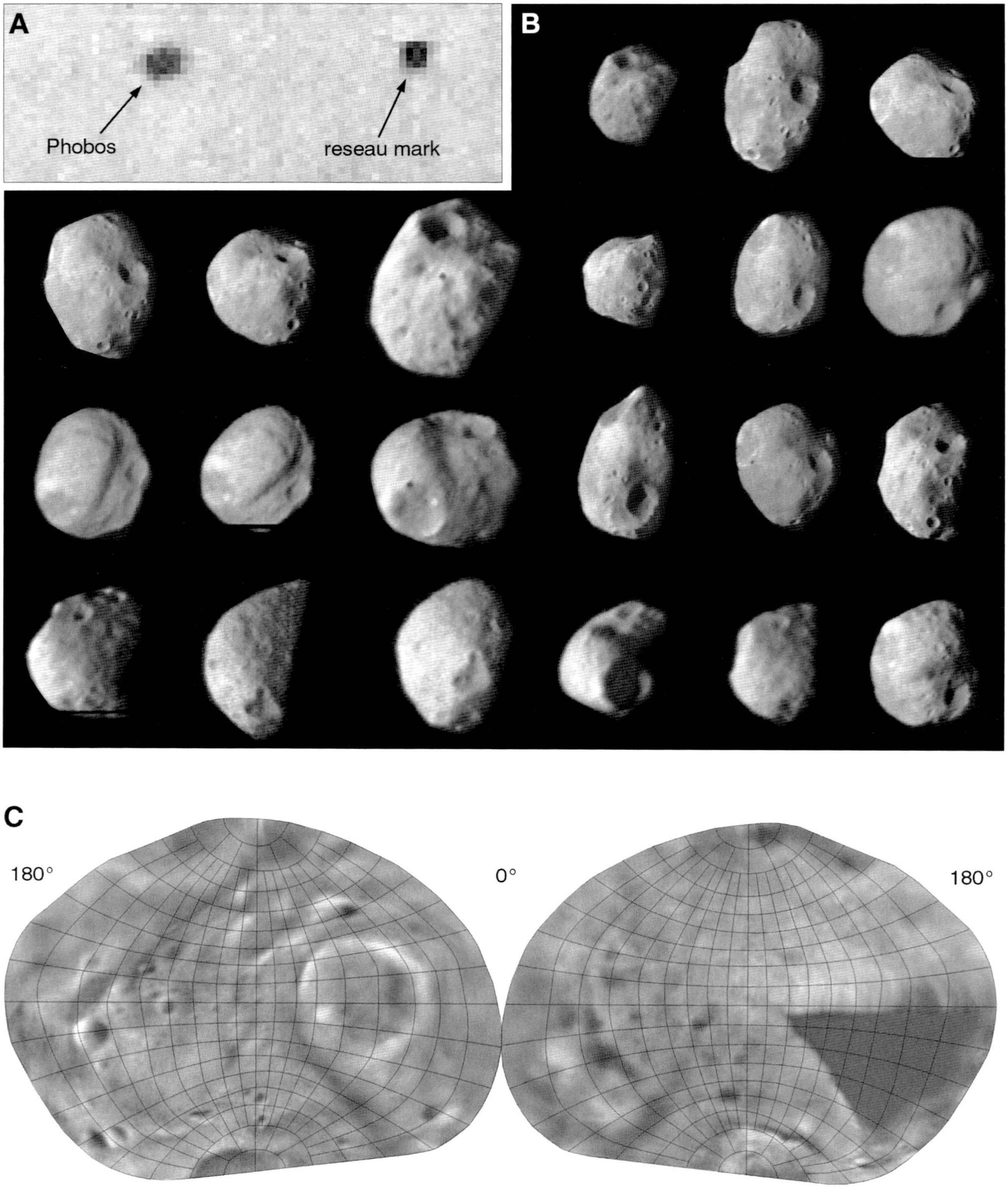

A

Phobos

reseau mark

B

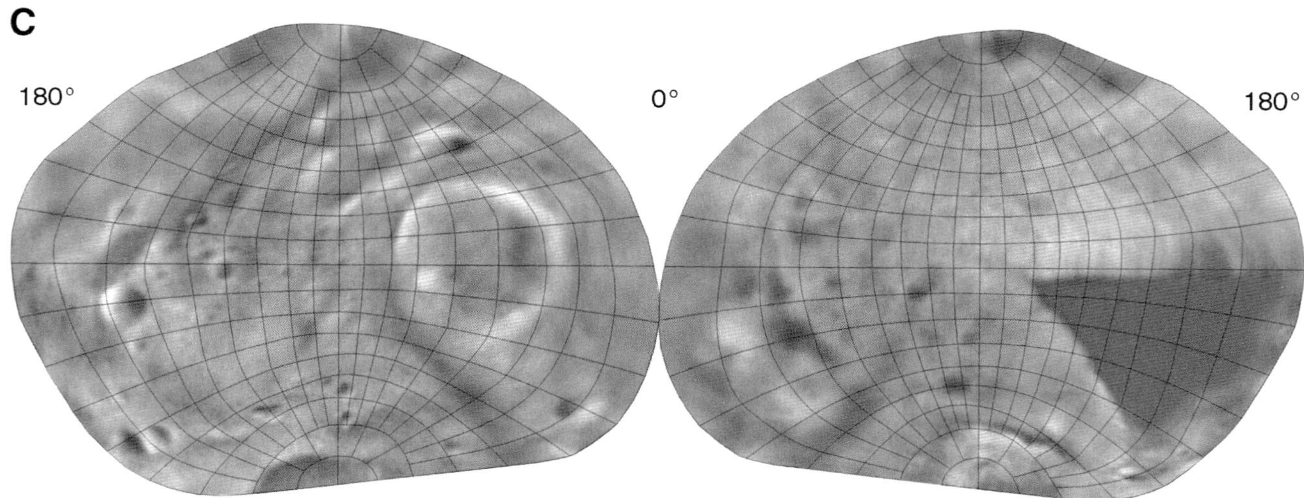

Figure 196 A: Mariner 7 image 7F91 of Phobos, also showing a nearby reseau mark, one of a grid of points added to the image for geometric analysis. **B:** Mariner 9 images of Phobos. The 21 Mariner 9 images are the complete Phobos imaging dataset from this mission. **C:** A photomosaic map made from the Mariner 9 images. The dark triangle at right on the map is the only region not observed by Mariner 9.

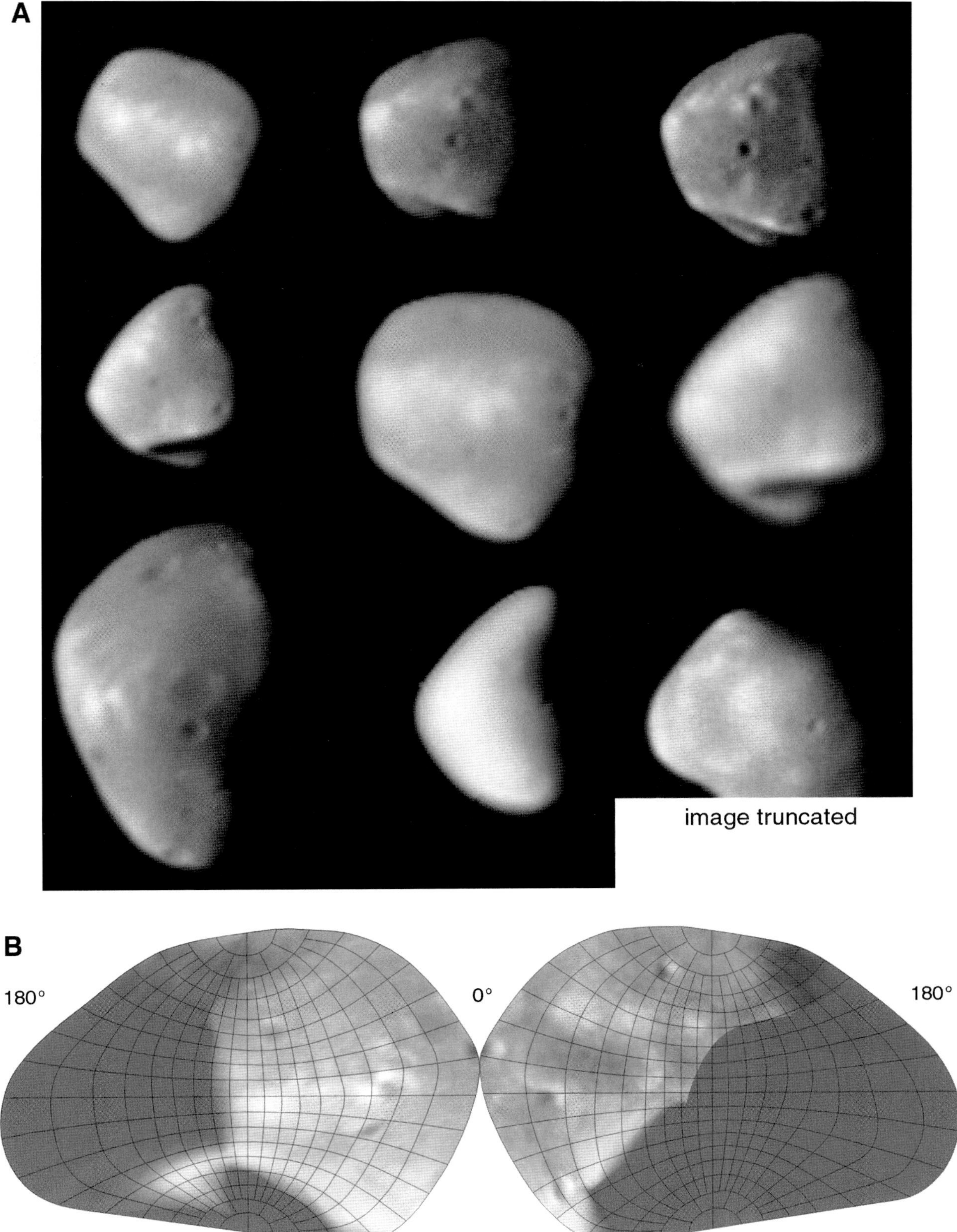

Figure 197 A: Mariner 9 images of Deimos. These are the only images of the satellite from this spacecraft. **B:** A photomosaic map of Deimos made from Mariner 9 images. The region in darkness in the map, about half of the surface, was not observed by this mission.

Using these new images, the first map of Phobos, in fact the first map of any irregularly shaped world, was compiled by Duxbury (1974), and a map and pictures of a globe of Deimos were published by Blunck (1977). These two maps were plotted on conventional map projections which ignored the irregular shapes of the moons, but the need for new mapping methods was becoming obvious. The first steps were taken by Ralph Turner, a scientific illustrator and modeller who had worked at the Lunar and Planetary Laboratory of the University of Arizona during the 1960s (Turner, 1978a, 1978b; Stooke, 1989). Mariner 9 images guided the sculpting of a physical model of Phobos, and its surface was mapped using an azimuthal (planar) projection in two halves. This type of projection produces circular maps of spherical worlds, such as those used for Mars throughout this atlas. Turner's maps were centred on the poles of Phobos with the equator forming the outer boundary, and because the equatorial cross-section of the satellite is elongated, the projection outline was modified to an ellipse to approximate the true shape.

1976: **Viking**

The Viking missions resulted in the first fully successful landings on the surface of Mars, but the two Viking orbiters were also able to make many observations of Phobos and Deimos (Figures 198, 199). Distant observations were made for global mapping and shape modelling, and close approaches provided fine details of the surfaces to permit studies of their history and geologic processes and to plan future landings on the satellites (Figures 200–202). The closest flyby of Deimos was on 15 October 1977 (MY 12, sol 649) when Viking Orbiter 2 passed only 28 km from the satellite. The closest Phobos flyby on 20 February 1977 (MY 12, sol 418) was at a distance of 89 km. The Viking results are presented by Thomas (1979) and Veverka and Burns (1980). Both satellite surfaces were now completely covered by images, almost everywhere at far greater resolution than Mariner 9 had provided. A completely unexpected system of long narrow valleys and crater chains, commonly called grooves, was discovered on Phobos. A few grooves are also present on Saturn's moon Pandora and asteroids 243 Ida, 433 Eros and 951 Gaspra, and larger numbers on 4 Vesta and 21 Lutetia (Figure 198). Deimos has no grooves at all, and its smooth surface is covered with a very thick regolith which fills craters and apparently slides downhill even in the very low gravity of this tiny satellite, forming bright streaks around higher points and ridges (Figure 198). It remains unclear why the moons look so different.

Thomas (1979) made improved maps of both satellites using the new images, reverting to the use of conventional map projections based on a sphere. Later, Simonelli *et al.* (1993) used Viking images to construct a new high-resolution shape model and photomosaic of each satellite, and the new image and shape data spurred the development of several novel map projections. Snyder (1985) devised a method for accurate conformal mapping of an ellipsoid approximating the shape of Phobos on a new type of cylindrical projection. This would make circular craters appear circular on a map, a characteristic often preferred for planetary maps, but it would only give precise results for the ellipsoid and not for the less regular shape of Phobos itself. A similar approach was described by Bugaevsky *et al.* (1992), who modified a cylindrical projection similar to Mercator by varying the spacing of its meridians and parallels to match their relative spacing on the Phobos ellipsoid. The parallels of latitude assume a sinusoidal shape, more noticeable away from the equator. Neither Snyder nor Bugaevsky used conventional planetocentric coordinates in their equations, and as a result their projections have not been widely adopted. Stooke (2003) attempted to overcome this problem with a simplified cylindrical projection called Prismographic, easy to construct but no longer precisely conformal.

Azimuthal projections are another broad class of map projection which might be modified for use with distinctly nonspherical objects. Turner's projection described above was the first to do this. Stooke (1986) modified azimuthal projections by allowing the radius factor in the projection equations to vary from point to point across the map rather than remaining constant. The result could be used with any shape model of the body, including a triaxial ellipsoid, the actual topography or the convex hull of the topography. The latter was generally found to be preferable (Stooke, 1998) and is used in Figures 194 and 195. These projections gave an indication of the shape of the object and so were named Morphographic, and they could be made in approximately conformal, equal area and equidistant versions, roughly preserving shapes, areas and distances, respectively. The equidistant version is used for global satellite maps in this atlas (*e.g.* Figures 194, 195).

Precise shape, area and distance representations were not possible using Stooke's method, but other approaches have been devised to improve these characteristics. Cheng and Lorre (2000), Nyrtsov (2000), Nyrtsov and Stooke (2002) and Berthoud (2005) have modified existing projections or created complex analytical solutions for detailed shape models to achieve more accurate representations than Stooke obtained. Although accuracy is improved, all these methods produce maps with irregular outlines and graticules which may deter some potential users. Clark *et al.* (2008) also devised maps with irregular outlines but with topological characteristics designed to reveal relationships between regions.

1970s: **Viking Phobos Mission Studies**

Martin Marietta, builder of the Viking landers, examined the possibility of using additional Viking spacecraft to explore Phobos and Deimos (Martin Marietta Corp., 1972). Similar studies continued for several years (Martin Marietta Corp., 1977b), and a variety of missions were suggested, but none were funded.

The 1972 study assumed that the Viking Phobos/Deimos missions would launch in 1979 and 1981, when the planetary geometry would require minimum energy. The 1979 mission would be a lander, the 1981 mission a sample return. Several designs were considered, the first consisting of a modified 482-kg lander and a 3600-kg orbiter with larger fuel tanks than the original Mars design. On arrival the orbiter would first enter an elliptical equatorial orbit, then transfer to an 15-hour elliptical orbit. The high point of this orbit would reach Deimos and the low point would be lower than Phobos, allowing multiple passes of both satellites and gathering data to help decide which satellite to land on. The spacecraft would then match its orbit with the chosen satellite. The lander would touch down and deploy instruments including a seismometer, a drill and a camera. The lander could either hop to new locations by firing its thrusters or drive on wheels mounted on each leg. An alternate mission design would see the orbiter itself land on the satellite. A Phobos rover option was also considered. It would land near 45° N, but at an unspecified longitude, at the beginning of October 1980 and drive to 60° S by the end of December.

The 1981 sample return mission would involve a 3374-kg spacecraft incorporating an orbiter with four legs and a 260-kg Earth return module derived from a Venus Pioneer probe design. The orbiter would land, gather a 2-kg sample and transfer it to the return capsule. The Earth return module would launch from the landed orbiter and enter a 1500 by 95 000 km orbit about Mars. Then it would adjust its orbital inclination and depart for Earth.

Another variation would combine a Mars lander with a satellite lander. The 4150-kg spacecraft would launch in 1979. The orbiter would enter an elliptical equatorial orbit around Mars with a 97-hour period, then release its Mars lander, whose landing sites would be limited to within 12° of the equator because of the need for an equatorial orbit. The orbiter would then transfer to a 15-hour orbit to study the satellites, and as described previously, select one satellite for a landing, and touch down itself.

1989: **Phobos 2**

The twin spacecraft Phobos 1 and Phobos 2 were launched to Mars and Phobos in July 1988. Phobos 1 failed on its way to Mars and flew silently past the planet on 25 January 1989 (MY 18, sol 647) (Zakharov, 1988), but Phobos 2 achieved orbit successfully on 29 January 1989 (MY 18, sol 651). The spacecraft also observed Mars (Figures 121, 122). Phobos approaches were made on 21 and 28 February and 25 March, with successful imaging each time at altitudes of 800–1100 km, 310–430 km and 190–220 km, respectively (Avanesov *et al.*, 1991).

During another flyby on 27 March 1989 (MY 19, sol 37), the spacecraft turned away from Earth to face Phobos, and its signal was never reacquired. A computer control error was blamed for the problem (Sagdeev and Zakharov, 1989). Although there was no landing, some useful data were obtained, including images (Figure 200A) which improved mapping of grooves in areas seen poorly by Viking and infrared spectral data on the

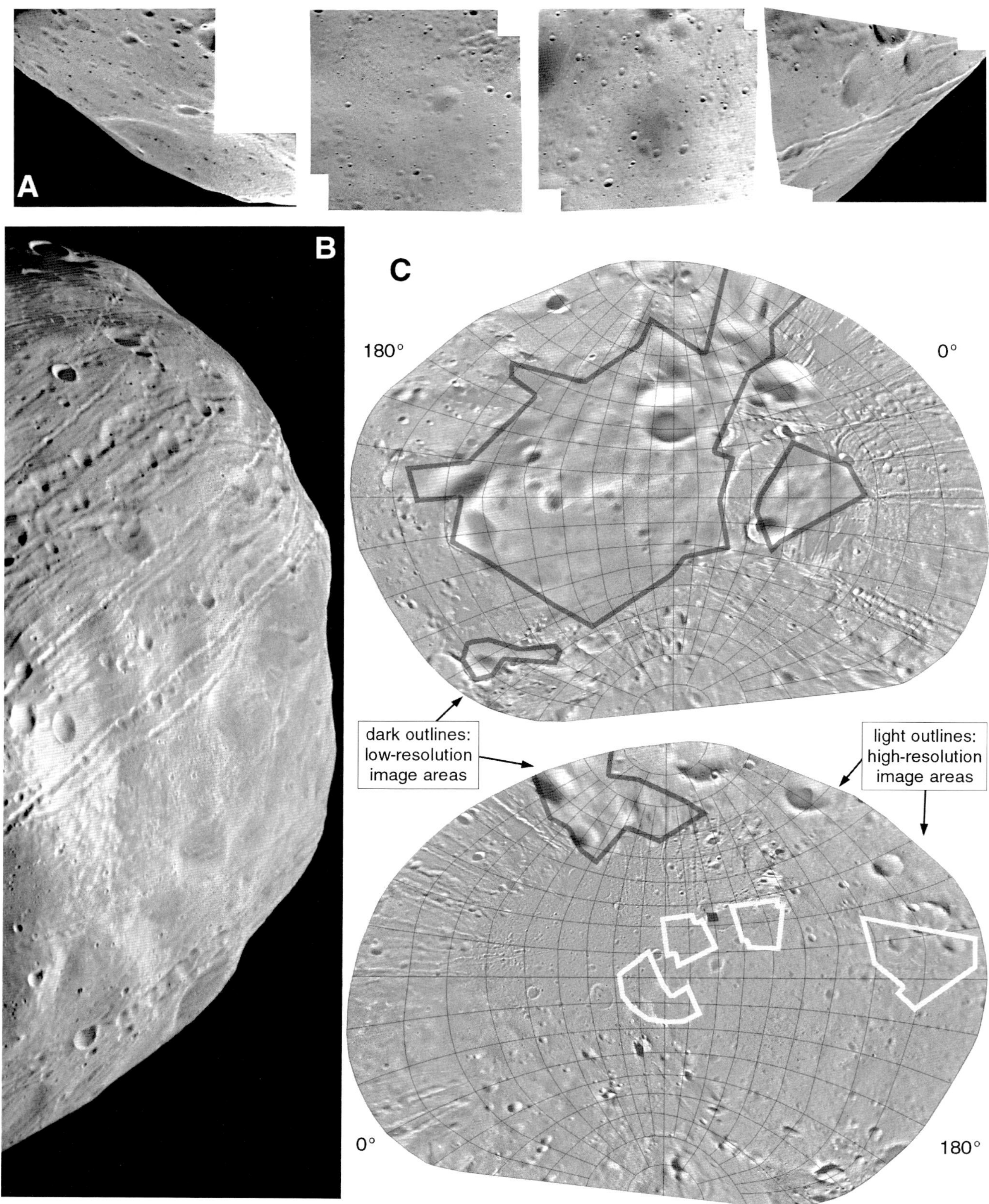

Figure 198 Viking images and photomap of Phobos.
A: The highest resolution images of Phobos (244A02-09). **B:** Mosaic of frames 343A06-17 showing grooves east of Stickney. **C:** Mosaic of Viking images modified from Simonelli *et al.* (1993). Areas of high-resolution images (Figure 198A) and the remaining areas seen only at low resolution are outlined.

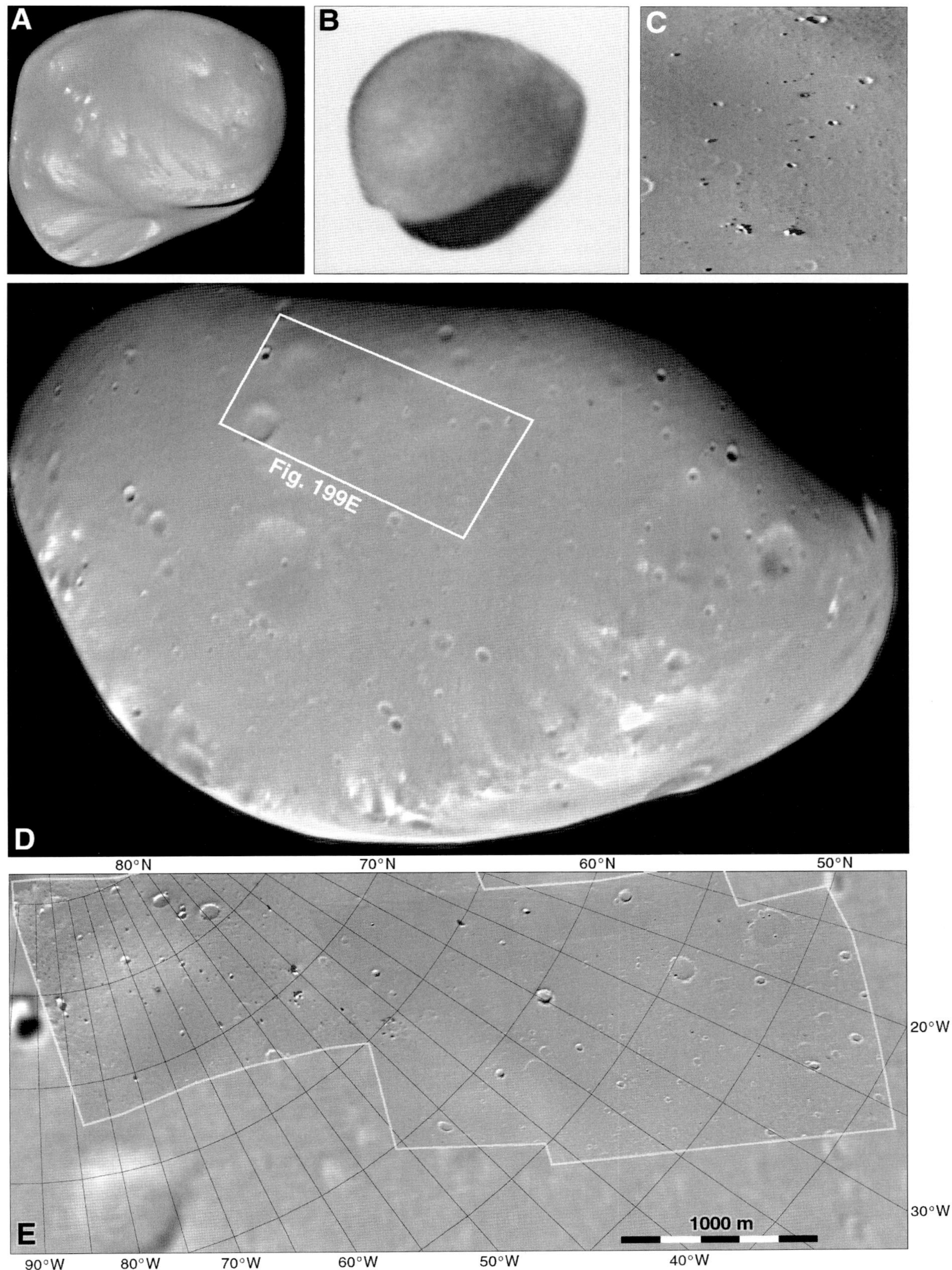

Figure 199 Viking images and photomap of part of Deimos.
Figure 195 illustrates the global Viking coverage with only small additions from other missions. **A:** Image 507A01 showing the region facing away from Mars. **B:** Distant view of Deimos transiting Mars (564A04). **C:** Part of the 423B highest-resolution mosaic. **D:** Mosaic from the 428B sequence, the best regional coverage from Viking 2. **E:** The highest-resolution mapping coverage of Deimos from Viking 2, orbit 423B. A very small area of additional high-resolution coverage occurs near 55° N, 160° W.

composition of Phobos. One infrared dataset from the ISM on 25 March, from 190-km altitude, covered an area near the equator and about 180° longitude, but it could not be calibrated due to light from Mars seen over the limb of Phobos. A second observation from 220 km was successful, and the area it covered is shown in Figure 200B. An ultraviolet and visible spectrometer (KRFM) obtained two data tracks across the surface, also shown in Figure 200B (Murchie and Erard, 1996; Murchie *et al.*, 1999). Compositional data did not match the dark asteroid types sometimes associated with Phobos.

If the mission had not failed at this point, it would have made a very low pass over the side of Phobos opposite Mars in early April 1989. The plan was to drop to within 50 m of the surface, holding that altitude by thrusting upwards or downwards as needed as the spacecraft drifted over the surface for about 15 minutes. Very high-resolution images would be obtained, showing features as small as 6 cm across. A laser would have vaporized small amounts of surface material which could reach the spacecraft and be analyzed directly. Another instrument was to emit krypton ions to dislodge surface ions for analysis, and neutron, gamma-ray and x-ray spectra would also reveal surface composition. The ground track would have been a little south of the equator, beginning west of the prominent crater Todd (Figure 195) at about 10° S, 165° W (Sagdeev and Zakharov, 1989) and extending westwards to about 4° S, 194° W. The small hopping probe would be dropped there, and shortly after that the long-term autonomous lander would be deployed at 3° S, 201° W. The track and landing site locations (Figure 200C) are illustrated by Sagdeev and Zakharov (1989), but they are uncertain by 1 or 2 km in latitude and perhaps more in longitude depending on the final trajectory. Before and after the closest approach, a radar sounder would have probed the interior of Phobos. This location near the anti-Mars region of Phobos maximized periods of sunlight and direct communication with Earth, because activities near the sub-Mars region would be interrupted during occultations by the planet. After the deployment of the landers, Phobos 2 would have been moved into a higher orbit to prevent a later impact on Phobos. Because that did not happen, a subsequent impact on Phobos remains a possibility.

The battery-powered hopping lander carried an x-ray spectrometer to measure surface composition and a magnetometer. It would fall to the surface, right itself with swivelling rods, and spend about 20 minutes taking data. Then a spring-loaded device would push the lander up, making it hop 20 to 40 m to a new location. That process might repeat five or six times over four hours to allow multiple measurements of composition. The solar-powered long-term autonomous lander would operate for a year, anchored to the surface by a harpoon-like penetrator. It would image the surface and operate a seismometer and spectrometers to measure surface composition, and its radio link would be used to monitor the orbit of Phobos more precisely than ever before. Phobos is being slowed by tidal effects, so its orbit will decay until it breaks up to form a ring, and the new observations would increase understanding of that process.

1997: **Mars Pathfinder**

Mars Pathfinder landed on Mars on 4 July 1997 (MY 23, sol 304) and operated for three months. Its cameras took images of both satellites, but as Phobos was only three pixels across and Deimos one pixel across, the images were effectively unresolved and were mainly useful for their spectral data (Murchie *et al.*, 1999). Images were taken of Phobos on 30 August and 12 September 1997, and of Deimos on 7 and 18 July. The spectra suggest that the two satellites were not dark C-type asteroids as had been suggested in the past, but either dark red D-type asteroids or weathered basalt-like material which might be basin ejecta from Mars. Because the cameras were on Mars, only the region on each satellite facing Mars could be imaged, with some variation depending on the position of the moon in the sky and the location of its terminator. Maps of the areas covered were published by Murchie *et al.* (1999), but they did not take the irregular shapes of the satellites into account. In Figure 201A the areas covered by the unresolved spectral data are shown, based on the actual topography. Just before the 30 August multispectral imaging, Phobos was observed passing out of eclipse, providing information on the atmosphere of Mars and refining its orbit (Thomas *et al.*, 1999).

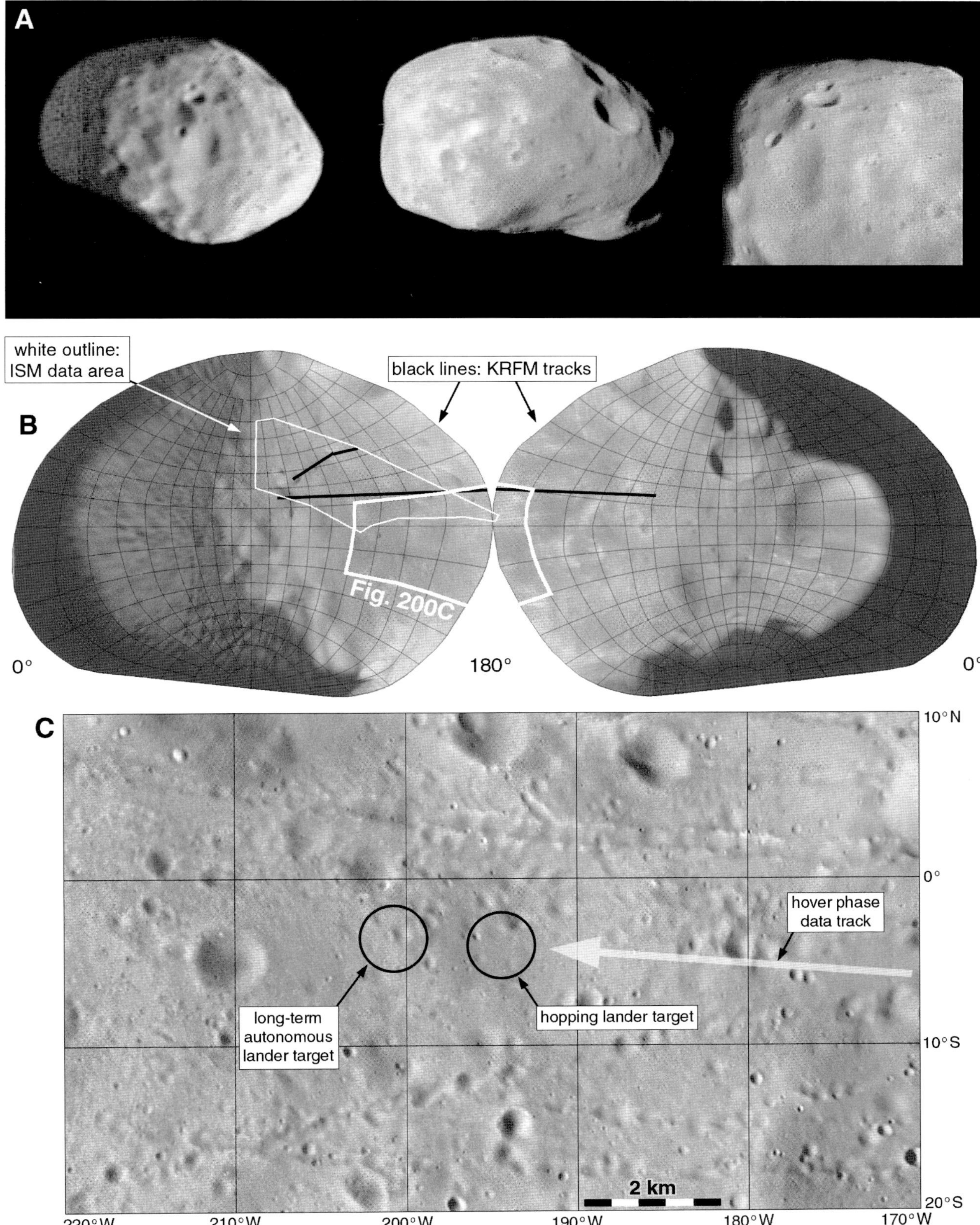

Figure 200 **A:** Representative Phobos 2 images, courtesy T. Stryk. **B:** Photomap showing the extent of image coverage and of ISM infrared data coverage, and the two KRFM data tracks. **C:** Locations of the Phobos 2 lander targets and low-elevation hover phase data collection area (Sagdeev and Zakharov, 1989). A and B show a region illuminated by light reflected off Mars between longitudes 110° W and 160° W.

1990s: **Aladdin**

Aladdin, a sample return mission to Phobos and Deimos, was proposed under the Discovery program in both 1996 and 1998. In 1997 Aladdin was shortlisted, but the solar wind sampler Genesis and the multi-comet flyby mission CONTOUR were selected in its place. In 1999 Aladdin was again shortlisted, but the Mercury orbiter MESSENGER and the Deep Impact comet mission were chosen instead. The second version of the mission (Mueller *et al.*, 2003) is described here.

Launch would have been on 30 May 2003, with arrival at Mars on 24 December 2003 (MY 26, sol 598). After spending 2004 near the Sun-Mars L2 Lagrange Point to ensure proper timing for the return journey, the satellite sampling phase would run from 4 February to 28 June 2005. Mars departure would occur on 28 June 2005 (MY 27, sol 467), with a return to Earth on 6 January 2006. The sampling phase would consist of 19 eccentric equatorial orbits with a 7.68-day period, a periapsis of 9000 km, slightly below the orbit of Phobos, and an apapsis of about 150000 km. The orbit period is 24 times that of Phobos and six times that of Deimos, allowing multiple encounters without using much fuel. Sampling was accomplished by firing a projectile at the surface during a very low flyby and scooping up particles as the spacecraft flew through the ejecta plume. Some orbits were used for satellite remote sensing, some for navigation and five for sampling (Table 79). Although three samples were to be taken at Phobos, only two sites were specified (Figure 201B, 202B).

Sample sites were chosen based on the Phobos 2 spectral data, which showed a distinct difference between the leading and trailing sides of the satellite (Murchie and Erard, 1996; Murchie *et al.*, 1999). An irregular boundary between the two areas, spectrally redder on the trailing side and bluer on the leading side, is indicated in Figure 201B. The corresponding boundary on the Mars-facing side was not observed by Phobos 2 but was seen later by Mars Reconnaissance Orbiter. The bluer material fills and surrounds the large crater Stickney and might consist of

Table 79. *Aladdin Orbital Activities (Mueller et al., 2003)*

Orbit	Satellite	Hemisphere	Activity	Closest distance (km)
1	Phobos	Trailing	Remote sensing	1000
2	Phobos	Trailing	Navigation	50
3	Phobos	Trailing	Navigation	50
4	Phobos	Trailing	Sampling	1.5
5	Deimos	Leading	Remote sensing	900
6	Phobos	Leading	Remote sensing	500
7	Phobos	Leading	Remote sensing	900
8	Phobos	Trailing	Navigation	50
9	Phobos	Trailing	Sampling	1.5
10	Phobos	Trailing	Spare	
11	Deimos	Leading	Navigation	50
12	Deimos	Leading	Navigation	50
13	Deimos	Leading	Sampling	1.5
14	Deimos	Leading	Navigation	50
15	Deimos	Leading	Sampling	1.5
16	Phobos	Leading	Navigation	50
17	Phobos	Leading	Sampling	1.5
18	Phobos	Leading	Spare	
19	Departure activities (rotate orbit plane, lower periapsis)			

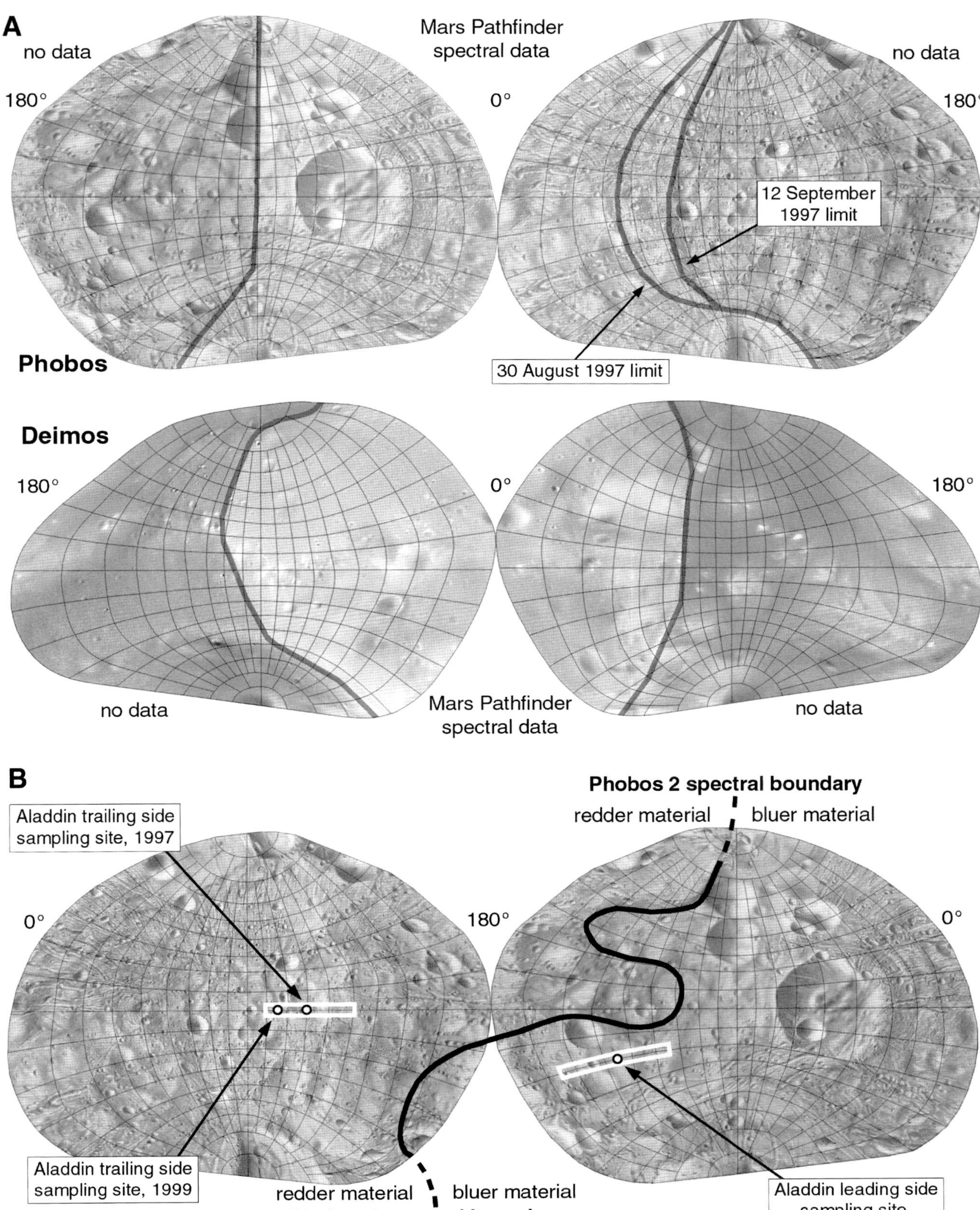

Figure 201 **A:** Areas on Phobos and Deimos covered by Mars Pathfinder spectral data, modified from Murchie *et al.*, 1999. **B:** Aladdin sampling areas on Phobos. The black line separates spectrally redder areas (left) from spectrally bluer areas, thought to be Stickney ejecta (Murchie *et al.*, 1999). Aladdin would sample both types of material. The white boxes would be covered by the highest resolution remote sensing data, and the centrally located dots would be targeted for sampling (Pieters *et al.*, 1999).

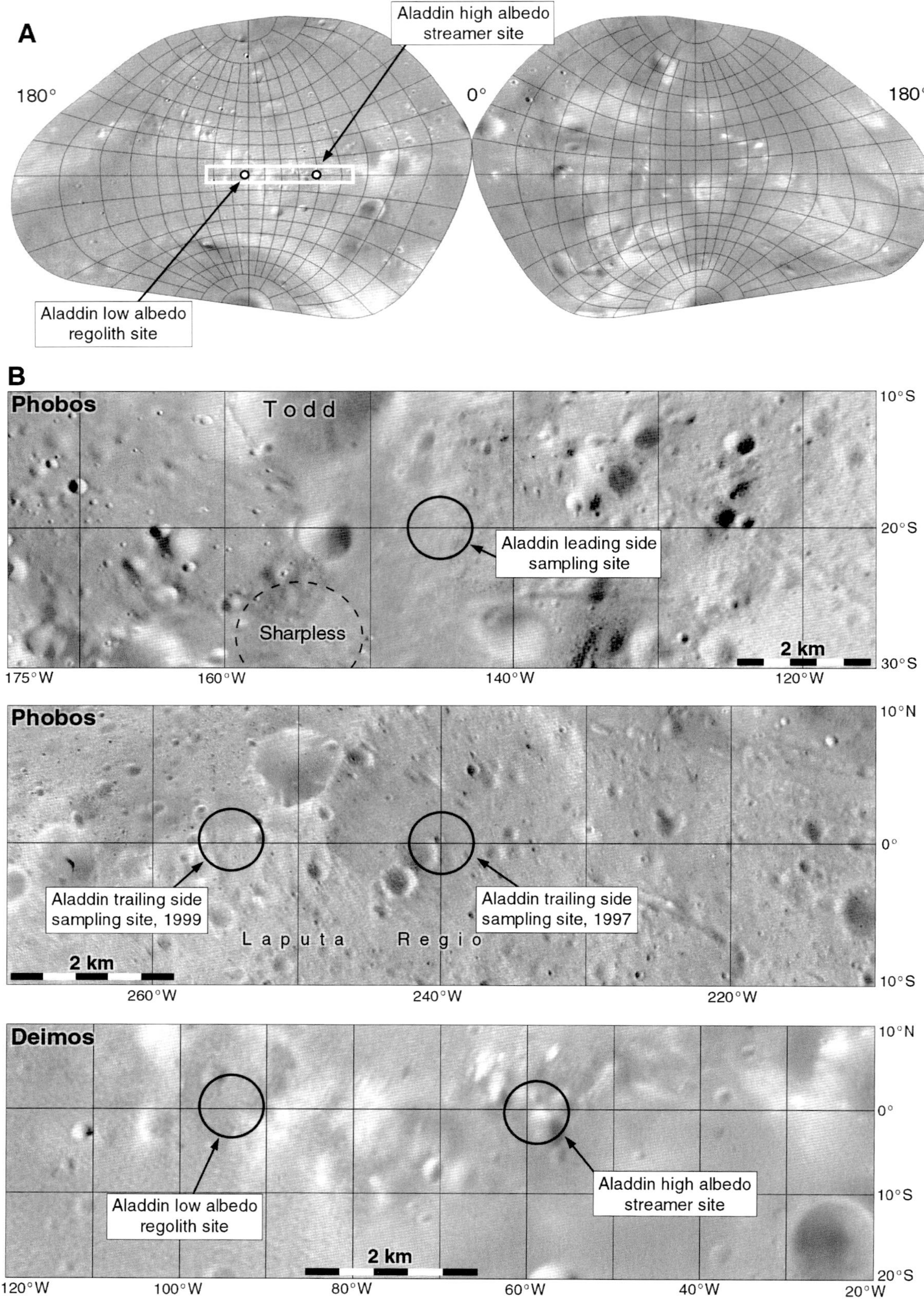

Figure 202 A: Aladdin sampling area on Deimos. The white box would be covered by the highest resolution remote sensing data, and the dots indicate sampling targets. **B:** Large-scale maps of the Aladdin sites on Phobos and Deimos.

its ejecta. The crater interior had already been suggested as a sampling target by Botbyl (1993) because the ejecta would provide access to material from the interior of Phobos. Samples of both sides would be obtained by Aladdin (Pieters *et al.*, 1997, 1999). During the low-altitude flyby, about 1500 m above the surface, detailed remote sensing would take place in swaths containing the landing sites, including observations of the sampling event and the resulting crater.

Pieters *et al.* (1997) and the contemporary mission website at the Applied Physics Laboratory of the Johns Hopkins University illustrated the remote sensing swaths and sampling sites (Figures 201B, 202A). The Phobos swaths extended about 5° N to S and 40° E to W, at 20° S and 125° W to 165° W on the leading side and on the equator and 220° W to 260° W on the trailing side. The sampling sites were at 20° S, 145° W and 0°, 240° W (Figure 202B). Two years later (Pieters *et al.*, 1999), the Phobos trailing side site had been moved to the west, probably to avoid sampling a topographically rough region. The remote sensing swath was still on the equator but now extended from 235° W to 275° W, with sampling at 0°, 255° W. Only the older swath is shown in Figure 201B.

On Deimos a single swath on the leading hemisphere was targeted, extending from 35° W to 120° W along the equator. Sampling would target a high albedo streamer and a low albedo regolith area. The streamers (Thomas, 1979, 1993), common on Deimos but not seen on Phobos, were apparently caused by slow movement of fine regolith particles from topographic ridges into depressions. Sampling sites (Figure 202B) were at roughly 0°, 58° W and 0°, 94° W.

1997: **Mars Global Surveyor**

Mars Global Surveyor (MGS) arrived at Mars on 12 September 1997 (MY 23, sol 371), initially entering an elliptical orbit which was gradually adjusted by aerobraking until it could perform high-resolution mapping from a low circular orbit. On 7, 19 and 31 August and 12 September 1998 (MY 24, sols 23, 35, 46 and 58, respectively), MGS made four close flybys and obtained five images of Phobos (Thomas *et al.*, 2000). The closest flyby distance was 265 km, and the best image, taken from 400-km altitude, had a resolution of about 2 m/pixel, the highest resolution available for Phobos. The other images were taken from distances between 1050 and 1770 km. Figure 203 shows some of these images and a map of surface coverage. The high-resolution images show Stickney crater and its surroundings, and the single highest resolution image extends from 30° N, 25° W to 10° N, 342° W. All these high-resolution images also show areas illuminated by light reflected from Mars, including the south pole. No more close passes were possible after the mapping orbit was attained, but by rotating to look past the limb of the planet, MGS could take more distant satellite images. On 1 June 2003 (MY 26, sol 398), it obtained a medium-resolution image of Phobos from 6000 km, and on 10 July 2006 (MY 28, sol 165), a low-resolution view of Deimos from 23 000 km was acquired. This was the only image of Deimos taken by MGS.

In addition to the images, two other instruments on MGS obtained data on Phobos. A MOLA track made on 12 September 1998 (MY 24, sol 58) was useful, especially for improving knowledge of the orbit of Phobos (Banerdt and Neumann, 1999). After MOLA ceased operating as an altimeter, it functioned as a passive radiometer, observing the brightness of the spot beneath it. In this mode it could detect the shadow of Phobos, which helped refine knowledge of the satellite's orbit (Bills *et al.*, 2005). Compositional data from TES suggested similarities to Mars rather than asteroids, adding to a growing feeling that the satellites might consist of Mars impact basin ejecta rather than having been captured from solar orbits (Roush and Hogan, 2000).

Russian Missions

Russia experienced severe economic difficulties in the 1990s but continued to suggest future missions to Mars and its satellites. Galeev *et al.* (1996) proposed a Phobos Sample Return mission, which eventually developed into the Phobos-Grunt mission launched in 2011. Its landing and sampling site would be near the anti-Mars point, on the equator at 180° W. A mission to Phobos and Deimos using ion thrusters was suggested by Rahe *et al.* (1999) and summarized by Mukhin *et al.* (2000) at a meeting on Mars exploration concepts held in Houston in July 2000. The small spacecraft could undertake remote sensing of both satellites and possibly a landing on Phobos.

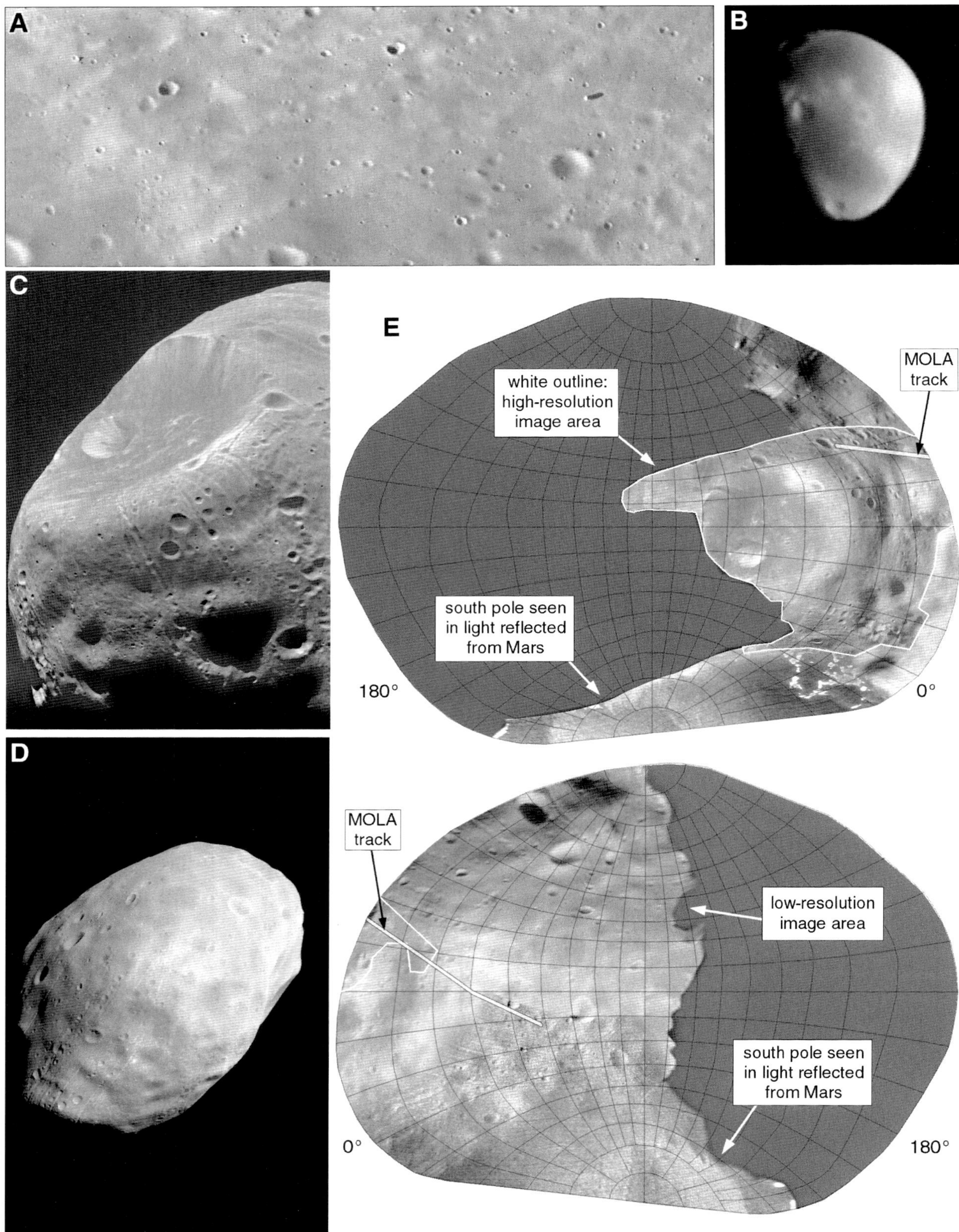

Figure 203 Mars Global Surveyor images and map of.
A: Highest-resolution image of Phobos (SP255103). **B:** MOC image of Deimos (S2000602). **C:** MOC image SP252603 showing Stickney crater. **D:** The Mars-facing side of Phobos (R0600044). **E:** Phobos image coverage by Mars Global Surveyor, including the MOLA data track of 12 September 1998, which coincides with the highest-resolution image (Banerdt and Neumann, 1999).

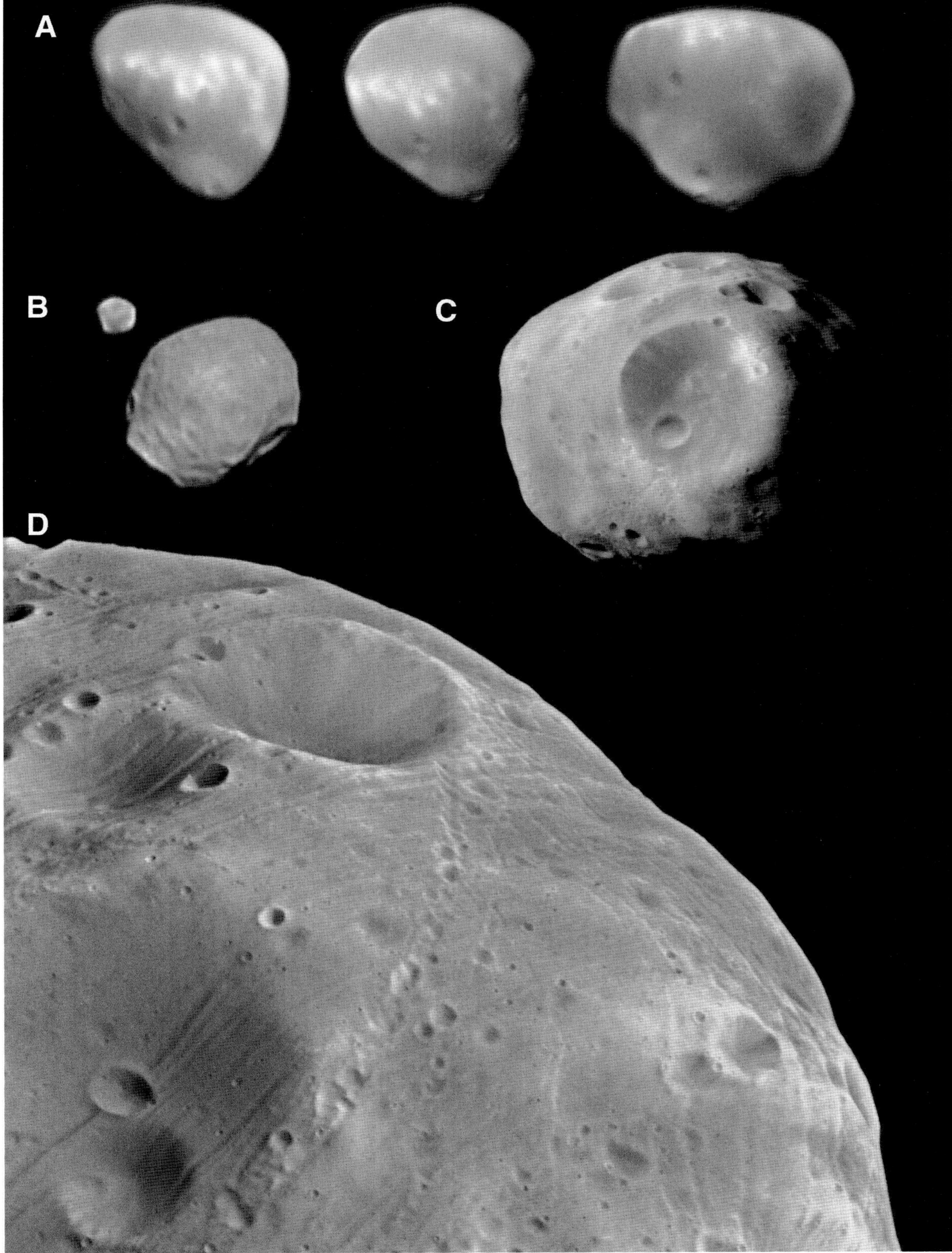

Figure 204 Mars Express images of Phobos and Deimos.
A: HRSC images h3196_0005_sr2, h0973_0002_sr2 and h1222_0003_sr2 of Deimos. **B:** HRSC image h7492_0054_sr2 showing Phobos and Deimos together in a single frame. **C:** HRSC image h2747_0000_s22 of Phobos centred on Stickney. **D:** High-resolution frame h5870_0000_nd2 of Phobos with Gulliver and Roche in the lower left corner.

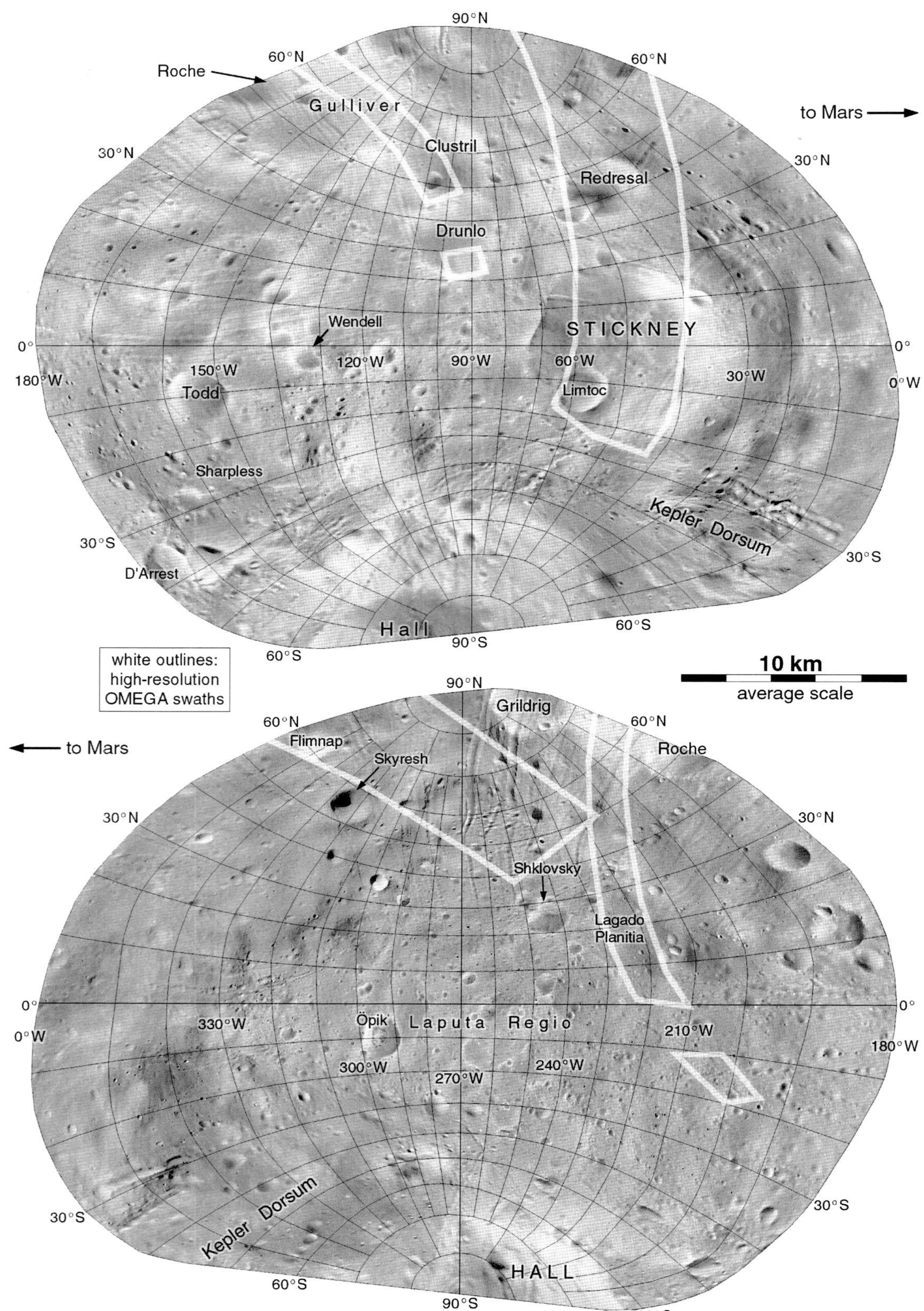

Figure 205 Global photomap of Phobos consisting mainly of Mars Express images.
Viking and Mars Reconnaissance Orbiter images are used where they provide improved resolution. Feature positions are based on a new coordinate system (Wählisch *et al.*, 2010) and differ slightly from previous maps. High-resolution OMEGA swaths are shown as white outlines.

2003: **Mars Express**

The European Space Agency's Mars Express mission arrived at Mars on 25 December 2003 (MY 26, sol 599). Its near-polar orbit for most of the mission extended from 300 km to between 10 000 and 14 000 km above the planet, allowing it to pass close to Phobos when their orbits intersected at intervals of about five months. No close passes to Deimos were possible. Phobos was imaged frequently during the mission from distances on the order of 5000 km and occasionally from as close as 1000 km. Very closes passes occurred on 22 August 2004 (149 km), 23 March 2006 (460 km), 17 July 2008 (274 km), 3 March 2010 (77 km) and 9 January 2011 (108 km). These dates are MY 27, sol 165, MY 28, sol 61, MY 29, sol 215, MY 30, sol 124 and MY 30, sol 428, respectively. The particularly close flyby on 3 March 2010 allowed an estimate of the mass of Phobos, and on 7 March the radar sounding system (MARSIS) was used to look for internal structures in the little moon. MARSIS was used again in January 2011. HRSC images covered the whole surface of Phobos, and in the leading hemisphere and polar regions they were superior to Viking coverage. Figure 204 shows representative images of Phobos and three distant views of the Mars-facing side of Deimos taken from roughly 11 000 km. These images were used to create a new global shape model and map of Phobos (Wählisch *et al.*, 2010). Positions of features in this map differ from those calculated by Simonelli *et al.* (1993), as can be seen by comparing Figures 195 and 205. Figure 205 is a map consisting largely of Mars Express images, but using Viking and Mars Reconnaissance Orbiter images in parts of the trailing side where they are superior in resolution or illumination. The photomosaic was compiled by P. Stooke using an early version of the Wählisch *et al.* (2010) map for positional control. The OMEGA spectrometer obtained compositional data, especially in the northern hemisphere (Figure 205).

Human Exploration of Phobos and Deimos

Singer (1981) described a study of human exploration of Phobos and Deimos undertaken for the Marshall Space Flight Center. A crew of eight people, including two medical specialists, would launch sometime after 1990, orbit Mars and land on Deimos. They would stay for three to six months on the satellite, using the time not just to study Deimos, but also to make a side trip to Phobos and to control up to 20 rovers with sample return capability on Mars itself. Deimos moves slowly across the Martian sky, making it better for rover communications than Phobos. This communication role requires at least some infrastructure, if not the entire Deimos habitat, on the Mars-facing side of the satellite. Singer imagined that two of the crew would control the Mars exploration, two more would explore and sample Deimos, and two would fly to Phobos to land and sample it. The medical team would take care of health issues if they occurred and conduct medical and psychological research. The full return trip would take two years.

Adelman and Adelman (1984) developed this idea further, proposing Phobos as the site for a base supporting an extended program of Mars exploration. They imagined a main base complex, a spaceport from which shuttles would fly to and from Mars, an observatory, and communication facilities, including satellites at the Mars-Phobos Lagrange Points 4 and 5 (60° ahead of and behind Phobos in its orbit). The spaceport would be on a ridge near the main base, and the observatory on a ridge, probably on Kepler Dorsum or the smaller ridges in the vicinity of Todd crater. A Mars communication system would be set up on the 0° meridian and an Earth communication facility near the 180° meridian. At the time Phobos was thought to be a volatile-rich C-type asteroid which might provide water or other resources.

Nash *et al.* (1989) surveyed human activities at the Moon, Mars, Phobos and asteroids, concluding that useful work could be done by astronauts at all of these potential targets. Three targets for geological exploration of Phobos were identified. Stickney, on the equator at 59° W, would provide samples from the small moon's interior. The structure and origin of grooves would be examined at an unspecified location, and typical Phobos regolith would be explored along the Kepler Ridge (Kepler Dorsum), between 28° and 30° S, extending from 60° W through the 0° meridian to 330° W.

3 Mars mission data

Table 80. *Mars Impact and Landing Events*

Spacecraft	Event	Arrival date	Mars date*	L_s	Location
Zond 2**	Impact?	6 August 1965	MY 6, sol 326	154	Unknown
Mars 2	Impact	27 November 1971	MY 9, sol 563	300	45° S, 302° W
Mars 3	Landing	2 December 1971	MY 9, sol 568	303	45° S, 158° W
Mars 6	Impact	12 March 1974	MY 11, sol 40	19	24° S, 19.5° W
Viking Lander 1	Landing	20 July 1976	MY 12, sol 209	97	22.48° N, 47.97° W
Viking Lander 2	Landing	3 September 1976	MY 12, sol 253	118	47.97° N, 225.74° W
Mars Pathfinder	Landing	4 July 1997	MY 23, sol 304	143	19.33° N, 33.55° W
Mars Climate Orbiter	***	23 September 1999	MY 24, sol 424	211	34° N, 170° W
Mars Polar Lander	Impact	3 December 1999	MY 24, sol 494	256	76.3° S, 194.5° W
Deep Space 2 – Amundsen	Impact	3 December 1999	MY 24, sol 494	256	75.0° S, 195.6° W
Deep Space 2 – Scott	Impact	3 December 1999	MY 24, sol 494	256	75.0° S, 195.6° W
Beagle 2	Impact	25 December 2003	MY 26, sol 599	322	11.6° N, 269.6° W
MER-A lander	Landing	4 January 2004	MY 26, sol 609	327	14.57° S, 184.52° W
Spirit (end of drive)	Drive	31 May 2010			14.52° S, 184.45° W
MER-B lander	Landing	25 January 2004	MY 26, sol 630	339	1.95° S, 5.53° W
Opportunity (end of drive)	Drive	31 May 2010			2.06° S, 5.52° W
Phoenix	Landing	25 May 2008	MY 29, sol 164	77	68.22° N, 125.75° E

*Mars dates are based on a calendar proposed by Clancy *et al.* (2000).
**Zond 2: possible impact, location unknown, suggested by Murray *et al.* (1967).
***Mars Climate Orbiter: approximate location at closest approach, or fragments if any reached the ground.

Table 81. *Mars Flyby and Orbital Events*

Spacecraft	Event	Arrival date	Mars date	L_s
Mars 1	Flyby	19 June 1963	MY 5, sol 237	110°
Mariner 4	Flyby	15 July 1965	MY 6, sol 305	143°
Zond 2	Impact or flyby	6 August 1965	MY 6, sol 326	154°
Mariner 6	Flyby	31 July 1969	MY 8, sol 405	200°
Mariner 7	Flyby	5 August 1969	MY 8, sol 410	203°
Mariner 9	Orbit	14 November 1971	MY 9, sol 550	292°
Mars 2	Orbit	27 November 1971	MY 9, sol 563	300°
Mars 3	Orbit	2 December 1971	MY 9, sol 568	303°
Mars 4	Flyby	10 February 1974	MY 11, sol 10	5°
Mars 5	Orbit	12 February 1974	MY 11, sol 12	6°
Mars 7	Flyby	9 March 1974	MY 11, sol 36	18°

Table 81. *(cont.)*

Spacecraft	Event	Arrival date	Mars date	L_s
Viking Orbiter 1	Orbit	19 June 1976	MY 12, sol 178	83°
Viking Orbiter 2	Orbit	7 August 1976	MY 12, sol 226	105°
Phobos 1	Flyby	25 January 1989	MY 18, sol 647	348°
Phobos 2	Orbit	29 January 1989	MY 18, sol 651	350°
Mars Observer	Flyby	24 August 1993	MY 21, sol 268	125°
Mars Global Surveyor	Orbit	12 September 1997	MY 23, sol 371	179°
Nozomi	Distant Flyby	28 August 1999	MY 24, sol 399	196°
Mars Climate Orbiter	Flyby, possible entry	23 September 1999	MY 24, sol 424	211°
Deep Space 1	Distant flyby	10 November 1999	MY 24, sol 471	240°
2001 Mars Odyssey	Orbit	24 October 2001	MY 25, sol 497	258°
Nozomi	Flyby	14 December 2003	MY 26, sol 588	316°
Mars Express	Orbit	25 December 2003	MY 26, sol 599	322°
Mars Reconnaissance Orbiter	Orbit	10 March 2006	MY 28, sol 46	23°
Rosetta	Flyby	25 February 2007	MY 28, sol 389	190°
Dawn	Flyby	18 February 2009	MY 29, sol 425	212°

Table 82. *Mars Exploration Chronology by Mars Year*

Mars year	Events
1	Sol 483: First well-observed global dust storm begins (McKim, 1996).
2	Sol 214: Sputnik 1 in orbit.
3	
4	Sol 129: Yuri Gagarin (Vostok 1) in orbit.
5	Sol 237: Mars 1 flyby.
6	Sol 305: Mariner 4 flyby; sol 326: Zond 2 flyby or impact.
7	
8	Sol 394: Apollo 11 lunar landing; sol 405: Mariner 6 flyby; sol 410: Mariner 7 flyby.
9	Sol 550: Mariner 9 in orbit; sol 563: Mars 2 impact, orbit; sol 568: Mars 3 landing, orbit.
10	Sol 156: Mars 2 and 3 end of mission; sol 220: Mariner 9 end of mission.
11	Sol 10: Mars 4 flyby; sol 12: Mars 5 orbit; sol 36: Mars 7 flyby; sol 40: Mars 6 impact.
12	Sol 178: Viking 1 in orbit; sol 209: Viking 1 landing; sol 226: Viking 2 in orbit; sol 253: Viking 2 landing.
13	Sol 255: Viking Orbiter 2 end of mission.
14	Sol 197: Viking Lander 2 end of mission; sol 311: Viking Orbiter 1 end of mission.
15	Sol 448: Viking Lander 1 end of mission.
16	
17	
18	Sol 647: Phobos 1 flyby; sol 651: Phobos 2 in orbit.
19	Sol 37: Phobos 2 end of mission.
20	
21	Sol 268: Mars Observer flyby.
22	

Table 82. *(cont.)*

Mars year	Events
23	Sol 304: Mars Pathfinder landing; sol 371: Mars Global Surveyor in orbit; sol 396: Mars Pathfinder end of mission.
24	Sol 399: Nozomi distant flyby; sol 424: Mars Climate Orbiter flyby or atmospheric entry; sol 470: Deep Space 1 distant flyby; sol 494: Mars Polar Lander and Deep Space 2 impacts.
25	Sol 497: Mars Odyssey in orbit.
26	Sol 588: Nozomi flyby; sol 599: Mars Express orbit and Beagle 2 impact; sol 609: MER-A (Spirit) landing; sol 630: MER-B (Opportunity) landing.
27	Sol 94: Opportunity enters Endurance crater; sol 100: Spirit reaches West Spur, Columbia Hills; sol 558: Spirit reaches summit of Husband Hill; sol 592: Opportunity reaches Erebus crater (Olympia).
28	Sol 18: Spirit reaches Home Plate; sol 46: Mars Reconnaissance Orbiter in orbit; sol 244: Opportunity arrives at Victoria crater; sol 277: Mars Global Surveyor end of mission; sol 389: Rosetta flyby; sol 583: Opportunity enters Victoria crater.
29	Sol 164: Phoenix landing; sol 257: Opportunity exits Victoria crater; sol 317: Phoenix end of mission; sol 425: Dawn flyby.
30	Sol 145: MER-A (Spirit) end of mission; sol 407: Opportunity arrives at Santa Maria crater; sol 636: Opportunity arrives at Endeavour crater (Cape York).
Anticipated events	
31	Sol 319: MSL (Curiosity) landing.
32	c. sol 400: MAVEN in orbit.
33	c. sol 475: ExoMars Trace Gas Orbiter in orbit, EDM landing.
34	c. sol 365: ExoMars 2018 rover landing.

Bibliography

Academy of Sciences of the USSR, 1969. *First Panoramas of the Lunar Surface*. Volume 2. Moscow: Akademiya Nauk.

Acuña, M. H., Connerney, J. E. P., Wasilewski, P., *et al.*, 2001. Magnetic Field of Mars: Summary of Results From the Aerobraking and Mapping Orbits. *Journal of Geophysical Research*, v. 106, no. E10, pp. 23403–17.

Adelman, S. J. and Adelman, B., 1984. The Case for Phobos. Paper AAS 84–165, in *The Case for Mars II*, AAS Science and Technology Series, v. 62, pp. 245–52. San Diego: Univelt Inc.

Albee, A. L., Arvidson, R. E., Palluconi, F. and Thorpe, T., 2001. Overview of the Mars Global Surveyor Mission. *Journal of Geophysical Research*, v. 106, no. E10, pp. 23291–316.

Albertz, J., Attwenger, M., Barrett, J., *et al.* and the HRSC CoI-Team, 2005. HRSC on Mars Express – Photogrammetric and Cartographic Research. *Photogrammetric Engineering & Remote Sensing*, v. 71, no. 10, pp. 1153–66.

Amrine, J. M., Berger, J. M., Blaufuss, D. J., *et al.*, 1992. *Project Ares, A Systems Engineering and Operations Architecture for the Exploration of Mars*. Technical Report AD-A258 082. Department of the Air Force, Air University, Department of Operational Sciences, Air Force Institute of Technology, Wright-Patterson Air Force Base, OH, 20 March 1992.

Anonymous, 1987. French/U.S. Team Spots Viking Crater on Martian Surface. *Aviation Week and Space Technology*, v. 126, 22 June, pp. 84–5.

Arnold, D. R., Bean, E. E. and Hughes, W. R., 1964. Study of the Mariner Mars 66 Spacecraft. Contractor Report no. NASA-CR-83726. Washington, DC: National Aeronautics and Space Administration.

Arvidson, R. E., Guinness, E. A., Moore, H. J., Tillman, J. and Wall, S. D., 1983. Three Mars Years: Viking Lander 1 Imaging Observations. *Science*, v. 222, no. 4623, pp. 463–8.

ASCONT, 1998. *Atlas of Mars by the TERMOSCAN Radiometer Data*. Moscow: Association for the Advancement of Space Science and Technology (ASCONT).

Avanesov, G., Zhukov, B., Ziman, Ya., *et al.*, 1991. Results of TV Imaging of Phobos (Experiment VSK-FREGAT). *Planetary and Space Science*, v. 39, nos. 1–2, pp. 281–95.

Avco Corporation, 1963a. *Voyager Design Studies. Volume 1: Summary*. Technical Report RAD-TR-63–34, 15 October 1963. Avco Research and Advanced Development Division. Wilmington, MA: Avco Corporation.

Avco Corporation, 1963b. *Voyager Design Studies. Volume 2: Scientific Mission Analysis*. Technical Report RAD-TR-63–34, 15 October 1963. Avco Research and Advanced Development Division. Wilmington, MA: Avco Corporation.

Banerdt, B., Chicarro, A. F., Coradini, M., *et al.*, 1996. Intermarsnet Phase-A Study Report. ESA Publication D/SCI(96)2, April 1996.

Banerdt, W. B. and Neumann, G. A., 1999. The Topography (and Ephemeris) of Phobos from MOLA Ranging. 30th Lunar and Planetary Science Conference, Houston, TX, 15–19 March 1999. Abstract no. 2021.

Barlow, N. G., 2000. "Following the Water" on Mars: Where Is It, How Much Is There, and How Can We Access It? Concepts and Approaches for Mars Exploration, Lunar and Planetary Institute, Houston, TX, 18–20 July 2000. Abstract no. 6013.

Barth, C. A. and Hord, C. W., 1971. Mariner Ultraviolet Spectrometer: Topography and Polar Cap. *Science*, v. 173, no. 3993, pp. 197–201.

Beer, W. and Mädler J. H., 1831. Physische Beobachtungen des Mars bei seiner Opposition im September 1830. *Astronomische Nachrichten*, v. 191, pp. 447–56.

Beer, W. and Mädler J. H., 1841. Beiträge zur Physischen Kenntniss der Himmlischen Körper im Sonnensysteme. *Weimar.*

Belton, M. J. S. and Hunten, D. M., 1969. Spectrographic Detection of Topographic Features on Mars. *Science*, v. 166, pp. 225–7.

Berthoud, M. G., 2005. An Equal-Area Map Projection for Irregular Objects. *Icarus*, v. 175, no. 2, June 2005, pp. 382–9, doi:10.1016/j.icarus.2004.11.021.

Bibring, J.-P., M. Combes, M., Langevin, Y. *et al.*, 1989. Results from the ISM Experiment. *Nature*, v. 341, pp. 591–3. doi:10.1038/341591a0.

Bibring, J.-P., Langevin, Y., Mustard, J. F., *et al.*, and the OMEGA team, 2006. Global Mineralogical and Aqueous Mars History Derived from OMEGA/Mars Express Data. *Science*, v. 312, pp. 400–4. doi:10.1126/science.1122659.

Bills, B. G., Neumann, G. A., Smith, D. E. and Zuber, M. T., 2005. Improved Estimate of Tidal Dissipation Within Mars from MOLA Observations of the Shadow of Phobos. *Journal of Geophysical Research*, v. 110, E07004, doi:10.1029/2004JE002376.

Blunck, J., 1977. *Mars and its Satellites: A Detailed Commentary on the Nomenclature*. 1st Edition. Hicksville, NY: Exposition Press.

Blunck, J., 1982. *Mars and its Satellites: A Detailed Commentary on the Nomenclature*. 2nd Edition. Smithtown, NY: Exposition Press.

Blunck, J., Zögner, J. and Jöns, H.-P., 1993. *Der Rote Planet im Karten Bild: 200 Jahre Marskartographie von Herschel, Beer und Mädler bis zur CD*. Gotha: Justus Perthes Verlag, 1993.

Botbyl, G., 1993. *Final Report for a Robotic Exploration Mission to Mars and Phobos. Project Aeneas*. Response to RFP Number ASE274L.0893, NASA/USRA Advanced Design Program. Argos Space Endeavours, 29 November 1993.

Bridges, J. C., Seabrook, A. M., Rothery, D. A., *et al.*, 2003. Selection of the Landing Site in Isidis Planitia of Mars Probe Beagle 2. *Journal of Geophysical Research*, v. 108 (E1), 5001, doi:10.1029/2001JE001820.

Budden, N. A. and Duke, M. B. (eds.), 1998. *HEDS-UP Mars Exploration Forum*. LPI Contribution no. 955. Houston, TX: Lunar and Planetary Institute.

Budney, C. J., Miller, S. L. and Cutts, J. A., 2000. Mars Stratigraphy Mission. Presented at the Concepts and Approaches for Mars Exploration meeting, Lunar and Planetary Institute, Houston, TX, 18–20 July 2000. Abstract no. 6035.

Bugaevsky, L. M., Krasnopevtseva, B. V. and K. B. Shingareva, 1992. Phobos Map and Phobos Globe. *Advances in Space Research*, v. 12, no. 9, pp. 17–21, doi:10.1016/0273–1177-(92)90314-N.

Cabrol, N. A. and Grin, E. A., 1999. Evolution of Lacustrine Environments on Mars and their Significance: The Case for the Brazos Lakes and East Terra Meridiani Basins as Landing Sites for Surveyor 2001. Presented at the Workshop on Mars 2001: Integrated Science in Preparation for Sample Return and Human Exploration, Lunar and Planetary Institute, Houston, TX, 2–4 October 1999.

Caplinger, M. A. and Malin, M. C., 2001. Mars Orbiter Camera Geodesy Campaign. *Journal of Geophysical Research*, v. 106, no. E10, pp. 23595–606.

Cheng, Y. and Lorre, J. J., 2000. Equal Area Map Projection for Irregularly Shaped Objects. *Cartography and Geographic Information Science*, v. 27, no. 2, April 2000, pp. 91–100.

Chicarro, A. F., 1988. Sampling the Ancient Volatile-Rich Areas of Mars. *Workshop on Mars Sample Return Science*. LPI Technical Report 88–07, pp. 57–58. Houston, TX: Lunar and Planetary Institute.

Chicarro, A. F., 1993. MARSNET: A European Network of Stations on the Surface of Mars. *Mars Past, Present and Future. Results from the MSATT Program*. Part 1, pp. 12–14. Houston, TX: Lunar and Planetary Institute.

Chicarro, A. F., Coradini, M., Fulchignoni, M., *et al.*, 1993. *MARSNET: Report on the Phase-A Study*, ESA SCI(93)2, ESA, 119 pp.

Christensen, E. J., 1975. Martian Topography Derived from Occultation, Radar, Spectral and Optical Measurements. *Journal of Geophysical Research*, v. 80, no. 20, pp. 2909–13.

Christensen, P. R., Bandfield, J. L., Hamilton, V. E., *et al.*, 2001. Mars Global Surveyor Thermal Emission Spectrometer Experiment: Investigation Description and Surface Science Results. *Journal of Geophysical Research*, v. 106, no. E10, pp. 23823–71.

Christensen, P. R., Jakosky, B. M., Kieffer, H. H., *et al.*, 2004. The Thermal Emission Imaging System (THEMIS) for the Mars 2001 Odyssey Mission. *Space Science Reviews*, v. 110, pp. 85–130.

Clancy, R. T., Sandor, B. J., Wolff, M. J., *et al.*, 2000. An Intercomparison of Ground-Based Millimeter, MGS TES, and Viking Atmospheric Temperature Measurements: Seasonal and Interannual Variability of Temperatures and Dust Loading in the Global Martian Atmosphere. *Journal of Geophysical Research*, v. 105, pp. 9553–71.

Clark, B. C., Baird, A. K., Weldon, R. J., Tsusaki, D. M., Schnabel, L. and Candelaria, M. P., 1982. Chemical Composition of Martian Fines. *Journal of Geophysical Research*, v. 87, no. B12, pp. 10059–67, doi:10.1029/JB087iB12p10059.

Clark, P. E., Clark, C. S. and Stooke, P. J., 2008. Using Boundary-based Mapping Projections for Morphological Classification of Small Bodies. Presented at the 39th Lunar and Planetary Science Conference, League City, TX, 10–14 March 2008. Abstract no. 1371.

Connerney, J. E. P., Acuña, M. H., Ness, N. F., *et al.*, 2005. Tectonic Implications of Mars Crustal Magnetism. *Proceedings of the National Academy of Sciences*, v. 102, no. 42, pp. 14970–5.

Conrath, B., Curran, R., Hanel, R., Kunde, V., Maguire, W., Pearl, J., Pirraglia, J. and Welker, J., 1973. Atmospheric and Surface Properties of Mars Obtained by Infrared Spectroscopy on Mariner 9. *Journal of Geophysical Research*, v. 78, no. 20, pp. 4267–78.

Costard, F., Achache, J., Bibring, J. P., *et al.*, 1993. A reference Martian mission for a long range rover. *Missions, Technologies, and Design of Planetary Mobile Vehicles*. CNES, pp. 129–138 (SEE N94–23373 06–91). Toulouse, France: Cépaduès-Éditions.

Costard, F., Mangold, N., Masson, Ph., Mege, D. and Peulvast, J. P., 1999. Melas Chasma: Potential Landing Site for the Mars 2001 Mission. Presented at the Workshop on Mars 2001: Integrated Science in Preparation for Sample Return and Human Exploration, Lunar and Planetary Institute, Houston, TX, 2–4 October 1999.

Darnell, W. L. and Wessel, V. W., 1974. A Conceptual Design and Operational Characteristics for a Mars Rover for a 1979 or 1981 Viking Science Mission. NASA TN-D-7462. Washington, DC: National Aeronautics and Space Administration, February 1974.

Davies, M. E., 1972. Coordinates of Features on the Mariner 6 and 7 Pictures of Mars. *Icarus*, v. 17, pp. 116–67.

De Angelis, G. and Chicarro, A. F., 1995. *Intermarsnet Mission – A Catalogue of Potential Landing Sites*. ESA/ ESTEC report no. ESA-X-1502.

De Angelis, G. and Chicarro, A. F., 1996. A Catalogue of Potential Landing Sites for the Intermarsnet Mission. *Planetary and Space Science*, v. 44, no. 2, pp. 1325–46.

Development Sciences Inc., 1978. *A Concept Study of a Remotely Piloted Vehicle for Mars Exploration. Final Technical Report.* NASA Contractor Report no. NASA-CR-157942. City of Industry, CA: Development Sciences Inc.

Dohm, J. M., Tanaka, K. L., Lias, J. H., Hare, T. M., Anderson, R. C. and Gulick, V. C., 1998. Warrego Valles and Other Candidate Sites of Local Hydrothermal Activity within the Thaumasia Region, Mars. Presented at the 29th Lunar and Planetary Science Conference, Lunar and Planetary Institute, Houston, TX, 16–20 March 1998. Abstract no 1669.

Dohm, J. M., Baker, V. R., Anderson, R. C., *et al.*, 2000. Martian Magmatic-driven Hydrothermal Sites: Potential Sources of Energy, Water and Life. Presented at the Concepts and Approaches for Mars Exploration meeting, Lunar and Planetary Institute, Houston, TX, 18–20 July 2000. Abstract no 6040.

Drake, M. J., Greeley, R., McKay, G. A., *et al.* (eds), 1988. *Workshop on Mars Sample Return Science*. LPI Tech. *Rpt.* 88–07. Houston, TX: Lunar and Planetary Institute, 196 pp.

Duxbury, T. C., 1974. Phobos: Control Network Analysis. *Icarus*, v. 23, pp. 290–9.

Duxbury, T. C. and Ocampo, A. C., 1988. Mars: Satellite and Ring Search From Viking. *Icarus*, v. 76, pp. 160–2.

Edgett, K. S., 1995. Physical Properties of the Ares Valles (Primary) and Trouvelot Crater (Backup) Landing Sites for Mars Pathfinder: Thermal Inertia and Rock Abundance from Viking IRTM observations. Lunar and Planetary Science XXVI, pp. 353–4. Houston, TX: Lunar and Planetary Institute.

Edgett, K. S., 1997. Around the Lander in 80 Sols. *EOS, Trans. IAU.*, v. 78, no. 40, p. 429.

European Space Agency (ESA), 1990. 'Mission to Mars: Report of the Mars Exploration Study Team', ESA SP-1117.

European Space Agency (ESA), 1991. *MARSNET: A Network of Stations on the Surface of Mars: Report on the Assessment Study.* Report no. ESA SCI(91)6, January 1991, ESA D/Sci Archives.

European Space Agency (ESA), 2000. Beagle 2 Team Assesses Landing Sites on Mars. ESA Press Release dated Wednesday, January 26, 2000.

Ezell, E. C. and Ezell, L. N., 1977. Selection and Certification: The Vikings' Search for Landing Sites on Mars. Unpublished draft manuscript dated 12 January 1977, Northeast Planetary Data Center, Brown University, Providence, RI.

Ezell, E. C. and Ezell, L. N., 1984. *On Mars: Exploration of the Red Planet 1958–1978*. The NASA History Series, NASA SP-4212. NASA Scientific and Technical Information Branch. Washington, DC: National Aeronautics and Space Administration.

Farmer, J., Des Marais, D., Greeley, R., Landheim, R. and Klein, H., 1995. Site Selection for Mars Exobiology. *Advances in Space Research*, v. 15, no. 3, pp. (3)157–(3)62.

Farmer, J., Nelson, D., Greeley, R., Klein, H. and Kuzmin, R., 1999. Site Selection for the MGS '01 Mission: an Astrobiological Perspective. Abstract in Gulick (1999), pp. 30–2.

Farrell, W. M., Clifford, S. M., Milkovich, S. M., *et al.*, 2008. MARSIS Subsurface Radar Investigations of the South Polar Reentrant Chasma Australe. *Journal of Geophysical Research*, v. 113, E04002, doi:10.1029/2007JE002974.

Ferri, F., P. H. Smith, M. Lemmon, and N. O. Rennó, 2003. Dust devils as observed by Mars Pathfinder. *Journal of Geophysical Research*, v. 108, no. E12, 5133, pp. 1–10, doi:10.1029/2000JE001421, 2003.

Flammarion, C., 1892. *La planète Mars et ses conditions d'habitabilité*. Paris: Gauthier-Villars et Fils.

Flammarion, C., 1909. *La planète Mars et ses conditions d'habitabilité*, Vol. 2. Paris: Gauthier-Villars et Fils.

Foley, C. N., Economou, T. and Clayton, R. N., 2003. Final Chemical Results from the Mars Pathfinder Alpha Proton X-Ray Spectrometer. *Journal of Geophysical Research*, v. 108, no. E12, pp. 8096–116, doi:10.1029/2002JE002019.

Folkner, W. M., Yoder, C. F., Yuan, D. N., Standish, E. M. and Preston, R. A., 1997. Interior Structure and Seasonal Mass Redistribution of Mars from Radio Tracking of Mars Pathfinder. *Science*, v. 278, pp. 1749–52.

Friedlander, A. L., Hartmann, W. K. and Niehoff, J. C., 1974. *Pioneer Mars 1979 Mission Options*. Report SAI 120-M1.

Chicago, IL: Science Applications, Inc., 29 January 1974.

Frumkin, M. and Riley, D., 2004. Optimizing Mars Airplane Trajectory with the Application Navigation System. Presented at Parallel and Distributed Computing Systems (PDCS04), Boston, MA, 9–11 November 2004.

Galeev, A. A., Moroz, V. I., Linkin, V. M., *et al.*, 1996. Phobos Sample Return Mission. *Advances in Space Research*, v. 17, no. 12, pp. 31–47.

Gangale, T. and Dudley-Rowley, M., 2005. Issues and Options for a Martian Calendar. *Planetary and Space Science*, v. 53, pp. 1483–95.

General Electric, 1963. *Voyager Design Study, Volume I, Design Summary*. Document no. 63SD801. Philadelphia, PA: General Electric Spacecraft Department, 15 October 1963.

General Electric, 1964. *Voyager Spacecraft System Study (Phase I - Titan IIIC Launch Vehicle), Final Report, Volume IIa*. Document no. 64SD933. Philadelphia, PA: General Electric Spacecraft Department, 7 August 1964.

Goldstein, R. M. and Gillmore, W. F., 1963. Radar Observations of Mars. *Science*, v. 141, pp. 1171–2.

Golombek, M. (ed.), 1994a. *Mars Pathfinder Landing Site Workshop*. LPI Tech. *Rpt*. 94–04. Houston, TX: Lunar and Planetary Institute, 49 pp.

Golombek, M., 1994b. Letter to Mars Pathfinder Landing Site Workshop Participants, June 20, 1994. Available at http://tes.asu.edu/PATHFINDER/p_f_landingsite_letter.html (obtained June 2006).

Golombek, M. P., Cook, R. A., Moore, H. J., and Parker, T. J., 1997. Selection of the Mars Pathfinder Landing Site. *Journal of Geophysical Research*, v. 102, pp. 3967–88.

Golombek, M.P., Anderson, R.C., Barnes, J.R., *et al.*, 1999. Overview of the Mars Pathfinder Mission: Launch through landing, surface operations, data sets, and science results. *Journal of Geophysical Research*, v. 104, pp. 8523–53.

Greeley, R. (ed.), 1990. *Mars Landing Site Catalog*. NASA Ref. Pub. 1238, August 1990. Washington, DC: National Aeronautics and Space Administration.

Greeley, R. and Batson, R. M., 1997. *The NASA Atlas of the Solar System*. Cambridge: Cambridge University Press.

Greeley, R. and Thomas, P. E. (eds.), 1995. *Mars Landing Site Catalog*, NASA Ref. Pub. 1238, 2nd Edition. Washington, DC: National Aeronautics and Space Administration.

Green, N., 1879. Observations of Mars, at Madeira, in August and September 1877. *Memoirs of the Royal Astronomical Society*, v. 44, pp. 123–40.

Guinness, E. A., Leff, C. E. and Arvidson, R. E., 1982. Two Mars Years of Surface Changes Seen at the Viking Landing Sites. *Journal of Geophysical Research*, v. 87, no. B12, pp. 10051–8.

Gulick, V., 1998. *Mars Surveyor 2001 Landing Site Workshop*. Abstracts of a meeting at NASA Ames Research Center, Moffett Field, CA, 26–27 January 1998.

Gulick, V. (ed.), 1999. *Second Mars Surveyor Landing Site Workshop*. Abstracts of a meeting at SUNY Buffalo, NY, 22–23 June 1999.

Haberle, R. M., 1995. Science Rationale for a Micro-Met Mission to Augment InterMarsNet. Presented at the International Workshop on InterMarsNet, Capri, Italy, 28–29 September 1995. Abstract.

Haberle, R. M. and Catling, D. C., 1996. A Micro-Meteorological Mission for Global Network Science on Mars: Rationale and Measurement Requirements. *Planetary and Space Science*, v. 44, no. 11, pp. 1361–83.

Hall, A., 1877. Notes. The Satellites of Mars. *The Observatory*, v. 1, pp. 181–5.

Hall, D. W., Parks, R. W. and Morris, S., 1997. *Airplane for Mars Exploration. Conceptual Design of the Full-Scale Vehicle Design, Construction and Test of Performance and Deployment Models*. Report submitted to NASA Ames on 27 May 1997. Sunnyvale, CA: David Hall Consulting.

Herr, K. C., Horn, D., McAfee, J. M. and Pimental, G. C., 1970. Martian Topography from the Mariner 6 and 7 Infrared Spectra. *Astronomical Journal*, v. 75, pp. 883–94.

Herschel, W., 1784. On the Remarkable Appearances at the Polar Regions of the Planet Mars, the Inclination of Its Axis, the Position of Its Poles, and Its Spheroidical Figure; With a Few Hints Relating to Its Real Diameter and Atmosphere. *Philosophical Transactions of the Royal Society of London*, v. 74, pp. 233–73.

Hibbs, A. R. (ed.), 1959. Exploration of the Moon, the Planets and Interplanetary Space. Report no. 30–1. Pasadena, CA: Jet Propulsion Laboratory, 30 April 1959.

Hoffman, S. J. and Kaplan D. (eds.), 1997. *Human Exploration of Mars: The Reference Mission of the NASA Mars Exploration Study Team*. NASA Special Publication 6107. Houston, TX: NASA Johnson Space Center.

Hoffman S. J., 1998. *The Mars Surface Reference Mission: A Response to the Human Exploration of Mars: The Reference Mission of the NASA Mars Exploration Team*. EX13–98-065.

Hoffman S. J., 2001. *The Mars Surface Reference Mission: A Description of Human and Robotic Surface Activities*. NASA/TP-2001–209371.

International Astronomical Union, 1960. *Transactions of the International Astronomical Union*. Moscow, 12–20 August 1958. V. 10, pl. 1, p. 262. Cambridge, MA: Cambridge University Press.

James, G., Chamitoff, G. and Barker, D., 1998. *Resource Utilization and Site Selection for a Self-Sufficient Martian Outpost*. NASA TM-98-206538. Washington, DC: National Aeronautics and Space Administration.

Jaumann, R., Neukum, G., Behnke, T., *et al.*, and the HRSC Co-Investigator Team, 2007. The High-Resolution Stereo Camera (HRSC) Experiment on Mars Express: Instrument Aspects and Experiment Conduct From Interplanetary Cruise Through the Nominal Mission. *Planetary and Space Science*, v. 55, pp. 928–52.

Jet Propulsion Laboratory (JPL), 1974. *Mariner Mars 1971 Television Picture Catalog*. NASA Technical Memorandum TM-33–585. Pasadena, CA: Jet Propulsion Laboratory.

Jet Propulsion Laboratory (JPL), 1978. Viking-Mars: Anatomy of Success. Viking Mission Status Bulletin no. 46. Pasadena, CA: Jet Propulsion Laboratory.

Jet Propulsion Laboratory (JPL), 2004. Spirit Looks Down Into Crater After Reaching Rim. Press release, 11 March 2004. Pasadena, CA: Jet Propulsion Laboratory.

Jones, K. L., Arvidson, R. E., Guinness, E. A., *et al.*, 1979. One Mars Year: Viking Lander Imaging Observations. *Science*, v. 204, no. 4395, pp. 799–806.

Kazarian-Le Brun, V., 1996. The MARS'96/BALTE Balloon Mission to Mars: Preliminary Results of Numerical Simulations. *Astrophysics and Space Science*, v. 239, pp. 197–211.

Kemmerly, J. E., 1961. *Summary of Mariner B Capsule Study*. Report U-1494, NASA CR-59715. Newport Beach, CA: Aeronutronic Division, Ford Motor Company.

Killeen, T. L., 1994. MUADEE: *Mars Upper Atmosphere Dynamics, Energetics, and Evolution Discovery Mission*. Final Report. Ann Arbor, MI: Michigan University, Ann Arbor, Dept. of Atmospheric, Oceanic and Space Sciences.

Killeen, T. L. and Brace, L., 1995. MUADEE: A Discovery-Class Mission for Exploration of the Upper Atmosphere of Mars. *Acta Astronautica*, v. 35, pp. 377–86.

Klein, H. P., Lederberg, J., Rich, A., Horowitz, N. H., Oyama, V. I. and Levin, G. V., 1976. The Viking Mission Search for Life on Mars. *Nature*, v. 262, no. 5563, pp. 24–7.

Kliore, A., Cain, D. L., Levy, G. S., Eshleman, V. R., Fjeldbo, G. and Drake, F. D., 1965. Occultation Experiment: Results of the First Direct Measurement of Mars's Atmosphere and Ionosphere. *Science*, v. 149, pp. 1243–8.

Kliore, A., Fjeldbo, G., Seidel, B. L. and Rasool, S. I., 1969. Mariners 6 and 7: Radio Occultation Measurements of the Atmosphere of Mars. *Science, New Series*, v. 166, no. 3911 (12 December 1969), pp. 1393–7.

Koroshetz, J. and Barlow, N. G., 1998. Possible Near-Surface Ice Reservoir South of Valles Marineris, Mars. 29th Lunar and Planetary Science Conference, Houston, TX, 16–20 March 1998. Abstract no. 1390.

Kuzmin, R. O, Landheim, R. and Greeley, R., 1994. Potential Landing Sites for Mars Pathfinder (abstract). In Golombek, M., ed., 1994. *Mars Pathfinder Landing Site Workshop*. LPI Tech. Rpt. 94–04, Lunar and Planetary Institute, Houston, TX, pp. 30–1.

Leighton, R. B., Murray, B. C., Sharp, R. P., Denton, A. J. and Sloan, R. K., 1965. Mariner IV Photography of Mars: Initial Results. *Science*, v. 149, no. 3684, pp. 627–30.

Ley, W. and von Braun, W., 1956. *The Exploration of Mars*. London: Sidgwick and Jackson.

Liwshitz, M., 1966. Landing Site Selection for MSSR. Bellcomm, Inc., Report No. NASA-CR-157878, 4 October 1966.

Lowell, P., 1895. *Mars*. New York: Houghton-Mifflin.

Lowell, P., 1906. *Mars and Its Canals*. New York: Macmillan.

Lowell, P., 1910. *Mars as the Abode of Life*. New York: Macmillan.

Lusignan, B. B., 1996. The Mars Surface Rover Marsokhod. Proceedings, 20th International Symposium on Space Technology and Science, Nagarawara, Japan, 19–25 May 1996, v. 2, pp. 1116–22.

Malin, M. C. and Edgett, K. S., 2000. Evidence for Recent Groundwater Seepage and Surface Runoff on Mars. *Science*, v. 288, no. 5475, pp. 2330–5.

Malin, M. C. and Edgett, K. S., 2001. Mars Global Surveyor Mars Orbiter Camera: Interplanetary Cruise Through Primary Mission. *Journal of Geophysical Research*, v. 106, no. E10, pp. 23429–70.

Malin, M. C., Edgett, K. S., Cantor, B. A., *et al.*, 2010. An Overview of the 1985–2006 Mars Orbiter Camera Science Investigation. *Mars*, v. 5, pp. 1–60, doi:10.1555/mars.2010.0001.

Malin Space Science Systems, 1998. Airplane Proposed for Mars Survey on Centennial of Wright Brothers' First Flight. Press release dated 20 July 1998. San Diego, CA: Malin Space Science Systems, Inc.

Malin Space Science Systems, 1999. Mars Climate Orbiter MARCI Approach Image. MGS MOC Release No. MARCI-1, 10 September 1999.

Malin Space Science Systems, 2004. Locating Landers on Mars. MGS MOC Release No. MOC2–595, 4 January 2004.

Malin Space Science Systems, 2005a. MGS Finds Viking Lander 2 and Mars Polar Lander (Maybe). MGS MOC Release No. MOC2–1082, 5 May 2005.

Malin Space Science Systems, 2005b. Mars Polar Lander NOT Found. MGS MOC Release No. MOC2–1253, 17 October 2005.

Malin Space Science Systems, 2006. Viking 1's 30th! MGS MOC Release No. MOC2–1529, 20 July 2006.

Manning, L. A. (ed.), 1977. *Mars Surface Penetrator: System Description*. NASA-TM-73243, June 1977.

Markun, C. D., 1988. Martian Sediments and Sedimentary Rocks. *Workshop on Mars Sample Return Science*. LPI Technical Report 88–07, pp. 117–118. Houston, TX: Lunar and Planetary Institute.

Marov, M. Ya. and Petrov, G. I., 1973. Investigations of Mars from the Soviet Automatic Stations Mars 2 and 3. *Icarus*, v. 19, no. 2, pp. 163–79.

Mars Science Working Group, 1977. *A Mars 1984 Mission. Report of the Mars Science Working Group*. NASA Tech. Memo. TM-78419, July 1977.

Mars Science Working Group, 1980. *Detailed Reports of the Mars Sample Return: Site Selection and Sample Acquisition Study*. JPL report 715–23, Volumes 1–10. Pasadena, CA: Jet Propulsion Laboratory, November 1980.

Marsal, O., Harri, A.-M., Lognonné, P., Rocard, F. and Counil, J.-L., 1999. Netlander: The First Scientific Lander Network on the Surface of Mars. Paper IAF-99-Q.3.03, 50th International Astronautical Congress. American Institute of Aeronautics and Astronautics.

Martin Marietta Corp., 1972. *A Study of System Requirements for Phobos/Deimos Missions*. Final Report. NASA Contractor Report CR-112077, 1 June 1972. Denver, CO: Martin Marietta Corp.

Martin Marietta Corp., 1977a. *Viking Lander System Primary Mission Performance Report*. NASA Contractor Report CR-145148, April 1977. Denver, CO: Martin Marietta Corp.

Martin Marietta Corp., 1977b. *Viking III Program, Mobile Lander Technical Studies*. Final Report S76–44594-001, January 1977. Denver, CO: Martin Marietta Corp.

Masursky, H., 1973. An Overview of Geological Results from Mariner 9. *Journal of Geophysical Research*, v. 78, no. 20, pp. 4009–30.

Masursky, H., Boyce, J. M. and Dial, A. L., 1974. *Viking Rover Studies*. NASA report CR-140391, July 1974.

Masursky, H. and Crabill, N. L., 1976. Search for the Viking 2 Landing Site. *Science*, v. 194, no. 4260, pp. 62–8.

Masursky, H. and Crabill, N. L., 1981. *Viking Site Selection and Certification*. NASA-SP-429. Hampton, VA: NASA Langley Research Center.

Masursky, H., Dial, A. L., Morris, E. C., Strobell, M. E., Applebee, D. J. and Chapman, M. G., 1988a. *Workshop on Mars Sample Return Science*. LPI Technical Report 88–07, pp. 119–120. Houston, TX: Lunar and Planetary Institute.

Masursky, H., Dial, A. L., Strobell, M. E. and Applebee, D. J., 1988b. Geology of Six Possible Martian Landing Sites. *MEVTV Workshop on Nature and Composition of Surface Units on Mars*. LPI Technical Report 88–05, pp. 85–87. Houston, TX: Lunar and Planetary Institute.

Masursky, H. and Strobell, M. H., 1976a. *Interagency Report: Astrogeology 59*. Geologic Maps and Terrain Analysis data for Viking Mars '75 Landing Sites Considered in December 1972. US Geological Survey open-file report 76–431.

Masursky, H. and Strobell, M. H., 1976b. *Interagency report: Astrogeology 60*. Geologic maps and terrain analysis data for Viking Mars 1975 landing sites considered in February and April 1973. U.S. Geological Survey open-file report 76–432.

Mayo, A. P., Blackshear, W. T., Tolson, R. H., *et al.*, 1977. Lander Locations, Mars Physical Ephemeris, and Solar System Parameters: Determination from Viking Lander Tracking Data. *Journal of Geophysical Research*, v. 82, pp. 4297–303.

McKim, R. J., 1996. The Dust Storms of Mars. *Journal of the British Astronomical Association*, v. 106, no. 4, pp. 185–200.

Merrihew, S. C., Haberle, R. M. and Lemke, L. G., 1996. A Micro-Meteorological Mission for Global Network Science on Mars: A Conceptual Design. *Planetary and Space Science*, v. 44, no. 11, pp. 1385–93.

Metzger, S., Carr, M., Johnson, J. R., Parker, T. J. and Lemmon, M. T., 1999. Dust Devil Vortices Seen by the Mars Pathfinder Camera. *Geophysical Research Letters*, v. 26, no. 18, pp 2781–4.

Mocquet, A., 1999. A Search for the Minimum Number of Stations Needed for Seismic Networking on Mars. *Planetary and Space Science*, v. 47, pp. 397–409.

Moore, H. J., 1994. A perspective of landing-site selection). In Golombek, M. (ed.), 1994. *Mars Pathfinder Landing Site Workshop*. LPI Tech. Rpt. 94–04. Houston, TX: Lunar and Planetary Institute, pp. 31–32. Abstract.

Moore, H. J., Spitzer, C. R., Bradford, K. Z., Cates, P. M., Hutton, R. E. and Shorthill, R. W., 1979. Sample Fields of the Viking Landers, Physical Properties, and Aeolian Processes. *Journal of Geophysical Research*, v. 84, no. B14, pp. 8365–77.

Moore, H. J., Hutton, R. E., Clow, G. D. and Spitzer, C. R., 1987. *Physical Properties of the Surface Materials at the Viking Landing Sites on Mars*. US Geological Survey Professional Paper 1389. Reston, VA: US Geological Survey.

Moore, J., 1970. *An Exploratory Investigation of a 1979 Mars Roving Vehicle Mission*. JPL Report 760–58. Pasadena, CA: Jet Propulsion Laboratory, December 1, 1970.

Morris, E. C., Jones, K. L. and Berger, J. P., 1978. Location of Viking 1 Lander on the Surface of Mars. *Icarus*, v. 34, no. 3, June 1978, pp. 548–55.

Morris, E. C. and Jones, K. L., 1980. Viking 1 Lander on the Surface of Mars: Revised Location. *Icarus*, v. 44, no. 1, October 1980, pp. 217–22.

Mueller, J. T., Guo, Y., von Mehlem, U. I. and Cheng, A. F., 2003. Aladdin Mission Concept. *Acta Astronautica*, v. 52, pp. 211–18.

Mukhin, L., Sagdeev, R., Karavasili, K. and Zakharov, A., 2000. Phobos, Deimos Mission. Presented at the Concepts and Approaches for Mars Exploration meeting, Houston, TX, 18–20 July 2000. Abstract no. 6096.

Murchie, S. and Erard, S., 1996. Spectral Properties and Heterogeneity of Phobos from Measurements by Phobos 2. *Icarus*, v. 123, pp. 63–86.

Murchie, S. L., Britt, D. T., Head, J. W., *et al.*, 1991. Color Heterogeneity of the Surface of Phobos: Relationships to Geologic Features and Comparison to Meteorite Analogs. *Journal of Geophysical Research*, v. 96, no. B4, pp. 5925–45.

Murchie, S., Thomas, N., Britt, D., Herkenhoff, K. and Bell, J. F., 1999. Mars Pathfinder Spectral Measurements of Phobos and Deimos: Comparison with Previous Data. *Journal of Geophysical Research*, v. 104, no. E4, pp. 9069–79.

Murray, B. C., Davies, M. E. and Eckman, P. K., 1967. Planetary Contamination II: Soviet and U.S. Practices and Policies. *Science*, v. 155, no. 3769, pp. 1505–11.

Murray, J. E. and Tartabini, P. V., 2001. *Development of a Mars Airplane Entry, Descent and Flight Trajectory*. NASA TM-2001-209035. NASA Dryden Flight Research Center, January 2001.

Mutch, T. A., Arvidson, R. E., Binder, A. B., Guinness, E. A., and Morris, E. C., 1977. The Geology of the Viking Lander 2 Site. *Journal of Geophysical Research*, v. 82, pp. 4452–67.

NASA, 1964. *Proceedings of the Symposium on Manned Planetary Missions 1963/1964 Status*. NASA Technical Memorandum TM X-53049. Huntsville, AL: NASA George C. Marshall Space Flight Center, 12 June 1964.

NASA, 1966. *Planetary Exploration Utilizing a Manned Flight System*. Washington, DC: National Aeronautics and Space Administration, Office of Manned Space Flight, 3 October 1966.

NASA, 1967. *Study of a Mars Surface Sample and Return Probe Launched from a Manned Mars Flyby*. NASA TR-293-7-132, Vol. 1, Summary Results and Conclusions; Vol. 2, Technical Analyses. Huntsville, AL: Northrop Space Laboratories, February 1967.

NASA, 1968a. *Mars Hard Lander Capsule Study*. NASA Contractor Report CR-66678-2, Volume 2, Mission and Science Definition. Washington, DC: National Aeronautics and Space Administration, 31 July 1968.

NASA, 1968b. *Mariner-Mars 1964, Final Project Report*. NASA SP-139. Washington, DC: National Aeronautics and Space Administration.

NASA, 1969. *Mariner-Mars 1969, A Preliminary Report*. NASA SP-225. Washington, DC: National Aeronautics and Space Administration.

NASA, 1973. *Mariner Mars 1971 Project Final Report. Volume 1. Project Development Through Launch and Trajectory Correction Maneuver*. NASA TR-32-1550. Pasadena, CA: Jet Propulsion Laboratory, 1 April 1973.

NASA, 1974. *Pioneer Mars Mission Study. Executive Summary*. NASA-TM-108688. Moffett Field, CA: NASA Ames Research Center, 1 August 1974.

NASA, 1978. *The Martian Landscape*. The Viking Lander Imaging Team. NASA SP-425. Washington, DC: Scientific and Technical Information Office, National Aeronautics and Space Administration.

NASA, 1987. *A Preliminary Study of Mars Rover/Sample Return Missions*. Washington, DC: The Mars Study Team, Solar System Exploration Division, NASA Headquarters, January 1987.

NASA, 1998. *Reference Mission Version 3.0; Addendum to the Human Exploration of Mars: The Reference Mission of the NASA Mars Exploration Study Team*. Drake, B. G. (ed.) NASA/SP-6107-ADD, EX13-98-036. Houston, TX: NASA Johnson Space Center.

Nash, D. B., Plescia, J., Cintala, M., *et al.*, 1989. *Science Exploration Opportunities for Manned Missions to the Moon, Mars, Phobos, and an Asteroid*. NASA CR-186295, JPL Publication 89–29, 30 June 1989.

Nickle, N. (ed.), 1980. *Mars Sample Return: Site Selection and Sample Acquisition Study*. JPL Publication 80–59. Pasadena, CA: Jet Propulsion Laboratory, 1 November 1980.

Nicks, O. W., 1967. *A Review of the Mariner IV Results*. NASA SP-130. Washington, DC: NASA.

Niehoff, J. C. and Friedlander, A. L., 1974. Pioneer Mars 1979 Mission Options. Paper AIAA-1974–783, American Institute of Aeronautics and Astronautics, Mechanics and Control of Flight Conference, Anaheim, CA, 5–9 August 1974, 13 pp.

Nyrtsov, M. V., 2000. *Developing Map Projections of Actual Surfaces of Celestial Bodies and Methods of Their Study*. Ph.D. Thesis, Moscow State University of Geodesy and Cartography (MIIGAiK).

Nyrtsov, M. V. and Stooke, P. J., 2002. The Mapping of Irregularly-Shaped Bodies at Planetary Scale. Proceedings from the International Conference InterCarto 8. Helsinki–St. Petersburg, 28 May–1 June 2002, pp. 433–436.

Oberst, J., Zeitler, W. and Kuschel, M., 2000. Where is Viking Lander 2? 31st Lunar and Planetary Science Conference, Houston, TX, 13–17 March 2000. Abstract no. 1612.

Otterman, J., 1967. Surface Pressures in the Martian Highlands and Lowlands. *Nature*, v. 213, pp. 1000–1.

Parker, T. J., Kirk, R. L. and Davies, M. E., 1999. Location and Geologic Setting for the Viking 1 Lander. 30th Lunar and Planetary Science Conference, Houston, TX, 15–19 March 1999. Abstract no. 2040.

Parker, T. J., and Kirk, R. L., 1999. Location and Geologic Setting for the Three US Mars Landers. Presented at The Fifth International Conference on Mars, July 18–23, 1999, Pasadena, CA. Abstract no. 6124. LPI Contribution No. 972, Lunar and Planetary Institute, Houston (CD-ROM).

Parker, T. J., McEwen, A. S., Kirk, R. L. and Bridges, N. T., 2007. HiRISE Captures the Viking and Mars Pathfinder Landing Sites. XXXVIII Lunar and Planetary Science Conference, League City, TX, 12–16 March 2007. Abstract no. 2368.

Perminov, V. G., 1999. *The Difficult Road to Mars: a Brief History of Mars Exploration in the Soviet Union*. Monographs in Aerospace History, no. 15. Washington, DC: National Aeronautics and Space Administration.

Pettengill, G. H., Counselman, C. C., Rainville, L. P. and Shapiro, I. I., 1969. Radar Measurements of Martian Topography. *Astronomical Journal*, v. 74, no. 1, pp. 461–82.

Pieters, C., Murchie, S., Cheng, A., *et al.*, 1997. Aladdin: Phobos-Deimos Sample Return. 28th Lunar and Planetary Science Conference, Houston, TX, 17–21 March 1997. Abstract no. 1113.

Pieters, C., Calvin, W., Cheng, A., *et al.*, 1999. Aladdin: Exploration and Sample Return of Phobos and Deimos. 30th Lunar and Planetary Science Conference, Houston, TX, 14–19 March 1999. Abstract no. 1155.

Platoff, A., 2001. *Eyes on the Red Planet: Human Mars Mission Planning, 1952–1970*. NASA/CR-2001–208928. Washington DC: National Aeronautics and Space Administration.

Pollack, J. B., Veverka, J., Noland, M., *et al.*, 1972. Mariner 9 Television Observations of Phobos and Deimos. *Icarus*, v. 17, pp. 394–407.

Pollack, J. B., Colburn, D. S., Flasar, F. M., Kahn, R., Carlston, C. E. and Pidek, D., 1979. Properties and Effects of Dust Particles Suspended in the Martian Atmosphere. *Journal of Geophysical Research*, v. 84, no. B6, pp. 2929–45.

Portree, D. S. F., 2001. *Humans to Mars: Fifty Years of Mission Planning*. Monographs in Aerospace History, no. 21. NASA SP-2001–4521. Washington, DC: National Aeronautics and Space Administration.

Proctor, R. A., 1888. Maps and Views of Mars. *Scientific American*, supplement, v. 26, July-December 1888, pp. 10659–60.

Pullan, D., Sims, M. R., Wright, I. P., Pillinger, C. T. and Trautner, R., 2004. Beagle 2: the Exobiological Lander of Mars Express. In Wilson, A. (ed), 2004. *Mars Express – The Scientific Payload*. European Space Agency Special Publication SP-1240, pp. 165–206.

Rahe, J., Mukhin, L., Sagdeev, R., *et al.*, 1999. Low Cost Mission to Phobos and Deimos. *Acta Astronautica*, v. 45, nos. 4–9, pp. 301–9.

Rayman, M. D., Chadbourne, P. A., Culwell, J. S. and Williams, S. N., 1999. Mission Design for Deep Space 1: A Low-Thrust Technology Validation Mission. *Acta Astronautica*, v. 45, nos. 4–9, pp. 381–8.

Rayman, M. D. and Varghese, P., 2001. The Deep Space 1 Extended Mission. *Acta Astronautica*, v. 48, nos. 5–12, pp. 693–705.

Rea, D. G., Carr, M., Bourke, R., *et al.*, 1988. *Mars Rover Sample Return: Results of Pre-Phase A Study*. Pasadena, CA: Jet Propulsion Laboratory, 4 October 1988.

Roush, T. L. and Hogan, R. C., 2000. Mars Global Surveyor Thermal Emission Spectrometer Observations of Phobos. XXXI Lunar and Planetary Science Conference, Houston, TX, 13–17 March 2000. Abstract no. 1598.

Sagan, C., Pollack, J. B. and Goldstein, R. M., 1966. *Radar Doppler Spectroscopy of Mars. I. Elevation Differences Between Bright and Dark Areas*. SAO Special Report no. 221. Cambridge, MA: Smithsonian Astrophysical Observatory.

Sagan, C., 1994. *Pale Blue Dot*. New York: Random House, Inc.

Sagdeev, R. Z. and Zakharov, A. V., 1989. Brief History of the Phobos Mission. *Nature*, v. 341, pp. 581–5.

Saunders, R. S., Arvidson, R. E., Badhwar, G. D., *et al.*, 2004. 2001 Mars Odyssey Mission Summary. *Space Science Reviews*, v. 110, pp. 1–36.

Schiaparelli, G. V., 1877. Osservazioni Astronomiche e Fisiche Sull'asse di Rotazione e Sulla Topografia del Pianete Marte. Atti della R. Accademia del Lincei, Memoria della cl. di Scienze Fisiche. Memoria 1, Ser. 3, Vol. 2, 1877–78, pp. 308–439.

Scott, D. H., 1988. Mars Sample Return: Recommended Sites. *Workshop on Mars Sample Return Science*. LPI Technical Report 88–07, pp. 151–3. Houston, TX: Lunar and Planetary Institute.

Scott, D. H. and Tanaka, K. L., 1988. Mars Sample Sites: Examples Based on a Global Geologic Perspective. *Workshop on Mars Sample Return Science*. LPI Technical Report 88–07, pp. 154–155. Houston, TX: Lunar and Planetary Institute.

Seiff, A., Haberle, R. and Murphy, J., 1994. Landing Site Considerations for Atmospheric Structure and Meteorology (abstract). In Golombek, M., ed., 1994. *Mars Pathfinder Landing Site Workshop*. LPI Tech. Rpt. 94–04. Houston, TX: Lunar and Planetary Institute, p. 38.

Service, R. W., 1909. *Ballads of a Cheechako*. Toronto: Ryerson Press.

Sheehan, W. and McKim, R. J., 1994. The Myth of Earth-Based Martian Crater Sightings. *Journal of the British Astronomical Association*, v. 104, no. 6, pp. 281–6.

Sidorenko, A. V. (ed.), 1980. *Poverkhnost Marsa*. Moscow: Nauka.

Shorthill, R. W., Moore, H. J., Hutton, R. E., Scott, R. F. and Spitzer, C. R., 1976. The Environs of Viking 2 Lander. *Science*, v. 194, no. 4271, pp. 1309–18.

Simonelli, D. P., Thomas, P. C., Carcich, B. T. and Veverka, J., 1993. The Generation and Use of Numerical Shape Models for Irregular Solar System Objects. *Icarus*, v. 103, pp. 49–61.

Simpson, R. A. and Tyler, G. L., 1981. Viking Bistatic Radar Experiment: Summary of First-Order Results Emphasizing North Polar Data. *Icarus*, v. 46, pp. 361–89.

Sims, M. R. (ed.), 2004. *Beagle 2 Mars, Mission Report*. Leicester, UK: Lander Operations Control Centre, University of Leicester.

Singer, S. F., 1981. The Ph-D Proposal. A Manned Mission to Phobos and Deimos. Paper AAS 81–231, in *The Case for Mars*, AAS Science and Technology Series, v. 57, pp. 39–65. San Diego: Univelt Inc.

Smith, B., 1970. Phobos: Preliminary Results from Mariner 7. *Science*, v. 168, no. 3933, pp. 828–30.

Smith, D. E., Zuber, M. T., Frey, H. V., *et al.*, 2001. Mars Orbiter Laser Altimeter: Experiment Summary After the First Year of Global Mapping of Mars. *Journal of Geophysical Research*, v. 106, no. E10, pp. 23689–722.

Snyder, C. W., 1977. The Missions of the Viking Orbiters. *Journal of Geophysical Research*, v. 82, no. 28, pp. 3971–3983.

Snyder, C. W., 1979. The Extended Mission of Viking. *Journal of Geophysical Research*, v. 84, no. B14, pp. 7917–33.

Snyder, C. W. and Evans, N., 1981. The Final Phases of the Viking Mission to Mars. *Icarus*, v. 45, pp. 2–24.

Snyder, J. P., 1985. Conformal Mapping of the Triaxial Ellipsoid. *Survey Review*, v. 28, no. 217, pp. 130–48.

Soderblom, L. A., Kreidler, T. J. and Masursky, H., 1973. Latitudinal Distribution of a Debris Mantle on the Martian Surface. *Journal of Geophysical Research*, v. 78, pp. 4117–22.

Soderblom, L. A. and Yelle, R. V., 2001. Near-Infrared Reflectance Spectroscopy of Mars (1.4–2.6 mm) from the New Millennium Deep Space 1 Miniature Integrated Camera and Spectrometer (MICAS). Presented at the 32nd Lunar and Planetary Science Conference, Houston, TX, 12–16 March 2001. Abstract no. 1473.

Soffen, G. A., 1976. Status of the Viking Missions. *Science*, v. 194, no. 4260, pp. 57–9.

Solomon, S. C., Anderson, D. L., Banerdt, W. B., *et al.* (eds), 1991. *Scientific Rationale and Requirements for a Global Seismic Network on Mars. Workshop, Morro Bay, California, 7–9 May 1990*. LPI Technical Report 91–02. Houston, TX: Lunar and Planetary Institute.

Space Technology Laboratories, Inc., 1962. Study of an Able-Mars Encounter Mission. NASW-246. Redondo Beach, CA: Thompson Ramo Wooldridge, Inc.

Squyres, S., 2005. *Roving Mars: Spirit, Opportunity, and the Exploration of the Red Planet*. New York: Hyperion.

Stallkamp, J. A., Herriman, A. G. and the Mariner Mars 1969 Experimenters, 1971. *Mariner Mars 1969 Final Project Report. Volume* 3. *Scientific Investigations*. JPL Technical Report 32–1460. Pasadena, CA: Jet Propulsion Laboratory, 15 September 1971.

Steinbacher, R. H. and Haynes, N. R., 1973. Mariner 9 Mission Profile and Project History. *Icarus*, v. 18, pp. 64–74.

Steinhoff, E. A., 1963. Use of Extraterrestrial Resources for Mars Basing. In Morgenthaler, G. (ed.), 1963. *Exploration of Mars. Proceedings of the American Astronautical Society Symposium on the Exploration of Mars*. Denver, CO, 6–7 June 1963, pp. 468–500.

Stooke, P. J., 1986. Automated Cartography of Non-Spherical Worlds. Proceedings of the Second International Symposium on Spatial Data Handling, Seattle, WA, 5–10 July 1986, pp. 523–36.

Stooke, P. J., 1989. Sizing Up Phobos. *Sky and Telescope*, v. 77, pp. 477–9.

Stooke, P. J., 1997. Horizon Topography and the Location of Viking Lander 2. *Earth, Moon and Planets*, v. 76, nos. 1 & 2, pp. 47–65.

Stooke, P. J., 1998. Mapping Worlds with Irregular Shapes. *The Canadian Geographer*, v. 42, no. 1, pp. 61–78.

Stooke, P. J., 1999. Revised Viking 1 Landing Site. XXX Lunar and Planetary Science Conference, Houston, TX, 15–19 March 1999. Abstract no. 1020.

Stooke, P. J., 2003. *Stooke Small Body Maps*. NASA Planetary Data System dataset EAR-A-3-RDR-STOOKEMAPS-V1.0.

Stooke, P. J., 2004. Viking 2 Landing Site in MGS/MOC Images. XXXV Lunar and Planetary Science Conference, 2004. Abstract no. 1074.

Stooke, P. J. and Keller, C. P., 1990. Map Projections for Non-Spherical Worlds. *Cartographica*, v. 27, no. 2, pp. 82–100.

Surkov, Yu. A. and Kremnev, R. S., 1998. Mars-96 Mission: Mars Exploration with the Use of Penetrators. *Planetary and Space Science*, v. 46, no. 11/12, pp. 1689–96.

Surkov, Yu. A., 1990. *Exploration of Terrestrial Planets from Spacecraft: Instrumentation, Investigation, Interpretation.* New York: Ellis Horwood.

Swan, P. R. and Sagan, C., 1965. Martian Landing Sites for the Voyager Mission. *Journal of Spacecraft and Rockets*, January-February 1965, pp. 18–25.

Swan, P. R., Hanselman, R. B., Ryan, R. L. and Suitor, R. F., 1966. *Manned Mars Surface Operations. A Volume of Technical Papers Presented at the AIAA/AAS Stepping Stones to Mars Meeting*, pp. 69–86; paper presented in Baltimore, MD, 28–30 March 1966.

Swift, J., 1726. *Travels into Several Remote Nations of the World, in Four Parts, by Lemual Gulliver*. London: Benjamin Motte.

Thakoor, S., Cabrol, N., Lay, N., *et al.*, 2003. Review: The Benefits and Applications of Bioinspired Flight Capabilities. *Journal of Robotic Systems*, v. 20, no. 12, pp. 687–706.

Thomas, N., Britt, D. T., Herkenhoff, K. E., *et al.*, 1999. Observations of Phobos, Deimos, and Bright Stars With the Imager for Mars Pathfinder. *Journal of Geophysical Research*, v. 104, no. E4, pp. 9055–68.

Thomas, P. C., 1979. Surface Features of Phobos and Deimos. *Icarus*, v. 40, pp. 223–43.

Thomas, P. C., 1993. Gravity, Tides, and Topography on Small Satellites and Asteroids: Application to Surface Features of the Martian Satellites. *Icarus*, v. 105, pp. 326–44.

Thomas, P. C. and Gierasch, P. J., 1985. Dust Devils on Mars. *Science*, v. 230, pp. 175–7.

Thomas, P. C., Veverka, J., Sullivan, R., *et al.*, 2000. Phobos: Regolith and Ejecta Blocks Investigated with Mars Orbiter Camera Images. *Journal of Geophysical Research*, v. 105, no. E6, pp. 15091–106.

Tombaugh, C. W., 1965. *Provisional Map of Mars, Mariner 4 Region*. NASA report CR-64772, TN-701-66-8. Las Cruces, NM: New Mexico State University, July 1965.

Turner, R. J., 1978a. A Model of Phobos. *Icarus*, v. 33, pp. 116–140.

Turner, R. J., 1978b. Modelling and Mapping Phobos. *Sky and Telescope*, v. 56, pp. 299–303.

Udrea, B., Delory, G., Landis, G., *et al.*, 2002. WOBBLE – A Proposed Mission to Characterize Past and Present Water on Mars. Paper AIAA IAF-02-Q.3.2.03, presented at the 53rd International Astronautical Congress, Houston, TX, 10–19 October 2002.

U.S. Geological Survey (USGS), 1984. Viking Lander Morning Scene – Camera 1. Atlas of Mars, Viking Lander Mosaic Series, Map I-1515. Reston, VA: US Geological Survey.

Vasavada, A. R., Williams, J.-P., Paige, D. A., *et al.*, 2000. Surface Properties of Mars' Polar Layered Deposits and Polar Landing Sites. *Journal of Geophysical Research*, v. 105, no. E3, pp. 6961–69.

Veverka, J. and Burns, J., 1980. The Moons of Mars. *Annual Review of Earth and Planetary Sciences*, v. 8, pp. 527–58.

Viking Data Analysis Team, c. 1972. *Viking 75 Project, Viking Data Analysis Team Report*, M75–144-0, NASA-TM-108221. Hampton, VA: NASA Langley Research Center, Viking Project Office, undated (internal references suggest a date late in 1972).

Viking Project, 1972. Viking 75 Project Science Requirements, Objectives and Constraints on Mission Design. Report M75–143-0. 15 June 1972. Hampton, VA: NASA Langley Research Center, Viking Project Office.

von Braun, W., 1953. *The Mars Project*. Urbana, IL: University of Illinois Press.

Wählisch, M., Willner, K., Oberst, J., *et al.*, 2010. A New Topographic Image Atlas of Phobos. *Earth and Planetary Science Letters*, v. 294, nos. 3–4, pp. 541–6.

Wells, R. A., 1969. Martian Topography: Large-Scale Variations. *Science*, v. 166, no. 3907, pp. 862–5.

Wells, R. A., 1972. Analysis of Large-Scale Martian Topography Variations – I. *Geophysical Journal of the Royal Astronomical Society*, v. 27, no. 1, pp. 101–33.

Wilson, A. (ed.), 2004. *Mars Express: A European Mission to the Red Planet*. ESA SP-1240. Noordwijk, The Netherlands: ESA Publications Division.

Zakharov, A. V., 1988. Close Encounters with Phobos. *Sky and Telescope*, v. 76, pp. 17–21.

Zeitler, W. and Oberst, J., 1999. The Mars Pathfinder Landing Site and the Viking Control Point Network. *Journal of Geophysical Research*, v. 104, no. E4, pp. 8935–42.

Zezin, R. B., Karyagin, V. P., Mamoshina, I. P., *et al.*, 1975. Analysis of Terrain Conditions in the Landing Site of the Mars-6 Automatic Interplanetary Station (in Russian). *Kosmicheskie Issledovaniya*, v. 13, no. 1, pp. 99–107.

Index

Aeneas, Project, 237
aerobraking, 10, 190, 248, 257, 282, 309, 315
Aeronautical Chart and Information, 1, 4, 20, 33
aircraft, 6, 81, 240, 241, 242, 251, 323, 324
Amazonian, 39, 318
Ambroise, Valerie, 257
Amundsen (probe), 290
Apollo 11, 20, 119
Apollo 16, 37
Apollo landing sites, 52, 55
Argyre, 23, 33, 39, 77, 196, 224, 228, 240, 243
Arnold, Jim, 175
array, 172
Arvidson, Ray, 178
Astronomical Unit, 18
Athena, 300, 305
Athena Precursor Experiment, 305
Avco Corporation, 5, 9, 11, 21, 25

Babakin, G. N., 45
balloon, 195, 197, 200, 225, 227, 236, 238, 239, 241, 243, 250, 252, 282, 285, 287, 289, 315
Banerdt, Bruce, 222
Barnard, Edward E., 4
Barsukov, V. L., 80, 90
Basilevsky, A. T., 31
Baum, William, 55, 57, 61, 74
Baumfree, Isabella, 259
Beer, Wilhelm, 1, 2, 17
Berman, Daniel, 298
Binder, Alan, 55, 57, 58
Blamont, Jacques, 195, 224, 227, 236, 239, 252
Borrelly (comet), 288
Braille (asteroid), 288
Bunch, T. E., 172
Burroughs, Edgar Rice, 4
canals, 1, 4, 9
Carr, Mike, 74, 117
Centre National des Etudes, 195, 224, 237, 241, 252
Ceres (asteroid), 195
Chryse, 14, 29, 57, 62, 63, 66, 69, 71, 73, 77, 81, 116, 128, 140, 156, 169, 175, 176, 178, 188, 190, 196, 224, 225, 230, 240, 243, 245, 248, 250, 255, 256, 257, 262, 263, 283, 295, 318, 320
Clarke, Arthur C., 4
Columbus, Christopher, 243
Crabill, Norman, 66, 74, 117
Cross, Charles, 25,
Crumpler, Larry, 122

Darwin, Charles, 320
Davies, M. E., 25, 83, 123, 133, 132
de Goursac, Olivier, 83
Deimos, 6, 37, 77, 95, 119, 126, 138, 140, 247, 267, 275, 282, 288, 306, 308, 318, 325, 330, 339
Discovery, 256, 257, 240
Discovery Program, 230, 238, 240, 315
Duke, Mike, 175
dust devil, 269, 275
dust storm, 33, 37, 95, 99, 116, 117, 119, 126, 147, 152, 169, 173, 202, 233, 247, 259

Eshleman, Von, 74
European Space Agency, 181, 202, 221, 234, 228, 315
exobiology, 243, 244, 255, 298

Farmer, Crofton, 73
Ford Motor Co., 9

Gale, 3, 302, 305, 312
Galileo (Jupiter mission), 176
Garvin, James, 83
General Electric, 9, 11, 19, 20, 21, 24
Geological Society of America, 275
Gliozzi, James, 58
Golombek, Matthew, 253
Google, Inc., 247
Greeley, Ronald, 172, 175
Green, Nathaniel, 3, 19
grooves (on Phobos), 330, 331, 332
gullies, 315, 316 , 322, 323
Gusev, 3, 188, 196, 202, 224, 228, 230, 240, 242, 243, 245, 251, 255, 257, 305, 306, 318, 324

Haberle, Robert, 224
Hall, Asaph, 325
Hargraves, Bob, 73
Hellas, 5, 23, 25, 26, 28, 29, 30, 31, 32, 33, 41, 42, 55, 57, 58, 73, 77, 139, 140, 152, 156, 167, 169, 173, 184, 188, 190, 198, 200, 201, 224, 225, 235, 239, 240, 249, 251, 285
Herschel, William, 1, 2, 15
Hesperian, 39, 318
Hess, Seymour, 124
Hubble, Edwin, 196
Hubble Space Telescope, 196, 221, 232, 234, 529
Huygens, Christiaan, 1
hydrothermal sites, 294, 297, 310

Ibragimov, N. B. O., 45
In-Situ Propellant Production, 305
In Situ Resource Utilization, 235, 275, 296, 304, 305, 310
Institute for Cosmic Research, 250
International Astronomical Union, 1, 69

Jet Propulsion Laboratory, 8

Kieffer, Hugh, 55, 57, 69, 122
King, Elbert, 176, 178, 180, 191
Klein, Harold, 117

Lederberg, Joshua, 58, 62, 63, 117
Lee, Gentry, 74, 120

Lewis, C. S., 4
Lowell, Percival, 1, 39
Luna 13, 31, 33, 52
Luna 19, 39
Luna 22, 39
Luna 3, 8, 29

Mädler, Johann, 1, 2, 17
magnetic anomaly, 11, 260, 251, 308
Malaska, Mike, 6, 319, 323
Malin Space Science Systems, 250
map projections, 325, 330
Marie Curie (rover), 298, 305
Mars day, 1
Mars Scout, 238, 241, 315
Mars Surveyor, 288, 298, 315
Marsokhod, 238, 250, 251, 260, 261, 285
Martian year, 1
Martin, Jim, 55, 74, 117, 122
Martin Marietta, 331
Masursky, Harold, 37, 55, 57, 58, 61, 71, 73, 74, 88, 117, 119, 120, 122, 124, 173, 175, 178, 180 , 191
Mellish, John, 1
Meridiani, 11, 14, 23, 69, 71, 197, 200, 247, 298, 308, 318, 320, 324
Moore, Henry, 69, 73, 74, 122, 178
Moroz, V. I., 45
Mountains of Mitchel, 5, 25
Mutch, Thomas, 52, 58, 88, 124, 172, 173

network mission, 167, 175, 195, 202, 222, 223, 228, 229, 232, 234, 235, 236, 239, 242, 244, 249, 294
New Millennium, 288, 290
Nix Olympica, 23, 26, 27, 30, 34, 55, 57, 69, 138, 139, 140, 152, 156, 167, 169
Noachian, 39, 318
Northrop, 5
NPO Lavochkin, 31

Oberth, Hermann, 196
occultation, 12, 23, 25, 69, 81, 95, 116, 119, 129, 138, 139, 140
Olympus Mons, 4, 69, 77, 196, 200, 201, 218, 223, 224, 225, 226, 230, 235, 240, 243, 248
Opportunity (rover), 247
Owen, Tobias, 57

Parker, Tim, 259, 264
penetrator, 124, 167, 168, 169, 172, 173, 175, 178, 179, 184, 195, 197, 200, 235, 241, 250, 252, 256, 282, 283, 285, 287, 288, 289, 290, 292, 294, 301, 302
Phobos, 5, 6, 25, 37, 77, 81, 95, 104, 116, 119, 126, 129, 138, 139, 191, 192, 205, 207, 235, 237, 240, 247, 248, 256, 257, 269, 275, 282, 288, 317, 318, 320, 325, 339
Phobos-Grunt, 318
Phoenix, 290, 294, 300, 315
Phyllosian, 318
Pilcher, Carl, 175
Planetary Society, 257
Porter, Jim, 58
prime meridian, 1
Proctor, Richard, 1, 3, 4, 19

radar, 6, 69, 71, 73, 74, 77, 126, 138, 140, 167, 238, 315, 317, 319, 343
radiation hazard, 308
RAND Corporation, 25, 83
Recant, George, 58
Robbins, Howard, 58
Rodionova, J. F., 192, 194, 207, 209
rover, 31, 33, 52, 72, 80, 172, 174, 175, 176, 178, 181, 183, 184, 185, 186, 188, 190, 195, 197, 200, 237, 238, 243, 250, 253, 259, 282, 285, 294, 298, 300, 306, 308, 324
rover (Phobos), 331
rover traverse, 169, 170, 171, 175, 177, 178, 179, 180, 181, 183, 185, 190, 191, 193, 196, 205, 206, 219, 220, 227, 229, 232, 238, 264, 271, 273, 274, 276, 279, 281, 284, 296 , 298, 304, 310, 313

Sagan, Carl, 33, 39, 55, 57, 66, 69, 88, 122, 264
sample return, 5, 169, 176, 178, 181, 182, 184, 185, 187, 188, 189, 191, 193, 194, 196, 197, 200, 203, 204, 205, 235
sample return (Deimos), 339
sample return (Phobos), 336, 339
Schiaparelli, Giovanni, 1, 4
Schmitz, Bob, 58
Scott (probe), 290
seasons, 1, 9, 20, 164
seismometer, 76, 88, 147, 172
Service, Robert, 73
Shapiro, Irwin, 71, 81
Siderikian, 318
Slocumb, Travis, 58
Smith, Bradford, 55, 57, 69
Soderblom, Larry, 73
Soffen, Gerald, 55, 124, 128
Sojourner, 238, 257, 264, 271, 275, 276, 277, 278, 279, 281, 282, 284, 286, 291, 292, 294, 298, 305, 306, 308
sol, 1
Space Shuttle, 172, 196, 241
Spirit, 240, 317, 324
Spirit (rover), 230, 240
Spitzer, Lyman, 196
Spudis, Paul, 122
Standish, Myles, 71, 81
Steinhoff, Ernst, 5
Stickney, Angelina, 325
Stickney (crater), 332, 339, 340, 341, 343
Strickland, E., 117
Stryk, Ted, 31, 36, 43, 46, 48, 49, 53, 56, 60, 193, 207, 208, 288, 293, 302, 303
Swift, Jonathan, 325
Syrtis Major, 1, 2, 5, 9, 12, 17, 20, 26, 29, 30, 42, 46, 57, 77, 169, 184, 224, 228, 231, 235, 239, 240, 243, 245, 249, 251, 252

Tharsis, 6, 21, 26, 29, 37, 38, 39, 43, 49, 77, 172, 184, 190, 193, 198, 200, 201, 202, 207, 208, 222, 224, 225, 228, 229, 230, 231, 233, 235, 238, 243, 248, 252, 256, 285, 295, 319, 323, 239

Theiikian, 318
Tombaugh, Clyde, 12, 17, 22
topography, 12, 23, 24, 25, 28, 29, 30, 39, 41, 43, 49, 238, 247
Truth, Sojourner, 259
Turner, Ralph, 330

U.S. Air Force, 1, 34, 43, 235
U.S. Army, 1, 4, 24, 28
U.S. Geological Survey, 3, 4, 39, 42, 46, 58, 178, 325
Utopia, 79, 84, 119, 126, 195, 201, 230, 243, 248, 252, 256, 257, 282, 283, 285, 294

Valles Marineris, 38, 42, 44, 45, 188, 190, 191, 194, 202, 205, 207, 209, 222, 224, 225, 230, 235, 240, 242, 247, 251, 285, 294, 324, 233
Vernadsky Institute, 80, 90
Vesta (asteroid), 195, 330
Viking Zone, 55
Vinogradov, A. P., 45
Vishniac, Wolf, 25, 55
von Braun, Wernher, 4
Voyager (outer planet mission), 76, 176

wave of darkening, 4, 5, 9, 10, 20
Wise, Don, 120
Withers, Paul, 290

Young, Thomas, 55, 61, 63, 66, 74